The Moths and Butterflies of Great Britain and Ireland

Volume 9

Volume 9

Sphingidae – Noctuidae
(Noctuinae and Hadeninae)

Editors: John Heath
A Maitland Emmet

Associate Editors:
D S Fletcher, E C Pelham-Clinton
W G Tremewan

Artists: Brian Hargreaves
Maureen Lane

THE MOTHS AND BUTTERFLIES OF GREAT BRITAIN AND IRELAND

Harley Books

Harley Books (B. H. & A. Harley Ltd.),
Martins, Great Horkesley,
Colchester, Essex, CO6 4AH, England

All rights reserved. No part of this publication
may be reproduced, stored in a retrieval system,
or transmitted, in any form or by any means,
electronic, mechanical, photocopying, recording
or otherwise without the prior permission of
the publisher.

Text set in Monotype Plantin (110)
Printed at The Curwen Press, London
Bound at The University Printing House, Cambridge

Book and jacket designed by James Shurmer

*The Moths and Butterflies of Great Britain
and Ireland* Volume 9
© John Heath; Harley Books, 1983

(Previously published by Curwen Books, 1979,
under ISBN 0 902068 07 5)

ISBN 0 946589 04 6   Volume 9

Volume 9: Contents

7	Preface
9	Eversible Structures, M. C. Birch
19	Systematic Section
20	Sphingidae, W. L. R. E. Gilchrist
39	Notodontidae, C. G. M. de Worms
65	Thaumetopoeidae, J. Heath
66	Lymantriidae, C. G. M. de Worms
78	Arctiidae, C. G. M. de Worms
111	Ctenuchidae, J. Heath
112	Nolidae, R. J. Revell
120	Noctuidae: Noctuinae and Hadeninae, R. F. Bretherton, B. Goater and R. I. Lorimer
281	Index

Plates A, B and C are between pages 40 and 41

Plates 1–13 follow page 280

Preface by John Heath and A. Maitland Emmet

In this volume, the first to be published in *The Moths and Butterflies of Great Britain and Ireland* dealing solely with Macrolepidoptera, we have made some changes in the format of the species text. The rearrangement of the section on *Life history* to include the descriptions of the early stages has eliminated some repetition which occurred with the earlier format.

Colour photographs of the imagines and larvae of typical representatives of each of the subfamilies treated in this volume and Volume 10 are included to amplify the introductory text to the families and subfamilies.

A consistent treatment of genera has proved impossible. Where recent taxonomic revisions have been made, *e.g.* in the Notodontidae, generic descriptions have been given, but in many instances even this has not been possible. Many of the early authors did not define their genera and definitions of these cannot be prepared until world-wide revisions of the groups concerned have been made. Such revisions are outside the scope of this book.

Three genera differ from those in Kloet & Hincks (1972). *Furcula* Lamarck, 1816 replaces *Harpyia* Ochsenheimer, 1810; *Harpyia* Ochsenheimer, 1810 replaces *Hybocampa* Lederer, 1853 and *Papestra* Sukhareva, 1973 has been introduced for *Phalaena biren* Goeze previously placed in *Lacanobia*.

Dot distribution maps have been provided by the Institute of Terrestrial Ecology, Biological Records Centre for the majority of species. These maps, which were prepared under contract from the Nature Conservancy Council, include all records received up to May 1978, or later for a few species (solid dots represent post-1960 records; open circles earlier records). For some of the migrant species special 'graded' maps devised by R. F. Bretherton are included. In many instances these maps give a good indication of the direction of the migrations they illustrate.

Species presumed not to be resident in the British Isles and of casual or reputed occurrence are represented by a brief paragraph in the text but most are omitted from the Keys. Regularly immigrant species, however, are treated as British.

We wish to acknowledge the continuing help of the Trustees of the British Museum (Natural History) and, in particular, Dr. P. Freeman, Keeper of Entomology, for making available the collections and library of that department; to his staff we are indebted for advice and assistance always so readily given. On behalf of the various authors we wish especially to thank D. J. Carter, A. H. Hayes and Miss Pamela Gilbert.

For their assistance with records from the Rothamsted Insect Survey we are indebted to R. A. French and I. Woiwod and, for transferring the light-trap data on to record cards prior to mapping, to Miss A. Sempkins.

We are grateful to Dr. J. D. Bradley, G. Dickson, G. Hyde, J. E. Knight, Dr. J. Mason, M. W. F. Tweedie and to the Trustees of the British Museum (Natural History) for the colour photographs reproduced on Plates A, B and C.

Chapter 1: M. C. Birch

EVERSIBLE STRUCTURES

Males of many species of Lepidoptera have scent-organs located on the abdomen, thorax, legs or wings. The structures vary greatly from simple scales and hair-tufts to complex eversible structures with storage areas and intricate mechanisms for expansion. All systems, however, involve some form of scale-tracts associated with gland-cells.

The existence of these structures has been known for a long time, but they have escaped mention in almost all popular systematic works. Varley (1962), McColl (1969) and Birch (1972) give references to the earlier literature. In this country, Eltringham (1925) first described the detailed morphology of abdominal brushes on the abdomen of male noctuid moths and also examined many other male structures (Eltringham, 1927; 1934; 1935). Swinton (1908) proposed a radial 'family tree' of butterflies and moths based on their scent-organs, but his system has little bearing on currently accepted systematic divisions and is, in any case, based on a character found in one sex only. Pierce (1909) emphasized the sporadic occurrence of male brush-organs in the Noctuidae with reference to generic relationships, an opinion which has since prevailed. It seems likely, however, that variations in structure have been overemphasized, since different types of structures do conform to broad taxonomic groups (McColl, 1969; Birch, 1972).

The male eversible structures of British species of Sphingidae, Arctiidae and Noctuidae are described in this chapter, but similar structures probably occur in some species of every major group of Lepidoptera. Some idea of the range and variety of these structures may be found in Varley (1962).

SPHINGIDAE

Müller (1878) traced the musky scent of *Agrius convolvuli* (Linnaeus) to pale brushes which spread out on either side of the second abdominal segment. Similar brushes had been discovered even earlier in *Acherontia atropos* (Linnaeus) (Nordmann, 1838).

Male *Deilephila elpenor* (Linnaeus) will readily erect a bundle of white hairs on each side of the base of the abdomen when disturbed (fig.1a, p.10); the white hairs are easily visible against the red abdominal scales. They are normally concealed by the pleural fold (fig.1b). The brush is everted by muscles twisting the base of the pocket outwards to spread the individual hairs, which are up to 3mm long, hollow, with longitudinal ridges and numerous pores. A complex gland opens through a duct and tiny pore into the pleural fold where the brushes normally lie. All this structural detail indicates that the brushes are disseminating structures for chemicals produced by the gland, but no obvious scent is produced by *D. elpenor*, nor have any chemicals been isolated from the brushes. *Manduca sexta* (Johansson) has similar brushes, also everted on disturbance, with a simpler gland structure. No scent is associated with them either (Grant & Eaton, 1973).

In contrast, a strong scent is emitted by *Acherontia atropos*. Swinton (1908) described

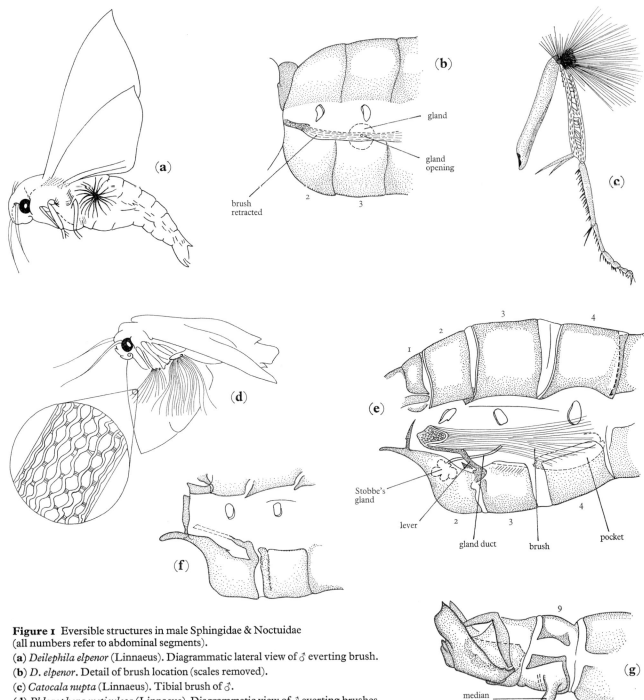

Figure 1 Eversible structures in male Sphingidae & Noctuidae (all numbers refer to abdominal segments).

(**a**) *Deilephila elpenor* (Linnaeus). Diagrammatic lateral view of ♂ everting brush.

(**b**) *D. elpenor*. Detail of brush location (scales removed).

(**c**) *Catocala nupta* (Linnaeus). Tibial brush of ♂.

(**d**) *Phlogophora meticulosa* (Linnaeus). Diagrammatic view of ♂ everting brushes. Inset: surface of one brush-hair greatly enlarged.

(**e**) *P. meticulosa*. Lateral view of anterior of ♂ abdomen with scales removed.

(**f**) *Apamea remissa* (Hübner). Lateral view of anterior of ♂ abdomen: partial brush-organ, lever only present.

(**g**) *Ochropleura plecta* (Linnaeus). Lateral view of posterior of ♂ abdomen with scales removed.

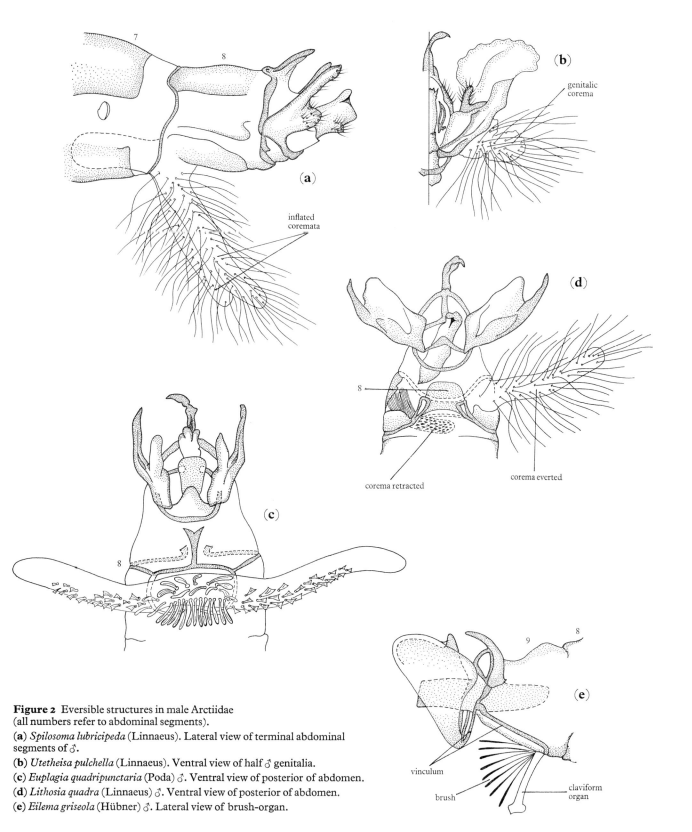

Figure 2 Eversible structures in male Arctiidae (all numbers refer to abdominal segments).

(**a**) *Spilosoma lubricipeda* (Linnaeus). Lateral view of terminal abdominal segments of ♂.
(**b**) *Utetheisa pulchella* (Linnaeus). Ventral view of half ♂ genitalia.
(**c**) *Euplagia quadripunctaria* (Poda) ♂. Ventral view of posterior of abdomen.
(**d**) *Lithosia quadra* (Linnaeus) ♂. Ventral view of posterior of abdomen.
(**e**) *Eilema griseola* (Hübner) ♂. Lateral view of brush-organ.

how a male held in the hand squeaked and spread out 'large, whirling, orange-coloured fans' which distributed a scent smelling first of jasmine, later becoming 'aluminous and disagreeable'. Apparently some believed that the scent and dust stirred up by *A. atropos* in flight caused blindness, though Swinton (1877) remarked that his moth 'had not that power'. Rothschild (1964) described an African species, *Lophostethus demolinii* (Angas), which produced a strong scent of 'burnt or bitter chocolate' perceived even when the insect was in flight.

British species with brush-organs similar to those of *Deilephila elpenor* are: *D. porcellus* (Linnaeus), *Sphinx ligustri* Linnaeus, *Daphnis nerii* (Linnaeus), *Hyles euphorbiae* (Linnaeus), and *H. lineata livornica* (Esper). The occasional or doubtful British species (Kloet & Hincks, 1972) with brush-organs are: *Agrius cingulata* (Fabricius), *Manduca quinquemaculatus* (Haworth), *M. sexta* and *Sphinx drupiferarum* Smith. Species with no trace of brushes are *Laothoe populi* (Linnaeus), *Mimas tiliae* (Linnaeus) and *Smerinthus ocellata* (Linnaeus). The remaining British species have not been examined.

Treatment of these structures in systematic works is terse. In their monumental work on the Sphingidae, Rothschild & Jordan (1903) state only that brushes are present in all species; this has now been shown to be incorrect. Hodges (1971) merely states that a pair of scent-scales is located on the second abdominal segment of males, but does not indicate in which species it is present.

ARCTIIDAE

Most of the British species of Arctiidae have eversible structures in the male. In many species these are coremata, eversible sacs covered with scales or hairs, on the posterior ventral part of the abdomen. Pagden's (1957) photographs of an Indo-Malaysian species, *Creatonotus gangis* (Linnaeus), show how large the structures can be and that the moth is capable of everting them.

Spilosoma lubricipeda (Linnaeus) has a good example of coremata (fig.2a, p.11). Pressure on the abdomen of a male will inflate a pair of coremata from between abdominal sternites 7 and 8. Light pressure exposes the tips of the hairs, further pressure forces air from tracheal sacs into the coremata and inflates them. Excessive pressure will fill the coremata with haemolymph. If the abdominal scales are brushed off, the base of the coremata is seen to arise from a sclerotized bar extending in a circle round the abdomen and fused to the tergite of segment 8.

The method of eversion by the moth is unknown, but there is a large air-sac in the posterior part of the abdomen connected both to the lateral tracheal trunks and directly to the coremata. Air could be pumped into this sac and into the coremata, or if everted in flight, the ventilatory air flow often associated with flight activity could be diverted into the coremata. The coremata collapse due to inherent elasticity of the wall when internal pressure is removed. Muscle action invaginates the collapsed coremata between sternites 7 and 8.

Coremata are absent from *S. lutea* (Hufnagel), but present in *S. urticae* (Esper). The coremata of *Phragmatobia fuliginosa* (Linnaeus) are similar to those of *Spilosoma lubricipeda* and *S. urticae*. *Diaphora mendica* (Clerck) lacks distinctive coremata, but a group of elongate scales is present on the anterior sclerotized arms of sternite 8. *Diacrisia sannio* (Linnaeus) and *Coscinia cribraria* (Linnaeus) have slightly larger groups of scales (brushes) on very small coremata, but the sclerotized structures of both species are greatly reduced compared with those of *Spilosoma lubricipeda*.

There are no coremata in *Arctia caja* (Linnaeus), *A. villica* (Linnaeus), *Parasemia plantaginis* (Linnaeus), *Spiris striata* (Linnaeus), *Callimorpha dominula* (Linnaeus) or *Tyria jacobaeae* (Linnaeus).

There are no abdominal coremata in *Utetheisa pulchella* (Linnaeus), but the male genitalia differ widely from those of other species of the subfamily. The sacculus region of each valve is membranous and covered with long setae and would, when everted, form a genitalic corema (fig.2b, p.11).

Male *Euplagia quadripunctaria* (Poda) have large coremata on sternite 8 (fig.2c). One sclerotized 'ring' incompletely circles the segment, and the coremata are everted through a second complete ring on the ventral surface. The coremata have scales only on their ventral surface and these scales are broad and triangular, especially towards the tips of the coremata.

In the subfamily Lithosiinae, no trace of brushes or coremata occur in *Thumatha senex* (Hübner), *Setina irrorella* (Linnaeus), *Nudaria mundana* (Linnaeus), *Atolmis rubricollis* (Linnaeus) or *Pelosia muscerda* (Hufnagel). However, brush-organs are found in several species of *Eilema* and both *Miltochrista miniata* (Forster) and *Lithosia quadra* (Linnaeus) have prominent coremata.

The coremata of *Miltochrista miniata* are everted through a ventral ring-shaped sclerite linked to the genitalia. Each corema is a short tube with hairs present only over the bulbous tip. The coremata of *Lithosia quadra* are longer and are covered with hairs (fig.2d). They extend from the pleural region of segment 8 which is surrounded by a complex system of sclerites.

The brush-organ of *Eilema griseola* (Hübner) is a central tuft of hairs with club-shaped structures (claviform organs—Eltringham, 1935) on either side (fig.2e). All are located on an extension of the vinculum and would be everted as the genitalia are extended. The clubs are formed of several scales with expanded tips fused together. The clubs are reduced in *E. lurideola* ([Zincken]). Four other species of *Eilema* which were examined had brushes but lacked clubs.

Judged by structural differences alone, coremata and brushes appear to have evolved independently many times in the Arctiidae, possibly as many as seven times in the few British species. The coremata of *Spilosoma lubricipeda*, *S. urticae* and *Phragmatobia fuliginosa* are similar in structure and location. Those of *Utetheisa pulchella*, *Coscinia cribraria* and *Eilema* represent three types radically different in structure and location. The coremata of *Euplagia quadripunctaria*, *Miltochrista miniata* and *Lithosia quadra* could have been modified from a single ancestral structure, but are different enough to suggest that independent origins are more likely. Even though our arctiid fauna is not large, it shows very well the lability and diversity of male eversible structures in Lepidoptera.

NOCTUIDAE

Male eversible structures in the Noctuidae are of many morphological types and can occur on the legs, thorax or, more often, on the abdomen. Some of the most complex scent-distributing systems are found in this family. The abdominal brush in Noctuinae,

Hadeninae, Cuculliinae and Amphipyrinae is of special interest, but other structures occur throughout the family.

Males of several species have a single eversible brush on the eighth abdominal sternite although the occurrence is sporadic between species. *Emmelia trabealis* (Scopoli) (Acontiinae) has such a structure, but other species examined in the subfamily, for example, *Lithacodia pygarga* (Hufnagel), *Eustrotia uncula* (Clerck), and *Deltote bankiana* (Fabricius), do not. In the Chloephorinae, *Pseudoips fagana* (Fabricius) has a pair of brushes at the base of the genitalia, but *Earias clorana* (Linnaeus) does not. *Colocasia coryli* (Linnaeus) (Pantheinae) has no male brushes, nor have any of the British species of the Ophiderinae.

There are brushes of hairs on the middle tibiae of all the British species of *Catocala* (fig.1c, p.10) except *C. sponsa* (Linnaeus). The brushes lie in a groove of the tibia and are concealed by enlarged scales. In the Hypeninae, e.g., *Paracolax derivalis* (Hübner), males have tufts of hair on the coxae, tibiae and femora of the forelegs. Brushes also occur on the legs of males of many species of Geometridae and other families (Swinton, 1908; Varley, 1962).

In the Plusiinae, the brush-organs of *Trichoplusia ni* (Hübner) have been described in detail (Grant, 1971) and Gothilf & Shorey (1976) recently photographed males using the brushes during courtship. Both British *Plusia* species and *Diachrysia chrysitis* (Linnaeus) have similar brushes which are easily everted by squeezing the abdomen of a male: the hair-tufts appear and suddenly spread out into a fan. *Abrostola trigemina* (Werneburg) and *A. triplasia* (Linnaeus) each have a single posterior brush. *Polychrysia moneta* (Fabricius) has no obvious male eversible structure. The remaining British Plusiinae have not been examined.

All the British Acronictinae, except *Acronicta auricoma* ([Denis & Schiffermüller]), were examined and lack male brushes or coremata.

An unusually complex organ occurs in the Noctuinae, Hadeninae, Cuculliinae and Amphipyrinae, the structure of which is remarkably constant in all species in which it is found. The system consists of paired brushes, scent-glands (Stobbe's glands) and storage-pockets. Each brush is attached to a sclerotized lever arising from the posterior angle of the second (apparent first) abdominal sternite (figs 1d,1e). Details of the method of operation have been described for *Phlogophora meticulosa* (Linnaeus) (Birch, 1970a). Clearwater (1975) gave an excellent description of the ultrastructure and development of the system in several noctuid species. The brush is extended by muscles acting at the base of the lever and the hairs are fanned out by a muscle at their base. Scent-secretion originates from Stobbe's glands in the second abdominal segment which are active only in the pharate adult. The scent is stored on the brushes within the pockets until mating occurs. The lattice-like structure of the brush-hairs provides a huge surface area both for storage of scent and for its evaporation when the brushes are everted (fig.1d). After locating a female, a male everts both brushes towards her and immediately attempts to copulate (Birch, 1970b).

The scent from many species of these subfamilies has now been chemically characterized. In all cases, the compounds are small, volatile and easily perceived by the human nose. The chemicals show some relationship to taxonomic groupings (Birch, 1974).

Some 40 per cent of 300 species of British Noctuidae surveyed have either fully developed brush-organs as described, or some of the structures involved. The remaining 60 per cent have no trace of them (Birch, 1972). A brush-organ of the 'typical' noctuid type is present in only one of 52 species of Noctuinae examined, 20 of 59 species of Hadeninae, 37 of 56 species of Cuculliinae, and 41 of 89 species of Amphipyrinae. About 20 per cent of the species with brush-organs have only a partial structure present (17 of 99 species).

In many genera, otherwise conceived on sound morphological criteria, some species have brush-organs, others have only a trace of the sclerotized structures, and the remainder show no trace of them at all. For example, in the genus *Mythimna*, most species have the 'typical' brush-structure. However, *M. pudorina* ([Denis & Schiffermüller]) and *M. loreyi* (Duponchel) have no brush-organs, and three species, *M. obsoleta* (Hübner), *M. comma* (Linnaeus) and *M. putrescens* (Hübner), have large inflatable coremata where the brushes would normally be.

The presence or absence of brushes is unfortunately of little value for separating critical species since in most cases both of the pair of species which may be confused either have or do not have the structures. *Apamea remissa* (Hübner) and *A. furva* ([Denis & Schiffermüller]) are exceptions (Birch, 1970c). Complete brush-organs are present in *A. furva*; only the sclerotized lever is present in *A. remissa* (fig.1f, p.10).

In the Noctuinae, only *Peridroma saucia* (Hübner) has any trace of the 'typical' noctuid brush-organs, but in all other respects it appears to be closely related to other genera in this subfamily (Birch, 1972). Curiously, *Ochropleura plecta* (Linnaeus) has a single median abdominal corema, with a red brush (fig.1g). Unlike the coremata of Arctiidae, Geometridae or the three species of *Mythimna*, this corema is apparently inflated by haemolymph.

The Noctuidae present a good example of the consistency and the lability of brush-organ structures. Consistency is demonstrated by the unique complex structure found in many species throughout four subfamilies, exhibiting very little variation. Lability is emphasized by the total absence of the structure in some species, partial loss in others and the evolution of completely new eversible organs in others.

Function of Eversible Structures

The significance of eversible structures in the life of butterflies and moths is far from understood. Brushes and coremata are usually concealed in pockets or folds and cannot be seen or studied unless they are artificially everted, which usually means destroying the perfection of an intended cabinet specimen. In fact many collectors are unaware of the existence of brush-organs.

Most moths are nocturnal and their behaviour is difficult to observe. It is perhaps not surprising that the behaviour evaluated in most detail is that of the Danaidae which are not only large and colourful, but are active and mate in the daytime.

A series of observations and experiments with the Danaidae has shown that the male brushes produce compounds during the courtship flight which induce the female to alight so that mating can occur. Brower *et al.* (1965) and Pliske & Eisner (1969) described the basic work on the Danaidae. In all but one or two other species the function of the structures is unverified by experiment.

Cardé *et al.* (1975) have recently described how males of the oriental fruit-moth,

Cydia molesta (Busck) (Tortricidae) always evert their abdominal hair-pencils before copulation. In the pyralid moth, *Plodia interpunctella* (Hübner), Grant & Brady (1975) have shown that the males release a scent from wing-glands and brushes located at the base of each forewing. The scent induces the female to remain stationary and to adopt an acceptance posture. Both these species are diurnal.

In the Noctuidae, the brushes of *Phlogophora meticulosa* (Birch, 1970b), *Trichoplusia ni* (Grant, 1970; Gothilf & Shorey, 1976) and *Heliothis virescens* (Fabricius) (Hendricks & Shaver, 1975) are everted as an integral part of courtship. In *Trichoplusia ni*, however, scent is not essential for successful mating, so the role of the male brushes is still unclear.

The account by Varley (1962) of males of *Creatonotus gangis* expanding their coremata appears to be the only one describing the natural eversion of coremata in Arctiidae and, apart from showing that they are everted, their exact role is no clearer. Although *Spilosoma lubricipeda* pairs readily in captivity, eversion of the coremata has not been described.

The use of brushes by male sphingids may have a different function. They are everted readily when a male is handled or disturbed and may form part of the prominent defensive display, which is particularly obvious in *Deilephila elpenor*. Other species with brushes appear to behave similarly. However, it is difficult to explain the adaptive advantage of a defensive structure that occurs in males alone.

Ford's (1955) hope that biologists and chemists would get together and investigate moth scents has been realized so well that the identity of chemicals secreted by male structures is better known than their function. It is possible to speculate endlessly on the evolutionary significance of male brushes and coremata, their role in species-isolation, and so on, but speculation is of little value without precise knowledge of how and when the structures are used. Many species in the Sphingidae, Arctiidae and Noctuidae are common, easy to rear, and very suitable for simple detailed observations. For example, it could be very interesting to analyse the courtship behaviour of species such as the arctiids *Spilosoma lubricipeda* and *S. lutea*, *i.e.*, one which has large coremata and one which has none, or of the noctuids with large brushes, *e.g.*, *Apamea monoglypha* (Hufnagel), and related species without brushes.

Techniques

Since the most needed information is on living insects, it would be best to rear insects so that mating behaviour can be observed first in captivity and then in natural or semi-natural conditions.

In the Arctiidae and Geometridae it is easy to check for the presence or absence of coremata or other structures near the genitalia by simply squeezing the abdomen, which causes at least partial eversion. In the Noctuidae, if the base of the lever (*i.e.*, either corner of the second abdominal segment) is pressed, the brush springs from the pocket, often releasing its distinctive odour.

To preserve any of these eversible structures for a collection, they can be inflated and dried. A fine glass tube is inserted into the thorax or anterior part of the abdomen of a fresh-killed specimen and is blown down until the brushes evert or the coremata inflate with air. The structures can be rapidly dried over a light-bulb and preserved in an extended position. Permanent preparations of abdomens with brush-organs can be made from fresh or dried specimens, using the techniques described by Reid (*MBGBI* 1: 117).

The abdomen is removed, cleared in KOH, stained, dehydrated, cut along the mid-dorsal line and mounted as described.

References

Birch, M. C., 1970a. Structure and function of the pheromone-producing brush-organs in males of *Phlogophora meticulosa* (L.) (Lepidoptera: Noctuidae). *Trans. R. ent. Soc. Lond.* **122**: 277–292.

———, 1970b. Pre-courtship use of abdominal brushes by the nocturnal moth, *Phlogophora meticulosa* (L.) (Lepidoptera: Noctuidae). *Anim. Behav.* **18**: 310–316.

———, 1970c. Critical species of Lepidoptera: *Apamea remissa* Hübn. and *Apamea furva* Schiff. (Noctuidae). *Entomologist's Gaz.* **21**: 262–264.

———, 1972. Male abdominal brush-organs in British noctuid moths and their value as a taxonomic character. *Entomologist* **105**: 185–205, 233–244.

———, 1974. Aphrodisiac pheromones. *In* Birch, M. C., (ed.) *Pheromones*, 115–134. North-Holland.

Brower, L. P., Brower, J. V. Z. & Cranston, F. P., 1965. Courtship behaviour of the queen butterfly, *Danaus gilippus berenice* (Cramer). *Zoologica, N.Y.* **50**: 1–39.

Cardé, R. T., Baker, T. C. & Roelofs, W. L., 1975. Ethological function of components of a sex attractant system for oriental fruit moth males, *Grapholitha molesta* (Lepidoptera: Tortricidae). *J. chem. Ecol.* **1**: 475–491.

Clearwater, J. R., 1975. Structure development and evolution of the male pheromone system in some Noctuidae (Lepidoptera). *J. Morph.* **146**: 129–176.

Eltringham, H., 1925. On the abdominal brushes of certain male noctuid moths. *Trans. ent. Soc. Lond.* **1925**: 1–5.

———, 1927. On the brush-organs in the noctuid moth *Laphygma frugiperda* Sm. and Ab. *Ibid.* **75**: 143–146.

———, 1934. On the brush-organs of the male ermine moth, *Spilosoma menthastri* Esper. *Ibid.* **82**: 41–42.

———, 1935. On the brush-organs of the male *Lithosia griseola* Hübn. (Lepidoptera). *Ibid.* **83**: 7–9.

Ford, E. B., 1955. *Moths*, 266 pp. London.

Gothilf, S. & Shorey, H. H., 1976. Sex pheromones of Lepidoptera: examination of the role of male scent brushes in courtship behavior of *Trichoplusia ni*. *Envir. Ent.* **5**: 115–119.

Grant, G. G., 1970. Evidence for a male sex pheromone in the noctuid *Trichoplusia ni*. *Nature, Lond.* **227**: 1345–1346.

———, 1971. Scent apparatus of the male cabbage looper, *Trichoplusia ni*. *Ann. ent. Soc. Am.* **64**: 347–352.

——— & Brady, U. E., 1975. Courtship behavior of phycitid moths. I. Comparison of *Plodia interpunctella* and *Cadra cautella* and role of male scent glands. *Can. J. Zool.* **53**: 813–826.

——— & Eaton, J. L., 1973. Scent brushes of the male tobacco hornworm *Manduca sexta* (Lepidoptera: Sphingidae). *Ann. ent. Soc. Am.* **66**: 901–904.

Hendricks, D. E. & Shaver, T. N., 1975. Tobacco budworm: male suppressed emission of sex pheromone by the female. *Envir. Ent.* **4**: 555–558.

Hodges, R. W., 1971. *The moths of America north of Mexico*, Fasc. 21 Sphingoidea, 158 pp., 19 figs., 14 col. pls. London.

Kloet, G. S. & Hincks, W. D., 1972. A check list of British insects (Edn 2). *Handbk Ident. Br. Insects* **11** (2): viii, 153 pp.

McColl, H. P., 1969. The sexual scent organs of male Lepidoptera. M.Sc. Thesis (University College of Swansea, Wales), 214 pp.

Müller, F., 1878. Wo hat der Moschusduft der Schwärmer seinen Sitz? *Kosmos, Stuttg.* **3**: 84–85. (A translation is given in Longstaff, G. B., 1912. *Butterfly hunting in many lands*. London.)

Nordmann, A. von, 1838. Über die Entdeckung des Stimmapparates bei dem Totenkopfschwärmer (*Sphinx* oder *Acherontia atropos*). *Bull. Acad. St. Petersb.* **3**: 164–194.

Pagden, H. T., 1957. The presence of coremata in *Creatonotus gangis* (L.) (Lepidoptera: Arctiidae). *Proc. R. ent. Soc. Lond.* (A) **32**: 90–94.

Pierce, F. N., 1909. *The genitalia of the group Noctuidae of the Lepidoptera of the British Islands*, xii, 88 pp., 32 pls. Liverpool.

Pliske, T. E. & Eisner, T., 1969. Sex pheromone of the queen butterfly: biology. *Science N.Y.* **164**: 1170–1172.

Reid, J., 1976. Techniques. *In* Heath, J. (ed.), *The Moths and Butterflies of Great Britain and Ireland*, **1**: 117–134.

Rothschild, M., 1964. A note on the evolution of defensive and repellant odours of insects. *Entomologist* **97**: 276–280.

Rothschild, W. & Jordan, K., 1903. A revision of the lepidopterous family Sphingidae. *Novit. zool.* **9**: Suppl., cxxxv, 972 pp., 67 pls.

Swinton, A. H., 1877. On stridulation in the genus *Acherontia*. *Entomologist's mon. Mag.* **13**: 217–220.

———, 1908. The family tree of moths and butterflies traced in their organs of sense. *Societas ent.* **23**: 99–101, 114–116, 124–126, 131–132, 140–141, 148–150, 156–157, 162–165.

Varley, G. C., 1962. A plea for a new look at Lepidoptera with special reference to the scent distributing organs of male moths. *Trans. Soc. Br. Ent.* **15**: 29–40, 2 figs.

SYSTEMATIC SECTION

Scheme of Classification

The scheme of classification adopted throughout this work is that detailed in Kloet & Hincks (1972) but modified where necessary. The families will be treated in the ten volumes (volume number indicated below) according to the following plan:

ZEUGLOPTERA

Micropterigoidea
    Micropterigidae   1

DACNONYPHA

Eriocranioidea
    Eriocraniidae   1

EXOPORIA

Hepialoidea
    Hepialidae   1

MONOTRYSIA

Nepticuloidea
    Nepticulidae   1
    Opostegidae   1
    Tischeriidae   1

Incurvarioidea
    Incurvariidae   1
    Heliozelidae   1

DITRYSIA

Cossoidea
    Cossidae   2

Zygaenoidea
    Zygaenidae   2
    Limacodidae   2

Tineoidea
    Psychidae   2
    Tineidae   2
    Ochsenheimeriidae   2
    Lyonetiidae   2
    Hieroxestidae   2
    Gracillariidae   2
    Phyllocnistidae   2

Yponomeutoidea
    Sesiidae   2
    Glyphipterigidae   2
    Douglasiidae   2
    Heliodinidae   2
    Yponomeutidae   3
    Epermeniidae   3
    Schreckensteiniidae   3

Gelechioidea
    Coleophoridae   3
    Elachistidae   3
    Oecophoridae   4
    Ethmiidae   4
    Gelechiidae   4
    Blastobasidae   4
    Stathmopodidae   4
    Momphidae   4
    Scythrididae   4

Tortricoidea
    Tortricidae   5
    Cochylidae   5

Alucitoidea
    Alucitidae   6

Pyraloidea
    Pyralidae   6

Pterophoroidea
    Pterophoridae   6

Hesperioidea
    Hesperiidae   7

Papilionoidea
    Papilionidae   7
    Pieridae   7
    Lycaenidae   7
    Nemeobiidae   7
    Nymphalidae   7
    Satyridae   7
    Danaidae   7

Bombycoidea
    Lasiocampidae   7
    Saturniidae   7
    Endromidae   7

Geometroidea
    Drepanidae   7
    Thyatiridae   7
    Geometridae   8

Sphingoidea
    Sphingidae   9

Notodontoidea
    Notodontidae   9
    Thaumetopoeidae 9

Noctuoidea
    Lymantriidae   9
    Arctiidae   9
    Ctenuchidae   9
    Nolidae   9
    Noctuidae
        Noctuinae   9
        Hadeninae   9
        Cuculliinae to Hypeninae   10
    Agaristidae   10

NOTE. The distribution maps included in this volume which have been provided by the Biological Records Centre in advance of the publication of the *Atlas of the Lepidoptera of the British Isles* must be regarded as provisional. Records from some localities mentioned in the text may not be shown on the maps when it has not been possible to localize them to a 10 km grid square, or when the records have been received too late for inclusion on the maps. Similarly records may be shown on the maps from localities not mentioned specifically in the text, *e.g.* the Isle of Man, the Channel Islands.

SPHINGIDAE
W. L. R. E. Gilchrist

The family Sphingidae consists of about 1,000 species of medium- to large-sized moths of world-wide distribution with most species occurring in the tropics. A few are cosmopolitan and some are wide-ranging migrants. There are 13 genera and 17 species placed in two subfamilies, the Sphinginae and the Macroglossinae, resident in, or regularly immigrant to, Great Britain and Ireland.

Imago (Pl.A, figs 1,3). Haustellum long and well developed in most species; antenna about half length of forewing, in male fasciculate or pectinate, in female simple, thickened medially, rapidly tapering apically to a pointed tip; labial palpus long, three-segmented; eye large, often lashed; ocelli and chaetosemata absent. Thorax with large prominent tegulae. Legs with spurs always present on midtibia, frequently also on fore- and hindtibia; a midtarsal 'comb' of rather long, slender setae. Forewing long and slender. Hindwing almost triangular; frenulum and retinaculum usually present, somewhat reduced in some species. Venation as in figure 3, the combination of R_1 crossing to Sc from Rs in the hindwing and the absence of 1A from both wings characterizing the family within the Macrolepidoptera (Hodges, 1971). Abdomen with paired scent glands on the second sternum of male; apices of segments 2–8 in male and 2–7 in female with modified setae, usually best developed on tergum.

Many species are crepuscular and nocturnal but some are exclusively diurnal. May often be found hovering in front of tubular flowers whilst feeding on the nectar.

Ovum. Rounded, often slightly flattened with very little surface sculpturing, green. The micropyle is lateral. Usually laid singly on the foodplant. Most British species are figured in Stokoe & Stovin (1958).

Larva (Pl.A, figs 2,4). Cylindrical or tapering anteriorly; head rounded or triangular; abdominal segment 8 with a caudal horn, sometimes much reduced. Many species have large eye-like warning markings laterally on thorax and assume a characteristic attitude when disturbed, with the head retracted into the raised thoracic segments. The family name is derived from this sphinx-like attitude. Many species are illustrated in Buckler (1887), Stokoe & Stovin (1958) and Newman (1965).

Pupa. Fusiform; haustellum often enlarged basally or free; cremaster prominent, usually triangular, often slightly bifurcate. In an earthen cell below the surface of the soil, or on the surface in a loosely spun cocoon. All the resident British species overwinter in this stage.

Much interesting detailed information on the occurrence and life history of the British species is given in Allan (1947). Williams (1971) tabulates the occurrence of the migrant species to 1962 and the work was continued by French (1965, 1966, 1968, 1971, 1973) to 1968 and (pers.comm. 1976) to 1975. For some of the migrant species special graded maps are included.

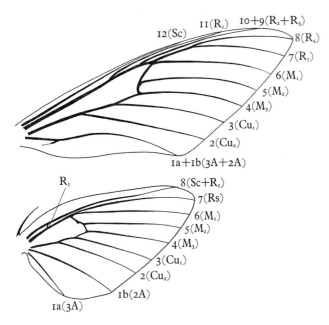

Figure 3 *Sphinx ligustri* Linnaeus, wing venation

Key to species (imagines) of the Sphingidae

1	Naked area on inner surface of first segment of labial palpus with a patch of fine, dense setae	(Macroglossinae) 2
–	Naked area on inner surface of first segment of labial palpus lacking a patch of fine setae	(Sphinginae) 11
2(1)	Wings with large areas without scales	3
–	Wings entirely covered with scales	4
3(2)	Borders of wings, especially hindwing, broad; black discal bar	*Hemaris fuciformis* (p. 30)
–	Borders of wings narrow; no discal bar	*H. tityus* (p. 29)
4(2)	Wingspan less than 60mm	5
–	Wingspan greater than 60mm	6
5(4)	Forewing dark grey or grey-brown; hindwing yellow	*Macroglossum stellatarum* (p. 31)

–	Fore- and hindwing pink and yellow *Deilephila porcellus* (p. 37)
6(4)	Fore- and hindwing intensely suffused pink *D. elpenor* (p. 36)
–	Wings not suffused pink or suffusion confined to hindwing 7
7(6)	Wings and body predominantly green or olive-green; very large species *Daphnis nerii* (p. 32)
–	Wings and body without green coloration; hindwing pinkish; forewing with prominent streak from dorsum to apex 8
8(7)	Forewing with very narrow streak; hindwing with prominent pink basal patch *Hippotion celerio* (p. 38)
–	Forewing with broad streak; hindwing with white basal patch 9
9(8)	Forewing with veins prominently white *Hyles lineata* (p. 35)
–	Forewing without prominently white veins 10
10(9)	Forewing with continuous broad, brown or olive-brown costal shade *H. gallii* (p. 34)
–	Forewing with costal shade interrupted forming basal and discal patches *H. euphorbiae* (p. 33)
11(1)	Wings with terminal margins heavily scalloped; very broad *Laothoe populi* (p. 28)
–	Wings with terminal margins smooth; elongate 12
12(11)	Hindwing with prominent eyespot *Smerinthus ocellata* (p. 27)
–	Hindwing without eyespot 13
13(12)	Abdomen and/or hindwing suffused pink 14
–	Abdomen and hindwing coloured otherwise 15
14(13)	Abdomen and hindwing suffused pink *Sphinx ligustri* (p. 24)
–	Abdomen only suffused pink *Agrius convolvuli* (p. 21)
15(13)	Wings brown, flecked with pale grey; abdomen banded brown and pale grey *Hyloicus pinastri* (p. 25)
–	Wings and body coloured otherwise 16
16(15)	Hindwing bright yellow; abdomen banded black and yellow; thorax with prominent skull-like marking; very large species *Acherontia atropos* (p. 23)
–	Termen of forewing with deep indentations; green or olive-green; prominent discal and dorsal patches; much smaller species *Mimas tiliae* (p. 26)

Sphinginae

AGRIUS Hübner
Agrius Hübner, [1819], *Verz.bekannt.Schmett.*: 140.
Herse Agassiz, [1846], *Nom.Syst.Gen.Lepid.*: 35.

A small genus with five species distributed throughout the world.

Characterized by the presence of a subapical fovea roofed over with scales on the inner surface of the second segment of the labial palpus.

AGRIUS CINGULATA (Fabricius)
The Sweet-potato Hornworm
Sphinx cingulata Fabricius, 1775, *Syst.Ent.*: 545.
Sphinx druraei Donovan, 1810, *Br.Ins.* **14**: 1, pl.469.
Type locality: America.

This Neotropical species occurred as accidental importations in London in 1778 and 1826 (Barrett, 1895, **2**: 35). It is similar in appearance to *A. convolvuli* (Linnaeus), but the forewing has conspicuous transverse markings and the abdomen and hindwing are suffused with brighter pink. The larva feeds on sweet-potato (*Ipomoea batatas*).

AGRIUS CONVOLVULI (Linnaeus)
The Convolvulus Hawk-moth
Sphinx convolvuli Linnaeus, 1758, *Syst.Nat.* (Edn 10) **1**: 490.
Type locality: not stated.

Description of imago (Pl.2, fig.1)
Wingspan 94–120mm. Head grey; antenna in male pectinate, in female simple, apically slightly hooked; haustellum *c.*75mm long. Thorax grey sprinkled whitish, especially in male. Forewing grey sprinkled whitish; male with whitish, broad, indistinct ante- and postmedian fasciae; long black streaks on veins in disc, at apex and basally on hind margin, especially in female. Hindwing pale grey; subbasal and median fasciae and terminal shade dark grey. Abdomen grey, banded with pink and black, edged white on each segment; broad dorsal stripe.
Similar species. *Sphinx ligustri* Linnaeus which has the hindwing pinkish white.

Life history
Ovum. Round and smooth; chorion bright green. Usually laid singly on the foodplant.

Larva. Full-fed *c.*100mm long. There are two distinct colour forms. (**a**) Head brownish green with black stripes. Body bright apple-green; spiracles black; subdorsal lines blackish; oblique lateral stripes yellow-ochre or brown; horn brownish red with black apex, posteriorly concave. (**b**) Head ochreous with black stripes. Body blackish or purplish brown, oblique lateral stripes yellow-ochre or pinkish; horn brownish red with black apex. Both forms are variable; the oblique lateral stripes can be whitish or bluish black with dark spots and patches on the body. The light form appears to be more common. Green penultimate instar larvae sometimes produce brown last instar (R. F. Bretherton pers.comm.). Imagines resulting from the two forms are identical. The larva feeds on various species of *Convolvulus*, in Britain especially on field bindweed (*Convolvulus arvensis*), and *Ipomoea*. Rarely found in the wild in Great Britain and Ireland.

Pupa. 47mm long; brown; proboscis sheath forming a very large projecting loop. In a large, very fragile subterranean cocoon.

Imago. A regular immigrant. Numbers vary greatly and years in which they have exceeded 250 are 1859, 1875, 1887, 1898, 1901, 1917, 1945, 1950 and 1976. The greatest number recorded was 505 specimens in 1945 (Williams, 1958). Flies at dusk and again at dawn between July and November. Feeds from the tubular flowers of such garden plants as *Nicotiana*, *Petunia* and *Lonicera*.

Distribution (Map 1)

It has been recorded from almost all parts of Great Britain and Ireland, and occurs throughout most of Eurasia, Africa and the Pacific. Eastern specimens appear to be slightly smaller than those in the western part of its range.

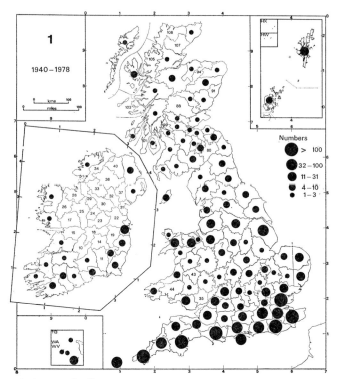

Agrius convolvuli

ACHERONTIA Laspeyres

Acherontia Laspeyres, 1809, *Jena. allg.Lit.-Ztg* **4**: 100.

A small genus with only one species in Europe and Africa which is a regular migrant to Great Britain and Ireland.

Characterized by the large size; brown and yellow coloration; short, thick, ciliate haustellum and thick straight antenna with terminal hook.

ACHERONTIA ATROPOS (Linnaeus)
The Death's-head Hawk-moth

Sphinx atropos Linnaeus, 1758, *Syst.Nat.* (Edn 10) **1**: 490.
Type locality: Europe.

Description of imago (Pl.2, fig.3)

Wingspan 102–135mm. Antenna short and stout, apically hooked; haustellum short. Thorax blackish brown with ochreous yellow markings resembling skull and crossbones. Forewing blackish brown with irregular areas of whitish scales; indistinct blackish brown lines and fasciae; veins blackish; discal spot white. Hindwing deep yellow; postmedian and terminal fasciae blackish brown; veins blackish brown from postmedian fascia to termen. Abdomen ochreous yellow, each segment with a black transverse band; dorsal stripe broad, grey-blue, interrupted by the black bands.

There is considerable minor variation in the forewing pattern.

Life history

Ovum. Oval; chorion slightly roughened, green. Laid singly on the upper surface of leaves of potato (*Solanum tuberosum*) and other Solanaceae.

Larva. Full-fed *c.*125mm long, stout. Head yellow with two black stripes. Body green or yellowish green, spotted dorsally with purple and with seven purple or purplish brown oblique stripes edged with yellow; spiracles black; true legs black with white spots; horn yellow, granulose, S-shaped. There are brown, white-freckled, and yellow forms. The brown form is predominantly nocturnal unlike the other forms which also feed during the day. An audible clicking noise can be made with the mandibles. In addition to various species of Solanaceae, including Duke of Argyll's teaplant (*Lycium barbarum*), it has also been found feeding on other shrubs and ash (*Fraxinus excelsior*). Has been found feeding from late July but most frequently in September and October.

Pupa. Brown, smooth; cremaster corrugated, with two points. In a large, very fragile cocoon 5–10cm below soil surface. Said to produce a squeaking noise shortly before emergence of the imago.

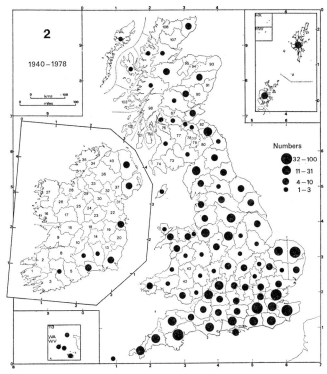

Acherontia atropos

Imago. A regular migrant to Great Britain and Ireland which is unable to survive the winter although early spring arrivals can produce a late autumn generation. Recorded in small numbers in most years, the greatest numbers being in 1931 (101) and 1956 (381); usually recorded from May onwards with most in August and September, when larvae are frequently found in potato fields. Known to feed on sap from trees and honey from beehives. Can emit a squeaking noise by forcing air through the proboscis.

Distribution (Map 2)

It has been recorded throughout Great Britain and Ireland. Occurs throughout Europe including the Canary Islands and Azores; also in the Middle East; Africa; Madagascar and the Seychelles.

MANDUCA Hübner
Manduca Hübner, [1807], *Samml.exot.Schmett.* **1**: pl. [170].

An entirely American genus, the majority of species Neotropical. It includes more than 40 large or very large species.

Characterized by the absence of spines on the foretibia and the strongly developed midtarsal comb.

MANDUCA QUINQUEMACULATUS (Haworth)
The Tomato Hornworm

Sphinx quinquemaculatus Haworth, 1803, *Lepid.Br.*: 59.

Type locality: England (presumably based on an imported specimen).

The occurrence of this Nearctic species in Britain is undoubtedly the result of accidental importations. It is somewhat similar to *Agrius convolvuli* (Linnaeus) but has the abdomen banded with yellow and black. A number of specimens are recorded as having been taken at Chelsea, Hull and Leeds in the nineteenth century (Barrett, 1895: 35). It occurs from Mexico northwards through Texas, the Mississippi valley, North Carolina, South Dakota and southern Ontario to Nova Scotia. The larva feeds on tomato (*Lycopersicum* spp.) and potato (*Solanum tuberosum*) (Hodges, 1971: 32).

MANDUCA SEXTA (Johansson)
The Tobacco Hornworm

Sphinx sexta Johansson, 1763, *Centuria Insect.rar.*: 27.
Sphinx carolina Linnaeus, 1764, *Mus.Lud.Ul.*: 346.

Type locality: U.S.A.; Carolina and Jamaica.

This Nearctic species is somewhat similar to *Agrius convolvuli* (Linnaeus) but has the abdomen banded with yellow and black as in *Manduca quinquemaculatus* (Haworth). It can be separated from the latter by the wavy, whitish subterminal line and more sharply chequered fringes. *M. sexta* has been recorded only once in Britain. This accidental importation is referred to in Curtis (1828) as follows, 'a pair were taken by Mr. Thompson (a friend of Mr. Plastead's), the 28th August 1796, at West Cowes, Isle of Wight, which we are so fortunate as to possess'.

The larva feeds on Solanaceae.

SPHINX Linnaeus
Sphinx Linnaeus, 1758, *Syst.Nat.* (Edn 10) **1**: 489.
Spectrum Scopoli, 1777, *Intr.Hist.Nat.*: 413.
Lethia Hübner, [1819], *Verz.bekannt.Schmett.*: 140.

A large genus, particularly Nearctic, with a few species occurring in the Palaearctic region, one of which is resident in Britain.

Characterized by the generally dull grey or grey-brown colour and the series of dorsolateral transverse bands of white and black on the abdomen.

SPHINX LIGUSTRI Linnaeus
The Privet Hawk-moth

Sphinx ligustri Linnaeus, 1758, *Syst.Nat.* (Edn 10) **1**: 490.

Type locality: not stated.

Description of imago (Pl.2, fig.6)

Wingspan 100–120mm. Antenna one-third length of forewing, apically hooked; haustellum well developed. Thorax black; tegulae white. Forewing brown, sprinkled white, basally pinkish-suffused; median fascia dark brown, narrow at apex, broad at dorsum; subterminal line black, edged white; veins in discal area black. Hindwing pale pinkish white; subterminal, median and subbasal fasciae black. Abdomen brown, banded with pink and black; dorsal line broad, grey with thin black central line.

Similar species. *Agrius convolvuli* (Linnaeus) q.v.

Life history

Ovum. Oval, flattened top and bottom, smooth; chorion pale green. Laid singly on leaves or stems of the foodplant.

Larva. Full-fed c.75mm long. Head green with black lateral stripes. Body green with seven oblique white stripes, each dorsally edged purple; spiracles reddish ochreous; claspers black; horn shiny black, ventrally yellow. Some have double purple stripes and others two or more horns. Turns brownish before pupation. Feeds on privet (*Ligustrum vulgare*), lilac (*Syringa vulgaris*) and ash (*Fraxinus excelsior*) in July and August.

Pupa. 70mm long; brown; proboscis sheath almost separate from thorax; cremaster with two apical and two subapical points. Subterranean at a greater depth than any other British sphingid. Overwinters, sometimes for two years.

Imago. Univoltine. Resident but not very common; reinforced by immigration from mainland Europe. Flies at dusk in June and July.

Distribution (Map 3)

Throughout south and south-eastern England being much scarcer in the Midlands and Wales. Old records in the

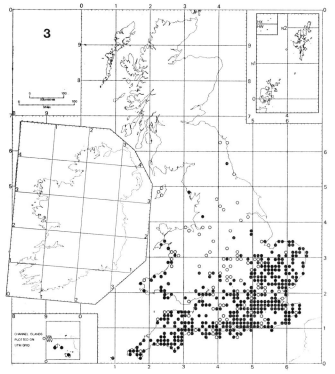
Sphinx ligustri

literature from northern England and Scotland possibly refer to migrants. It is said to be more common on the North Downs than on the Bagshot Sand or Greensand areas. Absent from Ireland. It occurs throughout most of the Palaearctic region.

SPHINX DRUPIFERARUM Smith
The Wild-cherry Sphinx

Sphinx drupiferarum Smith, 1797, *in* Smith & Abbot, *Nat.Hist.rarer lepid.Ins.Georgia* **1**: 71, pl.36.
Type locality: U.S.A.; Georgia.

This North American species has only once been recorded from Britain. A specimen was taken in a moth trap on 21 May 1970 at Weston-super-Mare, Somerset (Blathwayt, 1970). This was, presumably, an accidental importation.

It is similar in appearance to *S. ligustri* Linnaeus but has very little pink on the abdomen and hindwing.

HYLOICUS Hübner
Hyloicus Hübner, [1819], *Verz.bekannt.Schmett.*: 138.

A small genus with only one species in Great Britain.
Forewing long and narrow; hindwing with costa well arched; legs of uniform length, spurs on hindtibia very long.

HYLOICUS PINASTRI (Linnaeus)
The Pine Hawk-moth

Sphinx pinastri Linnaeus, 1758, *Syst.Nat.* (Edn 10) **1**: 492.
Type locality: not stated.

Description of imago (Pl.2, fig. 8)
Wingspan 72–80mm. Head dark brownish grey; antenna in male almost half length of forewing, in female shorter, tip hooked; haustellum well developed; eye lashed. Thorax dark grey, metathorax and lateral stripes black. Forewing dark grey densely irrorate white; three or four black dashes in disc; median and antemedian fasciae darker grey, indistinct; sometimes a black subbasal dash at dorsum and a black apical dash; fringe whitish, blackish-barred. Hindwing brownish grey, darker than forewing; veins dark grey; fringe whitish, blackish-barred. Abdomen grey, ringed black and whitish grey subdorsally; dorsal stripe grey with black central line.

Variation is confined to the intensity of the wing pattern, some specimens being almost unicolorous brown.

Life history
Ovum. Lozenge-shaped, yellow, turning grey before hatching. Laid in groups of two or three along pine needles.
Larva. Full-fed *c*.75mm long. In early instars body green with white stripes. In later instars dark green; dorsal line red-brown edged white or yellow; lateral marks brown with spiracular marks white and spiracles orange-red; horn dark brown with black spots. Feeds in August and September on needles of Scots pine (*Pinus sylvestris*) and Norway spruce (*Picea abies*).
Pupa. C.40mm long; brown, smooth; proboscis sheath short, rough, almost separate from thorax. Just subterranean or beneath fallen pine needles. Overwinters, sometimes for two years.
Imago. Univoltine. Flies at dusk from the end of June until August, feeding on the wing from flowers. Rests on pine trunks during the day.

Distribution (Map 4)
Almost entirely confined to Dorset, Hampshire, Surrey, Norfolk and Suffolk. Absent from Ireland. It has spread considerably since the 1930s owing to the increased planting of conifers. This species occurs throughout the pine forests of the Palaearctic region. It is sometimes a serious forest pest.

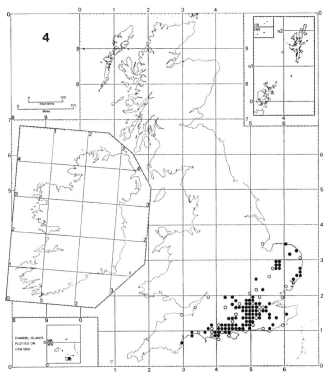
Hyloicus pinastri

MIMAS Hübner
Mimas Hübner, [1819], *Verz.bekannt.Schmett.*: 142.

A small genus with only one species in Great Britain.
   Haustellum short, poorly developed; tibiae with appressed scales and all spurs present.

MIMAS TILIAE (Linnaeus)
The Lime Hawk-moth
Sphinx tiliae Linnaeus, 1758, *Syst.Nat.* (Edn 10) **1**: 489.
Type locality: not stated.

Description of imago (Pl.1, figs 1–3)
Wingspan 70–80mm. Antenna less than half length of forewing; haustellum short. Thorax greenish grey with dorsal and lateral stripes darker. Forewing varying from light to dark brown, greyish-tinged; basal fascia obscure, dark greenish grey; median fascia dark olive-green, sometimes broken subdorsally, with a median terminal projection; terminal fascia broad, dark greenish grey; apical patch pale whitish green; termen irregularly dentate at tornus and medially. Hindwing ochreous brown; subterminal fascia and disc darker grey-brown. Abdomen greenish grey.
   Variation occurs in the ground colour, which is reddish brown in ab. *brunnea* Tutt (fig.2), and in the size and extent of the median fascia which may be reduced to a discal spot in ab. *centripuncta* Clark (fig.3).

Life history
Ovum. Ovate, pale shining green. Laid singly or in pairs on the underside of leaves of the foodplant in June or July.
Larva. Full-fed *c*.60mm long. Body tapering strongly towards head; green, turning brownish purple immediately before pupation; lateral stripes yellow; spiracles red; horn blue dorsally, red and yellow ventrally. Feeds from July to September on lime (*Tilia* spp.), English elm (*Ulmus procera*) and alder (*Alnus glutinosa*); probably also on birch (*Betula* spp.), oak (*Quercus* spp.) and *Prunus* spp. Feeds nocturnally, resting during the day on the underside of a leaf.
Pupa. *C*.35mm long; dark red-brown, roughened; proboscis sheath reaching wing-cases; cremaster rough with acute tubercles. In a loose cocoon usually just below the surface of the soil but sometimes in a crevice in the bark of a tree trunk. This stage lasts from August or September to May of the following year.
Imago. Univoltine. Flies at dusk in May and June. When at rest its wings resemble young leaves. Hybrids between *M. tiliae* and *Smerinthus ocellata* (Linnaeus) have been reported (Rothschild & Jordan, 1903).

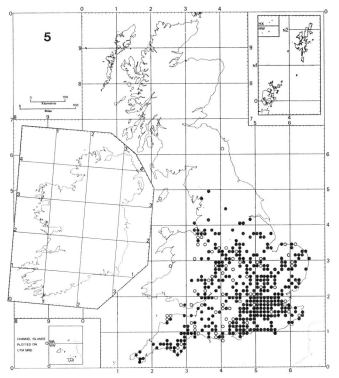
Mimas tiliae

Distribution (Map 5)
Throughout England as far north as a line from the Mersey to the Humber but nowhere very common; rare in Wales; absent from Scotland and Ireland. Abroad throughout Europe, except the Arctic, eastwards to Japan.

SMERINTHUS Latreille
Smerinthus Latreille, [1803], *in* Sonnini's Buffon, *Hist.nat. gén.part.Crust.Ins* **3**: 401.
Dilina Dalman, 1816, *K.svenska VetenskAkad.Handl.* **1816**: 205.

A small Holarctic genus with only one species in Great Britain and Ireland.
Costal margin of hindwing almost straight; foretibia with terminal spine.

SMERINTHUS OCELLATA (Linnaeus)
The Eyed Hawk-moth
Sphinx ocellata Linnaeus, 1758, *Syst.Nat.* (Edn 10) **1**: 489.
Type locality: not stated.

Description of imago (Pl.1, fig.9)
Wingspan 75–95mm. Antenna less than half length of forewing; haustellum weak, short. Thorax dark chocolate-brown, laterally pale brown. Frenulum reduced; retinaculum absent. Forewing pinkish brown; antemedian fascia fuscous; median and postmedian fasciae united medially, chocolate-brown; reniform stigma outlined brown; two broad subterminal lines with a median dark chocolate-brown blotch between them; terminal fascia forming a large triangular subapical mark; termen dentate at tornus. Hindwing rosy pink paling to ochreous at margin; a large black-edged bluish grey tornal eye-spot with a blackish centre. Abdomen pinkish brown.

Life history
Ovum. Ovate, smooth, yellowish green. Laid singly or in pairs on the underside of leaves or on petioles of the food-plant.
Larva. Full-fed *c.*70mm long. Body green, dotted white, with a yellow, blue or grey tinge; oblique lateral stripes yellow or violet, edged anteriorly with darker green; spiracles ringed red; sometimes red subdorsal or lateral spots or patches; horn light or grey-blue, sometimes green except at base. Feeds at night from June to September on many species of willow (*Salix*), especially small bushes, and on apple (*Malus* spp.) both wild and cultivated; will also feed on plum (*Prunus* spp.) and poplar (*Populus* spp.) in captivity. Sometimes 'sunbathes' during the day.
Pupa. *C.*40mm long, stout; blackish brown, smooth and glossy; proboscis sheath fused with thorax; cremaster broad basally, rough with lateral, pointed, tubercles. In a fragile cocoon just below the surface of the soil. This stage usually lasts from September to May of the following year.
Imago. Univoltine or occasionally partially bivoltine, with a second emergence in autumn. Flies at dusk from May onwards. Hybridizes with *Laothoe populi* (Linnaeus) and

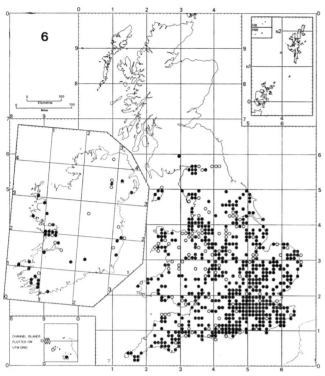

Smerinthus ocellata

Mimas tiliae (Linnaeus) but the offspring are normally infertile.

Distribution (Map 6)

Throughout England and Wales, becoming local in the north; absent from Scotland except for one record from Dumfriesshire; local in Ireland; also occurs in the Isle of Man and the Channel Islands. Abroad throughout Europe, except the Arctic, eastwards to west Siberia and Asia Minor.

LAOTHOE Fabricius

Laothoe Fabricius, 1807, *Magazin Insektenk.(Illiger)* **6**: 287.

A small Palaearctic genus with one species in Great Britain and Ireland.

Haustellum very short; forewing broad, termen dentate.

LAOTHOE POPULI (Linnaeus)
The Poplar Hawk-moth

Sphinx populi Linnaeus, 1758, *Syst.Nat.* (Edn 10) **1**: 489.
Type locality: not stated.

Description of imago (Pl.1, fig.6)

Wingspan 72–92mm. Antenna less than half length of forewing; haustellum short. Frenulum in male absent, in female vestigial; retinaculum absent. Forewing ashy grey-brown, rather variably suffused pinkish brown; veins pale ochreous; antemedian fascia narrow, dark brown; median fascia broad, dark brown, especially dorsally; subterminal fascia narrow, broader on dorsum, dark brown; terminal fascia forming a large triangular subapical mark, dark brown; reniform stigma whitish; termen scalloped. Hindwing colour and pattern as forewing but with a large, dull red, basal patch; margin scalloped. Thorax and abdomen grey-brown.

Specimens with the ground colour pale buff sometimes occur, especially in the female. Gynandromorphs are not uncommon (Tutt, 1902).

Life history

Ovum. Roundish, shining, pale yellowish green, finely reticulate. Laid in small groups on leaves of poplar (*Populus* spp.) and sallow (*Salix* spp.), generally on the underside.

Larva. Full-fed *c.*60mm long. Head triangular. Body green or blue-green, rough, spotted yellow; lateral stripes yellow; spiracles red; sometimes a series of red subdorsal spots; horn yellowish green, sometimes red-tipped. Occasionally two horns may be present. Feeds from July to September.

Pupa. *C.*35mm long; blackish brown, rather rough; proboscis fused with thorax. In a very slight earthen cocoon just below the surface of the soil. This stage usually lasts from September to May of the following year, but occasionally only until late autumn.

Imago. Univoltine, occasionally bivoltine. Flies from dusk onwards in May and June; a small second brood may occur in the autumn. When at rest it closely resembles dead leaves. Hybridization occurs occasionally with *Mimas tiliae* (Linnaeus) and *Smerinthus ocellata* (Linnaeus) and there is even an account of a male pairing in captivity with a crippled female *Saturnia pavonia* (Linnaeus) (Chinery, 1972).

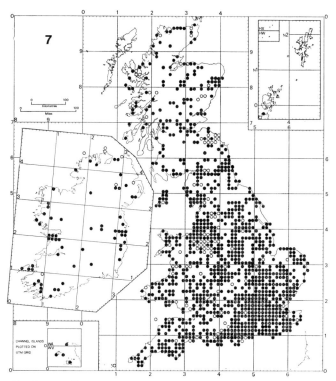
Laothoe populi

Distribution (Map 7)
Throughout Great Britain and Ireland and the Channel Islands. Western Palaearctic; throughout Europe, just reaching the Arctic, eastwards to the U.S.S.R. and Syria.

Macroglossinae

HEMARIS Dalman

Hemaris Dalman, 1816, *K.svenska VetenskAkad.Handl.* **1816**: 207.
Hemeria Billberg, 1820, *Enum.Ins.Mus.Blbg*: 82.
Haemorrhagia Grote & Robinson, 1865, *Proc.ent.Soc.Philad.* **5**: 173.

A moderate-sized Holarctic genus with two resident species in the British Isles.
Characterized by the wings having areas free of scales, which result from loose scales being shed as soon as the moths start to fly.

HEMARIS TITYUS (Linnaeus)
The Narrow-bordered Bee Hawk-moth
Sphinx tityus Linnaeus, 1758, *Syst.Nat.* (Edn 10) **1**: 493.
Sphinx bombyliformis sensu Esper, 1780, *Schmett.* **2** (14): 180.
Type locality: not stated.

Description of imago (Pl.2, fig.4)
Wingspan 41–46mm. Antenna more than half length of forewing, clubbed with thin, pointed, hooked tip; eye lashed; haustellum well developed. Thorax pale yellow-brown. Tibiae densely hairy. Forewing transparent except as follows: veins black; costa and dorsal basal patch blackish mixed ochreous yellow; terminal fascia broad at apex narrowing to tornus, blackish mixed ochreous yellow. Hindwing transparent except that veins are black; basal patch and narrow terminal border are blackish mixed ochreous yellow. Abdomen ochreous yellow with two narrow black bands; anal tuft large, almost black.

Similar species. *H. fuciformis* (Linnaeus) which has a dark brown discal spot, broader terminal fascia and a broad, reddish brown median band on abdomen.

Life history
Ovum. Ovate, smooth, shining, green. Laid singly on the underside of leaves of devil's-bit scabious (*Succisa pratensis*) or field scabious (*Knautia arvensis*).
Larva. Full-fed *c*.40mm long. Body green, sometimes whitish to bluish green, with whitish yellow dots bearing setae; dorsal lines yellowish with purplish or brownish red spots; spiracles within purplish or brownish red patches; ventral stripe purplish or brownish red; horn straight, rough, reddish brown. Turns reddish shortly before pupation. Feeds on the underside of leaves of the foodplant in July and August.

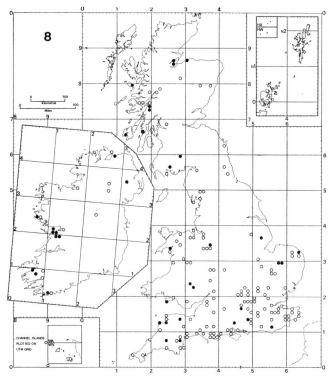
Hemaris tityus

Pupa. Dark brown. In a coarse, fragile cocoon just beneath the soil surface.

Imago. Univoltine. Diurnal, flying in May and June. In southern Britain it is a woodland species visiting flowers, especially bugle, for nectar. The flight is rapid and darting and it hovers in front of the flower when feeding. In flight has the appearance of a bumble-bee. In Ireland, especially the west, occurs on wet bogs where the foodplant grows and where it has been observed visiting, in numbers, the flowers of red valerian in adjacent gardens (A. M. Emmet pers.comm.). In Scotland also frequents bog and moorland.

Distribution (Map 8)

In Britain as far north as the Inner Hebrides and East Ross; Ireland and the Channel Islands; local and uncommon. It has decreased considerably during the last 25 years. Western Palaearctic; extending eastwards to Asia Minor and northern Iran.

HEMARIS FUCIFORMIS (Linnaeus)
The Broad-bordered Bee Hawk-moth
Sphinx fuciformis Linnaeus, 1758, *Syst.Nat.* (Edn 10) **1**: 493.
Sphinx bombyliformis sensu Ochsenheimer, 1808, *Schmett. Eur.* **2**: 189.
Type locality: Europe.

Description of imago (Pl.2, fig.2)

Wingspan 46–52mm. Antenna more than half length of forewing, clubbed with thin, pointed, hooked tip; eye lashed; haustellum well developed. Thorax pale yellow-brown. Tibiae densely hairy. Forewing transparent except for black veins, dark reddish brown costa and basal patch and broad, dark reddish brown terminal fascia narrowing towards tornus; discal spot dark brown. Hindwing transparent with veins black; basal patch and broad terminal border dark reddish brown. Abdomen ochreous yellow with broad, dark reddish brown median band; anal tuft large, almost black.

Similar species. H. tityus (Linnaeus) *q.v.*

Life history

Ovum. Roundish, bright green. Laid on the underside of leaves of honeysuckle (*Lonicera periclymenum*), bedstraw (*Galium* spp.) or snowberry (*Symphoricarpos rivularis*).

Larva. Full-fed *c*.50mm long. Head dark green. Body green, dorsally whitish green, ventrally reddish brown; dorsal lines broken, green and yellow; spiracles within reddish brown patches; horn slightly curved, red-brown, brown at tip, violet at base. Changes to purplish brown just before pupation. Feeds in July and August on the underside of leaves of the foodplant.

Pupa. Blackish brown, paler between segments; rough. In a fragile cocoon just beneath the surface of the soil.

Imago. Univoltine. Diurnal, flying in May and June, visiting flowers, especially bugle and rhododendron, for nectar. The flight is similar to that of *H. tityus*.

Distribution (Map 9)

Local and uncommon in woodland rides in England and Wales as far north as a line from the Mersey to the Humber. Absent from Scotland and Ireland. It has become much scarcer in recent years, possibly through heavy parasitization. Western Palaearctic; extending eastwards to northwest India.

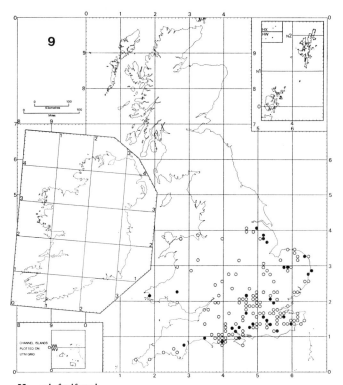
Hemaris fuciformis

MACROGLOSSUM Scopoli
Macroglossum Scopoli, 1777, *Intr.Hist.Nat.*: 414.

A large, almost entirely oriental genus with one western Palaearctic species which occurs as a regular immigrant in Great Britain and Ireland.

Characterized by the strongly clubbed antenna, forward-projecting palpus and absence of spines on the tibiae except for the short midtibial spur, which has a prominent comb of spines.

MACROGLOSSUM STELLATARUM (Linnaeus)
The Humming-bird Hawk-moth
Sphinx stellatarum Linnaeus, 1758, *Syst.Nat.* (Edn 10) **1**: 493.
Type locality: not stated.

Description of imago (Pl.2, fig.5)
Wingspan 50–58mm. Antenna more than half length of forewing, thickened distally with very short, thin, hooked tip; haustellum well developed; eye lashed. Thorax fuscous brown. Tibiae hairy. Forewing fuscous brown; subbasal line and antemedian lines faint; median and postmedian lines black, postmedian paler dorsally; subterminal line faint; terminal fascia dark fuscous brown; discal spot small, dark brown. Hindwing orange, fuscous-brown basally; terminal shade reddish brown narrowing towards, but not reaching, tornus. Abdomen fuscous-brown with median, lateral white patches and posterior black lateral and dorsal patches; anal tufts large, black.

Life history
Ovum. Roundish, smooth, glossy, green. Laid singly on bedstraw (*Galium* spp.) or wild madder (*Rubia peregrina*).
Larva. Full-fed c.60mm long. Body green or reddish brown with white dots; dorsal line darker; subdorsal line white; subspiracular line yellow; lines sometimes edged brown; spiracles black; horn blue, yellow at tip. Has been found in Britain feeding in July and August.
Pupa. Ochreous or brown, darker on wing-cases and spiracles; proboscis sheath forming a small beak-like projection; cremaster conical. In a loosely spun cocoon on the surface of the soil.
Imago. Immigrant. Diurnal, hovering like a humming-bird in front of the flowers at which it is feeding, the wings producing an audible high-pitched hum. Occurs almost every year with numbers varying greatly and has been recorded throughout the year; the winter immigrants probably originate in southern France. Believed to overwinter in the south-west in mild winters. Numbers recorded exceeded 1,000 in 1899, 1947 and 1955, the greatest number being 4,250 in 1947.

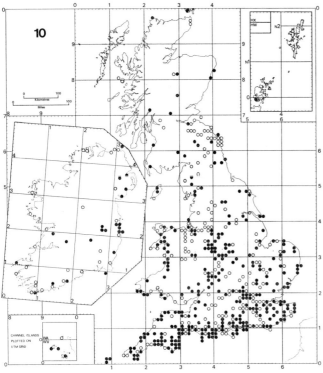
Macroglossum stellatarum

Distribution (Map 10)
Occurs irregularly throughout Great Britain and Ireland. Palaearctic.

DAPHNIS Hübner
Daphnis Hübner, [1819], *Verz.bekannt.Schmett.*: 134.

This genus contains three species, one Palaearctic, one Neotropical and one Oriental. *Sphinx nerii* Linnaeus was designated as the type-species by Curtis in 1837 (*Br.Ent.* **14**: 626).

DAPHNIS NERII (Linnaeus)
The Oleander Hawk-moth
Sphinx nerii Linnaeus, 1758, *Syst.Nat.* (Edn 10) **1**: 490.
Type locality: not stated.

Description of imago (Pl.1, fig.4)
Wingspan 72–115mm. Antenna with thin, hooked tip, in male pectinate, in female thickened distally; eye large; haustellum well developed. Thorax green with two oblique white stripes. Forewing deep olive-green; circular basal patch ochreous white with dark green centre; antemedian fascia pinkish white, strongly convex; median fascia pinkish white, united with antemedian fascia dorsally; a white dentate postmedian dorsal line reaching less than half across wing; subterminal fascia pinkish white obscured by a large grey blotch dorsally; apical streak white, dorsally bifurcate. Hindwing greyish green, basally fuscous; double subterminal line waved, inner line ochreous white, outer line dark green; a black subtornal spot. Abdomen greyish green, first segment with broad ochreous white band.

Life history
Ovum. Roundish, rather small, light green. Laid on either surface of leaves of the foodplant.
Larva. Full-fed c.100mm long. Body tapers anteriorly; yellowish green; subdorsal line on abdomen with white, black-encircled dots above and below from segment 2; white lines anteriorly on segments 2–5 with several white spots posterior to each line; large lateral ocellus on metathorax; horn yellowish orange, green at base. A rare form with the body colour deep golden orange occurs. Colour changes to olive-brown before pupation. Not recorded in Great Britain or Ireland; in southern Europe it feeds on greater periwinkle (*Vinca major*), lesser periwinkle (*Vinca minor*), oleander (*Nerium oleander*) and grape-vine (*Vitis vinifera*).
Pupa. C.25mm long; pale brown with black median line on thorax; small black dots on thorax and abdomen; spiracles black; cremaster long and thin, conical, with two short points. In a subterranean cocoon.
Imago. A rare immigrant, absent in many years. The greatest number recorded was 13 in 1953, recent records being of four in 1969 and one in 1975. Usually occurs between August and October.

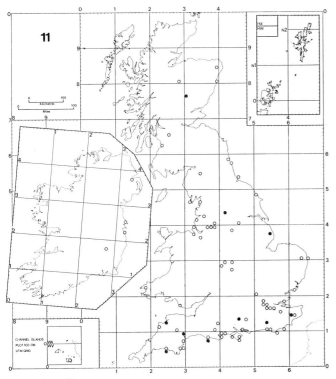
Daphnis nerii

Distribution (Map 11)

There are records from many parts of Great Britain and Ireland. A wide-ranging migrant known throughout the southern part of the western Palaearctic, Ethiopian and Oriental regions.

DAPHNIS HYPOTHOUS (Cramer)

Sphinx hypothous Cramer, 1780, *Papillons exotiques* **3**: 165, pl.285, fig.d.

Type locality: Indonesia; Amboina, Moluccas.

A specimen of this Asiatic species, which is similar to *Daphnis nerii* (Linnaeus), was taken at Crieff, Perthshire in July 1873 (South, 1961).

HYLES Hübner

Hyles Hübner, [1819], *Verz.bekannt.Schmett.*: 137.
Celerio Agassiz, [1846], *Nom.Syst.Gen.Lepid.*: 14.

A small genus of world-wide distribution with six Palaearctic species, three of which regularly migrate to Great Britain and Ireland.

Antenna broad near apex; first tarsal segment of foreleg with long, stout spines.

HYLES EUPHORBIAE (Linnaeus)
The Spurge Hawk-moth

Sphinx euphorbiae Linnaeus, 1758, *Syst.Nat.* (Edn 10) **1**: 492.

Type locality: not stated.

Description of imago (Pl.1, fig.7)

Wingspan 64–77mm. Antenna in male pectinate, in female simple, club-shaped with hooked tip; eye lashed; haustellum strongly developed. Forewing grey-brown, more or less pinkish-tinged, flecked with dark grey; basal blotch olive-brown, dorsally black; subterminal fascia triangular, wide dorsally, tapering to apex, dark olive-brown; discal spot large, dark olive-brown; termen pale olive-brown. Hindwing pale pinkish red, basal patch and postmedian fascia black; anal angle with rounded white patch. Abdomen with olive-brown dorsal stripe; anterior segments ochreous white with black bands; posterior segments olive-brown with dark brown bands.

Variation occurs in the extent of the pink suffusion and dark grey flecks.

Life history

Ovum. Roundish, smooth, bright green. Laid singly or in small batches on leaves of spurge (*Euphorbia* spp.).

Larva. Full-fed *c*.80mm long. Head crimson, dorsally black. Body black; dorsal stripe crimson; dorsal patches crimson, yellow or cream-coloured; many buff subdorsal and lateral dots giving an almost bronze appearance; occasionally large buff lateral spots alternate with dots; horn crimson with apex black. Feeds during August and September.

Pupa. *C*.35mm long; pale brown, head and thorax with black dots; proboscis sheath only feebly prominent; wing- and antenna-cases streaked blackish; abdominal segments with narrow black bands; spiracles black; cremaster bifid, black. In a fragile cocoon just below the soil surface.

Imago. A very occasional immigrant, absent in many years. Flies at dusk in June, visiting flowers.

Distribution (Map 12)

Most records are from southern England; not recorded

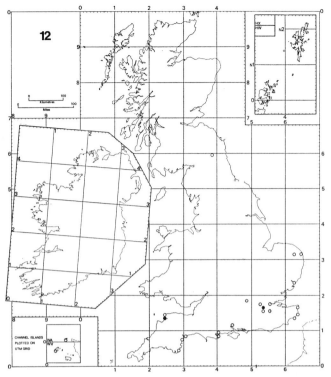

Hyles euphorbiae

from Scotland or Ireland. The greatest number recorded in any one year was seven in 1935. Abroad from the Canary Islands eastwards to northern Iran and north-west India; widely distributed in Europe from Spain to the Caucasus. It has been introduced into western Canada in an attempt to control certain *Euphorbia* weed species.

HYLES GALLII (Rottemburg)
The Bedstraw Hawk-moth

Sphinx gallii Rottemburg, 1775, *Naturforscher, Halle* 7: 107.
Sphinx galii [Denis & Schiffermüller], 1775, *Schmett.Wien.*: 42.
Type locality: Germany.

Description of imago (Pl.1, fig.5)

Wingspan 64–78mm. Antenna clubbed with a fine, hooked tip; eye lashed; haustellum well developed. Thorax dark olive-brown; tegulae white. Forewing dark olive-brown; a broad ochreous yellow stripe from basal half of dorsum to apex, with three costal projections and tapering to a point at apex; terminal fascia greyish brown. Hindwing pale ochreous; basal area and subterminal fascia black; median band pinkish red; anal patch white. Abdomen in anterior half ochreous white, banded black with olive-brown dorsal stripe; in posterior half dark olive-brown, banded ochreous white laterally; four small white dorsal spots.

Life history

Ovum. Roundish, slightly flattened, green. Laid on leaves of bedstraw (*Galium* spp.) and willowherb (*Epilobium* spp.).

Larva. Full-fed *c.*75mm long. Head pink. Body colour varies from greenish olive through reddish brown to black; ventral surface pink, except in black forms; large subdorsal buff, black-ringed spots, elongate on abdominal segment 8; spiracles white, ringed black; horn red. Feeds during August and September. In captivity will eat *Fuchsia*. Was reported in thousands in Lancashire and on the east coast in 1888 and from 1955 to 1958 in Norfolk.

Pupa. C.40mm long; reddish brown with black markings; cremaster long, bifid at tip. In a fragile subterranean cocoon in light soil.

Imago. An infrequent immigrant. Occasionally recorded as having become established for a few years, *e.g.* 1956–58 in north Norfolk, following large immigrations. Years when numbers have exceeded 50 are 1870, 1888, 1955 and 1973 when 65 were recorded; the greatest number was 71 in 1870. For a full account of the immigration of this species see de Worms (1975). Flies from May to July.

Distribution (Map 13)

This species has been recorded throughout Great Britain and Ireland as far north as Unst, Shetland Islands. Holarctic; from western Europe to Japan and the northern half of North America.

HYLES NICAEA (de Prunner)
The Mediterranean Hawk-moth

Sphinx nicaea de Prunner, 1798, *Lepid.Pedemont.*: 86.
Type locality: Italy; Piedmont.

This southern Palaearctic species very much resembles *H. euphorbiae* (Linnaeus) but differs in the pink tornal spots of the hindwing and in its usually greater size.

Two almost full-fed larvae were found feeding on common toadflax (*Linaria vulgaris*) on the Devon coast on 20 August 1954. These larvae produced imagines on 12 July 1955 (Wilkinson, 1956). This species is not normally migratory and the usual larval foodplant is spurge (*Euphorbia* spp.).

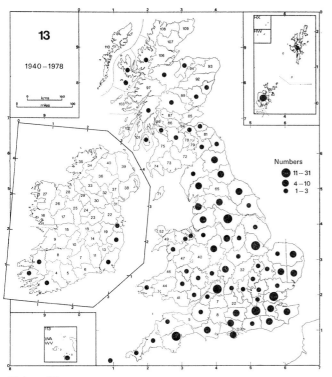

Hyles gallii

HYLES HIPPOPHAES (Esper)

Sphinx hippophaes Esper, 1793, *Schmett.* **2** (Suppl.) 2 Abs.: 6, pl.38, figs 1,2.

Type locality: Rumania; Wallachia.

There is no evidence to suggest that this Eurasiatic species has occurred in Great Britain and Ireland except possibly an accidental importation on one occasion. Stainton (1857, **1**: 91) writes '*Deilephila Hippophaes* is a *probable* British species, which should be looked for on the South Coast, where its foodplant, the sallow-thorn or sea-buckthorn (*Hippophaes Rhamnoides*) [sic] grows; *the green larva, with pink horn*, is said to feed in June and July, and again in September and October'. Barrett (1895, **2**: 51) comments 'Dr. Mason possesses from the collection of the late Mr. E. Brown, a specimen of *D. hippophaes*, Esp., a local species in Southern and Eastern Europe, labelled "Devonshire" '.

Hyles hippophaes is superficially similar to *H. euphorbiae* (Linnaeus) and *H. gallii* (Rottemburg) from which it differs in having the median fascia on the forewing grey.

HYLES LINEATA LIVORNICA (Esper)
The Striped Hawk-moth

Sphinx lineata Fabricius, 1775, *Syst.Ent.*: 541.
Sphinx livornica Esper, 1780, *Schmett.* **2** (13): 88.
Type locality: Germany.

Description of imago (Pl.1, fig.8)

Wingspan 78–90mm. Antenna clubbed with thin, hooked tip; haustellum well developed; eye lashed. Thorax dark ochreous brown with two lateral white stripes; tegulae white. Forewing dark ochreous brown, white basally; a broad, almost straight, pale ochreous stripe from dorsum basally to apex, tapering to apex; terminal fascia pale grey-brown, dark-edged; discal spot greyish white; veins white. Hindwing pinkish red, basal area and subterminal band black; anal patch whitish; terminal shade pale ochreous brown. Abdomen dark ochreous brown with broken black bands and ochreous white spots giving a chequered appearance.

Life history

Ovum. Ovate, light green. Laid singly on leaves of the foodplant.

Larva. Full-fed *c.*90mm long. Head black with red or yellow markings. Body colour variable from green to black, dotted yellow, sometimes dark-banded; dorsal and subdorsal lines yellow; subdorsal spots, ringed black, sometimes marked pink or black; subspiracular line, when present, white or pink; prolegs red; horn curved, red, tip black. Feeds in June and July on bedstraw (*Galium* spp.), dock (*Rumex* spp.),

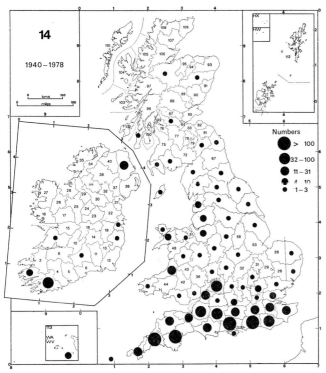
Hyles lineata

Fuchsia, grape-vine (*Vitis vinifera*) and other plants. Has been reported in Britain in years of large immigrations, *e.g.* 1943 and 1949.

Pupa. C.50mm long; brown; proboscis sheath enlarged basally; cremaster long, pointed. In a fragile subterranean cocoon.

Imago. A regular immigrant. Bivoltine. Flies in May and June and again in late summer, visiting flowers of such plants as red valerian at dusk, hovering while feeding. Occurs most years in small numbers, but occasionally there are large immigrations; in 1943, 543 were recorded and in 1949, 321. Has been much scarcer since 1950.

Distribution (Map 14)

This species has occurred throughout Great Britain and Ireland, especially in southern England. Cosmopolitan, occurring in most regions except in the Arctic and tropics.

DEILEPHILA Laspeyres

Deilephila Laspeyres, 1809, *Jena allg.Lit.-Ztg* **4**: 100.
Choerocampa Duponchel, 1835, *in* Godart & Duponchel, *Hist.nat.Lépid.Fr.* Suppl. **2**: 159.
Metopsilus Duncan, 1835, *Br.Moths*: 154.

A small Old World genus with two species occurring in Great Britain and Ireland.

Characterized by the spinose tarsi, midtarsus with posterior comb and unequal tibial spurs.

DEILEPHILA ELPENOR (Linnaeus)
The Elephant Hawk-moth
Sphinx elpenor Linnaeus, 1758, *Syst.Nat.* (Edn 10) **1**: 491.
Type locality: not stated.

Description of imago (Pl.2, fig.9)

Wingspan 62–72mm. Antenna less than half length of forewing, pectinate, thickening distally with thin, hooked tip; haustellum well developed; eye strongly lashed. Thorax olive-brown with four pink stripes; tegulae white. Forewing olive-brown; costa and termen pinkish red; dorsum white, black at base; median, subterminal and terminal fasciae olive-pink; discal dot white; veins paler. Hindwing bright pinkish red, basally black; cilia white. Abdomen olive-brown, laterally pink; dorsal stripe pink; lateral black patches on first segment and white lateral spots on each segment.

Life history

Ovum. Roundish, smooth and glossy, pale green. Laid singly on leaves of willowherb (*Epilobium* spp.), bedstraw (*Galium* spp.) and other introduced plants such as *Fuchsia* and *Impatiens*.

Larva. Full-fed c.85mm long. Body brown or green, freckled black except on thoracic segments which have a pale subdorsal line. Head and thoracic segments can retract into the first abdominal segment; abdominal segments 2 and 3 swollen with black, lateral, kidney-shaped ocelli with lilac centres; horn fairly short with black dorsal line sometimes indistinct. Feeds in July and August, sometimes sunning itself on the foodplant. When it is searching for food the thoracic segments can be extended like a 'trunk' and when it is disturbed they retract and dilate to display the ocelli as a defence response. Can 'swim' if it falls into water.

Pupa. C.45mm long, stout; medium brown freckled darker brown between segments; cremaster dark brown, conical. In a fragile cocoon on or just below the surface of the soil.

Imago. Univoltine, sometimes bivoltine. Flies at dusk in June, occasionally later in the second brood. Feeds at flowers, especially honeysuckle, whilst hovering. Fertile hybrids with *D. porcellus* (Linnaeus) have been reported.

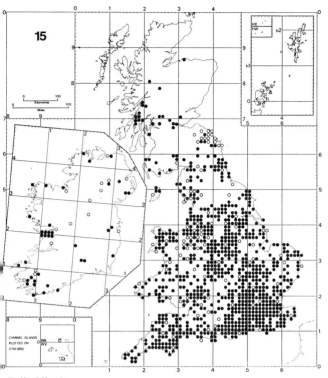

Deilephila elpenor

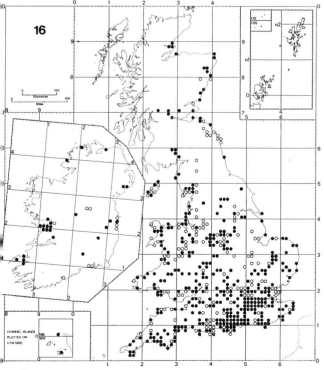

Deilephila porcellus

Distribution (Map 15)

Common and widely distributed throughout England and Wales; local in southern Scotland, but at present increasing its range; Ireland; Isle of Man. Palaearctic.

DEILEPHILA PORCELLUS (Linnaeus)
The Small Elephant Hawk-moth

Sphinx porcellus Linnaeus, 1758, *Syst.Nat.* (Edn 10) **1**: 492.
Type locality: not stated.

Description of imago (Pl.2, fig.7)

Wingspan 47–56mm. Antenna half length of forewing, thickening distally with a thin, hooked tip; haustellum well developed; eye lashed. Thorax pinkish red; metathorax olive-green; tegulae white. Forewing deep yellow-ochreous, antemedian, postmedian and subterminal lines brown; costal area irregularly pink; subterminal fascia pale yellow-ochreous; terminal fascia pinkish red. Hindwing deep yellow-ochreous, basally brownish grey; terminal shade pinkish red; cilia white. Abdomen pinkish red, dorsally deep ochreous yellow.

Variation occurs in the intensity of the lines on the forewing and the tint of pinkish red markings.

Life history

Ovum. Oval, glossy, green. Laid singly on bedstraw (*Galium* spp.).

Larva. Full-fed *c.*50mm long. Body greyish brown or green, spotted black; thoracic segments pinkish white dorsally except in final instar; abdominal segments 1 and 2 swollen, with lilac-coloured, brown-centred, black-edged round lateral ocelli; horn reduced to a double tubercle bearing setae. Feeds at night in July and August on flowers and leaves of bedstraw, also occasionally willowherb (*Epilobium* spp.) and purple loosestrife (*Lythrum salicaria*). When it is disturbed the head and thorax can be retracted into the first abdominal segment to display the threatening ocelli.

Pupa. *C.*25mm long, slender; pale ochreous brown, spotted darker brown; abdominal segments with small reddish brown hooks. In a fragile cocoon, on, or just beneath, the surface of the soil. The winter is passed in this stage.

Imago. Univoltine. Flies at dusk from May to July. Feeds at flowers especially honeysuckle and rhododendron whilst hovering.

Distribution (Map 16)

Most frequent in chalk and limestone areas. It is widely distributed in England and Wales as far north as the Lake District; very local in Scotland and Ireland; Isle of Man; the Channel Islands. Abroad from western Europe eastwards to west Siberia and Asia Minor.

HIPPOTION Hübner
Hippotion Hübner, [1819], *Verz.bekannt.Schmett.*: 135.

This genus contains 22 species, mostly Ethiopian and Oriental, of which only two occur in the Palaearctic region. One of these reaches Great Britain and Ireland as an immigrant.

Characterized by the apex of the first segment of the palpus being densely scaled internally and by the absence of an apical tuft of scales on the second segment.

HIPPOTION CELERIO (Linnaeus)
The Silver-striped Hawk-moth
Sphinx celerio Linnaeus, 1758, *Syst.Nat.* (Edn 10) **1**: 491.
Phalaena inquilinus Harris, 1776, *Exp.Engl.Ins.*: 93.
Type locality: not stated.

Description of imago (Pl.1, fig.10)
Wingspan 72–80mm. Antenna thickened distally with thin, hooked tip; haustellum well developed. Thorax light brown with two lateral ochreous white lines; tegulae silvery white. Forewing light brown suffused ochreous brown; basal streak, subterminal fascia and terminal line silvery white, edged black; dorsum silvery white. Hindwing pinkish red; basal and subterminal shades black, not reaching base; veins 3 (Cu_1) to 7 (Rs) black; cilia silvery white. Abdomen sharply tapered; light brown; dorsal line double, white; lateral lines double, broken, white.

Life history
Ovum. Ovate, small, green. Laid on either the upper or lower surface of a leaf of the foodplant.
Larva. Full-fed *c.*75mm long. Head small, pale brown. Body variably green or brown; dorsal line darker; subdorsal line pale, sometimes broken; a large oval black, white-ringed ocellus on first abdominal segment; a smaller similar white ocellus on second abdominal segment; supraspiracular dots red; spiracles white; horn short, straight, rough, brown or blackish brown. Head and thorax can be retracted into first abdominal segment, thereby dilating ocelli as a warning to predators. Turns darker brown before pupation. Feeds in June and from August to October on grape-vine (*Vitis vinifera*), bedstraw (*Galium* spp.), willowherb (*Epilobium* spp.), *Fuchsia* and Virginia creeper (*Parthenocissus* spp.).
Pupa. Light brown, glossy; proboscis sheath large, prominent, compressed; dorsal line darker; spiracles fuscous; wing-cases and abdomen below spiracles sprinkled fuscous; cremaster conical, long, thin, smooth, fuscous. In a fragile cocoon on or just below the surface of the soil.
Imago. A rare immigrant. Bivoltine. Flies in spring and from August to October. Most immigrants are of the autumn generation. Absent in many years; numbers ex-

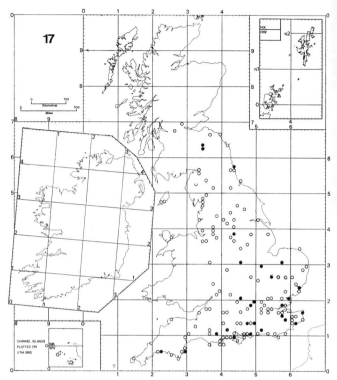

Hippotion celerio

ceeded ten in 1865, 1868, 1885 and 1963; 1885 with 41 was the record year; last recorded in 1978 when single specimens were taken in Hampshire, Sussex and Cambridgeshire.

Distribution (Map 17)
This species has been recorded from England, southern Scotland and the Channel Islands. Widely distributed abroad in the southern Palaearctic, Ethiopian and Oriental regions and Australia.

References
Allan, P. B. M., 1947. *A moth-hunter's gossip* (Edn 2), pp. 13–60. London.

Barrett, C. G., 1895. *The Lepidoptera of the British Islands*, **2**: 369 pp. London.

Blathwayt, C. S. H., 1970. A North American hawk moth in Britain. *Entomologist's Rec.J. Var.* **82**: 150.

Buckler, W., 1887. *The larvae of the British butterflies and moths*, **2**: xi, 172 pp., 35 col. pls. London.

Chinery, M., 1972. Cross-pairings of *Saturnia pavonia* (Linnaeus) (Lep., Saturniidae) and *Laothoe populi* (Linnaeus) (Lep., Sphingidae). *Entomologist's Gaz.* **23**: 119.

Curtis, J., 1828. *British Entomology*, **5**: 195. London.

French, R. A., 1965. Migration records, 1963. *Entomologist* **98**: 73–77.

——, 1966. Migration records, 1964. *Ibid.* **99**: 233–240.

——, 1968. Migration records, 1965. *Ibid.* **101**: 156–161.

——, 1971. Migration records, 1966 and 1967. *Ibid.* **104**: 204–218.

——, 1973. Migration records, 1968. *Ibid.* **106**: 256–263.

Hodges, R. W., 1971. *The moths of America north of Mexico*, Fasc. 21 Sphingoidea, 158 pp., 19 figs, 14 col. pls. London.

Newman, L. H., 1965. *Hawk-moths of Great Britain and Europe*, xv, 148 pp., figs. London.

Rothschild, W. & Jordan, K., 1903. A revision of the lepidopterous family Sphingidae. *Novit. zool.* **9**: Suppl. cxxxv, 972 pp., 67 pls.

South, R., 1961. *The moths of the British Isles* (Edn 4), **1**: 427 pp., 148 pls. London.

Stainton, H. T., 1857. *A manual of British butterflies and moths*, **1**: 91. London.

Stokoe, W. J. & Stovin, G. H. T., 1958. *The caterpillars of British moths*, **1**: 408 pp., 90 pls. London.

Tutt, J. W., 1902. *A natural history of the British Lepidoptera*, **3**: 452. London.

Wilkinson, C., 1956. *Celerio nicaea* (de Prun.): a hawk moth new to the British list. *Entomologist's Gaz.* **7**: 175–177.

Williams, C. B., 1958 (Reprinted 1971). *Insect migration*, xiii, 237 pp., 49 figs., 16 pls. London.

de Worms, C. G. M., 1975. A review of the immigration of *Hyles gallii* Rott. during 1973 with special reference to records for 1972 and 1974. *Entomologist's Rec. J. Var.* **87**: 232–239.

NOTODONTIDAE
C. G. M. de Worms

A world-wide family represented in the Palaearctic region by about 200 species and in Great Britain and Ireland by 25 large to moderate-sized species contained in 19 genera.

Imago (Pl.A, fig.5). Antenna about half length of forewing, in male usually bipectinate, in female pectinate or simple; haustellum usually developed. Metathorax with tympanal organs. Forewing triangular or ovate, more than twice as long as broad, a median tooth-like tuft of scales projecting from dorsum in many species; venation as in figure 4. Epiphysis present in male, absent in female.

Ovum. Hemispherical. Usually laid in small batches on the foodplant. Illustrated in Stokoe & Stovin (1958).

Larva (Pl.A, fig.6). Fleshy dorsal humps are usually present on one or more segments; anal segment with claspers modified into tail-like appendages with eversible flagella in some species. Head and anal segments are frequently raised when the larva is at rest. The larvae are figured by Buckler (1887, 89) and Stokoe & Stovin (1958). The British species all feed on deciduous trees and shrubs.

Pupa. In a hard cocoon on bark or subterranean in a soft, silken cocoon or in a chamber.

The generic descriptions which follow are based on those given by Kiriakoff (1967).

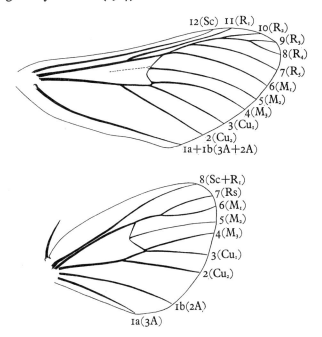

Figure 4 *Eligmodonta ziczac* (Linnaeus), wing venation

Key to species (imagines) of the Notodontidae

1	Forewing with median projecting tuft of scales on dorsum	2
–	Forewing without dorsal tuft of scales	16
2(1)	Forewing mainly white with yellow median fascia .... *Leucodonta bicoloria* (p. 57)	
–	Forewing otherwise, if white also with extensive dark markings	3
3(2)	Forewing with large postmedian lunule bordering an apical variegated pebble-like patch .... *Eligmodonta ziczac* (p. 48)	
–	Forewing markings otherwise	4
4(3)	Forewing ground colour white with black apical patch on costa and broad black dorsal streak	5
–	Forewing buff or brown	6
5(4)	Forewing with distinct, white, wedge-shaped tornal streak; hindwing without white subterminal line at anal angle .... *Pheosia gnoma* (p. 51)	
–	Forewing with tornal streak ill-defined; hindwing with white subterminal line intersecting dark anal patch .... *P. tremula* (p. 52)	
6(4)	Forewing with basal streak and subterminal line reddish .... *Notodonta dromedarius* (p. 46)	
–	Forewing otherwise	7
7(6)	Hindwing ground colour white or cream; wingspan greater than 50mm	8
–	Hindwing ground colour not white; wingspan less than 50mm	9
8(7)	Forewing with conspicuous dark grey basal and subapical areas; a small pale-margined orbicular stigma present in addition to the larger reniform stigma; subterminal line reddish brown; hindwing white; a rare vagrant .... *Tritophia tritophus* (p. 47)	
–	Forewing without dark grey patches; orbicular stigma absent; a subterminal series of brown dashes; hindwing cream; common species .... *Peridea anceps* (p. 50)	
9(7)	Forewing elongate with distinct transverse lines and fasciae; dorsal scale-tooth weak	10
–	Forewing triangular without distinct fasciae; dorsal scale-tooth prominent	12
10(9)	Forewing with distinct reniform mark .... *Drymonia ruficornis* (p. 60)	
–	Forewing without distinct reniform mark	11
11(10)	Forewing ante- and postmedian lines deeply dentate; common species .... *D. dodonaea* (p. 59)	
–	Forewing ante- and postmedian lines almost straight; a very rare vagrant .... *Gluphisia crenata* (p. 61)	
12(9)	Forewing pale greyish buff .... *Pterostoma palpina* (p. 56)	
–	Forewing reddish brown	13
13(12)	Forewing with a distinct white postmedian costal mark; dorsal half suffused whitish grey .... *Odontosia carmelita* (p. 55)	
–	Forewing without costal mark; dorsal half not suffused whitish grey	14
14(13)	Forewing dorsal half suffused white between subterminal line and termen .... *Ptilodontella cucullina* (p. 54)	
–	Forewing without white suffusion	15
15(14)	Hindwing with distinct black mark at anal angle .... *Ptilodon capucina* (p. 53)	
–	Hindwing without markings .... *Ptilophora plumigera* (p. 58)	
16(1)	Large species; wingspan 50mm or more	17
–	Smaller species; wingspan less than 50mm	19
17(16)	Forewing with large prominent yellowish buff apical patch .... *Phalera bucephala* (p. 41)	
–	Forewing without an apical patch	18
18(17)	Forewing ground colour white with prominent, acutely dentate transverse lines; veins on fore- and hindwings black .... *Cerura vinula* (p. 42)	
–	Wings ground colour brown or grey; forewing with subterminal series of darker spots and a darker median fascia .... *Stauropus fagi* (p. 45)	
19(16)	Wings ground colour white; forewing with a dark grey median fascia and apical patch	20
–	Forewing ground colour dark brown; hindwing pale or dark brown	22
20(19)	Forewing with inner margin of median fascia concave; fascia almost black .... *Furcula bicuspis* (p. 43)	
–	Forewing with inner margin of median fascia more or less straight; fascia grey	21
21(20)	Forewing with outer margin of fascia angled near costa, more or less dentate; fascia edged yellow .... *F. furcula* (p. 43)	
–	Forewing with outer margin of fascia smoothly concave; fascia edged black .... *F. bifida* (p. 44)	
22(19)	Forewing with a prominent silvery white, double reniform stigma .... *Diloba caeruleocephala* (p. 64)	
–	Forewing without silvery white reniform stigma	23
23(22)	Forewing without prominent apical patch .... *Clostera pigra* (p. 62)	
–	Forewing with large prominent, chocolate-brown apical patch	24
24(23)	Forewing with apical patch extending inwards beyond subterminal line; black spots present in tornal area .... *C. anachoreta* (p. 63)	
–	Forewing apical patch not extending inwards beyond subterminal line; tornal spots absent .... *C. curtula* (p. 63)	

Plate A

1 Sphingidae Sphinginae *Laothoe populi* (Linnaeus) Poplar Hawk-moth. Photo J. Mason. Page 28
2 Sphingidae Sphinginae *Smerinthus ocellata* (Linnaeus) Eyed Hawk-moth. Photo BMNH. Page 27
3 Sphingidae Macroglossinae *Deilephila elpenor* (Linnaeus) Elephant Hawk-moth. Photo J. Mason. Page 36
4 Sphingidae Macroglossinae *Deilephila elpenor* (Linnaeus) Elephant Hawk-moth. Photo M. W. F. Tweedie. Page 36
5 Notodontidae *Pterostoma palpina* (Clerck) Pale Prominent. Photo J. Mason. Page 56
6 Notodontidae *Notodonta dromedarius* (Linnaeus) Iron Prominent. Photo BMNH. Page 46
7 Lymantriidae *Lymantria monacha* (Linnaeus) Black Arches. Photo J. Mason. Page 76
8 Lymantriidae *Leucoma salicis* (Linnaeus) White Satin Moth. Photo BMNH. Page 75
9 Arctiidae Lithosiinae *Eilema lurideola* ([Zincken]) Common Footman. Photo G. Hyde. Page 94
10 Arctiidae Lithosiinae *Eilema lurideola* ([Zincken]) Common Footman. Photo G. Dickson. Page 94
11 Arctiidae Arctiinae *Parasemia plantaginis* (Linnaeus) Wood Tiger. Photo J. Mason. Page 98
12 Arctiidae Arctiinae *Spilosoma lutea* (Hufnagel) Buff Ermine. Photo BMNH. Page 103
13 Nolidae *Nola cucullatella* (Linnaeus) Short-cloaked Moth. Photo G. Hyde. Page 117

Plate B

1 Noctuidae Noctuinae *Noctua pronuba* (Linnaeus) Large Yellow Underwing. Photo J. Mason. Page 157
2 Noctuidae Noctuinae *Xestia agathina* (Duponchel) Heath Rustic. Photo BMNH. Page 188
3 Noctuidae Hadeninae *Orthosia gothica* (Linnaeus) Hebrew Character. Photo J. Mason. Page 255
4 Noctuidae Hadeninae *Hadena rivularis* (Fabricius) Campion. Photo BMNH. Page 228
5 Noctuidae Cucullinae *Xylocampa areola* (Esper) Early Grey. Photo J. Mason. (Vol.10)
6 Noctuidae Cucullinae *Cucullia absinthii* (Linnaeus) Wormwood. Photo BMNH. (Vol.10)
7 Noctuidae Acronictinae *Acronicta psi* (Linnaeus) Grey Dagger. Photo J. D. Bradley. (Vol.10)
8 Noctuidae Acronictinae *Acronicta leporina* (Linnaeus) Miller. Photo BMNH. (Vol.10)
9 Noctuidae Amphipyrinae *Phlogophora meticulosa* (Linnaeus) Angle Shades. Photo J. Mason. (Vol.10)
10 Noctuidae Amphipyrinae *Amphipyra pyramidea* (Linnaeus) Copper Underwing. Photo BMNH. (Vol.10)
11 Noctuidae Heliothinae *Heliothis peltigera* ([Denis & Schiffermüller]) Bordered Straw. Photo G. Hyde. (Vol.10)
12 Noctuidae Heliothinae *Helicoverpa armigera* (Hübner) Scarce Bordered Straw. Photo BMNH. (Vol.10)
13 Noctuidae Acontiinae *Eustrotia uncula* (Clerck) Silver Hook. Photo G. Hyde. (Vol.10)

Plate C

1 Noctuidae Chloephorinae *Bena prasinana* (Linnaeus) Scarce Silver Lines. Photo G. Dickson. (Vol.10)
2 Noctuidae Chloephorinae *Pseudoips fagana* (Fabricius) Green Silver Lines. Photo BMNH. (Vol.10)
3 Noctuidae Sarrothripinae *Nycteola revayana* (Scopoli) Oak Nycteoline. Photo BMNH. (Vol.10)
4 Noctuidae Pantheinae *Colocasia coryli* (Linnaeus) Nut-tree Tussock. Photo BMNH. (Vol.10)
5 Noctuidae Plusiinae *Autographa gamma* (Linnaeus) Silver Y. Photo G. Hyde. (Vol.10)
6 Noctuidae Plusiinae *Diachrysia chrysitis* (Linnaeus) Burnished Brass. Photo BMNH. (Vol.10)
7 Noctuidae Catocalinae *Catocala nupta* (Linnaeus) Red Underwing. Photo G. Hyde. (Vol.10)
8 Noctuidae Catocalinae *Catocala fraxini* (Linnaeus) Clifden Nonpareil. Photo G. Dickson. (Vol.10)
9 Noctuidae Ophiderinae *Scoliopterix libatrix* (Linnaeus) Herald. Photo J. Mason. (Vol.10)
10 Noctuidae Ophiderinae *Tyta luctuosa* ([Denis & Schiffermüller]) Four-spotted. Photo BMNH. (Vol.10)
11 Noctuidae Hypeninae *Hypena proboscidalis* (Linnaeus) Snout. Photo G. Hyde. (Vol.10)
12 Noctuidae Hypeninae *Hypena crassalis* (Fabricius) Beautiful Snout. Photo J. E. Knight. (Vol.10)

NOTE. The figures on Plates A, B and C are not to scale.

PHALERA Hübner

Phalera Hübner, [1819], *Verz.bekannt.Schmett.*: 147.
Hammatophora Humphreys & Westwood, 1843, *Br.Moths* **1**: 63.

A genus containing 15 Palaearctic species, one of which occurs in Great Britain and Ireland, besides others in the Oriental and Palaearctic regions.

Antenna in male strongly pectinate, in female weakly pectinate; haustellum reduced; palpi short; frons with scaly prominence. Thorax with a broad transverse crest. Posterior tibia with two pairs of spurs. Forewing with veins 3 (Cu_1) and 4 (M_3) approximated, 5 (M_2) from middle of discocellulars (dcs), one areole present, 6 (M_1) from areole, 7 (R_5), 8+9 (R_4+R_3), 10 (R_2) from apex of areole. Hindwing with veins 3 (Cu_1) and 4 (M_3) separate or forked, 5 (M_2) from middle of dcs, 6 (M_1) and 7 (Rs) stalked, 8 ($Sc+R_1$) approximated with cell near apex.

PHALERA BUCEPHALA (Linnaeus)
The Buff-tip

Phalaena (Noctua) bucephala Linnaeus, 1758, *Syst.Nat.* (Edn 10) **1**: 508.
Type locality: [Europe].

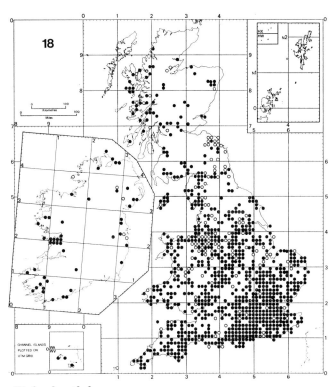

Phalera bucephala

Description of imago (Pl.3, fig.1)
Wingspan 55–68mm. Antenna in male strongly pectinate, in female weakly pectinate. Thorax yellow; tegulae silvery grey and reddish brown. Forewing violet-grey, dorsally flecked silver; subbasal, antemedian and terminal lines broken, black, edged reddish brown; subterminal line black, edged reddish brown, enclosing large yellowish buff patch at apex; discal spot pale. Hindwing pale yellow, basally lightly sprinkled grey. Abdomen yellowish buff.

Life history
Ovum. Hemispherical, white. Laid in large batches on the underside of leaves of deciduous trees such as oak (*Quercus* spp.), sallow (*Salix* spp.), elm (*Ulmus* spp.), hazel (*Corylus avellana*) and lime (*Tilia* spp.) in July or early August.

Larva. Full-fed *c.*80mm long. Head black, frons edged yellow. Body yellow; dorsal stripe broad, black, broken between segments; subdorsal, spiracular and subspiracular stripes grey, broken between segments; tufts of fine short setae on all segments. Feeds from July to October gregariously until nearly full-fed, becoming solitary in the last instar. Frequently defoliates entire branches of the trees on which it is feeding.

Pupa. Stout, abdomen tapering; shiny purple. In a chamber in the soil in which it overwinters; there is no cocoon.

Imago. Univoltine. Flies from late May to July. Comes freely to ultra-violet light, usually not until about 2 a.m. At rest it resembles remarkably a piece of dead wood.

Distribution (Map 18)
Widely distributed and common in England and Wales; more local in Scotland being most common in the western Highlands; local in Ireland; Isle of Man; the Channel Islands. Absent from Orkney and Shetland. Abroad throughout Europe eastwards to east Siberia.

CERURA Schrank

Cerura Schrank, 1802, *Fauna boica* **2**: 155.
Dicranura sensu auctt.

A world-wide genus with six Palaearctic species, one of which occurs in Great Britain and Ireland.

Antenna bipectinate; eyes glabrous; ocelli absent; palpi very short. Posterior tibia with one pair of spurs. Forewing with vein 5 (M_2) from middle of discocellulars (dcs); 6 (M_1) from areole near apex; 7 (R_5), 8+9 (R_4+R_3) and 10 (R_2) stalked from apex of areole. Hindwing vein 5 (M_2) from middle of dcs; 6 (M_1) and 7 (Rs) stalked; 8 ($Sc+R_1$) approximated to middle of upper margin of cell.

CERURA VINULA (Linnaeus)
The Puss Moth

Phalaena (Bombyx) vinula Linnaeus, 1758, *Syst.Nat.* (Edn 10) **1**: 499.

Type locality: [Sweden].

Description of imago (Pl.3, figs 6,7)
Wingspan 62–80mm. Antenna almost half length of forewing, bipectinate in both sexes, more strongly in male. Thorax yellowish grey, spotted black. Forewing in male very pale grey, almost white in terminal half; basal and subbasal fasciae reduced to black dots; other fasciae forming a series of acutely dentate dark grey lines; veins yellowish, edged with black; a terminal series of black dots between veins; in female ground colour darker grey and markings darker and heavier. Hindwing in male white, discal lunule and veins dark grey; in female brownish grey, veins brown. Abdomen pale grey with five dark grey bands; densely setose.

Variation is slight; Scottish females are darker than those from elsewhere.

Life history
Ovum. Hemispherical, reddish purple, finely reticulate with a slight yellow depression on top. Laid in batches of two or three, occasionally singly, on the upper surface of leaves of sallow (*Salix* spp.) or poplar (*Populus* spp.), from May to July, hatching in about ten days.

Larva. Full-fed *c.*70mm long. Head black. Body bright green, prominently 'humped' on metathorax; dorsal stripe a wide, white-edged purplish 'saddle', narrowing at metathorax and sometimes extending laterally almost to prolegs on abdominal segment 4; peritreme of spiracle reddish or black; anal claspers modified to form paired anal appendages with red flagella which are extruded and waved when it is alarmed. Becomes purplish prior to pupation. Feeds often initially in pairs, becoming full-fed in August or early September.

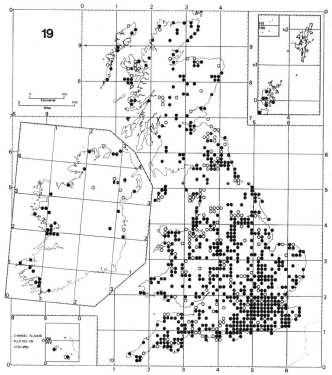

Cerura vinula

NOTE. Young larvae of this species are similar to those of the species of *Furcula*, but an easy identification is often provided by the egg-shells, reddish in this species but blackish in *Furcula*, which may be found on a nearby leaf.

Pupa. Rounded; black. In a very tough cocoon constructed from chewed-up wood and silk on a tree-trunk or fence-post. Immediately prior to emergence, it cuts through the end of the cocoon using a keel-like frontal structure. When the pupal case is ruptured the imago emits a fluid, said to be formic acid, which softens the cocoon in the area of the breach, thus facilitating its emergence. Overwinters from August to May.

Imago. Univoltine. Flies from May to July. Both sexes come freely to ultra-violet light.

Distribution (Map 19)
Widely distributed throughout the British Isles as far north as Orkney. Abroad throughout Europe eastwards to Japan.

FURCULA Lamarck

Furcula Lamarck, 1816, *Hist.nat.Anim.sans Vertèbres* **3**: 581.

A Holarctic genus containing 15 Palaearctic species, three of which occur in Great Britain and one of these in Ireland.

Antenna bipectinate to apex, pectinations shorter in female; palpi very short; haustellum reduced. Posterior tibia with one pair of very short spurs. Venation as in *Cerura* Schrank but vein 10 (R_2) free.

NOTE. Recent research has shown that *Phalaena furcula* Clerck was designated as the type species of the genus *Furcula* and that *Bombyx milhauseri* Fabricius was designated as the type species of the genus *Harpyia*. Therefore those species formerly in the genus *Harpyia* must be placed in the genus *Furcula*.

FURCULA BICUSPIS (Borkhausen)
The Alder Kitten

Phalaena (Bombyx) bicuspis Borkhausen, 1790, *Natgesch.eur. Schmett.* **3**: 380.

Type locality: Europe.

Description of imago (Pl.3, fig.5)
Wingspan 40–48mm. Head white; antenna pectinate, more strongly in male. Thorax dark grey-brown. Forewing white; a small black basal spot; antemedian line a series of black dots; median fascia broad, dark greyish brown or almost black, edged orange-yellow, inner and outer margins concave; subterminal line dentate, black; a black, almost triangular subapical patch edged orange-yellow inwardly, extending as a dentate line to dorsum; two black costal spots between median fascia and subapical patch; a black discal spot and a series of terminal black spots between veins. Hindwing white, discal lunule and veins grey-brown. Abdomen white, intersegmental bands grey.
Similar species. *F. furcula* (Clerck) and *F. bifida* (Brahm) which have the forewing markings paler grey.

Life history
Ovum. Hemispherical, purple. Laid in batches of two or three on the upper surface of leaves of birch (*Betula* spp.) or alder (*Alnus glutinosa*) in late May or June.
Larva. Full-fed *c*.35mm long. Head red-brown. Body bright green; intersegmental divisions incised; dorsal stripe a broad, purplish brown, white-edged 'saddle', tapering from head to metathorax where it is interrupted, thence to anal segment, with three lateral extensions, the largest on abdominal segment 4 where it almost reaches prolegs; anal prolegs modified to form paired anal appendages with eversible flagella. Feeds from late July to early September.

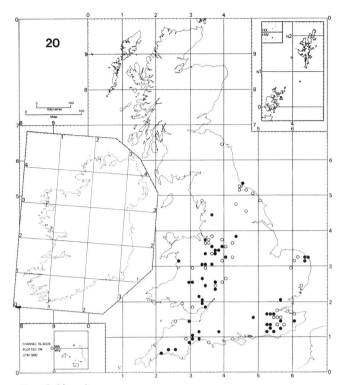

Furcula bicuspis

Pupa. Black. Overwinters in a small, tough cocoon usually constructed on the surface of a stem of birch or alder.
Imago. Univoltine. Flies in May and June. Male comes freely to ultra-violet light. Occasionally found at rest by day on tree-trunks.

Distribution (Map 20)
Very local; it occurs in western Britain from Cornwall northwards through Wales and the west Midlands to south Lancashire, and in south-east England, Norfolk and Co. Durham. Absent from Scotland and Ireland. Abroad throughout northern and central Europe to eastern Siberia.

FURCULA FURCULA (Clerck)
The Sallow Kitten

[*Phalaena*] *furcula* Clerck, 1759, *Icones Insect.rar.* **1**: pl.9, fig.9.

Type locality: [Sweden].

Description of imago (Pl.3, figs 2,3)
Wingspan 35–42mm. Head pale grey; antenna pectinate, more strongly in male. Thorax grey-brown. Forewing pale grey; a black basal spot; antemedian line a series of black spots; median fascia grey, edged black and orange-yellow,

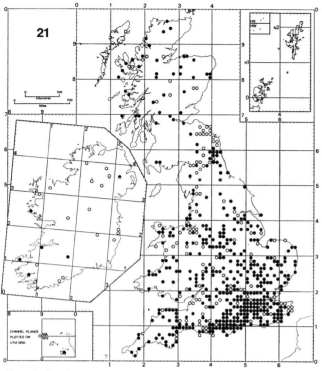

Furcula furcula

inner margin straight, outer margin rather ill-defined, angled near costa and more or less dentate; discal spot elongate, black; subapical patch ill-defined pale grey, edged black and orange-yellow inwardly, extending as a dentate line to dorsum; a series of terminal black spots between veins. Hindwing pale grey, discal spot and postmedian fascia darker grey; a series of terminal black spots between veins. Abdomen grey, banded dark grey.

Variation is slight, mainly confined to the intensity of the markings; pale forms (fig.3, a Surrey specimen) sometimes occur. Specimens from Scotland and Ireland tend to be larger than those from southern and central England.

Similar species. F. bifida (Brahm), a slightly larger species, the outer margin of the median fascia black and more evenly curved; F. bicuspis (Borkhausen) q.v.

Life history

Ovum. Hemispherical, deep purple. Laid in batches of two or three on the upper surface of leaves of sallow or willow (*Salix* spp.) in May or June and again in August, hatching after nine days.

Larva. Full-fed c.35mm long. Head purple-brown, retractable. Body pale green; dorsal stripe broad, pale grey, continuous, forming a 'saddle' extending almost to prolegs on abdominal segment 5, tapering to anal segment; 'hump' on metathorax slight; paired anal appendages with flagella, short, upright; intersegmental divisions not incised. Feeds in May and June and again in August and September. May be found often on isolated moorland sallow bushes.

Pupa. Dark purple-brown. In a tough cocoon constructed from a mixture of silk and chewed wood-pulp, placed in a recess hollowed out by the larva on a sallow stem. The cocoon has the appearance of a 'smudge'. The winter is passed in this stage.

Imago. Univoltine in Scotland and Ireland, bivoltine in England and Wales. Flies in late May and early June and again, in southern Britain, in early August. Comes freely to ultra-violet light. Occasionally found at rest on sallow stems.

Distribution (Map 21)

Widely distributed and fairly common throughout Great Britain and Ireland except the Isle of Man, Orkney and Shetland. Holarctic; extending throughout Europe eastwards to China; North America.

FURCULA BIFIDA (Brahm)
The Poplar Kitten

Phalaena bifida Brahm, 1787, *Neues Mag.Liebh.Ent.* **3** (2): 161.
Phalaena hermelina Goeze, 1781, *Ent.Beyträge* **3** (3): 227, nec Goeze 1781: 207.

Type locality: Germany; Mainz.

Description of imago (Pl.3, fig.4)

Wingspan 44–48mm. Head pale grey; antenna pectinate, very strongly in male. Thorax dark grey-brown; tegulae spotted yellow. Forewing pale grey; a small black basal dash; antemedian line a series of black dots; median fascia dark grey, irrorate whitish, inner margin straight, edged black and yellow, outer margin smoothly concave and sharply defined; discal spot black; two faint dentate postmedian lines; subapical patch dark grey, sharply edged, extending to dorsum as a pale grey subterminal line; a series of terminal black spots between veins. Hindwing very pale grey; discal spot large, dark grey; postmedian fascia grey; a series of terminal black spots between veins. Abdomen grey, banded dark grey.

Variation is slight but occasionally specimens occur with the median fascia very similar to that in *F. furcula* (Clerck).

Similar species. F. bicuspis (Borkhausen) and F. furcula q.v.

Life history

Ovum. Hemispherical, dark purple. Laid in batches of two or three on the upper surface of leaves of poplar or aspen (*Populus* spp.), from late May to early July.

Larva. Full-fed c.35mm long. Head purplish brown. Body bright green; 'saddle-like' dorsal stripe purplish brown,

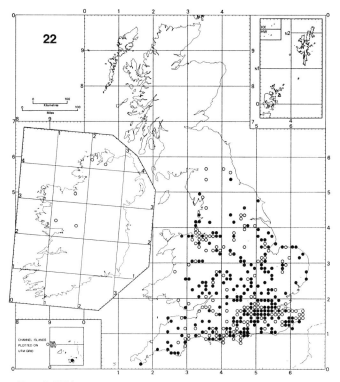

Furcula bifida

edged yellow, broken on metathorax, broadest on abdominal segment 4 where it extends laterally almost to the prolegs, then tapers to the anal segment; anal prolegs modified to form paired anal appendages with eversible flagella. Intersegmental divisions not incised. When it is at rest, the head and thorax are raised, forming a 'hump'. Feeds from July to September.

Pupa. Rounded; black. In a hard cocoon, covered with chewed pieces of wood and fragments of bark. The cocoon is spun on the surface or in chinks in the bark of the trunk of the host tree. Sometimes two winters are passed in this stage.

Imago. Univoltine. Flies from late May to July. Can be found at rest on trunks of poplar.

Distribution (Map 22)

Widely distributed in England south of a line from the Mersey to the Humber; local in Wales; old records from Ireland are not supported by specimens; Jersey. Abroad throughout Europe to the Urals.

STAUROPUS Germar
Stauropus Germar, 1811, *Syst.Gloss.Prod.*: 45.

A small genus of three species occurring in the Palaearctic and Oriental regions with one representative in Great Britain and Ireland.

Antenna in male bipectinate to two-thirds, in female filiform; eyes glabrous; ocelli absent. Legs clothed with long hairs; posterior tibia with one pair of spurs. Forewing elongate; tornus ill-defined; vein 5 (M_2) from middle of discocellulars (dcs); areole absent; 6 (M_1) and 7 (R_5) to 10 (R_2) arising from angle of cell; 7 (R_5) shortly stalked with 10 (R_2); 8+9 (R_4+R_3) on a long stalk. Hindwing veins 3 (Cu_1) and 4 (M_3) approximated; 5 (M_2) arising from middle of dcs; 6 (M_1) and 7 (Rs) shortly stalked; 8 ($Sc+R_1$) approximated to upper margin of cell beyond middle.

STAUROPUS FAGI (Linnaeus)
The Lobster Moth

Phalaena (*Noctua*) *fagi* Linnaeus, 1758, *Syst.Nat.* (Edn 10) **1**: 508.

Type locality: [Sweden].

Description of imago (Pl.3, figs 8,9)

Wingspan 55–70mm. Antenna in male strongly pectinate, in female simple. Dimorphic. One form has the wings ashy brownish grey (fig.8); forewing with broad median fascia deep brown, edged with dentate white lines; subterminal line a series of dark brown, whitish-edged spots between veins; a terminal series of dark brown spots between veins. Hindwing with whitish median costal mark. Thorax and abdomen brown-grey. The other form (fig.9) has the forewing deep blackish brown with prominent dentate white lines edging median fascia; subterminal and terminal spots less prominent. Hindwing with costal mark faint. Thorax and abdomen deep blackish brown. In some localities both forms occur in equal numbers.

In south-west Ireland, especially in Co. Kerry, a very large, pale, light grey-brown form occurs.

Life history

Ovum. Hemispherical, pale cream becoming purplish. Laid singly on leaves of various trees and shrubs such as beech (*Fagus sylvatica*), oak (*Quercus* spp.), birch (*Betula* spp.) and hazel (*Corylus avellana*), from May to July, hatching after about ten days.

Larva. Full-fed *c.*75mm long. Head large. Body ochreous brown; spiracular stripe interrupted, black; intersegmental divisions incised; abdominal segments 2–7 each with paired conical dorsal 'humps'; anal segment greatly enlarged; thoracic legs very long; prolegs well developed, anal pair modified into long, slender filaments; peritreme of spiracle white.

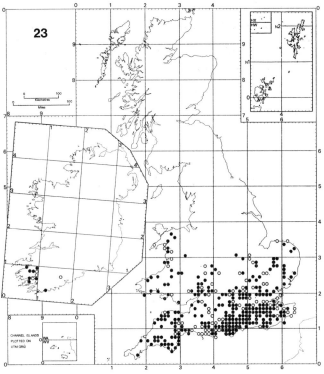
Stauropus fagi

When at rest its head, thorax and anal segment are raised, producing the characteristic 'lobster' appearance. On hatching, the first instar larva is ant-like and feeds solely on its eggshell, this instar being of nine days' duration. The larval stage lasts to September.

Pupa. Stout; blackish brown. In a flimsy cocoon spun amongst fallen leaves; overwinters from October to May of the following year.

Imago. Univoltine. Flies from mid-May to mid-July. When at rest on a small beech-stem with hindwings spread out below the forewings, it resembles a bunch of leaves.

Distribution (Map 23)
Widely distributed south of a line from Pembroke to the Wash; north-west Wales; Cos Kerry and Cork in Ireland. Abroad throughout Europe eastwards to Japan.

NOTODONTA Ochsenheimer
Notodonta Ochsenheimer, 1810, *Schmett.Eur.* **3**: 45.

A Holarctic genus with ten Palaearctic species, one of which is resident in Great Britain and Ireland, whilst a second has been recorded on a single occasion.

Antenna in male bipectinate to apex, pectinations very short on apical third, in female denticulate; haustellum rudimentary; palpi short; eyes hairy; ocelli absent. Posterior tibia with two pairs of spurs. Forewing with dorsal scale-tooth; veins 6 to 10 (M_1 to R_2) stalked.

NOTODONTA DROMEDARIUS (Linnaeus)
The Iron Prominent

Phalaena (Bombyx) dromedarius Linnaeus, 1767, *Syst.Nat.* (Edn 12) **1** (2): 827.

Type locality: not stated.

Description of imago (Pl.3, figs 10,11)
Wingspan 42–50mm. Antenna in male shortly pectinate, in female denticulate. Thorax dark brown. Forewing with dorsal scale-tooth; dark brownish fuscous, a whitish, ferruginous-marked basal patch; antemedian fascia whitish ochreous, edged dark fuscous; discal spot ferruginous, edged ochreous; postmedian fascia whitish ochreous, edged dark fuscous, broken medially; subterminal fascia ferruginous, ill-defined; fringe chequered black and grey. Hindwing pale brownish grey, darker at tornus; fringe chequered brown and pale grey. Abdomen dark brown.

In northern England and Scotland a form occurs (fig.11) with much darker wings and the markings almost obliterated.

Life history
Ovum. Hemispherical, small, bluish. Laid singly or in pairs on leaves of birch (*Betula* spp.) or, occasionally, alder (*Alnus glutinosa*) or hazel (*Corylus avellana*) in May and again in August.

Larva. Full-fed *c.*35mm long. Head large, yellow or brown. Body variable in colour, usually green; broad, dark brown dorsal stripe on thoracic and abdominal segments 1–4; subspiracular stripe pale yellow; ventral surface dark brown; abdominal segments 1–4 and 9 each with dorsal 'hump'; peritreme of spiracle whitish; anal claspers rudimentary. The anal segments are normally raised at an angle to the body. Feeds in June, July and August and sometimes as a second generation in September and October.

Pupa. Stout, rounded; glossy brown; cremaster with three small points. The larva constructs a lightly spun subterranean cocoon in which the pupa overwinters.

Imago. Univoltine, sometimes bivoltine. Flies in May, June and July, sometimes again in August and September or even later.

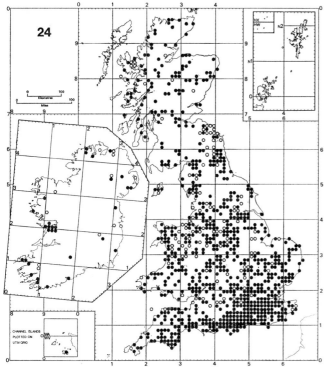

Notodonta dromedarius

Distribution (Map 24)

Widespread and common throughout Great Britain and Ireland; absent from Orkney and Shetland. Abroad throughout central and northern Europe.

NOTODONTA TORVA (Hübner)
The Large Dark Prominent

Bombyx torva Hübner, [1803], *Samml.eur.Schmett.* **3**: pl.7, fig.27.

Phalaena (Bombyx) tritophus sensu Esper, 1786, *Schmett.* **3**: 299, pl.60, figs 1,2.

Type locality: Europe.

Only one British specimen of this Palaearctic species is known. It was reared by Frank Norgate of Norfolk from either an egg or larva found by him in north Norfolk in July or August 1882. Originally placed doubtfully by him in his series of *Peridea anceps* (Goeze), it was recognized by Barrett (1896: 121–123). Though resembling *P. anceps* to some extent in colour and markings, it is similar to *N. dromedarius* (Linnaeus) in shape and size and could more probably be mistaken for that species in the field. Figured in South (1961: pl.22, fig.4).

TRITOPHIA Kiriakoff
Tritophia Kiriakoff, 1967, *Genera Insect.* **217** (B): 141.

A monotypic genus occurring in Europe, southern Asia and North America. The species occurs in Britain only as a rare immigrant.

Antenna in male shortly bipectinate, pectinations not more than three times diameter of shaft, gradually becoming denticulate on apical one-fifth, in female slightly denticulate; haustellum present; palpi very short, hairy, ascending; eyes hairy. Legs densely hairy, especially in male; anterior tarsus clothed with long hairs becoming shorter distally; other tarsi covered with short, smooth hairs; posterior tibia with two pairs of spurs. Forewing elongate, narrower in male, costa slightly concave in middle, then slightly arched; apex rounded; termen oblique, convex; tornus ill-defined; dorsum slightly convex with scale-tooth in middle; veins 3 (Cu_1) and 4 (M_3) well separated, 5 (M_2) from just beyond middle of discocellulars (dcs); 6 (M_1) and 7 to 10 (R_5 to R_2) arising from upper angle of cell; 7 (R_5), 8+9 (R_4+R_3) and 10 (R_2) stalked. Hindwing costa slightly arched; apex rounded; cilia long; veins 3 (Cu_1) and 4 (M_3) approximated, 5 (M_2) arising just beyond middle of dcs; 6 (M_1) and 7 (Rs) stalked at one-seventh; 8 ($Sc+R_1$) approximated to cell just beyond middle of upper margin.

TRITOPHIA TRITOPHUS ([Denis & Schiffermüller])
The Three Humped Prominent

Bombyx tritophus [Denis & Schiffermüller], 1775, *Schmett. Wien.*: 63.

Phalaena (Bombyx) phoebe Siebert, 1790, *in* Scriba, *Beitr. Insect.* **1**: 18.

Type locality: [Austria]; Vienna district.

Description of imago (Pl.3, fig.12)

Wingspan 54–62mm. Antenna in male weakly bipectinate, in female simple. Forewing with dorsal scale-tooth; dull grey-brown; median fascia indistinct, ochreous brown; discal spot large, reddish brown, preceded by a smaller spot, both outlined yellowish white; subterminal fascia narrow, ochreous brown, edged terminally yellowish white; fringe chequered grey and dark grey. Hindwing white, veins grey; fringe chequered white and grey, grey at tornus. Body dark fuscous.

Life history

Ovum. Hemispherical, almost white. Laid on leaves of poplar and aspen (*Populus* spp.).

Larva. Full-fed *c.*40mm long. Head and body brownish green becoming purple-brown immediately prior to pupation; dorsal and lateral stripes pale brown; ventral surface darker; abdominal segments 3–5 each with a prominent backward-pointing 'hump', and abdominal segment 9 with a forward-curving 'hump'; a series of black dorsal spots; peritreme of spiracle black. The anal segments are raised when the larva is walking. Feeds in June, July and August.

Pupa. Large; shining brown. Subterranean, in a silken cocoon; overwinters.

Imago. Flies from May to July on the Continent.

Occurrence and distribution

Only a few specimens of this rare immigrant have been reported: one male bred 10 August 1842 from a larva taken on aspen by J. W. Douglas in Essex; one in Suffolk by Mr. Garneys; one near Paisley by Morris Young; one at light in Bedford, May 1907; one at light taken by A. H. Sperring near Havant, Hampshire, 20 May 1920; one at Waterlooville, Hampshire, July 1920; one at Folkestone, Kent, 22 August 1953; and one in the Isle of Wight, August 1958. Also a number of larvae which failed to produce imagines have been taken over the years. Holarctic; throughout much of Europe and North America.

ELIGMODONTA Kiriakoff

Eligmodonta Kiriakoff, 1967, *Genera Insect.* **217** (B): 181.

A monotypic genus which occurs in Europe, southern and central Asia and throughout Great Britain and Ireland.

Antenna in male shortly bipectinate to apex, pectinations very short on apical fifth, in female denticulate, a tuft at base of antenna in both sexes; palpi short, hairy; eyes hairy. Legs clothed with long rough hairs; anterior tarsi rough-haired; other tarsi with short, smooth hairs; posterior tibia with two pairs of spurs. Forewing elongate; costa straight to three-quarters, then slightly arched; apex rounded; termen oblique, slightly convex; tornus forming an angle of $135°$; dorsum almost straight; median scale-tooth; forewing of female slightly broader than male; veins 3 (Cu_1) and 4 (M_3) well separated; 5 (M_2) from middle of discocellulars (dcs); 6 (M_1) and stalk of 7 (R_5), 8+9 (R_4+R_3) and 10 (R_2) from upper angle of cell. Hindwing costa very slightly convex; apex rounded; veins 3 (Cu_1) and 4 (M_3) arising together; 5 (M_2) from middle of dcs; 6 (M_1) and 7 (Rs) medially stalked; 8 (Sc+R_1) approximated to upper margin of cell for three-quarters of its length.

ELIGMODONTA ZICZAC (Linnaeus)
The Pebble Prominent

Phalaena (Bombyx) ziczac Linnaeus, 1758, *Syst.Nat.* (Edn 10) **1**: 504.

Type locality: [Europe].

Description of imago (Pl.3, figs 13–15)

Wingspan 42–52mm. Antenna in male finely bipectinate, in female denticulate. Thorax dark ochreous brown. Forewing with dorsal scale-tooth; ochreous brown, paler and greyish-tinged medially in costal half, terminally darker; a curved, black postmedian line extending towards apex outlining the 'pebble' mark; subterminal line paler; fringe chequered light and dark grey. Hindwing whitish brown, with large, dark brown discal lunule, darker in female. Abdomen dark ochreous brown.

There is some geographical variation. In Scottish specimens (fig.14) the forewing has black subbasal and antemedian costal marks and a prominent black line in the 'pebble' mark; the hindwing is paler and the body grey-brown. In Irish specimens (fig.15) the forewing is much lighter, with black subbasal and antemedian fasciae; the hindwing is paler in the male and the body is grey-brown.

Life history

Ovum. Hemispherical, pale bluish. Laid in small clusters on sallow and willow (*Salix* spp.) and occasionally poplar (*Populus* spp.) in May and June and again, in southern Britain, in August.

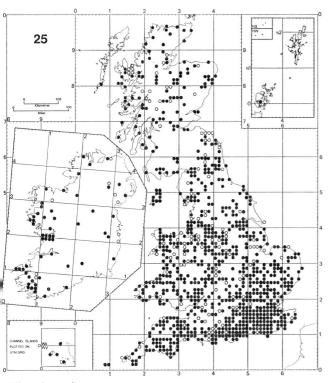

Eligmodonta ziczac

Larva. Full-fed *c.*40mm long. Head large, green. Body ochreous grey; abdominal segments 7–9 orange-brown; thoracic segments much more slender than abdominal segments; dorsal stripe yellow, edged brown; spiracular stripe greenish, whitish or purplish brown; abdominal segments 2, 3 and 9 each with a pyramidal 'hump'. The last three abdominal segments are usually held up at an angle to the rest of the body. Feeds in May and, in southern Britain, again in August.

Pupa. Reddish brown. In a fragile subterranean earthen cocoon; overwinters.

Imago. Univoltine, bivoltine in the south. Flies in May and June, and again, in southern Britain, in August with sometimes a third generation in October. Comes freely to light.

Distribution (Map 25)
Widespread and common throughout the British Isles except Shetland. Abroad throughout Europe to eastern Asia.

HARPYIA Ochsenheimer

Harpyia Ochsenheimer, 1810, *Schmett.Eur.* **3**: 19.
Hoplitis Hübner, [1819], *Verz.bekannt.Schmett.*: 147.
Hybocampa Lederer, 1853, *Verh.zool.-bot.Ver.Wien* **2**: 78.

A small Palaearctic genus of which one of the three species has been recorded on a single occasion in Britain.

Antenna bipectinate to three-quarters, pectinations very short in female; haustellum rudimentary; palpi short; eyes glabrous; ocelli absent. Legs with long hairs; posterior tibia with one pair of spurs. Forewing elongate and fairly narrow; broader in female; vein 2 (Cu_2) very short; 3 (Cu_1) and 4 (M_3) approximated; 5 (M_2) arising from middle of discocellulars (dcs); 6 (M_1) and 7 to 10 (R_5 to R_2) arising from angle of cell; stalk of 10 (R_2), 7 (R_5), 8+9 (R_4+R_3) of variable length. Hindwing veins 2 (Cu_2), 3 (Cu_1), 4 (M_3) and 5 (M_2) as in forewing; 6 (M_1) and 7 (Rs) stalked at one-quarter and one-third respectively; 8 ($Sc+R_1$) approximated to cell near base of upper margin.

See note under *Furcula* Lamarck (p. 43).

HARPYIA MILHAUSERI (Fabricius)

Bombyx milhauseri Fabricius, 1775, *Syst.Ent.*: 577.
Type locality: Germany; Dresden.

A single specimen of this Palaearctic species was taken at light in June 1966 at Aldwick Bay, Sussex (Pickering, 1966). There is no other record of this species from Great Britain and Ireland.

This is a species of the size and shape of *Pheosia tremula* (Clerck), but without the dorsal scale-tuft; it is whitish, much streaked and irrorate with grey, and with a blackish area along dorsum divided by the postmedian line.

PERIDEA Stephens
Peridea Stephens, 1828, *Ill.Br.Ent.* (Haust.) **2**: 32.

Of the 16 Palaearctic species in this genus only one occurs in Britain.

Antenna bipectinate almost to apex; palpi very short, hairy; eyes glabrous. Thorax slightly crested. Legs densely clothed with hairs as far as middle of tarsi. Forewing with dorsal scale-tooth; veins 6 (M_1) to 10 (R_2) stalked.

PERIDEA ANCEPS (Goeze)
The Great Prominent

Noctua anceps Goeze, 1781, *Ent.Beyträge* **3** (3): 207.
Bombyx trepida Esper, 1786, *Schmett.* **3** (24): 284.
Type locality: not stated.

Description of imago (Pl.3, figs 16,17)
Wingspan 52–72mm. Antenna in male pectinate, in female ciliate. Forewing with dorsal scale-tooth; pale grey-brown, irrorate ochreous; discal lunule, median dorsal mark and subterminal series of dashes between veins, dark brown, edged ochreous yellow. Hindwing pale yellowish white with pale grey apical patch; terminal line dark grey. Body grey-brown.

Variation is confined to the intensity of the brown ground colour of the forewing. Lake District specimens are frequently very dark and almost black examples occasionally occur.

Life history
Ovum. Hemispherical, large, bluish. Laid on the upper surface of leaves of oak (*Quercus* spp.) in May and June, hatching in ten days.
Larva. Full-fed *c.*40mm long. Head whitish green. Body olive-green; subdorsal and supraspiracular lines yellowish; pro- and mesothoracic segments with broad, reddish, lateral stripes; abdominal segments with oblique yellow lateral stripes posteriorly edged purple. When at rest it raises its head and thorax. Feeds from June to August.
Pupa. Elongate; reddish brown. In a subterranean cocoon covered with soil particles, well below the soil surface; overwinters.
Imago. Univoltine. Flies from late April until well into June. Whilst the male comes freely to ultra-violet light, the female only occasionally does so.

Distribution (Map 26)
Widely distributed and well established in oak woods in south and south-east England, the Lake District and Wales; in Scotland as far north as west Perthshire. Absent from Ireland. Abroad in central Europe from Spain to southern Fennoscandia and eastwards to the U.S.S.R.

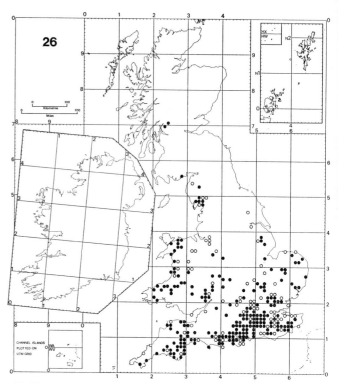

Peridea anceps

PHEOSIA Hübner

Pheosia Hübner, [1819], *Verz.bekannt.Schmett.*: 145.
Leiocampa Stephens, 1828, *Ill.Br.Ent.* (Haust.) **2**: 24.

Two of the eight Palaearctic species in this genus occur in Great Britain and Ireland.

Antenna bipectinate, pectinations very short in female; haustellum reduced; palpi very short; eyes naked; ocelli absent. Tegulae with long hairs. Posterior tibia with two pairs of spurs. Forewing elongate and rather narrow with dorsal scale-tooth; areole absent; vein 6 (M_1) from the same point as the stalk of veins 7 (R_5) to 10 (R_2). Hindwing with veins 6 (M_1) and 7 (Rs) on a long stalk.

PHEOSIA GNOMA (Fabricius)
The Lesser Swallow Prominent
Bombyx gnoma Fabricius, 1777, *Gen.Insect.*: 279.
Bombyx dictaeoides Esper, 1789, *Schmett.* **3**: 27, pl.84, fig.3.
Type locality: Germany; Hamburg.

Description of imago (Pl.3, figs 18,19)
Wingspan 46–58mm. Antenna in male pectinate, in female ciliate. Forewing elongate, rather narrow, with a small, dark brown dorsal scale-tooth; ochreous white; subapical elongate patch on costa dark brown; a broad, dark brown subdorsal streak from base to tornus with a white line near dorsum above scale-tooth; a very distinct white wedge-shaped tornal dash between veins 1b (2A) and 2 (Cu_2). Hindwing pale yellowish white with a diffuse dark brown tornal spot. Thorax and abdomen dark brown.

In specimens from the north of England and Scotland (fig.19) the ground colour of the forewing is clearer white and the thorax is either darker brown than in southern specimens or light grey, the two forms occurring in approximately equal numbers.

Similar species. *P. tremula* (Clerck) which is larger and has the distinct, white, wedge-shaped mark on the forewing replaced by a thin white line, and a thin white line in the tornal spot on the hindwing.

Life history
Ovum. Hemispherical, pale green. Laid in batches of two or three on the upper surface of leaves of birch (*Betula* spp.) in May and June and again in August and September.
Larva. Full-fed *c*.35mm long. Head and body purple-brown with a 'varnished' appearance; subspiracular stripe broad, yellow; peritreme of spiracle black; a 'hump' on abdominal segment 9. Feeds in June and July and again in September and October.
Pupa. Elongate; dark reddish brown. In a subterranean tough cocoon of silk intermixed with soil particles; overwinters.

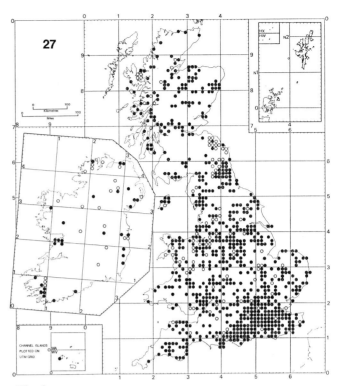

Pheosia gnoma

Imago. Bivoltine. Flies in May and June and again in August. The male especially comes freely to light.

Distribution (Map 27)
Widely distributed and generally common in birch woods throughout Great Britain and Ireland. Absent from Orkney and Shetland. Abroad throughout Europe eastwards to Siberia.

PHEOSIA TREMULA (Clerck)
The Swallow Prominent

[*Phalaena*] *tremula* Clerck, 1759, *Icones Insect.rar.* **1**: pl.9, fig.13.

Phalaena (*Bombyx*) *dictaea* Linnaeus, 1767, *Syst.Nat.* (Edn 12) **1** (2): 826.

Type locality: [Sweden].

Description of imago (Pl.3, figs 20,21)

Wingspan 50–64mm. Antenna in male pectinate, in female weakly ciliate. Forewing with dorsal scale-tooth; ochreous white; subapical patch on costa dark brown, rather ill-defined; base and broad dorsal streak dark brown, with a white line near dorsum above scale-tooth; white dashes on the terminal ends of veins at tornus, that between veins 1b (2A) and 2 (Cu$_2$) being long and very narrow. Hindwing pale yellowish white, with a small, dark brown tornal spot intersected by a white marginal line. Thorax and abdomen grey-brown.

Specimens from Scotland (fig.21) are paler and have the brown markings reduced and the thorax varies in the same way as in *P. gnoma* (Fabricius).

Similar species. *P. gnoma* q.v.

Life history

Ovum. Hemispherical, creamy white. Laid either singly or in pairs on the upper surface of leaves of poplar and aspen (*Populus* spp.) and willow (*Salix* spp.) in May and, in southern Britain, again in August.

Larva. Full-fed *c.*40mm long. Head and body green; dorsal line paler; subspiracular line reddish yellow; peritreme of spiracle black; a 'hump' on abdominal segment 9; thoracic legs red; prolegs and ventral surface bluish green. There is a grey-brown form in which the stripes are absent and the peritreme more prominent. Feeds in late June and July and again, in southern Britain, in September and October.

Pupa. Glossy reddish brown. The first generation pupates in a roughly spun cocoon amongst leaves; the second generation well beneath the soil, overwintering.

Imago. Univoltine, bivoltine in southern Britain. Flies in May and June and again, in southern Britain, in August. Comes freely to light.

Distribution (Map 28)

Widespread and common throughout Great Britain and Ireland where poplar and aspen occur. Absent from Orkney and Shetland. Abroad throughout Europe eastwards to Siberia.

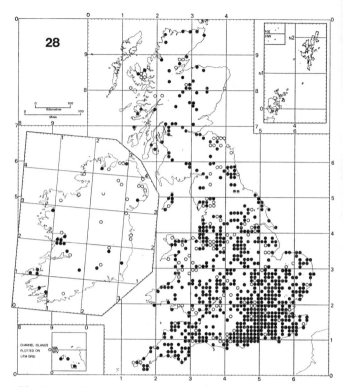

Pheosia tremula

PTILODON Hübner

Ptilodon Hübner, 1822, *Syst.-alph.Verz.*: 14, 15.
Lophopteryx Stephens, 1828, *Ill.Br.Ent.* (Haust.) **2**: 26.

A genus with 11 Palaearctic species, only one of which occurs in Great Britain and Ireland.

Antenna short, in male very slightly bipectinate, pectinations reduced towards apex, in female ciliate; palpi short, hairy, ascending; eyes slightly hairy; ocelli absent. Thorax crested. Posterior tibia with two pairs of spurs. Forewing with dorsal scale-tooth; vein 2 (Cu_2) twice as far from 3 (Cu_1) as 3 (Cu_1) is from 4 (M_3); 3 (Cu_1) and 4 (M_3) well separated; 5 (M_2) from middle of cell; areole short; 6 (M_1) from upper edge of cell; 7 (R_5) and 10 (R_2)+ 8+9 (R_4+R_3) from apex of areole. Hindwing veins 3 (Cu_1) and 4 (M_3) connate; 5 (M_2) from middle of cell; 6 (M_1) and 7 (R_s) stalked at one-third of 6 (M_1); 8 (Sc+R_1) approximated to cell just before apex.

PTILODON CAPUCINA (Linnaeus)
The Coxcomb Prominent

Phalaena (*Bombyx*) *capucina* Linnaeus, 1758, *Syst.Nat.* (Edn 10) **1**: 507.
Phalaena (*Bombyx*) *camelina* Linnaeus, 1758, *ibid.*: 507.
Type locality: Europe.

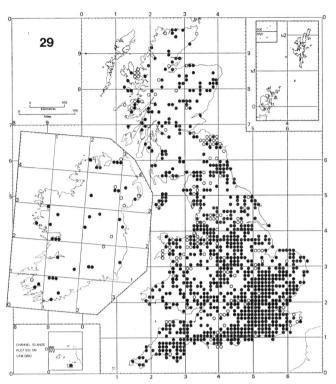

Ptilodon capucina

Description of imago (Pl.3, figs 22,23)

Wingspan 40–50mm. Antenna in male pectinate, in female ciliate. Thorax brown with distinct pale brown 'coxcomb-shaped' crest. Forewing with prominent dark dorsal scale-tooth; reddish brown; median fascia greyish brown; ante- and postmedian lines black, irregularly and acutely dentate, often obsolescent, the latter being sometimes absorbed into the median fascia; termen dentate. Hindwing pale ochreous; terminal shade darker brownish ochreous; tornal patch bluish black. Abdomen pale brown.

Variation is confined to the intensity of the ground colour; very dark forms (fig.23) occur with the ground colour purplish brown.

Similar species. *Ptilodontella cucullina* ([Denis & Schiffermüller]) which is smaller, paler and has a white subterminal line and white subterminal fascia.

Life history

Ovum. Hemispherical, white. Laid in batches of two or three on the upper surface of leaves of trees and shrubs such as birch (*Betula* spp.), poplar (*Populus* spp.), hazel (*Corylus avellana*) and willow (*Salix* spp.) in May and June and again in August and September.

Larva. Full-fed *c.*35mm long. Head green. Body bluish green, ventrally green; dorsal line dark green; subspiracular line yellow; two bright red dorsal projections on abdominal segment 8; a reddish spot posterior to each spiracle; a series of dorsal tufts of fine, short setae. There are several colour-forms with the body ochreous, yellowish or purplish coloured. Feeds in June and July and again, as a second generation, in September and October.

Pupa. Rounded, slightly elongate; purplish brown; cremaster with a single point. In a subterranean cocoon of silk mixed with fine particles of soil, often at a tree-root. The second generation overwinters in this stage.

Imago. Bivoltine. Flies in May and June and again in August and September. At rest the wings are folded downwards displaying the thoracic crest, thus producing a close resemblance to a dead leaf.

Distribution (Map 29)

Widely distributed and common throughout Great Britain and Ireland but absent from Orkney and Shetland. Abroad throughout Europe eastwards to Japan.

PTILODONTELLA Kiriakoff

Ptilodontella Kiriakoff, 1967, *Genera Insect.* **217** (B): 176.

A monotypic genus occurring throughout the Palaearctic region.

Antenna in male dentate-fasciculate, in female shortly dentate; palpi short, hairy, ascending; terminal segment short; eyes slightly hairy. Thorax with dorsal crest. Anterior femora with long, dense hairs; posterior tibia with two pairs of long spurs. Wings similar in shape to *Ptilodon* Hübner, but with termen less dentate. Forewing veins 2 (Cu_2), 3 (Cu_1) and 4 (M_3) equidistant and well separated; 5 (M_2) from middle of discocellulars (dcs); a long, narrow areole; 6 (M_1) from the upper angle of cell; 7 (R_5) and 10 (R_2)+8 (R_4)+9 (R_3) arising from apex of areole. Hindwing distance between veins 2 (Cu_2) and 3 (Cu_1) two and one-half times that between 3 (Cu_1) and 4 (M_3) which are well separated; 5 (M_2) from middle of dcs; 6 (M_1) and 7 (Rs) stalked beyond a half; 8 ($Sc+R_1$) close to cell to three-quarters.

PTILODONTELLA CUCULLINA ([Denis & Schiffermüller])

The Maple Prominent

Bombyx cucullina [Denis & Schiffermüller], 1775, *Schmett. Wien.*: 311.

Bombyx cuculla Esper, 1786, *Schmett.* **3**: 364.

Type locality: [Austria]; Vienna district.

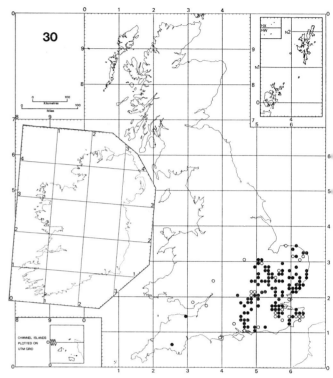

Ptilodontella cucullina

Description of imago (Pl.3, fig.24)

Wingspan 37–46mm. Antenna in male pectinate, in female finely dentate. Thorax reddish brown. Forewing with dark scale-tooth; reddish brown; costal half pale ochreous yellow from base to one-half; an irregular black line from middle of subterminal line to scale-tooth; subterminal fascia whitish, edged inwardly with a white line, not extending to apex; termen weakly dentate. Hindwing brown; tornal mark dark brown, divided by a yellow line. Abdomen brown.

Similar species. *Ptilodon capucina* (Linnaeus) q.v.

Life history

Ovum. Hemispherical, smooth and glossy, white. Laid in batches of two or three on the upper surface of leaves of field maple (*Acer campestre*) in late June and July.

Larva. Full-fed c.30mm long. Head pale ochreous brown. Body pale green; dorsal stripe dark green, broad on thoracic and first abdominal segments; subdorsal line pale yellow, edged green dorsally; spiracles pink; peritreme of spiracle black; 'hump' on abdominal segment 8 purplish. Feeds from July to mid-September; in captivity will accept sycamore (*Acer pseudoplatanus*).

Pupa. Purple-brown. In a thin silken cocoon amongst leaf-litter on the ground in mid-September; overwinters.

Imago. Univoltine. Flies in May and June. Both sexes come freely to light. At rest it resembles a piece of wood.

Distribution (Map 30)

Confined to the chalky soils of the south-eastern counties of England except for isolated records from Somerset and south Devon. Abroad throughout central Europe as far north as Denmark and eastwards to Siberia.

ODONTOSIA Hübner
Odontosia Hübner, [1819], *Verz.bekannt.Schmett.*: 145.

A small Holarctic genus containing four Palaearctic species, only one of which occurs in Great Britain and Ireland.

Antenna in male typically dentate-fasciculate, sometimes bipectinate, in female simple; haustellum rudimentary; palpi very short; eyes glabrous; ocelli absent. Legs densely clothed in long hairs; posterior tibia with two pairs of spurs. Forewing costa almost straight; apex pointed; termen oblique and dentate, especially at vein 4 (M_3); dorsum with scale-tooth; distance between veins 2 (Cu_2) and 3 (Cu_1) almost double that between 3 (Cu_1) and 4 (M_3) which are widely separated; 5 (M_2) from middle of discocellulars (dcs); areole present; 6 (M_1) arising from basal third of areole; 7 (R_5) and 8+9 (R_4+R_3) from apex of areole; 10 (R_2) from costal edge of areole, just before apex. Hindwing veins 2 (Cu_2), 3 (Cu_1), 4 (M_3) and 5 (M_2) as in forewing; 6 (M_1) and 7 (Rs) stalked for almost half length; 8 ($Sc+R_1$) approximated to cell just before apex.

ODONTOSIA CARMELITA (Esper)
The Scarce Prominent

Phalaena (Bombyx) carmelita Esper, 1799, *Schmett.* **3** Suppl. Abs. 3: 65, pl.91, fig.1.

Type locality: Germany.

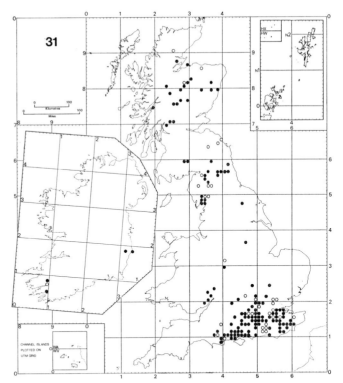

Odontosia carmelita

Description of imago (Pl.3, fig.25)
Wingspan 44–52mm. Antenna in male dentate-fasciculate, in female simple. Thorax dark red-brown. Forewing with a prominent, black dorsal scale-tooth; termen weakly dentate; purplish grey, dark brown costally; yellowish white costal marks at one-third and two-thirds; a postmedian series of black dots; fringe chequered black and yellowish white. Hindwing purplish grey; tornal patch dark grey divided by a pale, whitish line; fringe chequered black and yellowish white. Abdomen red-brown.

Variation is confined to the ground colour, Scottish specimens being darker grey.

Life history
Ovum. Hemispherical, pale blue. Laid on the underside of leaves of birch (*Betula* spp.) in batches of two or three at the end of April or in early May.

Larva. Full-fed *c.*35mm long. Head green. Body bright grass-green freckled with yellow; dorsal line dark green; spiracular line yellow, spotted with red. Feeds from May to mid-July.

Pupa. Deep purple-brown. In a tough silken cocoon covered with soil, usually under moss; overwinters in this stage from July to April.

Imago. Univoltine. Flies from early April to early May, somewhat later in the Scottish Highlands.

Distribution (Map 31)
Confined to birch woods; widely distributed in south-east England, in the Forest of Dean, Sherwood Forest, the Lake District, Northumberland and the Scottish Highlands; local in Cos Kerry and Wicklow in Ireland. Abroad in central and northern Europe.

PTEROSTOMA Germar

Pterostoma Germar, 1812, *Syst.Gloss.Prod.* **2**: 42.
Ptilodontis Stephens, 1828, *Ill.Br.Ent.* (Haust.) **2**: 28.

A small Palaearctic genus containing four species, only one of which occurs in Great Britain and Ireland.

Antenna in male bipectinate to apex, in female shortly pectinate; palpi as long as thorax; haustellum weakly developed; eyes glabrous; ocelli absent. Thorax with pointed dorsal crest. Legs clothed with long hairs; posterior tibia with two pairs of spurs. Forewing termen denticulate; dorsum with two scale-teeth; vein 6 (M_1) arising from upper angle of cell; areole elongate. Hindwing veins 6 (M_1) and 7 (Rs) stalked. Abdomen with prominent anal tuft in male.

PTEROSTOMA PALPINA (Clerck)
The Pale Prominent

[*Phalaena*] *palpina* Clerck, 1759, *Icones Insect.rar.* **1**: pl.9, fig.8.

Type locality: [Sweden].

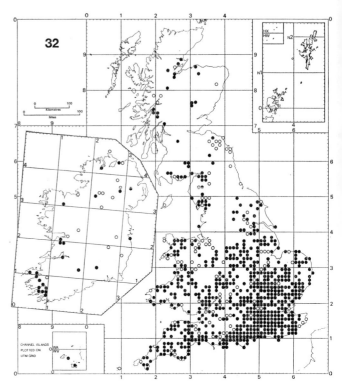

Pterostoma palpina

Description of imago (Pl.3, fig.26)
Wingspan 42–60mm. Antenna bipectinate, in male strongly, in female weakly. Thorax with dorsal crest; greyish buff. Forewing broad, greyish buff with prominent dorsal scale-tooth; termen dentate; a diffuse, greyish median fascia, sometimes obsolete; an irregular interrupted postmedian fascia consisting of black dots on the veins; a subterminal line of black dots; fringe chequered fuscous and buff. Hindwing greyish buff; postmedian fascia paler, often obscure or obsolete; terminal shade broad, grey. Abdomen buff with small dorsal and pronounced lateral crests.

Life history
Ovum. Hemispherical, whitish. Laid in small batches of two or three on the underside of leaves of poplar (*Populus* spp.), aspen (*Populus tremula*) or occasionally willow (*Salix* spp.) usually in late May.
Larva. Full-fed *c.*40mm long. Head and body bluish green; dorsal and subdorsal lines whitish; spiracular line yellowish, tinged with red, dorsally black. Feeds in June and July and again, in southern England, in the autumn.
Pupa. Stout, abdomen tapering; shining purplish brown; cremaster with four points. In a silken cocoon covered with soil particles amongst grass at the roots of the host tree; overwinters.
Imago. Univoltine; bivoltine in southern England. Flies in May and, in southern England, again in August. The male comes freely to light. When resting with the wings wrapped around the body it closely resembles a dead leaf.

Distribution (Map 32)
Widely distributed and common in southern England, becoming more local in the north; Wales; Scotland, in the Highlands amongst aspen; Ireland. Abroad throughout Europe eastwards to Japan.

LEUCODONTA Staudinger

Leucodonta Staudinger, 1892, *in* Romanoff, *Mém.Lépid.* **6**: 349.

A monotypic Palaearctic genus, now probably extinct in the British Isles.

Antenna in male dentate-fasciculate, in female shortly ciliate; haustellum rudimentary; palpi short; eyes glabrous; ocelli absent. Legs clothed in long hairs; posterior tibia with two pairs of spurs. Forewing apically rounded; termen smoothly arched; dorsum straight with median scale-tooth; vein 2 (Cu_2) from lower margin of cell at four-fifths; 3 (Cu_1) and 4 (M_3) closely approximated; 5 (M_2) from middle of discocellulars (dcs); areole narrow; 6 (M_1) from areole just beyond middle; 7 (R_5), 8 (R_4), 9 (R_3) stalked and 10 (R_2) from apex of areole. Hindwing rounded; veins 2 (Cu_2), 3 (Cu_1) and 4 (M_3) as in forewing; 5 (M_2) from upper one-third of dcs; 6 (M_1) and 7 (Rs) stalked on short stalks (one-fifth to one-sixth of length of veins); 8 ($Sc+R_1$) closely approximated to upper margin of cell as far as apex.

LEUCODONTA BICOLORIA ([Denis & Schiffermüller])
The White Prominent

Bombyx bicoloria [Denis & Schiffermüller], 1775, *Schmett. Wien.*: 49.

Type locality: [Austria]; Vienna district.

Description of imago (Pl.3, fig.27)

Wingspan 38–42mm. Antenna in male dentate-fasciculate, in female shortly ciliate. Forewing white; dorsal scale-tooth weak, speckled black; a black median basal spot; an orange median fascia extending from near dorsum to vein 11 (R_1) edged black basally and extending as a broad subdorsal streak almost to tornus; a dorsal patch of black spots below orange fascia; a subterminal line of black dots. Hindwing white. Thorax and abdomen pale yellowish white.

Life history

Ovum. Hemispherical, white. Laid on the underside of leaves of birch (*Betula* spp.) in mid-June.

Larva. Full-fed *c*.35mm long. Head and body pale green, whitish dorsally; dorsal and subdorsal lines green; subspiracular line yellow; peritreme of spiracle black within a white spot. Feeds from June to late July.

Pupa. Dark reddish brown. In a flimsy cocoon of silk and small leaf fragments usually amongst dead leaves of the foodplant; overwinters.

Imago. Univoltine. Flies in June. The male frequently comes to light and sometimes flies amongst birch by day.

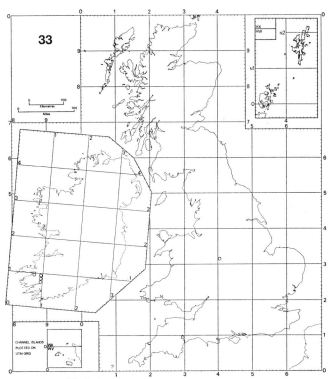

Leucodonta bicoloria

Occurrence and distribution (Map 33)

Presumed extinct. First recorded in the British Isles in June 1859 when it was found near Killarney, Co. Kerry, Ireland by P. Bouchard who obtained further examples in the same locality during the next few years. It was subsequently found sporadically in Co. Kerry until 1938. Since then none has been found despite much searching. The only other records from the British Isles are of a few taken in Burnt Wood, Staffordshire in 1861 and 1865 and of one taken near Exeter, Devon in 1880 (Barrett, 1896). Abroad throughout Europe eastwards to Japan.

PTILOPHORA Stephens

Ptilophora Stephens, 1828, *Ill.Br.Ent.* (Haust.) **2**: 29.

A small Palaearctic genus containing four species, only one of which occurs in England.

Haustellum rudimentary; eyes glabrous; ocelli absent. Anterior tibia with one short spine. Forewing vein 6 (M_1) arising from angle of cell; 7 (R_5), 8+9 (R_4+R_3), 10 (R_2) stalked. Hindwing veins 6 (M_1) and 7 (Rs) on a long stalk.

PTILOPHORA PLUMIGERA ([Denis & Schiffermüller])
The Plumed Prominent

Bombyx plumigera [Denis & Schiffermüller], 1775, *Schmett. Wien.*: 61.

Phalaena variegata Villers, 1789, *Linn.ent.* **2**: 160.

Type locality: [Austria]; Vienna district.

Description of imago (Pl.3, figs 28,29)

Wingspan 35–44mm. Antenna in male very strongly bipectinate, in female simple. Thorax reddish brown with weak anterior crest. *Male* dimorphic. One form with forewing reddish brown, darker basally; postmedian fascia narrow, broader at costa than at dorsum, pale yellow, veins fuscous; hindwing thinly scaled, pale yellowish buff. The other form with forewing more variegated with irregular greyish patches: hindwing pale grey basally. Fringes in both forms chequered buff and grey. *Female.* Wings reddish ochreous. Abdomen in both sexes reddish brown.

Life history

Ovum. Rounded, small, dark brown. Laid on a bare twig of field maple (*Acer campestre*) in November, not hatching until the following late April or May when the tree is in leaf.

Larva. Full-fed *c.*35mm long. Head and body delicate blue-green, the latter becoming almost transparent when the larva is about to pupate; dorsal stripe broad, green; narrow subdorsal stripe white with small dark dots. When resting on the underside of a leaf it curls its body into a loop. Feeds from late April till the end of May.

Pupa. Slender, abdomen tapering caudally; shining purplish brown. In a brittle cocoon amongst litter under bushes, suspended within the cocoon by a pair of anal hooks. The stage lasts from July till November.

Imago. Univoltine. Flies from the first warm spell in November until late December. The male comes freely to light about an hour after dusk whereas the female flies at dusk and again at about midnight.

Distribution (Map 34)

Very local and almost entirely confined to the chalk downs of south and south-east England, from Cranbourne Chase, Dorset to Suffolk; the Cotswolds. There are also old records from south Devon. Abroad throughout central Europe, including southern Fennoscandia, eastwards to Japan.

Ptilophora plumigera

DRYMONIA Hübner

Drymonia Hübner, [1819], *Verz.bekannt.Schmett.*: 144.
Chaonia Stephens, 1828, *Ill.Br.Ent.* (Haust.) **2**: 29.

A Holarctic genus containing eight Palaearctic species, two of which occur in Great Britain and Ireland.

Antenna in male shortly bipectinate to apex, in female simple; haustellum reduced; palpi short, hairy; eyes glabrous; ocelli absent. Posterior tibia with two pairs of spurs. Forewing frequently without areole; vein 6 (M_1) arising with 7 to 10 (R_5 to R_2), sometimes shortly stalked with them. Hindwing veins 6 (M_1) and 7 (Rs) on a long stalk.

DRYMONIA DODONAEA ([Denis & Schiffermüller])
The Marbled Brown

Bombyx dodonaea [Denis & Schiffermüller], 1775, *Schmett. Wien.*: 49.
Bombyx trimacula Esper, 1785, *Schmett.* **3**: 242.

Type locality: [Austria]; Vienna district.

Description of imago (Pl.3, figs 30–33)
Wingspan 39–44mm. Antenna in male bipectinate, in female simple. Thorax pale brown; tegulae greyish. Forewing less triangular than in other notodontids; pale whitish grey, with a silver lustre, variably suffused fuscous brown; median fascia broad at costa and dorsum, dark fuscous, edged whitish basally; a subapical triangular patch dark fuscous, sometimes extending to dorsum as a series of black dashes. Hindwing pale grey-brown. Abdomen pale brown.

There is considerable variation in the forewing pattern and the extent of the fuscous brown suffusion; in some forms (fig.31) this is much reduced, in others (figs 32,33) the pale ground colour is reduced to a whitish postmedian fascia; in f. *purpurascens* Cockayne the forewing is almost entirely black and the hindwing dark grey. The dark forms have the thorax uniformly dark brown.

Similar species. *D. ruficornis* (Hufnagel) from which it differs in lacking the dark crescent at cell apex.

Life history
Ovum. Hemispherical, pale bluish white. Laid in small clusters on the underside of leaves of oak (*Quercus* spp.), according to Stokoe & Stovin (1958) on silver birch (*Betula pendula*) and according to Allan (1949) on beech (*Fagus sylvatica*), between late May and early July.
Larva. Full-fed *c.*35mm long. Head green. Body bright green; dorsal stripe double, pale yellow; subspiracular stripe bright yellow; peritreme of spiracle reddish. Feeds during the summer till September.

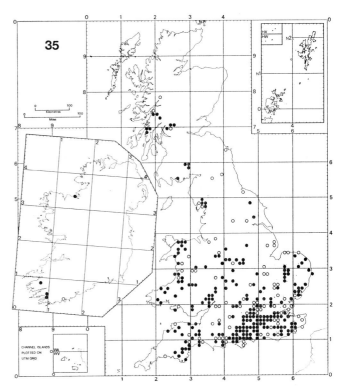

Drymonia dodonaea

Pupa. Reddish brown. In a light, hard, silken cocoon usually in the ground at the base of oak trees; overwinters.
Imago. Univoltine. Flies from late May to early July. The male comes freely to light but the female is seldom seen except occasionally at light or at rest on an oak trunk.

Distribution (Map 35)
An inhabitant of oak woods. Locally common throughout England, except the extreme south-west, as far north as the Lake District; Wales; west Scotland and south-west Ireland. Abroad ranging from central and southern Europe through Asia Minor to Japan.

DRYMONIA RUFICORNIS (Hufnagel)
The Lunar Marbled Brown

Phalaena ruficornis Hufnagel, 1766, *Berlin.Mag.* **2**: 424.
Bombyx chaonia [Denis & Schiffermüller], 1775, *Schmett. Wien.*: 49.
Noctua roboris Fabricius, 1777, *Gen.Insect.*: 283.
Type locality: Germany; Berlin.

Description of imago (Pl.4, figs 1–4)
Wingspan 38–46mm. Antenna in male pectinate, in female simple. Thorax greyish fuscous; tegulae paler. Forewing fuscous brown; basal spot whitish; ante- and postmedian lines black, dentate, edged basally and terminally with whitish yellow, sometimes enclosing a white fascia which partly (fig.2) or completely (fig.4) fills the area between the lines; a black crescent bordering apex of cell. Hindwing pale fuscous; postmedian fascia indistinct, grey. Fringe chequered grey and pale fuscous. Abdomen fuscous brown.

Irish specimens are usually larger and darker than those from Britain.

Similar species. *D. dodonaea* ([Denis & Schiffermüller]) q.v. and *Gluphisia crenata* (Esper), which is smaller, darker and has the forewing with a distinct black subterminal line.

Life history
Ovum. Hemispherical, pale blue. Laid on the upper surface of leaves of oak (*Quercus* spp.), usually in batches of two or three, in May.
Larva. Full-fed *c*.35mm long. Head green, mouth-parts pale yellow. Body bright grass-green, dorsally bluish green, ventrally greyish; subdorsal stripe pale cream; subspiracular stripe pale yellow, tinged with red dorsally; peritreme of spiracle black. Feeds from early June to July.
Pupa. Deep brown. In an ovate, loosely spun cocoon, at the base of an oak tree; overwinters.
Imago. Univoltine. Flies at the end of April and in early May. Both sexes come freely to ultra-violet light.

Distribution (Map 36)
Widely distributed and often common in southern and eastern England except the extreme south-west; more local in northern England; Wales; southern and central Scotland; Ireland in Co. Kerry, very local elsewhere. Abroad throughout northern and central Europe eastwards to Japan.

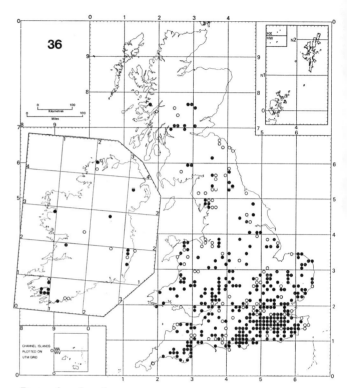

Drymonia ruficornis

GLUPHISIA Boisduval
Gluphisia Boisduval, 1828, *Eur.Lepid.Index method.*: 56.

A Holarctic genus containing two Palaearctic species, one of which has occurred in Britain as a rare vagrant.

Antenna bipectinate to apex, pectinations shorter in female; haustellum rudimentary; palpi short, ascending; eyes hairy; ocelli present. Thorax and tegulae densely hairy. Posterior tibia with one pair of spurs. Forewing fairly broad; costa weakly arched subapically; apex rounded; termen convex; tornus weak; dorsum straight except at base where it is slightly convex; distance between veins 2 (Cu_2) and 3 (Cu_1) twice that between 3 (Cu_1) and 4 (M_3) which are well separated; 5 (M_2) from middle of discocellulars (dcs); 6 (M_1) from upper angle of cell at same point as stalk of 7 (R_5), 8+9 (R_4+R_3), 10 (R_2). Hindwing veins 2 (Cu_2), 3 (Cu_1), 4 (M_3) and 5 (M_2) as in forewing; 6 (M_1) and 7 (Rs) stalked at slightly more than three-fifths; 8 (Sc+R_1) approximated to cell at upper angle.

GLUPHISIA CRENATA VERTUNEA Bray
The Dusky Marbled Brown

Bombyx crenata Esper, 1785, *Schmett.* **3**: 245.
Gluphisia crenata vertunea Bray, 1929, *Lambillionea* **29**: 18.
Type locality: Belgium; Virton.

NOTE. The name *vertunea* was originally given, by Derenne, to an aberration of *G. crenata* taken at light at Virton, Belgium on 17 June 1919 (*Revue mens.Soc.ent.namur.* **20**: 23). Subsequently Bray gave this form subspecific status, describing it as race *vertunea*.

Description of imago (Pl.4, fig.5)
Wingspan 30–38mm. Antenna in male bipectinate, in female ciliate. Forewing dark blue-grey; subbasal, antemedian, postmedian and subterminal lines darker, edged with white; antemedian concave, postmedian and subterminal weakly dentate. Hindwing grey, darker basally; median fascia indistinct, pale grey. Body grey-brown.

Similar species. *Drymonia ruficornis* (Hufnagel) *q.v.*

Life history
Ovum. Round, convex above, concave below, pale green with a pearly lustre. Laid in small clusters on leaves of poplar (*Populus* spp.) in June.
Larva. Full-fed *c.*26mm long. Head and body pale green; dorsal stripe thin, whitish; reddish brown dorsal spots on metathorax and abdominal segments 1 and 4–9; subdorsal line broad, yellow. Feeds from July to end of August.
Pupa. Glossy black. In a silken cocoon usually spun between leaves of poplar; overwinters.

Imago. Univoltine, possibly bivoltine in the southern part of its range. Flies in April and again in June and July.

Occurrence and distribution
It is doubtful whether more than three authentic specimens have been taken in the British Isles, all in the mid-1800s. The first, a female, was taken in Ongar Park Wood, Essex in June 1839, the second in the same locality in June 1841 and the third, a larva, was beaten from poplar by the Rev. Joseph Greene at Halton, Buckinghamshire in August 1853 (Barrett, 1896). Other allegedly British specimens extant in collections include three 'taken among aspens' on the Isle of Man in 1870. Abroad throughout northern and central Europe including northern Italy to east Siberia and Japan.

CLOSTERA Samouelle

Clostera Samouelle, 1819, *Ent.useful Compendium*: 247.
Ichthyura Hübner, [1819], *Verz.bekannt.Schmett.*: 162.
Pygaera sensu auctt.

A Holarctic genus containing nine Palaearctic species, three of which occur in Britain and one of these also in Ireland.

Antenna short, bipectinate to apex, pectinations shorter in female; haustellum reduced; palpi relatively long; eyes hairy; ocelli absent. Thorax with dorsal tuft; tegulae hairy. Posterior tibia with two pairs of spurs. Wings broad. Forewing with costa slightly convex basally, then slightly concave, straight apically; apex rounded; termen smoothly curved; tornus weak; dorsum slightly convex; veins 2 (Cu_2) and 3 (Cu_1) twice as far apart as 3 (Cu_1) and 4 (M_3) which are well separated; 5 (M_2) arising just beyond middle of discocellulars (dcs); 6 (M_1), 7 (R_5), 8+9 (R_4+R_3), 10 (R_2) stalked from apex of cell. Hindwing veins 3 (Cu_1) and 4 (M_3) arising together; 5 (M_2) from just beyond middle of dcs, slightly weaker than other veins; 6 (M_1) and 7 (Rs) stalked at just less than one-half; 8 ($Sc+R_1$) approximated to cell just before apex. Abdomen long, anal tuft in male prominent, bifid.

CLOSTERA PIGRA (Hufnagel)
The Small Chocolate-tip

Phalaena pigra Hufnagel, 1766, *Berlin.Mag.* **2**: 426.
Bombyx reclusa [Denis & Schiffermüller], 1775, *Schmett. Wien.*: 56.
Type locality: Germany; Berlin.

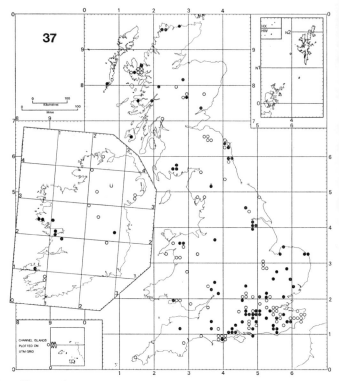

Clostera pigra

Description of imago (Pl.4, figs 6,7)
Wingspan 24–28mm. Antenna in male bipectinate, in female ciliate. Forewing pale brownish grey; antemedian fascia in the form of two white lines, sometimes joined dorsally; postmedian fascia similarly formed of two white lines enclosing darker area; apical patch rather ill-defined, chocolate-brown. Hindwing dark fuscous. Thorax and abdomen brown.

Similar species. *C. anachoreta* ([Denis & Schiffermüller]) and *C. curtula* (Linnaeus) which are much larger species with the 'chocolate-tip' sharply defined.

Life history
Ovum. Hemispherical, pale olive-green. Laid in irregular lines on leaves of creeping willow (*Salix repens*) or other species of *Salix* in May and also August in the south.
Larva. Full-fed *c.*30mm long. Head black. Body dorsally pale orange with fine setae, laterally dark grey; subdorsal line blue-grey; spiracular line broad, orange; pinacula on abdominal segments 1 and 8; legs black. Feeds from May to July and again in October (in the south) in a 'tent' of spun leaves.
Pupa. Bright red; cremaster with single point. In a cocoon among spun leaves in which it overwinters.
Imago. Bivoltine, except in northern Britain. Flies in May, and again in August in the south. Sometimes comes freely to light. Occasionally flies in sunshine.

Distribution (Map 37)
A wetland species which is widely distributed and common in southern England and more local throughout the remainder of Great Britain; widely distributed and locally very common in Ireland. Abroad throughout most of northern and central Europe, and south to the Balkans; Asia as far east as Tibet.

CLOSTERA ANACHORETA ([Denis & Schiffermüller])
The Scarce Chocolate-tip

Bombyx anachoreta [Denis & Schiffermüller], 1775, *Schmett.Wien.*: 56.

Type locality: [Austria]; Vienna district.

Description of imago (Pl.4, fig.8)

Wingspan 36–38mm. Antenna in male bipectinate, in female ciliate. Thorax brown with dark brown dorsal streak. Forewing pale grey-brown; apical quarter chocolate-brown; subbasal, antemedian, postmedian and subterminal lines indistinct, pale grey; postmedian line dentate, white in chocolate patch; discal spot black; two black subterminal spots between chocolate patch and tornus. Hindwing pale greyish brown. Abdomen dark brown.

Similar species. *C. pigra* (Hufnagel) *q.v.* and *C. curtula* (Linnaeus) from which it differs in having the forewing with the postmedian line within the chocolate patch and the two spots above the tornus.

Life history

Ovum. Hemispherical, small, white. Laid on the upper surface of leaves of poplar (*Populus* spp.).

Larva. Full-fed *c.*40mm long. Head black. Body blue-grey, black dorsally; dorsal stripe broad, orange; peritreme of spiracle yellow; spiracular stripe yellow; abdominal segment 1 with a reddish prominence with white pinacula on either side, and segment 8 with a reddish prominence; each segment with dorsal tufts of fine setae. Feeds, often gregariously when young but later singly, in spun leaves of the foodplant.

Pupa. Black, shiny. In spun leaves of the foodplant; overwinters on the ground after leaf-fall.

Imago. Bivoltine with an occasional third generation. Flies in May, August, and sometimes in October.

Occurrence and distribution (Map 38)

Little was known of this species in Britain until larvae were reported from near Folkestone, Kent in 1859 by H. G. Knaggs. It apparently became established there until 1912, since when only occasional specimens have been seen, most recently at Dungeness, Kent, August 1978 (E. Wild pers.comm.). Larvae were found in Sussex in 1893 and a few specimens have been taken on the Suffolk coast. Abroad throughout central and northern Europe eastwards to China and Japan.

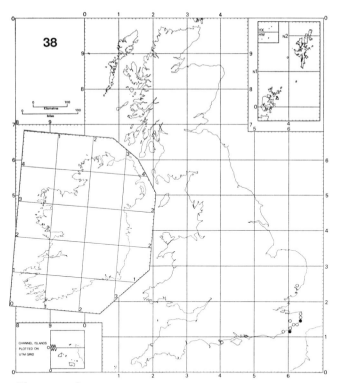
Clostera anachoreta

CLOSTERA CURTULA (Linnaeus)
The Chocolate-tip

Phalaena (*Bombyx*) *curtula* Linnaeus, 1758, *Syst.Nat.* (Edn 10) **1**: 503.

Type locality: [Europe].

Description of imago (Pl.4, fig.11)

Wingspan 36–38mm. Antenna in male bipectinate, in female ciliate. Thorax dark brown; tegulae grey. Forewing pale grey-brown speckled with black; antemedian, median and postmedian lines whitish; apical two-thirds of subterminal area chocolate-brown, sharply edged with white basally. Hindwing pale greyish brown. Abdomen dark brown.

Similar species. *C. pigra* (Hufnagel) and *C. anachoreta* ([Denis & Schiffermüller]) *q.v.*

Life history

Ovum. Hemispherical, small, blue-green. Laid in small batches on the upper surface of leaves of poplar (*Populus* spp.) and aspen (*Populus tremula*).

Larva. Full-fed *c.*35mm long. Head black. Body blue-grey; dorsal stripe yellow; a series of orange spiracular spots; abdominal segments 1 and 8 each with a raised black 'hump'; each segment sprinkled with black spots and bearing tufts

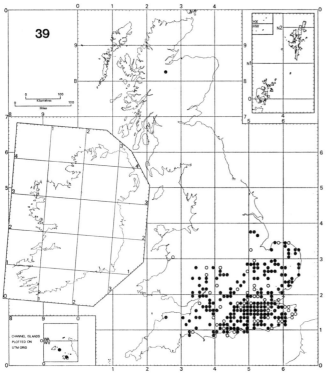

Clostera curtula

of light grey setae. Feeds in May and June, and again in August and September.

Pupa. Reddish brown. Between spun leaves of the foodplant; the second generation overwinters on the ground after leaf-fall.

Imago. Bivoltine with an occasional third generation. Flies in April and May; July and August; and sometimes September and October. Comes frequently to ultra-violet light.

Distribution (Map 39)
Widely distributed in England south of a line from the Severn to the Humber but absent from the south-west; south-east Wales; in Scotland recently recorded from Loch Ness, Inverness-shire (B. Skinner pers.comm.); the Channel Islands Absent from Ireland. Western Palaearctic.

DILOBA Boisduval
Diloba Boisduval, 1840, *Genera Index method.Eur.Lepid.*: 88.
Episema sensu auctt.

A monotypic genus occurring in Europe, including the British Isles, and Asia Minor. Formerly included in the Noctuidae (see Nye, 1975: 158).

DILOBA CAERULEOCEPHALA (Linnaeus)
The Figure of Eight Moth
Phalaena (Bombyx) caeruleocephala Linnaeus, 1758, *Syst. Nat.* (Edn 10) **1**: 504.
Type locality: [Sweden].

Description of imago (Pl.4, figs 9,10)
Wingspan 34–40mm. Antenna in male pectinate, in female ciliate. Thorax greyish brown. Forewing in male elongate and triangular, in female broader and more rounded; dark grey-brown; a short black median dash at base; antemedian and subterminal lines dark grey-brown; orbicular stigma in the shape of a figure 8 edged with yellowish white, contiguous with reniform stigma which extends to costa; a black tornal dash; fringe chequered grey and brown. Hindwing pale ochreous grey; median fascia pale brown; tornal spot dark grey; termen black; cilia grey.

Life history
Ovum. Domed, ribbed and finely reticulate. Laid in late autumn in small clusters on stems and twigs of the foodplant, usually blackthorn (*Prunus spinosa*), not hatching till May.
Larva. Full-fed *c.*40mm long. Head blue-grey. Body bluish grey dorsally with small black pinacula each bearing a small seta; broad, interrupted, dorsal stripe yellow; subspiracular stripe broad, yellow; ventral stripe dark greenish grey. Feeds exposed from May until July on blackthorn, hawthorn (*Crataegus* spp.), crab apple (*Malus sylvestris*) and cultivated apple, plum and other fruit trees.
Pupa. Pale purplish brown; cremaster with two small points. In a stout cocoon constructed from silk, earth and moss on the ground in a sheltered situation.
Imago. Univoltine. Flies in late autumn. Both sexes come readily to light.

Distribution (Map 40)
Widely distributed and common throughout England except the extreme south-west; north Wales; scarce and local in Scotland; scarce in Ireland (Baynes, 1964: 55). Abroad throughout Europe eastwards to Asia Minor.

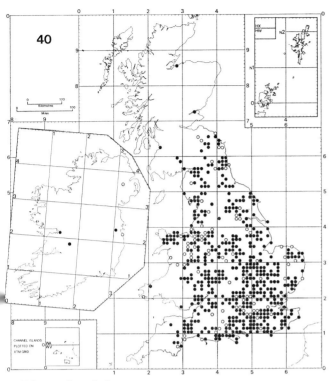

Diloba caeruleocephala

References

Allan, P. B. M., 1949. *Larval foodplants*, 126 pp. London.

Baynes, E. S. A., 1964. *A revised catalogue of Irish Macrolepidoptera*, 110 pp. Hampton.

Barrett, C. G., 1896. *The Lepidoptera of the British Islands*, **3**: 396 pp. London.

Buckler, W., 1887, 1889. *The larvae of the British butterflies and moths*, **2**: 172 pp., 35 col. pls., **3**: 79 pp., 53 col. pls. London.

Kiriakoff, S. G., 1967. Lepidoptera. Fam. Notodontidae. *Genera Insect.* **217** (B): 238 pp., 8 pls., 136 figs.

Nye, I. W. B., 1975. *The generic names of moths of the world*, **1**: 568 pp. London.

Pickering, R. R., 1966. Hoplitis milhauseri F. (Lep. Notodontidae), a species new to Britain. *Entomologist's Gaz.* **17**: 100.

South R., 1961. *The moths of the British Isles* (Edn 4), **1**: 427 pp., 148 pls. London.

Stokoe, W. J. & Stovin, G. H. T., 1958. *The caterpillars of British moths*, **1**: 408 pp., 90 pls. London.

THAUMETOPOEIDAE
J. Heath

Three species of doubtful status are listed in Kloet & Hincks (1972).

Two of these, *Thaumetopoea pityocampa* ([Denis & Schiffermüller]) (Pine Processionary Moth) and *T. processionea* (Linnaeus) (Oak Processionary Moth) were reported as having been bred from larvae and pupae found in a pine tree in Kent by T. Bachelor, a dealer, in 1872 and 1874 respectively. A full account of the circumstances of these discoveries is given in Allan (1943).

The third species, *Trichiocercus sparshalli* (Curtis), which occurs in Australia, is listed in Curtis (1830) as having been taken near Horning, Norfolk. No other record of the occurrence of this insect in the British Isles has been traced.

References

Allan, P. B. M., 1943. *Talking of moths*, 340 pp. Newtown.

Curtis, J., 1830. *A guide to an arrangement of the British insects*, (5): 143. London.

Kloet, G. S. & Hincks, W. D., 1972. A check list of British insects (Edn 2). *Handbk Ident.Br.Insects* **11** (2): viii, 153 pp.

LYMANTRIIDAE
C. G. M. de Worms

This family of about 2,500 species reaches its greatest development in the Old World tropics. It is represented in the Palaearctic region by about 200 species and in Great Britain and Ireland by 11 moderate-sized species contained in seven genera.

Imago (Pl.A, fig.7). Antenna less than half length of forewing, in male strongly bipectinate to apex, in female shortly bipectinate or ciliate; haustellum reduced or absent. Wings broadly triangular; forewing often with areole, figure 5. Females of some species brachypterous and flightless; in all other species female larger than male. Thorax and abdomen densely hairy; abdomen in female often with dense anal tuft.

All except the males of *Orgyia* are nocturnal, the males of some species being abundant at light.

Ovum. Hemispherical, rounded or subcylindrical. Laid in a cluster, usually covered with hair-scales from anal tuft. Illustrated in Stokoe & Stovin (1958).

Larva (Pl.A, fig.8). Conspicuously hairy; in some species long dense dorsal tufts or hair-pencils, called 'tussocks', present on abdominal segments 1–4 and a long hair-pencil on segment 8. Long setae present on all other segments. The British species mainly feed exposed on deciduous trees and shrubs. Care should be taken when handling these larvae as the hairs are urticating and are made up of numerous dart-shaped spicules which easily become detached and work their way into the skin, causing considerable irritation. Illustrated in Buckler (1889).

Pupa. Stout and hairy. In a silken cocoon which incorporates larval hairs.

Some species are of considerable economic importance abroad, being major forest pests.

The generic descriptions which follow are based, in part, on those given by Ferguson (1978).

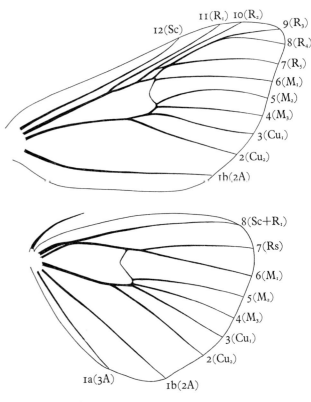

Figure 5 *Dasychira pudibunda* (Linnaeus), wing venation

Key to species (imagines) of the Lymantriidae

1	Wings vestigial	2
–	Wings fully developed	3
2(1)	Upper surface light ochreous grey; length not more than 16mm	*Orgyia antiqua* ♀ (p. 70)
–	Upper surface darker grey; length usually more than 16mm	*O. recens* ♀ (p. 69)
3(1)	Forewing ground colour white or cream	4
–	Forewing ground colour grey or brown	10
4(3)	Forewing unicolorous white or with a single black mark in disc or at tornus	5
–	Forewing with numerous black or grey transverse lines	9
5(4)	Forewing white with a black V-shaped mark in disc	*Arctornis l-nigrum* (p. 76)
–	Forewing unicolorous or with blackish tornal spot only	6
6(5)	Abdomen with anal tuft yellow or brown	7
–	Anal tuft concolorous with abdomen	8
7(6)	Anal tuft usually dark brown; abdomen mostly brown or grey dorsally	*Euproctis chrysorrhoea* (p. 73)
–	Anal tuft yellow; abdomen mostly white dorsally	*E. similis* (p. 74)
8(6)	Forewing and abdomen of ♂ whitish ochreous, ♀ creamy white; extinct species	*Laelia coenosa* (p. 68)
–	Forewing thinly scaled, satin white; tibiae and tarsi black with white rings	*Leucoma salicis* (p. 75)
9(4)	Thorax black-marked; abdomen tinged pink; hindwing pale grey	*Lymantria monacha* (p. 76)
–	Thorax not black-marked; abdomen not tinged pink; hindwing creamy white	*L. dispar* ♀ (p. 77)
10(3)	Forewing with white tornal spot; wingspan less than 40mm	11
–	Forewing without tornal spot; wingspan more than 40mm	12
11(10)	Forewing with white subapical spot	*Orgyia recens* ♂ (p. 69)
–	Forewing without white subapical spot	*O. antiqua* ♂ (p. 70)
12(10)	Forewing without markings	13
–	Forewing with grey or brown lines or fasciae	14
13(12)	Antenna of ♂ strongly bipectinate	*Dasychira pudibunda* ab. *concolor* (p. 72)
–	Antenna of ♀ weakly pectinate	*Lymantria monacha* f. *eremita* (p. 76)
14(12)	Forewing with four distinct blackish transverse lines	*L. dispar* ♂ (p. 77)
–	Forewing with not more than three distinct lines	15
15(14)	Lines of forewing mixed with orange-yellow	*Dasychira fascelina* (p. 71)
–	Lines of forewing not mixed with orange-yellow	*D. pudibunda* (p. 72)

LAELIA Stephens

Laelia Stephens, 1828, *Ill.Br.Ent.* (Haust.) **2**: 62.

Most species in this genus occur in the Indo-Malayan region. The one European species is now extinct in Britain.

LAELIA COENOSA (Hübner)
The Reed Tussock

Bombyx coenosa Hübner, [1808], *Samml.eur.Schmett.* **3**: 77, pl.51, fig.218.
Type locality: Europe.

Description of imago (Pl.4, figs 12,13)
Wingspan 46–50mm. Antenna in male strongly bipectinate, in female ciliate. Forewing in male broad, triangular, pale yellowish buff; in female narrower, creamy white, faintly buff-tinged. Hindwing creamy white. Abdomen in male yellowish buff, in female white.

Life history
Ovum. Cylindrical, top saucer-like with a central raised point. Laid on stems of branched bur-reed (*Sparganium erectum*), great fen-sedge (*Cladium mariscus*) and common reed (*Phragmites australis*) in July.
Larva. Full-fed *c*.25mm long. Body dusky brown with a black dorsal stripe; pinacula grey, bearing yellowish setae; a large, long tuft of forward-directed hairs on either side of the prothorax; tussocks of yellow hairs on abdominal segments 1–4 and a long dense pencil of backward-directed, brownish hairs on segment 8. Hatches in August and feeds until autumn and again in the spring, after hibernation, on reeds and sedges.
Pupa. Brownish. In a long silken cocoon attached to a stem of a reed or sedge; in late May.
Imago. Univoltine. Flies in July and August.

Occurrence and distribution (Map 41)
This insect was first reported from Whittlesea Mere in 1819 and was later found in Wicken, Burwell and Yaxley Fens. It was quite abundant in these latter localities, especially in the larval stage, till about 1865 when the larvae became increasingly difficult to find. The male moths were, however, attracted to strong lights at Wicken Fen till 1871 but the moth was on the decline and none has been recorded since 1879. Abroad it is widespread in France, Germany, Spain and the Balkans.

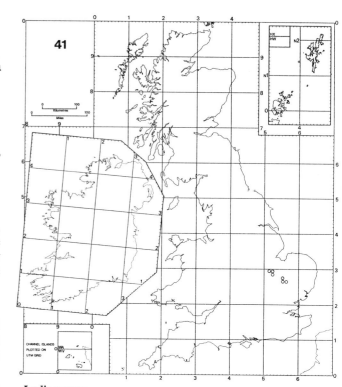

Laelia coenosa

ORGYIA Ochsenheimer
Orgyia Ochsenheimer, 1810, *Schmett.Eur.* **3**: 208.

A widely distributed genus containing about 60 species, mainly Holarctic, two of which occur in Britain. One of these also occurs in Ireland.

The females are flightless, with vestigial wings. After emergence the female remains on the outside of the cocoon on which the eggs are laid.

The males are usually day-flying with a characteristic wild, swooping flight.

ORGYIA RECENS (Hübner)
The Scarce Vapourer
Gynaephora recens Hübner, [1819], *Verz.bekannt.Schmett.*: 161.
Orgyia gonostigma sensu auctt.
Type locality: not stated.

Description of imago (Pl.4, figs 14,15)
Wingspan male 34–40mm, female with vestigial wings. Antenna short, in male bipectinate, in female serrate. *Male.* Forewing dark reddish brown, paler apically; pale grey basal and subbasal lines enclosing a subquadrate brown spot; reniform stigma outlined pale grey; a white subapical mark and white spot at tornus sometimes connected by a broken white subterminal line; a faint dark brown, broken, postmedian fascia. Hindwing rounded, deep brown. Fringes of both wings chequered. *Female.* Pale grey, very shortly winged, stout and up to 25mm long when full of eggs. Moths of the second generation are smaller than those of the first.
Similar species. *O. antiqua* (Linnaeus) from which it differs in having a white subapical spot.

Life history
Ovum. Somewhat cylindrical, rounded, with a slight depression at the apex, white. Laid in a batch of 400 to 800 on the cocoon from which the female has emerged.
Larva. Full-fed *c.*40mm long. Body blackish grey with interrupted reddish subdorsal and spiracular lines; long black and white hairs in tufts along the back and sides; a large, long pencil of forward-directed blackish hairs on either side of the prothorax; four large tussocks of deep brown hairs on abdominal segments 1–4 and a large, long, backward-directed pencil of hairs on segment 8.

Some larvae feed up rapidly on sallow (*Salix* spp.), hawthorn (*Crataegus* spp.), oak (*Quercus* spp.) and other deciduous trees and shrubs, producing a small second generation. The second generation and part of the first hibernate when quite small, feed up in the following spring and pupate in late May.

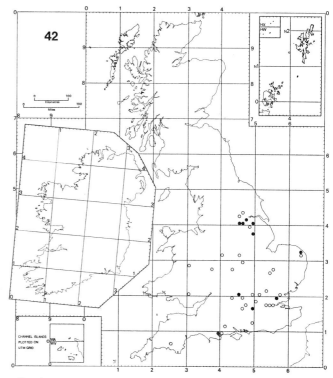

Orgyia recens

Pupa. Small, deep brown, covered with greyish hairs. In a tough silken cocoon often spun amongst leaves of the foodplant.
Imago. Partially bivoltine. First generation flies in June and July, and the small second generation from August to October. Diurnal.

Distribution (Map 42)
Exceedingly local, now almost entirely confined to a few localities in south-west Yorkshire, north Lincolnshire and the Norfolk Broads. Formerly much more widespread in southern England and in Wales. Abroad widespread in northern and central Europe, Spain and Italy.

ORGYIA ANTIQUA (Linnaeus)
The Vapourer

Phalaena (Bombyx) antiqua Linnaeus, 1758, *Syst.Nat.* (Edn 10) **1**: 503.
Phalaena gonostigma Scopoli, 1763, *Ent.Carn.*: 199.
Type locality: [Sweden].

Description of imago (Pl.4, figs 16,17)
Wingspan male 35–38mm, female with vestigial wings. Antenna short, in male bipectinate, in female shortly bipectinate. *Male.* Forewing slightly broader than in *O. recens* (Hübner), ochreous red; ante- and postmedian fasciae dark brown; a very distinct, comma-shaped, white tornal spot. Hindwing ochreous red. *Female.* Light ochreous grey, very short-winged.
Similar species. *O. recens q.v.*

Life history
Ovum. Rounded, flattened top and bottom with a small darker depression on the upperside, brownish. Laid in a batch of several hundreds on the cocoon of the female.
Larva. Full-fed *c.*40mm long. Body dark grey with a subdorsal series of raised red spots with pinacula bearing tufts of yellowish hairs; a large pencil of dark brown forward-directed hairs on either side of prothorax; four large tussocks of pale yellow hairs on abdominal segments 1–4 and a long, backward-directed pencil of hairs on segment 8. Female larvae are much larger than those producing males.

Hatches from the overwintering batch of eggs in the spring as soon as the foliage starts to appear. Polyphagous, feeding on most deciduous trees and shrubs; in Scotland almost always on birch (*Betula* spp.).
Pupa. Glossy black, hairy. In a silken cocoon often placed in crevices in the bark of trees, fences, *etc.*
Imago. Bivoltine with an occasional third generation. The first moths are on the wing in July and August, producing a further brood in the early autumn, often in October. Univoltine in the north. The flightless female never leaves her cocoon and pairing with the day-flying male takes place *in situ*. Occasionally males have been taken flying to light.

Distribution (Map 43)
Widespread throughout the British Isles to the north of Scotland, and in Ireland. It is sometimes a pest in the larval stage in cities such as London. Abroad throughout the Holarctic region.

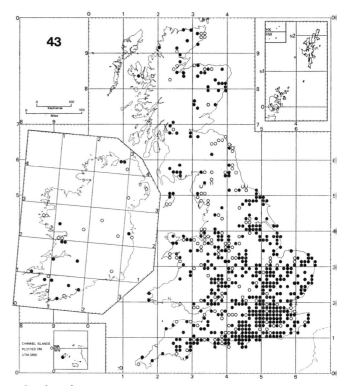

Orgyia antiqua

DASYCHIRA Hübner
Dasychira Hübner, [1809], *Samml.exot.Schmett.* **1**: pl.[178].

An extensive Old World genus of more than 400 species two of which occur in Great Britain and Ireland.

Female often almost twice as large as male. Antenna bipectinate with pectinations very long in male, short in female.

The recent rediscovery of the type-species *D. tephra* Hübner indicates that the Old World species at present included in *Dasychira* may be misplaced (Ferguson, 1978).

DASYCHIRA FASCELINA (Linnaeus)
The Dark Tussock

Phalaena (Bombyx) fascelina Linnaeus, 1758, *Syst.Nat.* (Edn 10) **1**: 503.

Type locality: [Sweden].

Description of imago (Pl.4, figs 21,22)
Wingspan 44–53mm. Antenna short, in male strongly bipectinate, in female shortly bipectinate. Forewing dull slaty grey in southern forms, but a much deeper grey in examples from the Scottish moors; blackish ante- and postmedian fasciae; a small blackish subapical costal mark and sometimes an indistinct broken subterminal fascia bounded outwardly by a pale line; dark markings sometimes entirely absent, sometimes very pronounced and yellowish. Hindwing paler grey, usually with an indistinct dark discal mark and sometimes an indistinct subterminal fascia.

An exceptionally large race occurs at Dungeness, Kent where the larva feeds on broom (*Sarothamnus scoparius*).

Life history
Ovum. Spherical, whitish. Laid in batches on stems and twigs of heather (*Calluna vulgaris*), hawthorn (*Crataegus monogyna*), sallow (*Salix* spp.) and broom where they are well concealed by a covering of hairs from the anal tuft of the female.

Larva. Full-fed *c*.45mm long. Head black. Body deep brown with a series of star-like tufts of hair along the body, yellow on thorax, greyer on abdomen; a pencil of long, grey hairs on either side of prothorax; four large tussocks of long, black-tipped white hairs on abdominal segments 1–4 and pencil of long grey hairs on segment 8.

Usually hatches in August and after feeding till the autumn goes into hibernation in a small silken envelope spun amongst leaves of the foodplant. It resumes feeding in late March and is usually full-fed in late May.

Pupa. Glossy black, densely covered with hairs which are interwoven into the cocoon which is spun amongst the foodplant.

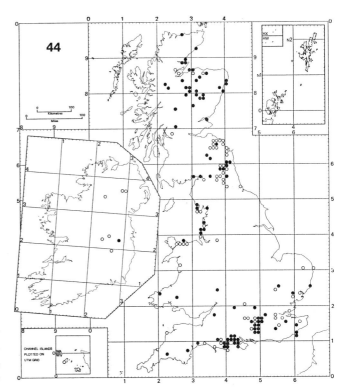

Dasychira fascelina

Imago. Univoltine. Flies during the latter half of July or sometimes in early August.

Distribution (Map 44)
Usually associated with moorland but also occurs coastally. Widespread and locally common in Surrey, Sussex, Hampshire, Dorset, the coastal sandhills of Cheshire and Lancashire, Northumberland; the Highlands of Scotland; very local in Ireland. Abroad north and central Europe to eastern Siberia.

DASYCHIRA PUDIBUNDA (Linnaeus)
The Pale Tussock

Phalaena (Bombyx) pudibunda Linnaeus, 1758, *Syst.Nat.*
(Edn 10) **1**: 503.
Type locality: [Sweden].

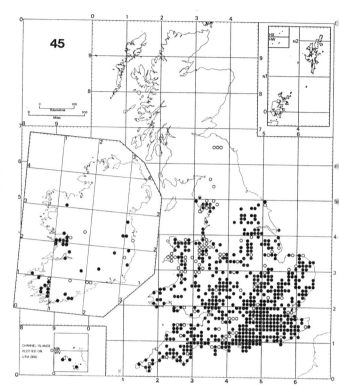

Dasychira pudibunda

Description of imago (Pl.4, figs 18–20)
Wingspan 50–70mm. Antenna short, in male strongly bipectinate, in female shortly bipectinate. Forewing grey, paler in female; subbasal, ante- and postmedian lines usually distinct, in male dark brownish grey, in female brown; reniform stigma dark-outlined, indistinctly in female; a subterminal fascia represented by two dark crescents outwardly whitish-edged; fringe strongly chequered. Hindwing in male pale grey, in female almost white; a faint, greyish basal patch in cell; postmedian fascia, more distinct at tornus, greyish brown.

In recent years melanism has been increasing in the male; occasionally the female has the forewing a deep unicolorous grey and the hindwing light grey (fig.19). The extreme ab. *concolor* Cockayne has both wings deep grey.

Life history
Ovum. Rounded, with a slight depression on top, white. Laid in batches on foliage of many deciduous trees and on hop (*Humulus lupulus*) in late May and June.
Larva. Full-fed *c.*45mm long. Body either deep green mottled with white, or yellow with greenish shading; four large concolorous brushes of hairs on abdominal segments 1–4 with a black band between each; smaller tufts of hairs on segments 5–7 which have black supraspiracular dashes; a large pencil of long reddish hairs on segment 8. Feeds from May/June till late summer. Locally known as the 'hop-dog'.
Pupa. Shiny brown, hairy. In a lightly spun silken cocoon; overwinters.
Imago. Univoltine. Flies in May and June. The male comes freely to light. Both sexes rest in a very characteristic manner with the forelegs stretched well forward.

Distribution (Map 45)
Throughout England and Wales to the Lake District; only doubtfully recorded from Scotland; local in Ireland. Abroad widespread in central and northern Europe; central Asia; Japan.

EUPROCTIS Hübner

Euproctis Hübner, [1819], *Verz.bekannt.Schmett.*: 159.
Liparis Ochsenheimer, 1810, *Schmett.Eur.* **3**: 186.
Nygmia Hübner, [1820], *Verz.bekannt.Schmett.*: 193.
Porthesia Stephens, 1828, *Ill.Br.Ent.* (Haust.) **2**: 65.

A large Old World genus containing more than 600 species. It is represented in Great Britain by two species, one of which also occurs in Ireland.

The female often has a wingspan twice that of the male. Both sexes of most species fly equally well.

EUPROCTIS CHRYSORRHOEA (Linnaeus)
The Brown-tail

Phalaena (Bombyx) chrysorrhoea Linnaeus, 1758, *Syst.Nat.* (Edn 10) **1**: 502.
Bombyx phaeorrhoeus Haworth, 1803, *Lepid.Br.*: 109.

Type locality: [Europe].

Description of imago (Pl.4, figs 23,24)
Wingspan 36–42mm. Antenna bipectinate, in male pectinations very long, in female short. Thorax white. Forewing glistening white usually devoid of all markings; occasionally specimens occur with a few black dots. Hindwing white. Abdomen white basally and ventrally, in male brown dorsally, in female greyish brown; an anal tuft of dark brown or, rarely, brownish yellow hairs, very large in female.
Similar species. *E. similis* (Fuessly) which has a completely white abdomen and yellow anal tuft.

Life history
Ovum. Round, flattened top and bottom, with a slight depression on top. Laid in a large batch on a twig of blackthorn (*Prunus spinosa*) or hawthorn (*Crataegus* spp.) and completely covered with hairs from the anal tuft of the female.
Larva. Full-fed *c.*45mm long. Body deep reddish brown with red dorsal pinacula on abdominal segments 1, 2, 7 and 8, each bearing a cluster of small brownish hairs; tufts of whitish hairs on either side of the pinacula; dorsal line edged with red and a red dorsal spot on abdominal segments 6 and 7.

The larvae hatch simultaneously from a batch of eggs and immediately start spinning a nest in which they live and feed gregariously during the late summer, after which they spin a much tougher nest in which to pass the winter. They start to feed again in the following April, becoming full-fed about the end of May when they disperse for pupation. Sometimes they do considerable damage to blackthorn and hawthorn hedgerows (Baker, *MBGBI* **1**: 83). The hairs are

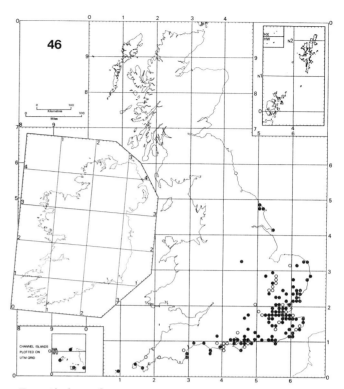

Euproctis chrysorrhoea

extremely irritating, causing rashes on exposed parts of the skin. At the end of the eighteenth century it was a great pest in the centre of London where measures were taken to destroy it (Curtis, 1782). This still proves necessary from time to time in Canvey Island, Essex and other localities.
Pupa. Deep blackish brown, covered with tufts of larval hair. In a lightly spun silken cocoon amongst foliage. Sometimes a number of cocoons are spun together into a small ball.
Imago. Univoltine. Flies in July and August. Both sexes come to light.

Distribution (Map 46)
This species, still periodically a local pest, is now almost entirely coastal, with only occasional records inland though apparently spreading into the east Midlands. It occurs from Yorkshire round the Thames estuary to Kent, and along the south coast to Cornwall; the Isles of Scilly; the Channel Islands. Palaearctic; accidentally introduced to the eastern U.S.A. where it is now a serious orchard pest.

EUPROCTIS SIMILIS (Fuessly)
The Yellow-tail

Phalaena similis Fuessly, 1775, *Verz.bekannt.schweiz.Ins.*: 35.
Bombyx auriflua [Denis & Schiffermüller], 1775, *Schmett. Wien.*: 52.
Bombyx chrysorrhoea sensu Haworth, 1803, *Lepid.Br.*: 109.
Type locality: Switzerland.

Description of imago (Pl.4, figs 25,26)
Wingspan 35–45mm. Antenna bipectinate, in male pectinations very long, in female short. Forewing ground colour pure white with silky sheen; in male almost always a small blackish brown tornal mark, sometimes double. Abdomen white with anal tuft yellow, very large in female.
Similar species. E. *chrysorrhoea* (Linnaeus) q.v.

Life history
Ovum. Spherical with flattened base and slight depression on top. Laid in a batch on a twig of hawthorn (*Crataegus* spp.), blackthorn (*Prunus spinosa*), oak (*Quercus* spp.), birch (*Betula* spp.), sallow (*Salix* spp.) and other trees, covered in hairs from the anal tuft of the female.
Larva. Full-fed *c*.35mm. long. Head black. Body black; dorsal stripe vermilion with a narrow black central line; tufts of whitish hairs and small white elongate markings on side of each segment; a pencil of whitish hairs on either side of prothorax; a small thick hump on abdominal segment 1 and a broad patch of red on segments 8 and 9.

Hatches in August and feeds gregariously until autumn when each constructs a small cocoon-like hibernaculum. In spring it remains solitary when it recommences feeding and is full-fed about the end of May.
Pupa. Brown, hairy. In a light brown cocoon containing larval hairs spun amongst foliage.
Imago. Univoltine. Flies in July and August. Often found at rest looking rather like a white feather with its forelegs outstretched along the length of a twig or leaf.

Distribution (Map 47)
Common throughout England but confined to the coast in Devon and Cornwall and absent from the north-east; Wales; very local in southern Scotland; Isle of Man; the Isles of Scilly; very local in Ireland; the Channel Islands. Abroad throughout Europe eastwards to China and Japan.

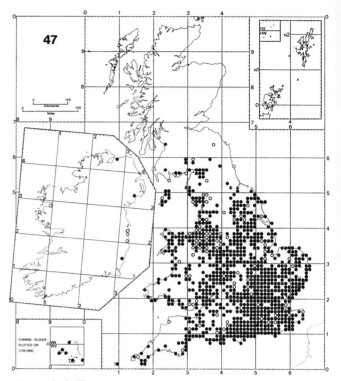

Euproctis similis

LEUCOMA Hübner

Leucoma Hübner, 1822, *Syst.-alph.Verz.*: 14–16, 18, 19.
Stilpnotia Humphreys & Westwood, 1843, *Br.Moths* 1: 90.

An Old World genus containing about 40 species, one of which occurs in Great Britain and Ireland.

Males and females are of similar size and the female flies almost as extensively as the male.

Some species are of considerable economic importance as forest pests.

LEUCOMA SALICIS (Linnaeus)
The White Satin Moth

Phalaena (Bombyx) salicis Linnaeus, 1758, *Syst.Nat.* (Edn 10) 1: 502.

Bombyx salicinus Haworth, 1803, *Lepid.Br.*: 107.

Type locality: [Sweden].

Description of imago (Pl.4, figs 27,28)
Wingspan 43–60mm. Antenna bipectinate, in male strongly, in female shortly. Legs white, ringed black. Wings thinly scaled, satin-white, immaculate with a distinct silky lustre when fresh; costa of forewing cream-coloured in some specimens.
Similar species. Arctornis l-nigrum (Müller) from which it differs in the much narrower forewing and the absence of the black V-mark.

Life history
Ovum. Hemispherical, top slightly depressed, dark-coloured. Laid in batches, with some female abdominal hairs adhering, on twigs of poplar (*Populus* spp.) and sallow (*Salix* spp.).
Larva. Full-fed *c.*45mm long. Head dark grey, dorsally black. Body russet brown; a brilliant white dorsal patch and a bright red spot on each segment; a subdorsal row of orange-red pinacula bearing tufts of whitish hairs; a bright blue subspiracular line and a row of orange pinacula.

Hatches in late August and hibernates when quite small, spinning a web in which to spend the winter. Feeding recommences as soon as the foodplant comes into leaf, when it feeds up rapidly, often gregariously, but still in its individual web. Periodically its huge numbers can defoliate large stands of poplar. Pupates at the end of June.
Pupa. Glossy black, covered with tufts of white hairs in dorsal and ventral rows. In a thin silken cocoon spun in crevices of the bark of the foodplant.
Imago. Univoltine. Emerges in July and August. Both sexes come to light. Sometimes sits on tree-trunks in great numbers.

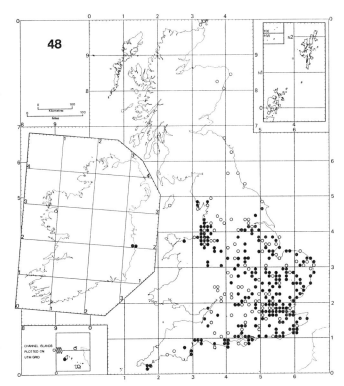

Leucoma salicis

Distribution (Map 48)
Coastal in the eastern and southern counties from Yorkshire to Cornwall and in Lancashire; also inland in eastern England and, although usually local and uncommon, it is sometimes especially abundant, particularly in parts of the London area; records from Scotland as far north as Orkney and Shetland are presumed to be migrants; very local in Ireland. Abroad throughout western Europe to Japan; also in North America where it was accidentally introduced; sometimes assuming pest proportions.

ARCTORNIS Germar
Arctornis Germar, 1811, *Syst.Gloss.Prod.*: 18.

One species in this genus has been recorded in Britain.

ARCTORNIS L-NIGRUM (Müller)
The Black V Moth

Phalaena (Bombyx) l-nigrum Müller, 1764, *F.Ins.Frid.*: 40.
Bombyx v-nigrum Fabricius, 1775, *Syst.Ent.*: 577.
Leucoma vau-nigra Stephens, 1828, *Ill.Br.Ent.* (Haust.) **2**: 64.

Type locality: Denmark; Copenhagen.

Description of imago (Pl.4, fig.29)
Wingspan 52–56mm. Antenna short, in male strongly pectinate, in female shortly. Legs alternately black- and white-marked. Wings light green, quickly fading to satin-white; forewing with a very prominent black V-shaped mark at end of cell.

Similar species. *Leucoma salicis* (Linnaeus) q.v.

Life history
Ovum. Laid in small batches on the bark of various deciduous trees in July.

Larva. Full-fed c.45mm long. Head black. Body black, laterally reddish; lateral hairs yellow; tussocks of white hairs on abdominal segments 1, 2, 6 and 7 and of reddish hairs on segments 3–5; very long tufts of whitish hairs on the thoracic segments and the last two abdominal segments.

Hatches in July and overwinters in the first instar. Feeding recommences in the spring on various trees, including lime (*Tilia* spp.), elm (*Ulmus* spp.) and sallow (*Salix* spp.), also sometimes on pine (*Pinus* spp.).

Pupa. Small, blackish with short hairs. In a silken cocoon spun amongst the foliage of trees at the end of May.

Imago. Univoltine. Emerges at the end of June.

Occurrence and distribution
Barrett (1895) reported this insect from Bromley in Kent in the early part of the nineteenth century and also from Sheffield, Yorkshire. In this century, prior to 1947 it seems to have visited the British Isles only on two occasions; a specimen was taken at Chelmsford, Essex in July 1904 and another at light on the Sussex coast in July 1946. From June 1947 to July 1960 specimens were taken at light annually at Bradwell-on-Sea, Essex, the maximum being 15 in 1953 (Dewick, 1978). A wild larva was also found and there can be no doubt that during this period it was breeding locally. Abroad it is widespread in central and southern Europe and in Scandinavia.

LYMANTRIA Hübner
Lymantria Hübner, [1819], *Verz.bekannt.Schmett.*: 160.
Porthetria Hübner, [1819], *ibid.*: 160.
Psilura Stephens, 1828, *Ill.Br.Ent.* (Haust.) **2**: 57.

An Old World genus containing some 150 species. This genus is represented in Great Britain by two species, one of which also occurs in Ireland.

The female frequently has a wing expanse twice that of the male. Although the female has fully developed wings, those of some species, *e.g. L. dispar* (Linnaeus), do not fly.

Some species are major forest pests.

LYMANTRIA MONACHA (Linnaeus)
The Black Arches

Phalaena (Bombyx) monacha Linnaeus, 1758, *Syst.Nat.* (Edn 10) **1**: 501.

Type locality: [Europe].

Description of imago (Pl.4, figs 32–35)
Wingspan 44–54mm. Antenna bipectinate, in male strongly, in female very shortly. Pro- and mesothorax white with black dorsal marks. Forewing ground colour white; one basal and a series of subbasal spots black; antemedian, median, postmedian and subterminal lines black, irregularly dentate; a series of black terminal spots extending into cilia; a black orbicular spot and reniform mark sometimes almost completely obscured by dark grey median shading. The extent and intensity of the shading varies considerably and in the extreme f. *eremita* Ochsenheimer (fig.35) the wing is unicolorous dark grey-brown. Hindwing grey-brown; terminal shade darker; fringe with a series of black marks. Abdomen tinged pink.

Life history
Ovum. Ovoid, rough and pitted. Laid singly or in pairs in crevices of the bark of oak (*Quercus* spp.) and other trees; overwinters.

Larva. Full-fed c.35mm long. Body dark grey; a black dorsal stripe, indented with whitish patches on abdominal segments 4–6; a black mark in a white patch on mesothoracic segment; red dots sometimes present in the whitish patches; spiracular line darker; pinacula bearing tufts of whitish hairs along dorsal and spiracular lines. Feeds from spring until June, almost exclusively on oak.

Pupa. Glossy brown with short hairs, almost metallic in appearance. In a light silken cocoon spun in crevices of the bark of trees, especially oak.

Imago. Univoltine. Emerges in August. The male flies freely at night and comes to light, whilst the sluggish female can be found sitting on tree-trunks, often during the day.

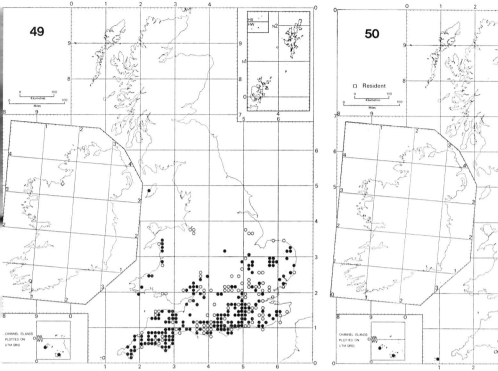

Lymantria monacha

Lymantria dispar

Distribution (Map 49)

Throughout England south of a line from the Wash to Pembroke; very local in Wales; no recent record from Ireland; the Isles of Scilly; the Channel Islands. Abroad throughout Europe to the Balkans; Japan.

Economic importance. In some European countries this species is a serious forest pest causing extensive defoliation of Norway spruce and Scots pine.

LYMANTRIA DISPAR (Linnaeus)
The Gipsy Moth

Phalaena (Bombyx) dispar Linnaeus, 1758, *Syst.Nat.* (Edn 10) **1**: 501.

Type locality: [Europe].

Description of imago (Pl.4, figs 30,31)

Wingspan 48–65mm. Antenna bipectinate, in male strongly, in female very shortly. Sexually dimorphic. *Male.* Head, thorax, wings and abdomen deep grey-brown. Forewing with antemedian, median, postmedian and subterminal lines black, irregularly dentate; a prominent crescent-shaped black reniform mark and a black orbicular spot; subterminal fascia darker grey-brown. Hindwing with darker terminal shade. *Female.* Much larger than male; ground colour pale creamy white. Forewing with transverse lines finer, brownish grey, subterminal fascia absent. Hindwing with pale brown postmedian fascia and discal spot. A series of black spots in fringes between veins in fore- and hindwings in both sexes. Abdomen with brown anal tuft.

The extinct British form was much larger than the continental form.

Life history

Ovum. Large, glossy, button-shaped. Laid in late summer in batches on foliage of the foodplant and covered with hairs from the anal tuft of the female.

Larva. Full-fed *c.*50mm long. Head pale yellow-brown, marked with black. Body greenish grey, dorsal line paler; prominent pinacula bearing tufts of hairs on each segment along the subdorsal and subspiracular lines; pinacula red-brown, except the subdorsals on thoracic and abdominal segments 1 and 2, which are blue.

In Britain it hatched in April and fed on bog-myrtle (*Myrica gale*) and creeping willow (*Salix repens*) in preference to forest trees, which are the foodplants on the Continent.

Pupa. Short and stout with tufts of short hairs. Spun up in a flimsy silken cocoon at the end of June.

Imago. Univoltine. Emerges in August. The male flies freely by day and pairing takes place almost immediately after the female has emerged. The female seldom moves far from its cocoon.

Occurrence and distribution (Map 50)

This species was abundant in the Fens in the early 1800s but apparently declined rapidly in the 1850s. The last certain record was from Wennington Wood, Huntingdon in 1907 (Heath, 1974). Sporadic occurrences from that date to the 1920s are thought to be of accidentally introduced specimens. Occasional males have been taken on the south and east coasts of England since 1945: large migrations of the male take place in eastern Europe from which these may originate. The female does not fly, but the species is spread by wind-borne early-instar larvae. Abroad it is abundant from Europe as far north as Finland eastwards to China; also in North America where it was accidentally introduced in the latter half of the nineteenth century. It is a major forest pest throughout much of its range.

References

Barrett, C. G., 1895. *The Lepidoptera of the British Islands*, **2**: 369 pp. London.

Buckler, W., 1889. *The larvae of the British butterflies and moths*, **3**: 79 pp., 53 col. pls. London.

Curtis, W., 1782. *A short history of the brown-tail moth*, 13 pp., 1 col. pl. London.

Dewick, A. J., 1978. *Arctornis l-nigrum* (Müller, O. P., 1764) (Lep.: Lymantriidae): The Black V Moth – a possible occasional resident species in Britain. *Entomologist's Rec. J. Var.* **90**: 223–225.

Ferguson, D. C., 1978. *The moths of America north of Mexico*, Fasc. 22.2 Noctuoidea Lymantriidae, 110 pp., 23 figs, 1 pl., 8 col. pls. London.

Heath, J., 1974. A century of change in the Lepidoptera. *In* Hawksworth, D. L., *The changing flora and fauna of Britain*, pp. 275–292. London.

Stokoe, W. J. & Stovin, G. H. T., 1958. *The caterpillars of British moths*, **1**: 408 pp., 90 pls. London.

ARCTIIDAE
C. G. M. de Worms

A world-wide family of some 10,000 species with many New World and tropical genera. Represented in the Palaearctic region by about 200 species and in Great Britain and Ireland by 32 species in two subfamilies.

Imago (Pl.A, figs 9,11). Antenna about half length of forewing, usually bipectinate or setose-ciliate in male, setose-ciliate or simple in female; haustellum frequently reduced. Forewing broad and often cryptically patterned in the Arctiinae; generally elongate and almost unicolorous in the Lithosiinae. Hindwing brightly coloured in the Arctiinae. Venation as in figures 6 and 7, p. 79.

In the Lithosiinae some species, when at rest, fold their wings flat over the body and others roll their wings downwards around the body. This is a useful field aid to the determination of closely related species.

Some arctiids are aposematic and secrete toxins. These have been studied by Rothschild (1972) and Marsh & Rothschild (1974).

Ovum. Hemispherical. Usually laid in large clusters on the foodplant, sometimes covered in hairs from the anal tuft of the female. Illustrated in Stokoe & Stovin (1958).

Larva (Pl.A, figs 10,12). Usually with dense tufts of setae arising from verrucae. All the British Arctiinae feed on herbaceous plants and the Lithosiinae on lichens, although in captivity some species will eat leaves of vascular plants. Figured by Buckler (1889), Cockayne (1944) and Stokoe & Stovin (1958).

Pupa. Glabrous; cremaster weak or absent. In a cocoon of felted larval hairs.

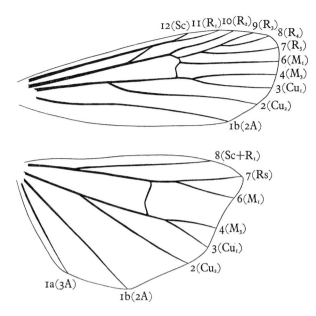

Figure 6 *Eilema lurideola* ([Zincken]), wing venation

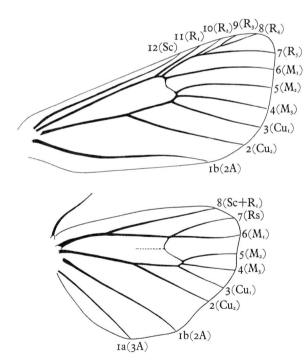

Figure 7 *Phragmatobia fuliginosa* (Linnaeus), wing venation

Key to the species (imagines) of the Arctiidae

1	Ocelli absent; forewing elongate; body slender .... (Lithosiinae) 2	
–	Ocelli present; forewing broad; body usually stout .... (Arctiinae) 23	
2(1)	Wing black; collar red, thorax and abdomen black, tip of abdomen red .... *Atolmis rubricollis* (p. 85)	
–	Wings not black 3	
3(2)	Wings pale grey, rounded; small species (wingspan 19–23mm) 4	
–	Wings otherwise coloured, or if grey, elongate; large species (wingspan 24–55mm) 5	
4(3)	Forewing with ante- and postmedian series of black dashes; hindwing with distinct black discal lunule .... *Thumatha senex* (p. 81)	
–	Forewing with faint cross-lines; hindwing without discal lunule; very thinly scaled .... *Nudaria mundana* (p. 84)	
5(3)	Forewing pale grey 6	
–	Forewing otherwise coloured 13	
6(5)	Forewing wholly pale grey 7	
–	Forewing with base or costa orange or yellow 8	
7(6)	Forewing very elongate; three black dots costally at two-thirds; two black dots dorsally at one-third .... *Pelosia muscerda* (p. 87)	
–	Forewing shorter; a postmedian curved line of black dots from costa to dorsum .... *P. obtusa* (p. 87)	
8(6)	Forewing grey; base yellow and basal third of costa black .... *Lithosia quadra* ♂ (p. 95)	
–	Forewing grey; costa orange or yellow 9	
9(8)	Hindwing whitish or grey 10	
–	Hindwing pale yellow 11	
10(9)	Hindwing whitish .... *Eilema caniola* (p. 90)	
–	Hindwing grey .... *E. deplana* ♀ (p. 93)	
11(9)	Costal streak narrowing to a point at apex .... *E. lurideola/sericea* (pp. 94, 92)	
–	Costal streak of even width 12	
12(11)	Wings long and narrow, costa weakly arched .... *E. complana* (p. 91)	
–	Wings broad, costa strongly arched .... *E. griseola* (p. 89)	
13(5)	Forewing yellow or orange without markings 14	
–	Forewing otherwise coloured or, if yellow, with markings 17	
14(13)	Abdomen grey; anal tuft yellow .... *E. sororcula* (p. 88)	
–	Abdomen yellow 15

15(14)	Costa strongly arched; hindwing very broad .... *E. griseola* f. *stramineola* (p. 89)	29(23)	Forewing almost without markings 30
–	Costa weakly arched; hindwing narrower 16	–	Forewing with distinct markings 31
16(15)	Hindwing uniformly yellowish ... *E. deplana* ♂ (p. 93)	30(29)	Forewing deep orange-yellow; hindwing with extensive black suffusion .... *Diacrisia sannio* ♀ (p. 101)
–	Hindwing suffused grey costally .... *E. pygmaeola* (p. 91)	–	Forewing reddish fuscous; hindwing pink or grey; wings rather thinly scaled; smaller species .... *Phragmatobia fuliginosa* (p. 106)
17(13)	Forewing yellow or orange with many black dots or striae; abdomen black .... *Setina irrorella* (p. 82)	31(29)	Forewing yellow with large dark discal spot, cilia pink .... *Diacrisia sannio* ♂ (p. 101)
–	Forewing otherwise 18	–	Forewing black or dark brown with contrasting markings 32
18(17)	Wings pink with strongly dentate postmedian line .... *Miltochrista miniata* (p. 83)	32(31)	Forewing blackish with red costal streak; two red terminal spots; hindwing red .... *Tyria jacobaeae* (p. 109)
–	Wings otherwise 19	–	Forewing brown or blackish irregularly streaked and/or spotted cream 33
19(18)	Forewing white or yellow with one costal and one subdorsal median black spot 20	33(32)	Hindwing yellow .... *Arctia villica* (p. 100)
–	Forewing with many markings 21	–	Hindwing red 34
20(19)	Forewing with spots dot-like; hindwing grey .... *Cybosia mesomella* (p. 86)	34(33)	Thorax uniformly dark brown .... *A. caja* (p. 99)
–	Forewing with spots large; hindwing yellow .... *Lithosia quadra* ♀ (p. 95)	–	Thorax black- and yellowish-coloured 35
21(19)	Forewing yellow, broadly striate with black on veins; hindwing yellow with broad black costal and terminal bands .... *Spiris striata* (p. 96)	35(34)	Abdomen yellow .... *Euplagia quadripunctaria* (p. 107)
–	Forewing white or greyish white 22	–	Abdomen black and crimson .... *Callimorpha dominula* (p. 108)
22(21)	Forewing white with many black and red spots and striae .... *Utetheisa pulchella* (p. 97)		
–	Forewing greyish white heavily spotted and suffused dark grey .... *Coscinia cribraria* (p. 96)		
23(1)	Fore- and hindwings more or less similarly patterned and coloured 24		
–	Fore- and hindwings contrastingly patterned and coloured 29		
24(23)	Wings white or cream heavily marked black; abdomen with broad black dorsal stripe .... *Parasemia plantaginis* (p. 98)		
–	Wings white, pale yellowish, or grey, lightly dotted .. 25		
25(24)	Wings white 26		
–	Wings yellowish, buff or grey 28		
26(25)	Forewing with two small black dots, hindwing without markings .... *Spilosoma urticae* (p. 104)		
–	Forewing with numerous black dots 27		
27(26)	Abdomen yellow with black dorsal spots .... *S. lubricipeda* (p. 102)		
–	Abdomen white .... *Diaphora mendica* ♀ and *D. mendica* ♂ f. *rustica* (p. 105)		
28(25)	Wings and body uniform grey .... *D. mendica* ♂ (p. 105)		
–	Wings yellowish buff, dotted or striate with black .... *Spilosoma lutea* (p. 103)		

Lithosiinae

THUMATHA Walker
Thumatha Walker, 1866, *List Specimens lepid.Insects Colln Br.Mus.* **35**: 1900.

Represented in Great Britain and Ireland by a single species.

THUMATHA SENEX (Hübner)
The Round-winged Muslin

Bombyx senex Hübner, [1808], *Samml.eur.Schmett.* **3**: pl.55, figs 236,237.
Nudaria rotunda Haworth, 1809, *Lepid.Br.*: 156.
Type locality: Europe.

Description of imago (Pl.5, fig.1)
Wingspan 20–22mm, female smaller than male. Antenna short, setose-ciliate. Forewing rounded, less so in female, thinly scaled, pale greyish white; veins dark grey; discal spot black, prominent; antemedian fascia an indistinct row of brown elongate spots; postmedian fascia a row of brown elongate spots; fringe chequered brown and pale grey. Hindwing pale greyish white; discal spot dark brown; fringe chequered brown and pale grey. Body pale greyish white.

Similar species. Nudaria mundana (Linnaeus) which differs in having the wings less rounded, more thinly scaled and almost transparent, the spots much less conspicuous, and the fringes uniform in colour.

Life history
Ovum. Globular, weakly reticulate, glossy yellow. When laid, the eggs adhere to the female's anal tuft and she detaches them by means of her two hind tarsi (Buckler, 1889). Oviposition takes place amongst lichens growing on the vegetation of damp, swampy places.
Larva. Full-grown *c.*15mm long, very stout. Head shining black. Body deep reddish with a wax-like appearance; verrucae with dense tufts of setae which conceal the larval skin, the setae being of two types, some pale brown with black tips, others black and slightly plumose, the 'plumage' being pale brown; legs glossy with black tips; prolegs pale grey.

Hatches in August or September and feeds for a short time before going into hibernation. Becomes active again in the spring and is full-grown in May. Feeds on several kinds of lichen, especially *Peltigera canina*, and also on mosses such as *Homalothecium sericeum* and *Dicranoweisia cirrata*.

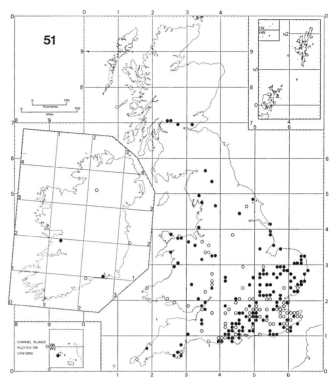

Thumatha senex

Pupa. Deep shining brown. In a thick brownish cocoon of silk mixed with larval hairs, spun amongst fallen leaves.
Imago. Univoltine. Flies in July and August. Flies for a brief period at dusk and comes readily to light at about midnight.

Distribution (Map 51)
An inhabitant of fens and marshes. It is widespread in south-east England and occurs more sparingly westwards to Wales and northwards to Inverness-shire; Guernsey; in Ireland it is recorded from scattered localities. It is easily overlooked and is probably more widespread than the records indicate. Abroad in northern Europe eastwards to the Urals.

SETINA Schrank

Setina Schrank, 1802, *Fauna boica* **2** (2): 165.
Endrosa Hübner, [1819], *Verz.bekannt.Schmett.*: 167.
Philea Zetterstedt, 1839, *Ins.Lapp.*: 931.

A small genus occurring in Europe and western Asia.
 The male has a large tympanal organ and, in most species, a tymbal organ which is capable of producing sound.

SETINA IRRORELLA (Linnaeus)
The Dew Moth

Phalaena (*Tinea*) *irrorella* Linnaeus, 1758, *Syst.Nat.* (Edn 10) **1**: 535.
Lithosia irrorata Fabricius, 1798, *Suppl.Ent.syst.*: 461.
Type locality: Europe.

Description of imago (Pl.5, figs 2–4)
Wingspan 26–32mm, female smaller than male. Head black; antenna setose-ciliate. Patagia orange. Forewing thinly scaled, pale orange-yellow, darker in female; antemedian, median and subterminal rows of three or four black spots. Hindwing pale ochreous yellow. Body black, much stouter in female; anal tuft yellow.
 In f. *signata* Borkhausen (fig.3) the spots on the forewing are united to form the figure ≦.

Life history
Ovum. Globular, purplish brown. Laid on various lichens growing on rocks.
Larva. Full-fed *c*.20mm long. Head black. Body blackish brown, greyer laterally; discontinuous dorsal stripe yellow; subdorsal stripe white, mixed with yellow; subspiracular stripe pale yellow, edged above with dark brown; verrucae white in subdorsal region, elsewhere blackish, bearing tufts of brown setae; legs dark brown.
 Hatches in August and soon goes into hibernation. Starts feeding again in the spring on various black and yellow lichens growing on rocks and stones and is full-grown in late May. Feeds exposed in bright sunshine but is hard to find because of its cryptic coloration.
Pupa. Short and stout; brown. In a loose cocoon among small stones or in a crevice in a rock.
Imago. Univoltine. Flies in late June and July. The male flies at dawn and again in the afternoon, but the female is seldom active before dark. Both sexes come to light. Sometimes to be seen by day hanging from blades of grass, especially the male.

Distribution (Map 52)
It occurs mainly in coastal areas, often just above high-water-mark, but is also recorded sparingly inland on chalk

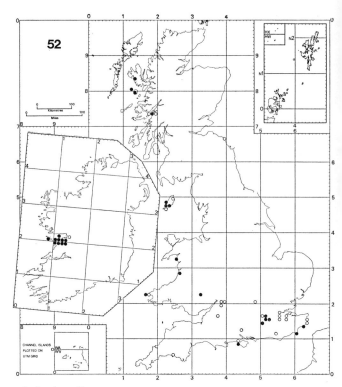

Setina irrorella

and limestone. It is known from the coastal region of Kent, especially Dungeness; Freshwater, Isle of Wight; south Devon; the coast of west Wales; Isle of Man; inland, the best-known locality is the neighbourhood of Boxhill, Surrey, but it has also been found in the Cotswolds; in Scotland on the east and west coasts; and in Ireland it is common in the Burren, Co. Clare. Abroad it occurs in Europe as a montane species.

MILTOCHRISTA Hübner

Miltochrista Hübner, [1819], *Verz.bekannt.Schmett.*: 166.
Calligenia Duponchel, 1844, *Cat.méth.Lépid.Eur.*: 59.

A genus with more than 100 species represented in Europe, tropical Asia and Australia.

MILTOCHRISTA MINIATA (Forster)
The Rosy Footman
Phalaena miniata Forster, 1771, *N.Sp.Ins.*: 75.
Bombyx rosea Fabricius, 1775, *Syst.Ent.*: 587.
Type locality: England.

Description of imago (Pl.5, fig.5)
Wingspan 25–33mm, female smaller than male. Head and thorax rosy; antenna setose-ciliate. Forewing rather elongate, ochreous pink, darker on costa and termen; waved antemedian line not reaching dorsum, acutely dentate postmedian line, and subterminal row of dots, blue-black. Hindwing very pale pink, darker terminally. Abdomen pink-fuscous.

Life history
Ovum. Fusiform, deep yellow. Laid in neat rows on several kinds of lichen, especially *Peltigera canina*, growing on stems of trees and bushes.
Larva. Full-fed *c.*15mm long, rather stout. Head brown. Body dark grey, densely covered with tufts of mouse-coloured setae, those on abdominal segments 5–10 being somewhat paler; dorsal setae are of two types, those on the mesothorax and abdominal segment 10 being simple, whereas those on the remaining segments are plumose.

Hatches in August and continues feeding throughout the winter, except in severe weather; is full-grown in May. In captivity it will accept withered leaves as well as lichens.
Pupa. Short, reddish brown. In an oval silken cocoon interwoven with larval setae which project upright, giving the cocoon a superficial resemblance to the larva.
Imago. Univoltine. Flies from mid-June throughout July. Sluggish by day and if disturbed flutters to the ground. Flies freely at night and comes to light and sugar.

Distribution (Map 53)
Occurs chiefly in woods and lanes and on well-timbered heathland. It is widely distributed and locally common in the southern half of England, the New Forest being a noted locality. It also occurs more sparingly northwards to Lincolnshire and Caernarvonshire; in Ireland in Cos Waterford and Galway; the Channel Islands. Palaearctic.

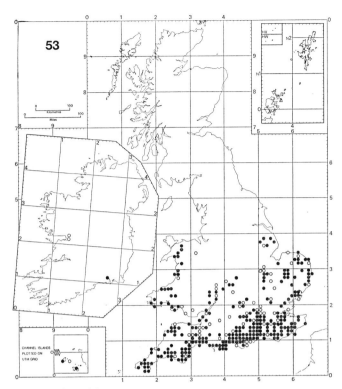
Miltochrista miniata

NUDARIA Haworth

Nudaria Haworth, 1809, *Lepid.Br.*: 156.
Nelopa Billberg, 1820, *Enum.Ins.Mus.Blbg*: 85.

A small Indo-Malayan and Australian genus with one species in Europe.

NUDARIA MUNDANA (Linnaeus)
The Muslin Footman

Phalaena (Tortrix) mundana Linnaeus, 1761, *Fauna Suecica* (Edn 2): 349.
Bombyx munda Fabricius, 1793, *Ent.syst.* **3**: 482.
Bombyx hemerobia Hübner, [1803], *Samml.eur.Schmett.* **3**: pl.17, fig.65.
Type locality: Sweden.

Description of imago (Pl.5, fig.6)
Wingspan 19–23mm. Antenna in male finely ciliate, in female simple; scape with tuft. Wings very thinly scaled, almost transparent and iridescent. Forewing pale brownish grey; veins darker; ante- and postmedian lines composed of faint grey dots; discal spot pale grey; cilia brownish grey. Hindwing unmarked with faint ochreous tinge. Body fuscous whitish.

Similar species. Thumatha senex (Hübner) *q.v.*

Life history
Ovum. Oval, weakly reticulate and glossy. Laid on small lichens, particularly on orange-coloured species growing on stone walls, or less often, palings or bushes.
Larva. Full-fed *c.*13mm long, rather stout. Head dark brown or blackish. Body greyish; a broad yellow dorsal stripe enclosing a fine greyish brown central line; subdorsal stripe blackish brown; ventral surface paler; abdominal segment 4 with a black velvety spot and pairs of sulphur-coloured spots on the remaining segments; verrucae with long, greyish setae.
 Hatches in late July and feeds until the autumn, when it goes into hibernation. In the spring it is readily found on stone walls, either feeding in sunshine or concealed under loose stones. Pupation takes place in May.
Pupa. Yellowish green. In a flimsy, network cocoon in which it is clearly visible, spun, sometimes communally, in a crevice.
Imago. Univoltine. Flies in late June and July. The male can sometimes be disturbed by day and comes readily to light, but the female is secretive and seldom seen in the wild.

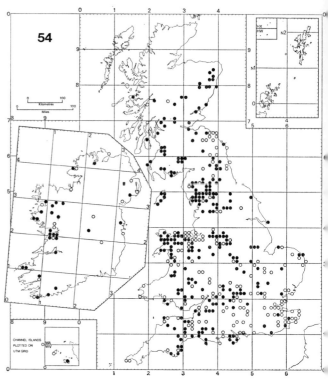

Nudaria mundana

Distribution (Map 54)
Widely distributed, mainly in stony terrain, throughout most of the British Isles as far north as the Scottish Highlands. Abroad in northern and central Europe eastwards to Asia Minor.

ATOLMIS Hübner

Atolmis Hübner, [1819], *Verz.bekannt.Schmett.*: 164.
Gnophria Stephens, 1829, *Ill.Br.Ent.* (Haust.) **2**: 98.

Represented in Great Britain and Ireland by a single species.

ATOLMIS RUBRICOLLIS (Linnaeus)
The Red-necked Footman

Phalaena (Noctua) rubricollis Linnaeus, 1758, *Syst.Nat.* (Edn 10) **1**: 511.
Type locality: Europe.

Description of imago (Pl.5, fig.7)
Wingspan 28–36mm, female slightly larger than male. Head black; antenna setose-ciliate. Thorax black; patagia red. Wings uniform velvety black. Abdomen black, terminal segments yellowish orange.

Life history
Ovum. Spherical, creamy white. Laid in neat batches on trunks of trees, especially conifers, oak and beech, which carry a good growth of small lichens.
Larva. Full-fed *c.*32mm long. Head intense black. Body greyish, irrorate with yellow; dorsal stripe narrow, whitish; subdorsal stripe blackish, interrupted on abdominal segments; verrucae reddish, bearing fairly long, brownish setae.

Hatches in July and feeds up quickly on lichens or algae, attaining full growth in October. Sometimes occurs in numbers on the same tree and is well concealed by its cryptic coloration as it rests in crevices of the bark.
Pupa. Shining red-brown. In a close-fitting cocoon of silk mixed with larval hairs, spun in a chink of bark or on the ground amongst leaf litter, in some cases communally; overwinters.
Imago. Univoltine. Occurs mainly in June and July, but sometimes as early as late May. In favourable years has been observed flying on sunny days in vast numbers round the tops of trees, usually oak, and is occasionally on the wing on dull days as well; may also be found at rest on low herbage. At night comes freely to light and sometimes to sugar.

Distribution (Map 55)
Occurs mainly in woodland. Its numbers vary greatly from year to year and occasionally it has appeared in profusion. It is found in most of the southern counties of England but occurs only sporadically farther north; Wales; the Channel Islands and the Isles of Scilly. In Ireland it used to occur in many areas, but there are few recent records. Abroad it is found over most of northern and central Europe, its range extending to eastern Asia.

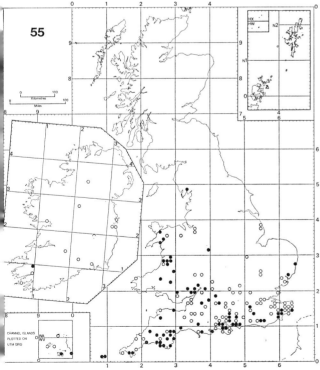
Atolmis rubricollis

CYBOSIA Hübner
Cybosia Hübner, [1819], *Verz.bekannt.Schmett.*: 167.

A monotypic genus.

CYBOSIA MESOMELLA (Linnaeus)
The Four-dotted Footman

Phalaena (Tinea) mesomella Linnaeus, 1758, *Syst.Nat.* (Edn 10) **1**: 535.

Type locality: [Europe].

Description of imago (Pl.5, figs 8,9)
Wingspan 29–34mm. Head ochreous orange; antenna in male shortly setose-ciliate, in female simple. Thorax whitish grey; patagia ochreous orange. Forewing broad, pale whitish grey, costa and termen suffused with orange-yellow; a prominent black median dot near costa and a second similar dot near dorsum. Hindwing very broad, ochreous grey; cilia pale orange. Abdomen whitish grey, terminally yellowish.

Variation occurs in the extent of the orange-yellow suffusion of the forewing. In one form (fig.9) the forewing is wholly orange-yellow; this form occurs throughout the range and is prevalent in east Kent.

Life history
Ovum. Spherical, glossy, greyish white with a faint ring of whiter spots round the micropyle. Laid in neatly arranged batches, in nature probably amongst lichens growing on tree-trunks or on stems of heather.

Larva. Full-fed *c.*25mm long, tapering slightly at the head and anal segments. Head shining black. Body velvety slate-black; a more intense velvety black patch on mesothorax; verrucae bearing dense tufts of dark brown or black setae which are plumose except on the mesothorax and below the spiracles where they are simple.

Hatches in late July or August and feeds till early autumn when it goes into hibernation; reappears in the spring and is full-grown in May. Besides eating lichens growing amongst heather or on bushes and trees, it will accept, in captivity, fresh or decaying leaves of sallow (*Salix* spp.) and heather (*Calluna vulgaris*).

Pupa. Pale brown. In a lightly spun, semitransparent cocoon amongst moss or herbage, or, more often, in a chink in the bark of a tree.

Imago. Univoltine. Occurs in July. May often be flushed from heather and is on the wing between sunset and dusk. Also flies after dark and comes freely to light.

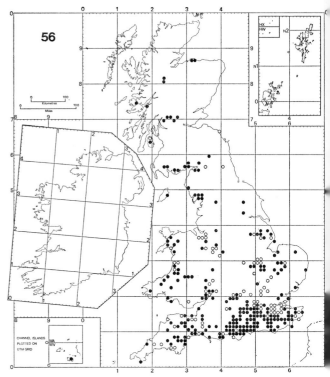

Cybosia mesomella

Distribution (Map 56)
Occurs on heaths, moorland and in open woodland and is widespread in southern England; Wales. In Scotland local as far north as Morayshire; Jersey. Absent from Ireland. Abroad it occurs throughout Europe, except for the extreme north, eastwards into Asia.

PELOSIA Hübner
Pelosia Hübner, [1819], *Verz.bekannt.Schmett.*: 165.

A small Palaearctic genus with two species in England.

PELOSIA MUSCERDA (Hufnagel)
The Dotted Footman
Phalaena muscerda Hufnagel, 1766, *Berlin.Mag.* **3**: 400.
Type locality: Germany; Berlin.

Description of imago (Pl.5, fig.10)
Wingspan 30–34mm, female slightly larger than male. Antenna finely and shortly setose-ciliate. Forewing very elongate, pale grey somewhat suffused with brown; an oblique line of three or four black postmedian dots from costa reaching half across wing; two black antemedian dots placed obliquely near dorsum. Hindwing broad, pale brownish grey. Body pale grey.

Similar species. *P. obtusa* (Herrich-Schäffer) which is smaller, has more rounded wings and a very curved median row of black dots from costa to dorsum.

Life history
Ovum. Hemispherical, glossy white. Probably laid upon lichens growing on bushes and trees.

Larva. Full-fed *c.*20mm long, stout, tapering slightly towards the head and anal segments. Head shining black. Body brownish black marbled with grey, velvety; dorsal and subdorsal stripes black; subspiracular stripe grey, narrow and interrupted; on prothorax and abdominal segment 9 a pair of dull red spots between the dorsal and subdorsal stripes; verrucae brown, the anterior pair on each segment ringed with reddish grey, all bearing dense tufts of short, brown setae; legs and prolegs reddish grey.

Hatches in August and soon goes into hibernation. Feeds again in spring, probably on lichens growing on sallow in the wetter parts of fens and also, possibly, on species of moss. In confinement it has accepted *Peltigera canina*, *Homalothecium sericeum* and *Weisia* spp. as well as decayed bramble and sallow leaves, but its habits in its natural habitat have not been studied.

Pupa. Dark chestnut-brown. In a double cocoon, the inner one of grey and the outer one of white silk, pupating in May.

Imago. Univoltine. Flies in late July and August. Seldom seen by day, but flies at dusk and dawn, as well as at midnight, when it comes readily to light. Has also been taken at sugar. At rest it folds its wings downwards along the sides of the abdomen.

Distribution (Map 57)
It is chiefly found in the wetter areas of marshes and fens,

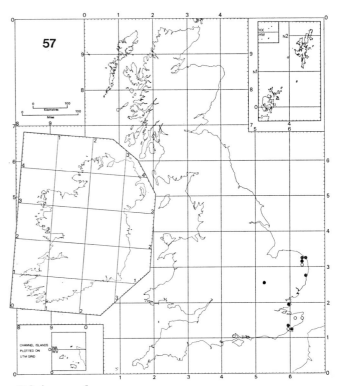

Pelosia muscerda

often in carrs of alder, sallow and buckthorn. It is very local, being found chiefly in the Norfolk Broads where it was discovered in 1829. Recently it has been recorded from Cambridgeshire, the Dungeness, Kent, area and Suffolk. There are old records from Matley Bog in the New Forest and from marshes in east Kent. Palaearctic.

PELOSIA OBTUSA (Herrich-Schäffer)
The Small Dotted Footman
Paidina obtusa Herrich-Schäffer, 1852, *Syst.Bearb.Schmett. Eur.* **2**: 156, 161.
Type locality: S. France.

Description of imago (Pl.5, fig.11)
Wingspan 26mm. Forewing rounded, brownish grey, veins darker; a very curved median row of black dots from costa to dorsum. Hindwing broad, pale brownish grey. Body brownish grey.

Similar species. *P. muscerda* (Hufnagel) q.v.

Life history
Ovum. Not described.
Larva. Stated to be very similar to that of *P. muscerda*. Hatches in late August and feeds for a short time probably

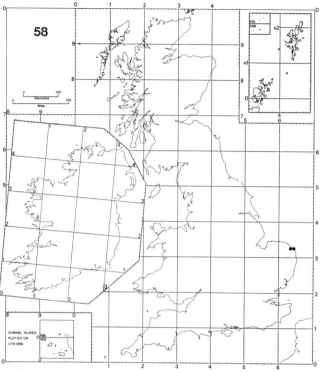

Pelosia obtusa

on lichens and mosses before going into hibernation. Resumes feeding in April and is full-fed in late June.
Pupa. Not described.
Imago. Univoltine. Flies in late July and early August.

Distribution (Map 58)

Two specimens only have been recorded in the Norfolk Broads, the first in August 1961. It is found in remote and thick reed-beds and is probably indigenous. Abroad it occurs in marshy areas through northern Europe eastwards to Russia and Siberia.

EILEMA Hübner

Eilema Hübner, [1819], *Verz.bekannt.Schmett.*: 165.
Systropha Hübner, [1819], *ibid.*: 166.
Piesta sensu auctt.

A large cosmopolitan genus.
Represented in Great Britain and Ireland by seven species, or eight if *E. sericea* (Gregson) is accepted as a valid species.

EILEMA SORORCULA (Hufnagel)
The Orange Footman

Phalaena sororcula Hufnagel, 1766, *Berlin.Mag.* **3**: 398.
Bombyx aureola Hübner, [1803], *Samml.eur.Schmett.* **3**: pl.24, fig.98.
Type locality: Germany; Berlin.

Description of imago (Pl.5, fig.12)
Wingspan 27–30mm. Head and thorax orange; antenna finely setose-ciliate. Forewing costa arched, orange-yellow. Hindwing pale orange-yellow. Abdomen greyish yellow, anal tuft yellow.
Similar species. *E. griseola* f. *stramineola* Doubleday which is larger, has the costa more strongly arched, and the thorax and abdomen yellow.

Life history
Ovum. Undescribed. Laid in late May or early June on various lichens growing on several species of tree, but most commonly on oak or beech.
Larva. Full-fed *c.*20mm long and rather slender. Head blackish brown. Body with the ground colour brownish grey except for the dorsum, which is white; dorsal, subdorsal and supraspiracular lines brown, interrupted on metathorax and abdominal segments 4 and 8 by blackish patches and on abdominal segment 5 by a conspicuous white patch; prothorax with reddish anterior and lateral markings; verrucae red except on the black patches, bearing dense tufts of brownish grey setae.
Hatches in July and is full-grown in September.
Pupa. Glossy dark brown. In a slight silken cocoon spun usually amongst moss; overwinters.
Imago. Univoltine. Flies in late May and throughout June. By day it rests on branches of oak or beech and may be beaten out, when it flutters to the ground. Flies from late dusk onwards and comes occasionally to sugar or light.

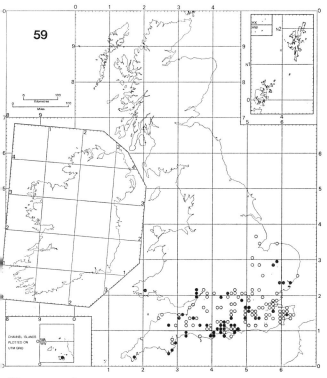

Eilema sororcula

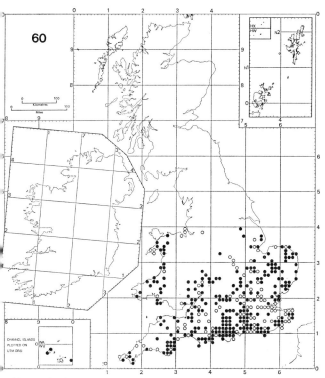

Eilema griseola

Distribution (Map 59)

Occurs locally in oak and beech woodland south of a line from the Wash to Pembrokeshire. Abroad it is widespread in northern and central Europe eastwards to southern Russia.

EILEMA GRISEOLA (Hübner)
The Dingy Footman

Bombyx griseola Hübner, [1803], *Samml.eur.Schmett.* **3**: pl.23, fig.97.
Lithosia plumbeolata Stephens, 1829, *Ill.Br.Ent.* (Haust.) **2**: 96.
Lithosia stramineola Doubleday, 1847, *Zoologist* **5**: 1914.
Type locality: Europe.

Description of imago (Pl.5, figs 13,14)

Wingspan 32–40mm. Head pale yellow; antenna setoseciliate. Thorax pale grey. Forewing elongate, costa strongly arched, dull slate-grey with yellow costal streak of even width from base to apex. Hindwing pale ochreous yellow. Abdomen pale yellow.

In f. *stramineola* Doubleday (fig.14) the thorax and forewing are entirely pale yellow. This form was formerly considered to be a distinct species and was named the Straw-coloured Footman.

Life history

Ovum. Hemispherical, weakly reticulate, glossy. Laid in clusters on the bark close to patches of lichen, especially *Peltigera canina*.

Larva. Full-fed *c.*25mm long, rather stout. Head glossy black. Body velvety black, slightly paler on ventral surface; dorsal stripe purplish, enclosing a narrow blackish line; subdorsal stripe orange with black verrucae, narrow but expanded to form orange dorsal blotches on pro- and mesothorax and abdominal segment 9, these orange markings being variable in development and intensity; the verrucae, except those on the subdorsal stripe, brown, all bearing tufts of brownish setae which are longer on the thoracic and anal segments.

Hatches in late July or August and feeds until it goes into hibernation in September, reappearing in the spring to complete its growth in May. Besides lichens growing on trees and bushes, it will eat, at any rate in captivity, various mosses, withered leaves of sallow and foliage of many herbaceous plants.

Pupa. Stout, reddish brown. In a silken cocoon intermixed with mosses or lichens, spun usually in a sheltered situation.

Imago. Univoltine. Occurs in July. Flies freely at dusk and at night when it comes to light, sugar and honey-dew.

Generally conceals itself by day, but may sometimes be seen resting on foliage with the wings held flat over the abdomen.

Distribution (Map 60)
Occurs in fens and marshy areas, especially around alder or sallow carr, and also sometimes in damp woodland. It is locally common in southern England; west Wales; the Channel Islands; unknown in Ireland. Palaearctic; f. *stramineola* is known only in Britain.

EILEMA CANIOLA (Hübner)
The Hoary Footman

Bombyx caniola Hübner, [1808], *Samml.eur.Schmett.* **3**: pl.51, fig.220.

Type locality: Europe.

Description of imago (Pl.5, fig.15)
Wingspan 28–35mm. Head orange; antenna shortly and finely setose-ciliate. Thorax whitish grey; patagia orange. Forewing elongate, pale silky whitish grey, costa pale yellow almost to apex. Hindwing very pale yellowish grey. Abdomen whitish grey, anal tuft pale yellow.

Variation is confined to the intensity of the grey coloration; in f. *lacteola* Boisduval the wings are almost entirely white.

Similar species. *E. complana* (Linnaeus) and *E. lurideola* ([Zincken]) which differ in having the forewing darker grey, the costal streak wider and much more intensely yellow and the hindwing pale ochreous yellow.

Life history
Ovum. Hemispherical, weakly reticulate, glossy whitish. Laid on or near lichens growing on coastal rocks.

Larva. Full-fed *c*.25mm long. Head grey-brown to black, with pale setae projecting forwards. Body greyish brown; dorsal stripe black; a subdorsal series of black-edged orange spots; spiracular and subspiracular stripes orange; verrucae brown, bearing tufts of short, greyish setae.

Hatches in August or September. Some larvae feed on lichens growing on rocks or even house-slates; others have been found on leguminous plants such as bird's-foot trefoil (*Lotus corniculatus*), kidney vetch (*Anthyllis vulneraria*) or white clover (*Trifolium repens*). According to Barrett (1895), it is reluctant to change from one class of foodplant to another. Overwinters when small and feeds again from April till the end of May or early June.

Pupa. Stout, shining brown. In a loose cocoon spun amongst lichens growing on rocks, or on the ground.

Imago. Univoltine. Occurs from late July till September. Flies freely at dusk and later comes to sugar and light. Stays concealed by day, but may occasionally be beaten

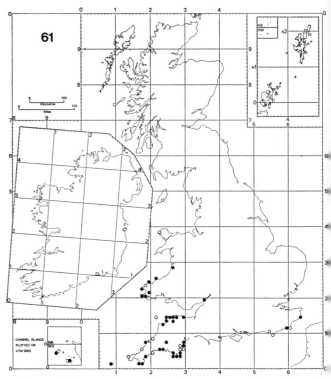

Eilema caniola

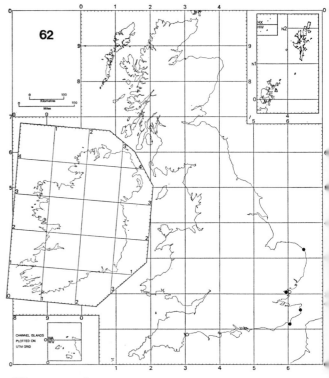

Eilema pygmaeola

from thick herbage at the foot of sea-cliffs, when it drops to the ground. Rests with its wings closely folded to the sides of the abdomen.

Distribution (Map 61)

In south-east England it has been found at Dover, in Romney Marsh and near Beachy Head; it occurs in many localities near the sea in the West Country; in Wales recorded in Pembrokeshire and formerly in Anglesey; in Ireland at Howth, Co. Dublin, where the first British specimens were taken in 1861 by C. G. Barrett, and in Co. Waterford; the Channel Islands. Abroad it is quite common inland in many parts of Europe; though predominantly a Mediterranean species and common in some Italian cities, it is also widespread in west Germany.

EILEMA PYGMAEOLA (Doubleday)
The Pigmy Footman

Lithosia pygmaeola Doubleday, 1847, *Zoologist* **5**: 1914.
Lithosia lutarella sensu Meyrick, 1895, *Handbk Br.Lepid.*: 26.

Type locality: England; Kent.

Description of imago (Pl.5, figs 16,17)

Wingspan 24–28mm, female considerably smaller than male. Head ochreous yellow; antenna simple. Forewing whitish ochreous or sometimes greyish. Hindwing pale yellow, costally suffused greyish. Abdomen pale yellowish grey.

Subsp. *pallifrons* (Zeller)

Lithosia pygmaeola pallifrons Zeller, 1847, *Stettin.ent.Ztg* **8**: 339.

(Fig.17) Head with frons grey. Body yellow. Forewing pale yellow.

Similar species. E. deplana (Esper) male, which has the antenna setose-ciliate, the costa of the forewing suffused with orange and is much larger.

Life history

Ovum. Hemispherical, minutely punctate, salmon or flesh-coloured, glossy. Laid loose (Buckler, 1889).

Larva. Short and rather stout (Buckler, *loc. cit.*). Head black. Body brown, paler ventrally; dorsal stripe black; subdorsal stripe dark brown; spiracular stripe whitish; verrucae with tufts of short, brown setae.

Hatches in August and soon goes into hibernation; resumes feeding in the spring until June. Little is known of the natural foodplants, but it is probable that these are lichens growing on or near the ground in coastal areas.

Pupa. Undescribed.

Imago. Univoltine. Occurs in late July and August. Flies at dusk and again late at night when it comes freely to light and sometimes to sugar. Rests on marram and other herbage with the wings folded against the sides of the abdomen, when it may easily be mistaken for a species of *Crambus* (Pyralidae).

Distribution (Map 62)

Occurs only on coastal sandhills. It is found on the east coast of Kent, where it was discovered in 1847 and where it still occurs plentifully; in the early 1930s it was found to be common at Dungeness, Kent; its only other known present locality is Waxham, Norfolk, but single specimens were taken at Bradwell-on-Sea, Essex (1958, 1961) and Exmouth, Devon in 1937 (Stidston, 1952). The Dungeness population is referable to subsp. *pallifrons*. Its status abroad is uncertain, since it has been confused with the very similar *E. lutarella* (Linnaeus).

EILEMA COMPLANA (Linnaeus)
The Scarce Footman

Phalaena (*Noctua*) *complana* Linnaeus, 1758, *Syst.Nat.* (Edn 10) **1**: 512.

Type locality: [Sweden].

Description of imago (Pl.5, fig.18)

Wingspan 30–36mm. Head orange; antenna in male weakly setose-ciliate, in female simple. Thorax grey; patagia orange. Forewing silvery grey, costal streak orange-yellow, extending in uniform width from base to apex. Hindwing pale ochreous yellow, very slightly suffused with grey at costa. Abdomen grey, terminal segments yellowish.

Similar species. E. sericea (Gregson) which has a narrower forewing with the costa more rounded, the costal streak narrower, often tapering before the apex, the hindwing more heavily suffused with grey. South (1961) states that another difference is the size of the costal tuft of scales on the forewing underside. For other differences see *E. sericea* (p. 92). *E. caniola* (Hübner) *q.v.* and *E. lurideola* ([Zincken]) which has the costal streak of the forewing tapering towards the apex.

Life history

Ovum. Hemispherical, glossy pale cream. Laid in batches on or near lichens growing on trees or bushes.

Larva. Full-fed *c*.25mm long, stout. Head glossy black. Body greyish brown, paler ventrally; dorsal line black, edged white; subdorsal line consisting of a series of alternate orange and white spots, the orange spots being very faint on pro- and metathorax; spiracular line orange, edged with brown above and below; verrucae brown, bearing tufts of short, greyish brown setae.

Hatches in August or September and overwinters when

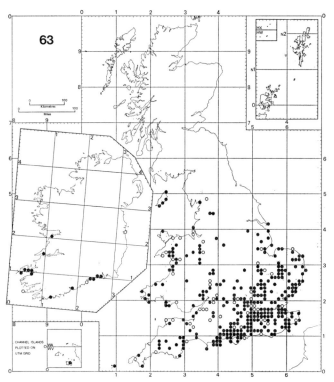

Eilema complana

still small; recommences feeding in the spring and is full-fed in June. Though lichens are the principal foodplant, it will also eat mosses and withered leaves, foliage of knotgrass (*Polygonum aviculare*) or hare's-foot clover (*Trifolium arvense*) and flowers of bird's-foot trefoil (*Lotus corniculatus*).

Pupa. Reddish brown. In a slight cocoon spun amongst lichens and mosses on trunks of trees or in a sheltered situation on the ground.

Imago. Univoltine. Appears in July and August. Remains concealed by day amongst heather and other low vegetation and if disturbed falls to the ground as if dead. Active at dusk and has been recorded flying in dozens round isolated bushes, perhaps assembling to a freshly emerged female. After dark it visits flowers such as those of thistle and field scabious and comes freely to light. Rests with its wings rolled tightly round its abdomen in a stiff, 'footman-like' manner, a character which distinguishes it from *E. lurideola* which folds its wings flat over its back.

Distribution (Map 63)
Inhabits mainly heathland and moors, though it is also found on sandhills and in woodland. It occurs in most of southern England being most prevalent in the eastern and south-eastern counties, extending northwards to the Humber and north Lancashire; it occurs commonly on the coast of Wales and the Isle of Man; in Ireland it is mainly coastal and is not uncommon in Cos Waterford and Kerry. Abroad it is found throughout Europe eastwards to Asia Minor.

EILEMA SERICEA (Gregson)
The Northern Footman

Lithosia sericea Gregson, 1860, *Ent.wkly Intell.* **9**: 30.
Lithosia molybdeola Guenée, 1861, *Annls Soc.ent.Fr.* (4) **1**: 50.

Type locality: England; [Warrington, Lancashire].

Status. The characters used by Gregson to separate this supposed species from *E. complana* (Linnaeus) are very variable, nor can the two be separated on the structure of their genitalia (Pierce & Beirne, 1941). Therefore it seems likely that they are conspecific.

Description of imago (Pl.5, fig.23)
Wingspan 34mm. The original description by Gregson is as follows:

'——the species we take on our mosses is distinct from the *Complana* on Mr. Doubleday's List, and ought to be placed between that species and his *L. Complanula*: from the first it differs in its more rounded costa, the costal streak being narrower and not carried out to the apex of the wing parallel, as in *L. Complana*: in its more silky appearance, and general narrower form; in the underwings always being suffused more or less, – sometimes . . . they are quite dark, without any of the yellow upon them. It also differs much on the under side; perhaps the best character there is one pointed out to me by Dr. Knaggs: in *Complana* there is always a costal streak on the underside of the inferior wing, from the apex to the base, of one uniform width, or nearly so; whereas in the species from our mosses [Warrington] the same streak is broad for nearly one-third of the wing, when it either disappears abruptly or is only continued to the base as a very faint narrow streak.

'From *Complanula* it differs in being less rounded on the costa, and also in the collar being continued unicolorous with the costal streak, as in *Complana*; the costal streak varies much in its breadth at the apex, sometimes being carried through of an equal width, as in *Complana*, at others approaching the form so invariably seen in *Complanula*; but it always wants the *ample* yellow under wings of that species'.

Similar species. *E. complana* q.v.

Life history
All the literature indicates that neither the larva nor the pupa can be distinguished from those of *E. complana*. The

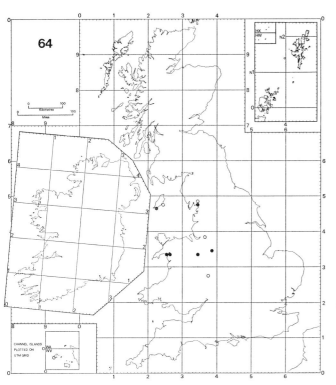

Eilema sericea

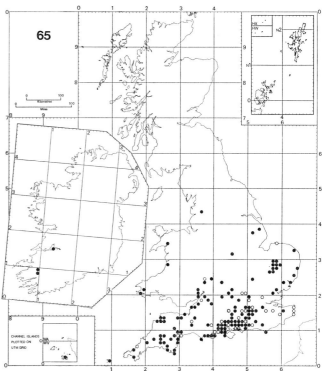

Eilema deplana

imago flies in July; it comes freely to light and is sometimes disturbed from vegetation or found at rest by day.

Distribution (Map 64)

The northern mosses of Shropshire, Cheshire and Lancashire; North Wales; Isle of Man. Many of these have been drained for urban development. Unknown abroad.

EILEMA DEPLANA (Esper)
The Buff Footman

Noctua deplana Esper, 1787, *Schmett.* **4**: 97.
Noctua depressa Esper, 1787, *ibid.* **4**: 98.
Bombyx helvola Hübner, [1803], *Samml.eur.Schmett.* **3**: pl.23, fig.95.
Lithosia helveola Ochsenheimer, 1810, *Schmett.Eur.* **3**: 133.
Type locality: Europe.

Description of imago (Pl.5, figs 19,20)

Wingspan 28–36mm. Head greyish orange; antenna in male strongly setose-ciliate, in female simple. Patagia greyish orange. Forewing in male ochreous grey, costal margin yellowish-tinged, in female yellowish to deep ochreous grey with distinct yellow costal streak tapering towards apex; cilia orange. Hindwing in male pale grey, in female darker. Abdomen yellowish grey, anal segments yellow.

In some localities the prevalent form is f. *unicolor* Bankes, which has the wings orange-buff with only slight grey suffusion on the hindwing.

Similar species. E. *pygmaeola* (Doubleday) *q.v.*

Life history

Ovum. Hemispherical. Laid in August in batches in the vicinity of coarse lichens growing on trunks of yew (*Taxus baccata*) and other trees.

Larva. Full-fed *c.*20mm long. Head dark brownish grey freckled with black. Body greenish grey; dorsal stripe yellowish, edged with black and enclosing a broken, dark central line; a black transverse bar on the metathorax and black triangular spots on abdominal segments 4 and 8; setae deep brown and short.

On hatching, feeds on lichens for a short period before hibernation; feeds again in the spring and is full-fed in early June. May readily be beaten from branches of the trees where it is feeding.

Pupa. Brown, smooth and glossy. In a closely woven, boat-shaped cocoon, spun amongst lichens on a branch or trunk.

Imago. Univoltine. Flies in late July and throughout August.

Comes freely to ultra-violet light. Rests with its wings folded flat over its back. By day it may be beaten from branches of yews and other trees, when it drops or flutters to the ground.

Distribution (Map 65)

Inhabits woodland, especially where there are yews, but is local and seldom numerous. It has been recorded from many localities in southern England, including the Isles of Scilly, the New Forest being a well-known locality. It has also been found in the eastern counties, the Wye Valley, Gwent and west Wales and there are records as far north as Lancashire. In Ireland it is confined to the south-west. Eurasiatic; in Europe it is widespread as far south as Italy.

EILEMA LURIDEOLA ([Zincken])
The Common Footman

Lithosia lurideola [Zincken], 1817, *Allg.Lit.-Ztg* **1817** (217): 68.

Lithosia complanula Boisduval, 1834, *Icon.hist.Lépid.Eur.* **2**: 97.

Type locality: not stated.

Description of imago (Pl.5, fig.24)

Wingspan 31–38mm. Head orange; antenna in male weakly setose-ciliate, in female simple. Thorax grey; patagia orange, greyish-centred. Forewing rather broad, leaden grey, costal streak orange-yellow, tapering well before and not reaching apex. Hindwing fairly broad, pale ochreous yellow. Abdomen grey, anal tuft yellow.

Similar species. *E. complana* (Linnaeus) and *E. sericea* (Gregson) q.v.

Life history

Ovum. Hemispherical, finely punctate and glossy white. Laid in batches, generally on bark.

Larva. Full-fed c.25mm long. Head glossy black. Body deep grey; dorsal and subdorsal lines blackish; abdominal segments with orange spiracular stripe; verrucae and setae black on dorsum and yellow on lateral surface.

Hatches in August or September and feeds for a short time before hibernation; resumes feeding early in the spring and attains full growth at the end of May. Lichens growing on trees and bushes are its normal diet; it will also accept foliage of sallow (*Salix* spp.), apple (*Malus* spp.), oak (*Quercus* spp.), dogwood (*Swida sanguinea*) and other trees and bushes.

Pupa. Short and stout, with the anal end bluntly rounded; deep brown. In a light silken cocoon spun usually in a crevice of the bark.

Imago. Univoltine. Flies from the end of June until August. Rests by day amongst foliage of trees and if beaten out

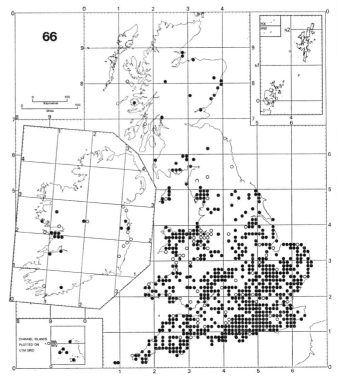

Eilema lurideola

falls as though dead to the ground. Flies from dusk onwards and visits sugar and flowers, especially those of traveller's-joy and thistles. Comes freely to light and rests with the wings folded flat over its back; this attitude readily distinguishes it from *E. complana* which rests with its wings rolled round its abdomen.

Distribution (Map 66)

Widespread and common throughout Britain to southern Scotland and farther north on the east coast, Invernessshire and Isle of Mull; the Channel Islands; in Ireland it has been found in many coastal areas and especially in the Burren, Co. Clare. Eurasiatic.

LITHOSIA Fabricius

Lithosia Fabricius, 1798, *Suppl.Ent.syst.*: 419.
Lithosis Billberg, 1820, *Enum.Ins.Mus.Blbg*: 91.
Oeonistis sensu auctt.

A small genus of about 20 species represented in Europe, temperate Asia including Japan, and Africa.

LITHOSIA QUADRA (Linnaeus)
The Four-spotted Footman

Phalaena (Noctua) quadra Linnaeus, 1758, *Syst.Nat.* (Edn 10) **1**: 511.
Type locality: [Europe].

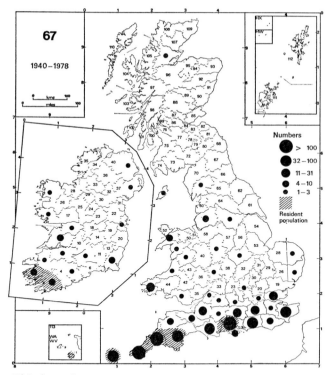

Lithosia quadra

Description of imago (Pl.5, figs 21,22)

Wingspan 35–55mm, female much larger than male. Sexually dimorphic. *Male.* Head orange; frons blue-black; antenna shortly setose-ciliate. Thorax orange. Forewing bluish grey, darker apically, slightly suffused with yellowish; basal patch orange; basal third of costa with a broad blue-black streak. Hindwing pale lemon-yellow, suffused costally with pale grey. *Female.* Head and thorax ochreous yellow; antenna weakly setose-ciliate. Forewing ochreous yellow; a large blue-black triangular spot on costa at just beyond one-half and a similar spot just before centre of wing below disc. Hindwing pale lemon-yellow.

Life history

Ovum. Hemispherical, minutely punctate, bright green when laid, becoming olive later and finally dark brown before hatching (Buckler, 1889). Laid in orderly batches near lichens on the bark of oak (*Quercus* spp.) or other trees.

Larva. Full-fed 37–40mm long. Head glossy black. Body blackish; a broad, yellow dorsal stripe, consisting of four undulating lines, the intervening spaces also punctate with yellow; black cruciform dorsal spots on metathorax and abdominal segments 4 and 8; verrucae dorsally red and laterally grey, each bearing mixed grey and black setae.

Hatches in September and feeds on lichens, especially *Peltigera canina*, on the trunks and branches of oak and other trees. Hibernates when small and resumes feeding in the spring, becoming full-grown in June or July. The full-fed larva has been observed sunning itself on the foliage. It has cannibalistic tendencies.

Pupa. Stout, glossy black. In a loose, grey cocoon of silk mixed with larval hairs, spun in a crevice or under bark or moss.

Imago. Univoltine. Occurs from early August till September. Often flies in bright sunshine as well as after dark, when it comes freely to light, sugar and flowers. May also be found resting on tree-trunks with its wings flat over its back.

Distribution (Map 67)

Inhabits well-established woodland, where its numbers fluctuate greatly; in some years both larvae and adults are observed in great plenty and it is probable that the population is from time to time reinforced by immigration. It has been recorded from most of the southern counties of England and has occurred sporadically as far north as Ross-shire and the Isle of Man; it is also found in the Channel Islands, the Isles of Scilly, west Wales and much of Ireland. Palaearctic.

Arctiinae

SPIRIS Hübner
Spiris Hübner, [1819], *Verz.bekannt.Schmett.*: 169.
Callopis Billberg, 1820, *Enum.Ins.Mus.Blbg*: 91.
Eulepia Curtis, 1825, *Br.Ent.* **2**: 56.
Emydia Boisduval, 1828, *Eur.Lepid.Index method.*: 39.

A monotypic genus.

SPIRIS STRIATA (Linnaeus)
The Feathered Footman
Phalaena (Bombyx) striata Linnaeus, 1758, *Syst.Nat.* (Edn 10) **1**: 502.
Phalaena (Bombyx) grammica Linnaeus, 1758, *ibid.*: 822.
Type locality: Germany.

Description of imago (Pl.5, figs 25,26)
Wingspan 33–42mm. Antenna in male bipectinate, in female simple. Thorax yellow, black dorsally. Forewing pale lemon-yellow; veins heavily shaded with black in male; discal spot black; subterminal fascia ill-defined, pale grey, absent in female. Hindwing deep yellow; terminal shade and discal spot black. Abdomen yellow, dorsal stripe black.

Life history
Ovum. Hemispherical, a large dark spot in area of micropyle; chorion punctate and glossy (Stokoe & Stovin, 1958).
Larva. Full-fed *c.*30mm long. Head black. Body blackish brown; dorsal stripe orange; subdorsal stripe white; verrucae yellowish, bearing reddish brown setae.
 Hatches in July and feeds on herbaceous plants. Hibernates gregariously in a web when still small and resumes feeding in the spring, becoming full-fed in May.
Pupa. Red-brown. In a flimsy, whitish cocoon.
Imago. Univoltine. Appears in June and July. Flies in sunshine and may readily be disturbed in dull weather.

Occurrence and distribution
On the Continent it is found in small colonies on heathland. Between 1815 and 1859 two specimens were taken in Berkshire, one in Essex, one in Yorkshire and three in north Wales; there has been no subsequent record. It is doubtful whether this species has ever been indigenous in Britain. Palaearctic; it is common in many parts of Europe, especially in the Mediterranean region.

COSCINIA Hübner
Coscinia Hübner, [1819], *Verz.bekannt.Schmett.*: 169.

A monotypic genus.

COSCINIA CRIBRARIA BIVITTATA (South)
The Speckled Footman
Phalaena (Bombyx) cribraria Linnaeus, 1758, *Syst.Nat.* (Edn 10) **1**: 507.
Phalaena (Bombyx) cribrum Linnaeus, 1761, *Fauna Suecica* (Edn 2): 302.
Emydia cribrum bivittata South, 1900, *Entomologist* **33**: 68.
Euprepia cribraria anglica Oberthür, 1911, *Études Lép.comp.* **5**: 171.
Type locality: England.

Description of imago (Pl.5, fig.27)
Wingspan 33–40mm, female larger than male. Antenna in male bipectinate, in female simple. Thorax white, spotted with brown. Forewing pale greyish white; antemedian, median, postmedian and terminal fasciae represented by rows of dark brownish black spots; discal area suffused with brownish grey, with greyish streaks above and below; fringe chequered grey and white. Hindwing grey, paler in discal area. Abdomen white, ringed with dark grey.
 Variation occurs in the intensity of the grey suffusion and a very dark form is known.
Subsp. *arenaria* (Lempke)
Euprepia cribraria arenaria Lempke, 1937, *Lambillionea* **37**: 150.
This continental subspecies has the forewing almost white.

Life history
Ovum. Hemispherical, glossy yellow when laid, becoming purplish brown before hatching. Has been found laid in a batch round a slender bare twig of bell-heather (*Erica cinerea*).
Larva. Full-fed *c.*30mm long and rather stout. Head blackish brown. Body purple-brown; dorsal stripe white; spiracular stripe brown; verrucae pale brown, small, bearing tufts of brown setae; legs brown.
 Hatches in August. Though the eggs are generally laid on heather, the young larva feeds mainly on dandelion (*Taraxacum officinale*) or other herbaceous plants. Overwinters when small and resumes feeding in March when it is often to be found sunning itself on the blades of grasses such as *Deschampsia cespitosa*. Full-fed in June.
Pupa. At first reddish, becoming deep brown. In a white, rather flimsy silken cocoon spun amongst grass or heather.

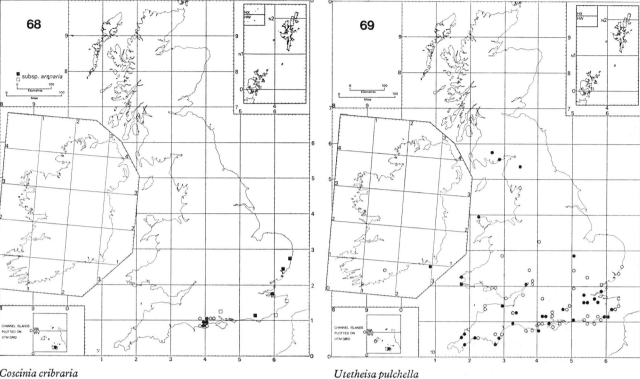

Coscinia cribraria

Utetheisa pulchella

Imago. Univoltine. Flies mainly in July though it has been recorded from late May until August. Rests with its wings tightly folded round its body and after dark is to be found freely at rest on heather. During the day it hides in the heather and when disturbed flies swiftly if the sun is shining, but drops to the ground in dull weather.

Distribution (Map 68)
Very local and confined to the heaths west of the New Forest from Ringwood and Verwood to Wareham. Occasionally migrant specimens of subsp. *arenaria* have been taken on the coasts of Kent and Suffolk and inland in Hampshire and Sussex. This subspecies occurs throughout Europe to the Mediterranean.

Conservation. This species has declined in recent years and according to Goater (1975) has not been recorded in Hampshire since 1960. The greatest restraint should be exercised in collecting specimens.

UTETHEISA Hübner

Utetheisa Hübner, [1819], *Verz.bekannt.Schmett.*: 168.
Deiopeia Stephens, 1829, *Ill.Br.Ent.* (Haust.) **2**: 92.

A small, cosmopolitan genus. Most species are distinctively patterned and brightly coloured, and some are migratory.

UTETHEISA PULCHELLA (Linnaeus)
The Crimson Speckled

Phalaena (Tinea) pulchella Linnaeus, 1758, *Syst.Nat.* (Edn 10) **1**: 534.
Noctua pulchra [Denis & Schiffermüller], 1775, *Schmett. Wien.*: 68.

Type localities: Europe and Mauritania.

Description of imago (Pl.5, fig.28)
Wingspan 28–40mm. Head orange; antenna in male setoseciliate, in female simple. Thorax white, spotted with black and orange. Forewing creamy white with five transverse series of crimson spots alternating with six similar series of black dots; fringe chequered grey and white. Hindwing brilliant white, terminal shade dark grey, broader apically; discal spot pale grey. Abdomen white.

Life history

Ovum. Hemispherical, greenish. Laid in batches on foliage of forget-me-not (*Myosotis* spp.), borage (*Borago officinalis*) and other herbaceous plants.

Larva. Full-grown *c.*30mm long and rather slender. Head reddish, edged with black. Body greyish; dorsal and spiracular stripes white; a transverse orange bar on each segment; verrucae black, bearing black dorsal setae and grey lateral setae.

The species is continuously brooded in subtropical regions. In temperate climates it hatches in the autumn and may either hibernate or feed at intervals till March; warmth and bright sunshine are necessary if it is to be reared successfully.

Pupa. Reddish brown. In a white silken cocoon spun amongst the foliage of the foodplant or on the ground.

Imago. Immigrant specimens have reached Britain between the months of May and October, but mostly in the autumn. The adult flies in sunshine.

Occurrence and distribution (Map 69)

The first specimen was taken in Hampshire, probably in 1818. Since then there have been occasional records from many areas of Britain as far north as southern Scotland, and in Ireland. At least 30 were reported in 1871 and a similar number in 1961, mostly in southern England; in the latter year a large number showing remarkable variation were reared from a specimen taken in Kent. Eurasiatic; it is widespread in the Mediterranean region, which is the probable source of British immigrants.

UTETHEISA BELLA (Linnaeus)

Phalaena (Tinea) bella Linnaeus, 1758, *Syst.Nat.* (Edn 10) **1**: 534.

Type locality: North America.

A specimen of this American species was taken on Skokholm Island, Pembrokeshire at the end of July 1948 (South, 1961, **2**: 72, pl.35, fig.6).

This species is similar to *U. pulchella* (Linnaeus) but has the ground colour of the hindwing pink.

PARASEMIA Hübner

Parasemia Hübner, [1820], *Verz.bekannt.Schmett.*: 181.
Nemeophila Stephens, 1828, *Ill.Br.Ent.* (Haust.) **2**: 72.

This genus contains only two species, one European and one Chinese.

PARASEMIA PLANTAGINIS (Linnaeus)
The Wood Tiger

Phalaena (Bombyx) plantaginis Linnaeus, 1758, *Syst.Nat.* (Edn 10) **1**: 501.

Type locality: [Sweden].

Description of imago (Pl.5, figs 29–33)

Wingspan 33–42mm. Antenna in male strongly bipectinate, in female ciliate. Thorax black; tegulae yellow. Forewing black; a broad yellow streak near dorsum from base almost to termen joined to a diagonal streak from tornus to costa at two-thirds meeting a similar streak from apex; a yellow costal patch at one-half. Hindwing in male orange-yellow; terminal shade black, interrupted; two black streaks from base and a black discal spot; in female terminal shade narrower; basal streaks very broad, sometimes uniting to cover almost half of wing. Abdomen black; lateral stripe in male yellow, in female red.

In northern England and Scotland males with the yellow markings replaced by white occur regularly (f. *hospita* [Denis & Schiffermüller], fig.30); in ab. *matronalis* Freyer (fig.31) the markings are cream with the hindwing almost completely black. In the female ab. *rufa* Tutt (fig.33) the forewing markings are white and the hindwing dusky rufous. Many other forms are known with great variation in the extent and colour of the markings which range from yellow or buff to red.

Subsp. *insularum* Seitz

Parasemia plantaginis insularum Seitz, 1910, *Gross-Schmett. Erde* **2**: 81, pl.16F.

Male hindwing ground colour darker; markings of cell and submarginal spots more extensive.

Life history

Ovum. Hemispherical, smooth, glossy and yellowish. Laid in small batches on bell-heather (*Erica cinerea*) or foliage of various herbaceous plants.

Larva. Full-fed *c.*35mm long. Head black. Body blackish, paler ventrally; verrucae greyish black, bearing tufts of blackish setae, except on abdominal segments 1–3 where the verrucae and setae are reddish or orange; the setae of the anal segments are greyish and much longer than the rest.

Hatches from July onwards and is polyphagous. Feeds slowly until the autumn when it hibernates half-grown.

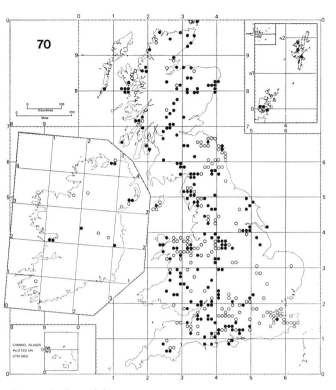

Parasemia plantaginis

Resumes feeding in the early spring and is full-fed by the end of April.

Pupa. Blackish brown. In a cocoon of silk mixed with larval hairs, spun in heather or amongst leaves of the larval foodplant.

Imago. Univoltine. Occurs from June until August. The male flies in sunshine, careering low over the ground; the female on hot, sunny afternoons and after dark. In captivity a second generation can be reared in the late autumn.

Distribution (Map 70)

Formerly widespread in open woodland and on heaths, moors and downs, but recently it has disappeared from many localities, especially in southern England. It occurs more freely from the Midlands and north Wales to the Shetlands and locally in Ireland. Subsp. *insularum* occurs in Scotland, Orkney and Shetland. Northern Palaearctic.

ARCTIA Schrank
Arctia Schrank, 1802, *Fauna boica* **2** (2): 152.
Eyprepia Ochsenheimer, 1810, *Schmett.Eur.* **3**: 299.
Zoote Hübner, [1820], *Verz.bekannt.Schmett.*: 181.
Epicallia Hübner, [1820], *ibid.*: 182.

A Holarctic genus containing about 12 species. Their brilliant colours are considered to be aposematic.

ARCTIA CAJA (Linnaeus)
The Garden Tiger
Phalaena (Bombyx) caja Linnaeus, 1758, *Syst.Nat.* (Edn 10) **1**: 500.
Type locality: [Sweden].

Description of imago (Pl.5, figs 34–41)
Wingspan 50–78mm. Antenna in male bipectinate, in female ciliate; shaft white, scape red. Thorax dark brown; patagia edged with red. Forewing dark brown; fasciae, streaks and spots anastomosing in disc, ochreous white, extremely variable in size and extent. Hindwing orange-red; about six large blue-black yellow-edged spots, variable in number and size. Underside of both wings white, variably suffused orange-red, with the upperside markings repeated in brown. Abdomen red, with black dorsal bars, these sometimes obsolescent on terminal segments; ventral surface brown with red intersegmental bands.

Exceedingly variable both in the form and extent of the markings of the forewing and the colour of the hindwing (figs 35–41). Cockayne (1949) and Smith (1958) discuss the variation of this species and illustrate many forms.

Life history
Ovum. Hemispherical, with the surface punctate and glossy, yellowish to green. Laid in large batches on the underside of leaves of herbaceous plants.

Larva. Full-fed *c.*60mm long. Head black. Body brownish; a spiracular series of white dots; verrucae on thoracic segments bearing long, black, white-tipped setae mixed with shorter, paler setae; on the abdominal segments dorsal and subdorsal verrucae white with long, pale setae; lateral verrucae dark brown with red-brown setae.

Very hairy and popularly known as the 'woolly bear'. It is polyphagous. The eggs hatch in late August and the larva goes into hibernation when still small. Starts to feed again in the early spring and is full-fed by the end of June, when it is often to be seen walking very rapidly over roads and footpaths. If kept warm in captivity, it may be induced to feed up rapidly without diapause and produce a moth in the autumn.

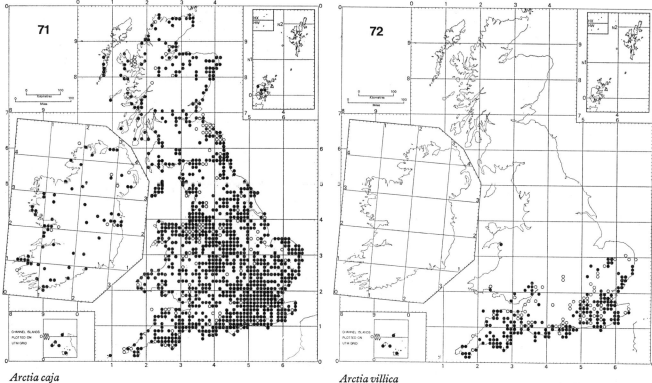

Arctia caja

Arctia villica

Pupa. Stout; deep bluish purple-brown. In a yellowish silken cocoon spun amongst detritus on the ground.

Imago. Univoltine. Flies in late July and August. Seldom seen by day but at night comes freely to light.

Distribution (Map 71)

Widespread and common in many habitats throughout the British Isles north to Orkney. Northern Palaearctic.

ARCTIA VILLICA BRITANNICA Oberthür
The Cream-spot Tiger

Phalaena (Bombyx) villica Linnaeus, 1758, *Syst.Nat.* (Edn 10) **1**: 501.

Arctia villica britannica Oberthür, 1911, *Études Lép.comp.* 5: 135.

Type locality: France; Cherbourg.

Description of imago (Pl.6, figs 1–4)

Wingspan 50–66mm. Head black; antenna in male bipectinate, in female simple. Thorax black; tegulae white. Forewing black; a basal blotch, two large antemedian spots, two smaller postmedian spots, two large subterminal spots and one or more subapical spots, creamy white. Hindwing orange-yellow; large discal spot, and several other spots of variable size, usually four in male, five in female, black; terminal shade black, incomplete and broken into spots. Underside of both wings similar to upperside but costa suffused red. Abdomen in male orange-yellow, terminally red, in female almost completely red; black dorsal and lateral spots frequently present on each segment in both sexes.

Variation is considerable (figs 2,3). In some specimens the pale markings of the forewing are reduced or totally absent, in others they are enlarged and coalesce to form blotches. The markings of the hindwing are equally variable; extreme forms may be unicolorous yellow or almost uniform black.

Life history

Ovum. Globular, glossy white. Laid in neatly arranged batches on foliage of various herbaceous plants.

Larva. Full-fed *c.*50mm long. Head bright red. Body glossy black; verrucae bearing large spreading clusters of reddish brown setae, those on the posterior segments being longer and directed slightly backwards; legs and prolegs bright red.

Polyphagous. Hatches in early July and feeds slowly until the autumn when it goes into hibernation. Resumes feeding in the early spring and is often full-fed by the end of March.

Pupa. Black. In a slight cocoon spun amongst leaves or detritus on or close to the ground.

Imago. Univoltine. Occurs from late May to early July. Only occasionally encountered flying by day, but after dusk it is active and comes readily to light.

Distribution (Map 72)

Locally common, principally in coastal districts, in the southern counties of England and Wales, though occasional specimens have been recorded from north Wales. Abroad subsp. *britannica* is found in northern France and the nominate subspecies throughout central and southern Europe.

DIACRISIA Hübner

Diacrisia Hübner, [1819], *Verz.bekannt.Schmett.*: 169.
Euthemonia Stephens, 1828, *Ill.Br.Ent.* (Haust.) **2**: 68.

A large Old World genus some members of which are of considerable economic importance in the tropics, being pests of such crops as cocoa.

DIACRISIA SANNIO (Linnaeus)
The Clouded Buff

Phalaena (*Bombyx*) *sannio* Linnaeus, 1758, *Syst.Nat.* (Edn 10) **1**: 506.
Phalaena (*Bombyx*) *russula* Linnaeus, 1758, *ibid.*: 510.
Type locality: Europe.

Description of imago (Pl.6, figs 5–8)

Wingspan 35–50mm, female much smaller than male. Sexually dimorphic, originally described by Linnaeus as two species. *Male.* Head bright yellow; antenna bipectinate. Forewing bright yellow; dorsum suffused with pinkish red; discal spot double-lobed, crimson with grey centre; apical half of costa and cilia pinkish red. Hindwing creamy white; subterminal shade and discal lunule dark grey; termen creamy white; cilia pinkish red. Abdomen creamy white, dorsal spot on each segment grey. *Female.* Head and thorax deep orange; antenna orange, dentate. Forewing deep orange; discal spot and costa darker; veins reddish. Hindwing orange, subterminal shade and extensive basal patch enclosing discal lunule dark grey.

In the male, variation is confined to the intensity of the grey suffusion on the hindwing; this is very extensive in ab. *maerens* Strand (fig.6) and completely absent in ab. *immarginata* Niepelt (see Watson, 1975a).

Life history

Ovum. Hemispherical, lightly punctate, glossy, whitish. Laid in small batches on foliage of bell-heather (*Erica cinerea*) and various herbaceous plants.

Larva. Full-fed *c.*40mm long. Head dark grey. Body reddish brown; broad dorsal stripe whitish, enclosing a yellow spot on each segment; subdorsal and spiracular stripes dark brown; peritreme of spiracles black; a subspiracular row of white spots; verrucae bearing thick tufts of short brown hairs.

Polyphagous. Hatches in late July and hibernates when small. Resumes feeding in early spring and is full-fed by the end of May or early June.

Pupa. Brown, streaked with grey. In rather a flimsy cocoon spun amongst detritus or herbage close to the ground.

Imago. Univoltine. Appears in late June and early July.

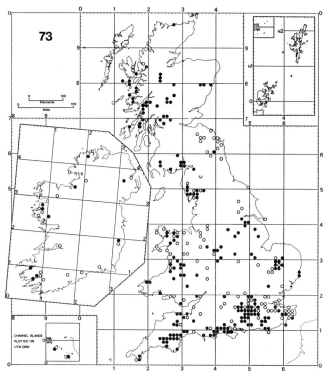

Diacrisia sannio

The male flies freely in sunshine over the heather, the female is less often seen by day but becomes active soon after dark and again shortly before dawn.

Distribution (Map 73)

Locally common on heaths and moors or, less frequently, downland. It is widely distributed over the British Isles, but is absent from the Isle of Man, Orkney and Shetland. In Europe, where it is mainly bivoltine, it is widespread and its range extends into Asia Minor.

SPILOSOMA Curtis

Spilosoma Curtis, 1825, *Br.Ent.* **2** (23): 92.

Primarily an Old World tropical genus with a few representatives in America and the temperate zones.

SPILOSOMA LUBRICIPEDA (Linnaeus)
The White Ermine

Phalaena (Bombyx) lubricipeda Linnaeus, 1758, *Syst.Nat.* (Edn 10) **1**: 505.
Bombyx menthastri [Denis & Schiffermüller], 1775, *Schmett. Wien.*: 54.

Type locality: [Sweden].

Description of imago (Pl.6, figs 10–14)

Wingspan 34–48mm. Head creamy white; antenna in male bipectinate, in female setose-dentate; labial palpus dark brown. Forewing pale ochreous white with black markings as follows: two dots on costa near base, two subbasal dots in centre of wing with another between them and dorsum, a dot antemedially on costa and another below it, an irregular median row of dots from dorsum almost to costa, a curved postmedian line of dots, a discontinuous subterminal line of dots, and a few terminal dots. Hindwing pale ochreous white; discal and one or more subterminal spots blackish grey. Abdomen yellow, paler ventrally, anal tuft white; black dorsal, lateral and subventral spots on each segment.

This species varies considerably in the extent of the black dotting of the forewing, from almost immaculate (fig.11) to the other extreme with the dots united into striations in ab. *godarti* Oberthür (fig.14), and in the ground colour of the forewing which can vary from white to dark fuscous-brown in ab. *brunnea* Oberthür (fig.12). In Ireland many specimens have the forewings and sometimes the hindwings buff (fig.13), being not unlike *S. lutea* (Hufnagel). Individuals with ochreous brown wings are known from Scotland.

Similar species. *S. urticae* (Esper) which has the wings pure white, with very few black dots on the forewing and the basal segment of the labial palpus yellow; *S. lutea* from which Irish specimens of *S. lubricipeda* differ in the pattern of the black dots; *Diaphora mendica* (Clerck) which has the abdomen white.

Life history

Ovum. Globular, smooth and glossy, yellowish white. Laid in large batches on the underside of leaves of various species of herbaceous plants.

Larva. Full-fed *c.*50mm long. Head small, black. Body dark brown; dorsal stripe red or orange; verrucae with tufts of long, brown setae; legs black.

Polyphagous on herbaceous plants without showing

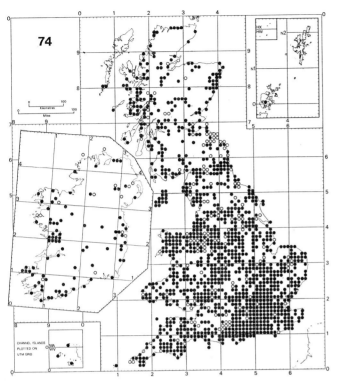

Spilosoma lubricipeda

any particular preference. Hatches in late July and feeds throughout August and September. Often seen walking rapidly over the ground; the scientific name is derived from this characteristic behaviour.

Pupa. Stout; shining purplish brown; cremaster with short blunt bristles. In a substantial cocoon of silk and larval hairs spun amongst debris on the ground or in a convenient crevice; overwinters.

Imago. Univoltine. Flies from late May till July; occasionally there is a small second brood in the early autumn. Comes freely to light, usually arriving about midnight.

Distribution (Map 74)

Widespread and common in many habitats throughout the British Isles except Shetland. Palaearctic.

SPILOSOMA LUTEA (Hufnagel)
The Buff Ermine

Phalaena lubricipeda lutea Hufnagel, 1766, *Berlin.Mag.* **2**: 412.

Spilosoma lubricipeda sensu auctt.

Type locality: Germany; Berlin.

Description of imago (Pl.6, figs 15–20)

Wingspan 34–42mm. Head yellowish buff; antenna in male bipectinate, in female setose-dentate. Forewing yellowish buff, paler in female, with variable black markings as follows: usually a large spot near base of costa, discal spot and line of spots from costa near apex to midpoint on dorsum repeated more distinctly on underside; sometimes also a subbasal line of dots and a curved postmedian line of elongate spots. Hindwing pale yellowish buff, pale yellow in female; a median spot near dorsum and sometimes a few tornal spots pale grey. Abdomen yellowish buff with black dorsal, lateral and subventral spots on each segment.

Very considerable variation occurs in the extent of the black markings: some specimens are devoid of all spots, others have prominent black striae, ab. *intermedia* Standfuss (fig.20), and in the extreme ab. *zatima* Stoll (figs 17,18) the wings are almost wholly suffused black with the veins yellow.

Similar species. Irish forms of *S. lubricipeda* (Linnaeus) *q.v.*

Life history

Ovum. Hemispherical, smooth, glossy, whitish green with darker annuli. Laid in large batches on leaves of various herbaceous plants such as dock (*Rumex* spp.) or dandelion (*Taraxacum officinale*).

Larva. Full-grown *c.*50mm long. Head glossy brown. Body brownish; dorsal stripe reddish, distinct; spiracular stripe yellowish or light grey; verrucae large, bearing thick tufts of long, blackish brown setae. In the first instar it is yellowish and in the second, greyish.

Polyphagous on a wide range of herbaceous plants. Hatches in July or August and completes its growth in September or October.

Pupa. Reddish brown, glossy. In a dingy grey cocoon of silk mixed with larval hairs spun amongst leaves or on the ground; overwinters.

Imago. Univoltine. Flies from late May to July. Comes freely to light.

Distribution (Map 75)

Widespread and common in gardens and most other habitats throughout England and Wales; in Scotland it is mainly western, but is absent from Orkney, Shetland and the Outer Hebrides; in Ireland it occurs widely on the coast and more locally inland. Palaearctic.

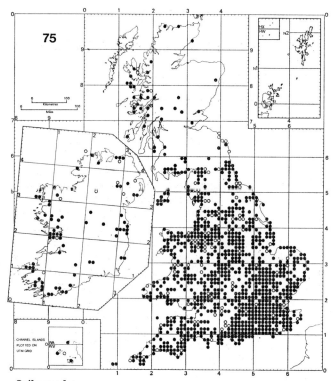

Spilosoma lutea

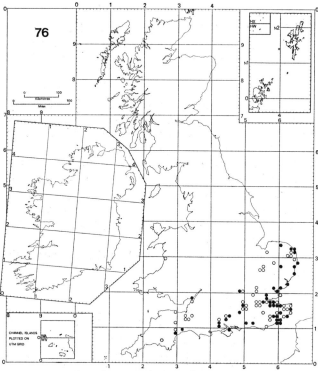

Spilosoma urticae

SPILOSOMA URTICAE (Esper)
The Water Ermine

Bombyx urticae Esper, 1789, *Schmett.* **3**: 20.
Phalaena (Bombyx) papyratia Marsham, 1791, *Trans.Linn. Soc.Lond.* **1**: 72.
Type locality: Europe.

Description of imago (Pl.6, fig.9)
Wingspan 38–46mm. Head white; antenna in male bipectinate, in female simple; labial palpus dark brown, basal segment yellow. Wings white; forewing with two small black discal spots and sometimes a short row of black apical spots. Abdomen yellow, paler ventrally, first and anal segments white; dorsal, lateral and subventral spots on each segment black.

Similar species. S. lubricipeda (Linnaeus) *q.v.*; *Diaphora mendica* (Clerck), from which it differs in having the abdomen yellow.

Life history
Ovum. Hemispherical, smooth, shining whitish green. Laid in batches on various herbaceous plants such as mint (*Mentha* spp.), yellow loosestrife (*Lysimachia vulgaris*), water dock (*Rumex hydrolapathum*) or lousewort (*Pedicularis sylvatica*).

Larva. Full-fed *c.*50mm long. Head black and glossy, mouth-parts yellow. Body purplish brown; verrucae large, black, bearing tufts of long, deep brown setae; peritreme of spiracles white; legs brown, prolegs yellow.

Hatches in midsummer and feeds until the end of August.

Pupa. Dark reddish brown. In a light cocoon of silk mixed with larval hairs, spun on the ground or in a folded leaf of the foodplant; overwinters.

Imago. Univoltine. Emerges in June. Comes freely to light.

Distribution (Map 76)
Very local in fens, marshes and water-meadows. It is found mostly in the counties south and east of a line from the Wash to the Severn. Palaearctic.

DIAPHORA Stephens

Diaphora Stephens, 1827, *Ill.Br.Ent.* (Haust.) **2**: 77.
Cycnia sensu auctt.

A monotypic genus.

DIAPHORA MENDICA (Clerck)
The Muslin Moth

[*Phalaena*] *mendica* Clerck, 1759, *Icones Insect.rar.* **1**: pl.3, fig.5.

Type locality: [Sweden].

Description of imago (Pl.6, figs 21–24)
Wingspan 30–43mm. Head grey or white; antenna in male bipectinate, in female ciliate. Forewing in male brown-grey to blackish grey except in Ireland where it is white or creamy yellow (f. *rustica* Hübner (figs 22,23)), in female pale creamy white; two subbasal dots and a few dots representing postmedian line black. Hindwing concolorous with forewing; small discal spot and subterminal series of dots, larger in female, black. Abdomen concolorous with the wings, with dorsal, lateral and subventral dark spots on each segment; these markings are obscure in the dark form of the male.

Variation occurs in the ground colour of the male, which is brownish grey in southern England and almost black in the north. In Ireland the rare ab. *venosa* Adkins has the veins prominently darker. In the female the spotting of the wings is sometimes replaced by heavy dark grey striations.

Life history
Ovum. Hemispherical, slightly rough and pitted, pale greenish white. Laid in batches, often in double rows, on leaves of various herbaceous plants such as dock (*Rumex* spp.), chickweed (*Stellaria* spp.), plantain (*Plantago* spp.) or dandelion (*Taraxacum officinale*).

Larva. Full-grown *c.*40mm long. Head glossy, pale chestnut-brown. Body greyish brown; dorsal stripe paler; spiracles with black peritreme; subspiracular series of whitish dots; verrucae pale brown, bearing tufts of yellow-brown setae. In the first instar it is whitish and in the second grey with black dots and black setae.

Hatches in late June and is full-fed by the end of the summer. In captivity the imagines sometimes emerge in September and give rise to a second generation in the autumn, but this seems hardly ever to occur under natural conditions.

Pupa. Black, glossy and finely punctate. In a close-fitting cocoon of silk intermixed with larval hairs spun amongst foliage or on the ground; overwinters.

Imago. Univoltine. Flies in May and early June. The male

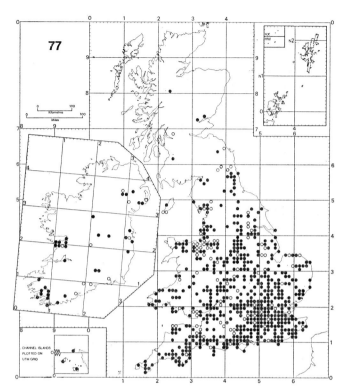

Diaphora mendica

comes freely to light, but it is doubtful if the female is ever so attracted, being more frequently seen flying in sunshine.

Distribution (Map 77)
Widespread especially in open woodland in England and Wales as far north as the Lake District and County Durham, thence extending much more locally to central Scotland. In Ireland it occurs commonly in the south-west and the Burren, Co. Clare and locally elsewhere. Abroad it is frequent in northern and central Europe as far east as Asia Minor.

PHRAGMATOBIA Stephens
Phragmatobia Stephens, 1828, *Ill.Br.Ent.* (Haust.) **2**: 73.

A genus of some 50 species represented in most regions except Australasia.

PHRAGMATOBIA FULIGINOSA (Linnaeus)
The Ruby Tiger

Phalaena (Noctua) fuliginosa Linnaeus, 1758, *Syst.Nat.* (Edn 10) **1**: 509.
Type locality: [Europe].

Description of imago (Pl.6, figs 25,26)
Wingspan 28–38mm. Head reddish brown; antenna greyish white, in male finely setose-ciliate, in female simple. Thorax reddish brown. Forewing bright reddish brown, thinly scaled in disc; discal marks black. Hindwing rose-pink, suffused with grey on costal half; discal spots and terminal spots, often united to form a band, dark grey; cilia deep pink. Abdomen red, dorsal stripe and lateral spots black.

In northern England the forewing is darker brown and the hindwing is more heavily suffused with grey, approaching the Scottish subsp. *borealis* (Staudinger).

Subsp. *borealis* (Staudinger)
Spilosoma fuliginosa var. *borealis* Staudinger, 1871, *in* Staudinger & Wocke, *Cat.Lepid.eur.Faunengeb.*: 59.
Forewing deep brown and the hindwing grey, except for a pink basal suffusion (fig.26).

Life history
Ovum. Hemispherical, white. Laid in rather large batches on leaves of various herbaceous plants such as dock (*Rumex* spp.), dandelion (*Taraxacum officinale*) and goldenrod (*Solidago virgaurea*).
Larva. Full-grown *c*.35mm long. Head glossy black. Body blackish; dorsal stripe and a subdorsal series of spots reddish; verrucae grey or black and very large, each bearing a star-like tuft of yellowish, reddish or blackish brown setae which tend to conceal the red markings. In early instars it is greyish or brownish and the red markings are more conspicuous.

Hatches in late May or June. Some feed up quickly and produce a small second brood of adults in September, but the majority feed more slowly and overwinter when full-fed; these reappear in the spring but do not feed again.
Pupa. Stout, black. In a closely spun cocoon of brown silk amongst foliage or ground debris; on moorland it is often placed amongst the twigs of heather. The cocoon of the first generation is spun in April; that of the smaller second generation in August.

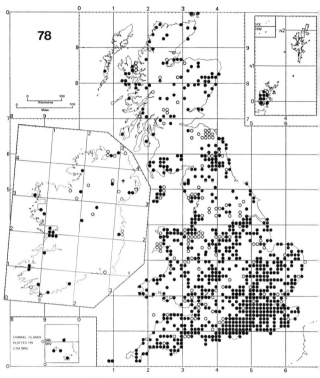

Phragmatobia fuliginosa

Imago. Mainly univoltine. Flies in May and June, but with a small second generation in September, rare in subsp. *borealis*. Both sexes come freely to light. Has occasionally been observed flying in sunshine or running rapidly over herbage.

Distribution (Map 78)
Widespread and locally common in most habitats throughout the British Isles as far north as the Orkneys. Holarctic.

PYRRHARCTIA Packard
Pyrrharctia Packard, 1864, *Proc.ent.Soc.Philad.* **3**: 120.

PYRRHARCTIA ISABELLA (Smith)
Phalaena isabella Smith, in Smith & Abbot, 1797, *Nat.Hist. rarer lepid.Ins.Georgia* **1**: 131.
Type locality: U.S.A.; Georgia.

Larvae of this species are recorded as having been found at Carnforth, Lancashire for several years prior and up to 1906 by H. Murray. An imago was reared from some of those larvae by Chapman (1906). In recent years larvae have been imported with American oak by an Edinburgh firm of coopers and three moths reared (E. C. Pelham-Clinton pers.comm.) (*cf. Ecpantheria deflorata* (Fabricius), p. 110).

HALISIDOTA Hübner
Halisidota Hübner, [1819], *Verz.bekannt.Schmett.*: 170.

HALISIDOTA MOESCHLERI Rothschild
Halisidota moeschleri Rothschild, 1909, *Novit.zool.* **16**: 280.
Type locality: Jamaica.

A single male of this West Indian species was taken in an m.v. light-trap at Charlton Kings, Cheltenham, Gloucestershire on 19 July 1961 (Ford, 1965).

EUPLAGIA Hübner
Euplagia Hübner, [1820], *Verz.bekannt.Schmett.*: 180.

A monotypic genus.

EUPLAGIA QUADRIPUNCTARIA (Poda)
The Jersey Tiger
Phalaena (Noctua) quadripunctaria Poda, 1761, *Ins.Mus. Graec.*: 89.
Phalaena (Noctua) hera Linnaeus, 1767, *Syst.Nat.* (Edn 12) **1** (2): 835.
Type locality: [Greece].

Description of imago (Pl.6, figs 27,28)
Wingspan 52–65mm. Head white, marked with yellow and black; antenna simple. Thorax whitish, dorsally black; tegulae and patagia edged black. Forewing black with creamy white markings as follows: a dorsal streak from base almost to tornus; a curved basal streak extending to at least one-fifth; a short, fine, oblique subbasal line from costa to middle of wing; an oblique antemedian fascia not reaching dorsum; a short narrow oblique median mark; an oblique postmedian fascia joining a subterminal fascia in centre of wing, reaching termen near tornus where it is orange-tinged and black-spotted; fringe chequered black and yellow. Hindwing red, with large blue-black discal, subapical and pretornal spots. Abdomen orange-yellow, dorsal, lateral and ventral spots black.

In ab. *lutescens* Staudinger the ground colour of the hindwing is yellow; intermediate forms between this and the typical form are known. Ab. *lutescens* occurs commonly but other variation is rare, though extreme forms are known in which the forewing is wholly black and others in which the white marks are extensively confluent.

Life history
Ovum. Hemispherical, pale yellow when laid, becoming deep violet before hatching. Laid in batches on leaves of herbaceous plants such as dandelion (*Taraxacum officinale*), plantain (*Plantago* spp.), dead-nettle (*Lamium* spp.) or ground-ivy (*Glechoma hederacea*), hatching in 15 or 16 days.
Larva. Full-fed *c.*50mm long. Head glossy black. Body deep brown; dorsal stripe broad, orange-yellow, irregular; a spiracular series of creamy white spots; verrucae brown, each bearing mixed brown and grey setae; spiracles black with white peritreme; ventral surface greyish.

Goes into hibernation soon after hatching; resumes feeding nocturnally early in the spring and is full-fed in May. In captivity it will feed through the winter and has sometimes been induced to complete its growth before the end of the year.

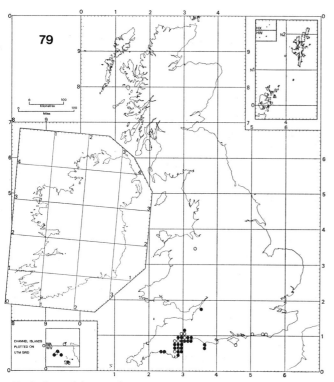
Euplagia quadripunctaria

Pupa. Stout; deep brown. In a flimsy, web-like cocoon spun amongst debris or moss.

Imago. Univoltine. Occurs from the end of July till early September. Flies in sunshine and can readily be disturbed from bushes in dull weather. At night it comes freely to ultraviolet light. In parts of the Continent, such as the island of Rhodes, it aestivates gregariously in huge numbers, but this habit has not been observed in Britain.

Distribution (Map 79)
Common in the Channel Islands and along the south coast of Devon from Seaton to Torquay, where it has been established since about 1880, and frequents gardens and parks. Individuals recorded from other localities along the south coast are probably migrants from the Channel Islands or the Continent; there was a record from north Wales in 1859. Abroad it is widespread in central and southern Europe and Asia Minor to Iran.

CALLIMORPHA Latreille
Callimorpha Latreille, 1809, *Gen.Crust.Ins.* **4**: 220.
Panaxia Tams, 1939, *Entomologist* **72**: 73.

A small genus represented in Europe and temperate Asia eastwards to Japan.

CALLIMORPHA DOMINULA (Linnaeus)
The Scarlet Tiger
Phalaena (*Noctua*) *dominula* Linnaeus, 1758, *Syst.Nat.* (Edn 10) **1**: 509.
Type locality: [Europe].

Description of imago (Pl.6, figs 31–34)
Wingspan 52–58mm. Head greenish black; antenna in male setose-ciliate, in female simple. Thorax greenish black with two orange streaks. Forewing metallic greenish black; a small orange subbasal streak on dorsum; two obliquely placed antemedian spots, costal orange, inner creamy white; a median orange spot near costa; a large, creamy white postmedian blotch extending from near costa to middle of wing; two or three subapical dots below costa, two subterminal spots and a large and a small pretornal spot, all creamy white. Hindwing scarlet; a prominent median costal blotch often extending to middle of wing, and a broad, interrupted subterminal shade, meeting termen medially, metallic blue-black; cilia dark greyish brown. Markings repeated on underside of both wings, but the pale markings of forewing are more extensively orange. Abdomen scarlet, dorsal stripe and ventral surface black.

Variation in the extent of both fore- and hindwing markings is considerable. In f. *bimacula* Cockayne (fig.32) the forewing is almost entirely black with only the two antemedian spots remaining, in f. *ocellata* Kettlewell (fig.33) the forewing markings are very extensively confluent and the hindwing markings much reduced, whilst in f. *rossica* Kolenati the hindwing is yellow. Kettlewell (1944) gives a detailed account of variation in this species.

Life history
Ovum. Hemispherical, glossy cream. Laid in batches on many species of herbaceous plants such as nettle (*Urtica* spp.), dock (*Rumex* spp.) and comfrey (*Symphytum* spp.); according to Stokoe & Stovin (1958), they are scattered at random and not attached to the foodplant.

Larva. Full-fed c.50mm long. Head glossy black. Body purplish black; discontinuous yellow dorsal and subdorsal stripes; verrucae black, bearing mixed grey and black setae.

Hatches in late July or August and feeds for a while before hibernation. Resumes feeding in April and is full-grown in late May. It is polyphagous on a wide range of

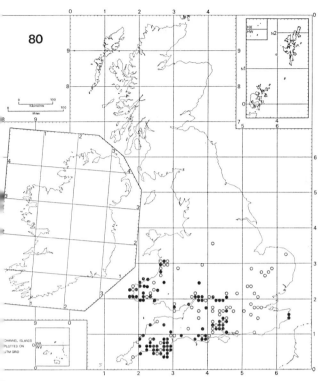

Callimorpha dominula

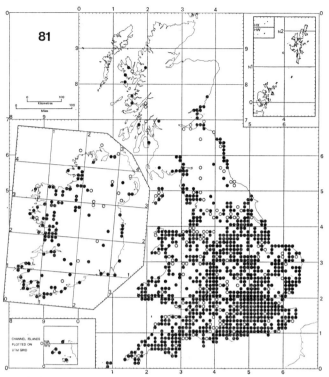

Tyria jacobaeae

herbaceous plants and is even found on trees such as blackthorn (*Prunus spinosa*), goat willow (*Salix caprea*) or oak (*Quercus* spp.), feeding fully exposed, often in sunshine.

Pupa. Deep red. In a light, silken cocoon spun amongst debris on the ground.

Imago. Univoltine. Occurs in June and July. Flies freely in sunshine.

Distribution (Map 80)

Very local, occurring on coasts and inland along river banks and in fens and water meadows. Except for colonies on the east coast of Kent, it no longer occurs in the eastern counties and is now confined to the southern counties from Hampshire to Cornwall, Berkshire, Gloucestershire and south and west Wales. Widespread throughout Europe.

TYRIA Hübner

Tyria Hübner, [1819], *Verz.bekannt.Schmett.*: 166.

A monotypic Eurasiatic genus.

TYRIA JACOBAEAE (Linnaeus)
The Cinnabar

Phalaena (*Noctua*) *jacobaeae* Linnaeus, 1758, *Syst.Nat.* (Edn 10) **1**: 511.

Type locality: [Sweden].

Description of imago (Pl.6, figs 29,30)

Wingspan 35–45mm. Antenna finely setose ciliate. Forewing brownish grey; a scarlet subcostal streak from base almost to apex, apically broader; subapical and tornal scarlet spots on termen; dorsum scarlet from base to just beyond one-half. Hindwing scarlet; costa and cilia brownish grey. Underside of both wings similar to upperside. Body black.

Variation is usually confined to the size of the scarlet markings which are sometimes confluent. A remarkable series of all-red forms has been reared in captivity (Watson, 1975b). An extremely rare form with the markings yellow (ab. *flavescens* Thierry-Mieg) is known, as is an all-black form (ab. *negrana* Cabeau).

Life history
Ovum. Hemispherical, bright, glistening yellow when laid, becoming greyish just prior to hatching. Laid in batches of 30–40 on the underside of the lower leaves of common ragwort (*Senecio jacobaea*) or occasionally groundsel (*S. vulgaris*); exceptionally as many as 150 may be found in a single cluster.

Larva. Full-fed *c.*25mm long. Head black. Body bright orange-yellow with a black band on each segment bearing short, blackish setae. In the first instar it is greyish green.

Feeds in July and August on the upper leaves of the foodplant, sometimes occurring in great numbers and devouring all the foliage.

Pupa. Dark brown ringed with reddish. In a slight, loose cocoon spun on or below the surface of the soil or amongst moss or grass-roots; overwinters.

Imago. Univoltine. Flies from late May till mid-July. Comes freely to light, usually late at night, and is readily disturbed by day.

Distribution (Map 81)
Inhabits mainly well-drained pasture, heathland and sand-dunes where the foodplant grows plentifully. The population ecology is described in detail by Dempster (1971). Widespread and common throughout southern and midland England and Wales, becoming local and mainly coastal in the north and in Scotland; in Ireland it is common in coastal districts but local inland. Eurasiatic; moths have been released in North America and Australia in an attempt to control ragwort.

ECPANTHERIA Hübner
Ecpantheria Hübner, [1820], *Verz.bekannt.Schmett.*: 183.

ECPANTHERIA DEFLORATA (Fabricius)
Bombyx deflorata Fabricius, 1794, *Ent.syst.* **3** (2): 127.
Type locality: India.

A larva of this species, from which a moth was bred on 8 August 1969, was imported with American oak by an Edinburgh firm of coopers (E. C. Pelham-Clinton pers. comm.).

References

Barrett, W., 1895. *The Lepidoptera of the British Islands*, **2**: 369 pp. London.

Buckler, W., 1889. *The larvae of the British butterflies and moths*, **3**: 79 pp., 18 col. pls. London.

Chapman, T. A., 1906. A new British Arctiid. *Entomologist's mon. Mag.* **42**: 100–101.

Cockayne, E. A., 1944. The larvae of the British Lithosiinae. *Proc. Trans. S. Lond. ent. nat. Hist. Soc.* **1942–43**: 75–78, 1 pl.

———, 1949. *Arctia caja* L.: its variation and genetics. *Ibid.* **1947–48**: 155–191, 2 col. pls.

Dempster, J. P., 1971. The population ecology of the Cinnabar moth, *Tyria jacobaeae* L. (Lepidoptera, Arctiidae). *Oecologia (Berl.)* **7**: 26–27.

Ford, M. L., 1965. A specimen of *Halisidota moeschleri* Rothschild collected in England. *Entomologist's Gaz.* **16**: 113.

Goater, B., 1975. *The butterflies and moths of Hampshire and the Isle of Wight*, xiv, 439 pp. Faringdon.

Kettlewell, H. B. D., 1944. A survey of the insect *Panaxia* (64) (*Callimorpha*) *dominula*, L. *Proc. Trans. S. Lond. ent. nat. Hist. Soc.* **1942–43**: 1–49, 5 pls.

Marsh, N. & Rothschild, M., 1974. Aposematic and cryptic Lepidoptera tested on the mouse. *J. Zool., Lond.* **174**: 89–122.

Pierce, F. N. & Beirne, B. P., 1941. *The genitalia of the British Rhopalocera and the larger moths*, 66 pp., 21 pls. Oundle.

Rothschild, M., 1972. Secondary plant substances and warning colouration in insects. *Symp. R. ent. Soc. Lond.* **6**: 59–83.

Smith, S. G., 1958. New aberrations of *Arctia caja* (Linn.) (Lep. Arctiidae). *Entomologist's Gaz.* **9**: 3–6, 3 pls.

South, R., 1961. *The moths of the British Isles* (Edn 4) **2**: 379 pp., 141 pls. London.

Stidston, S. T., 1952. *A list of the Lepidoptera of Devon Introduction and Part 1*, 74 pp. Torquay.

Stokoe, W. J. & Stovin, G. H. T., 1958. *The caterpillars of British moths* **1**: 408 pp., 90 pls. London.

Watson, R. W., 1975a. Aberrations of *Diacrisia sannio* Hübner (Lep.: Arctiidae). *Entomologist's Rec. J. Var.* **87**: 258, 1 col. pl.

———, 1975b. New aberrations of *Tyria jacobaeae* L. (Lep.: Arctiidae). *Ibid.* **87**: 267, 1 col. pl.

CTENUCHIDAE
J. Heath

A family of about 3,000 species, mostly tropical, but with a few species in North America and Europe. Many of the species are mimics of various wasps and lycid beetles.

Five species in four genera have been recorded as introductions to Britain but none is resident or immigrant.

SYNTOMIS Ochsenheimer
Syntomis Ochsenheimer, 1808, *Schmett.Eur.* **2**: 104.

SYNTOMIS PHEGEA (Linnaeus)
Sphinx phegea Linnaeus, 1758, *Syst.Nat.* (Edn 10) **1**: 494.
Type locality: Germany.

A specimen of this European species was taken on 24 June 1872 near Dover, flying in sunshine, by the dealer, T. Batchelor (Barrett, 1895). It is figured in South (1961).

DYSAUXES Hübner
Dysauxes Hübner, [1819], *Verz.bekannt.Schmett.*: 171.

DYSAUXES ANCILLA (Linnaeus)
Phalaena (*Noctua*) *ancilla* Linnaeus, 1767, *Syst.Nat.* (Edn 12) **1**: 835.
Type locality: Germany.

A specimen, said to have been captured near Worthing, Sussex, was exhibited at the Entomological Society in 1867 by E. Newman. No other British specimen is known (Barrett, 1895). It is figured in South (1961).

EUCHROMIA Hübner
Euchromia Hübner, [1819], *Verz.bekannt.Schmett.*: 121.

EUCHROMIA LETHE (Fabricius)
Zygaena lethe Fabricius, 1775, *Syst.Ent.*: 553.
Type locality: Africa.

This species is imported, from time to time, among consignments of bananas from Africa. It is figured in Laithwaite *et al.* (1975).

CERAMIDIA Butler
Ceramidia Butler, 1876, *Trans.Linn.Soc.Lond.* (Zool.) **12**: 412.

CERAMIDIA VIRIDIS (Druce)
Antichloris viridis Druce, 1884, *in* Godman & Salvin, *Biol. centr.-amer.*, Insecta.Lepid.-Het. **1**: 67, pl.7, fig.25.
Ceramidia musicola Cockerell, 1910, *Can.Ent.* **42**: 60.
Type locality: Panama.

Occurrences of this West Indian species in widely scattered localities in Britain are the result of accidental introductions, probably in consignments of bananas (South, 1961).

CERAMIDIA CACA (Hübner)
Antichloris caca Hübner, 1818, *Zuträge Samml.exot.Schmett.* **1**: 24.
Type locality: Brazil.

Although listed in Kloet & Hincks (1972) no other reference to the occurrence of this species in Britain has been traced.

References

Barrett, C. G., 1895. *The Lepidoptera of the British Islands*, **2**: 369 pp. London.

Kloet, G. S. & Hincks, W. D., 1972. A check list of British insects (Edn 2). *Handbk Ident. Br. Insects* **11** (2): viii, 153 pp.

Laithwaite, E., Watson, A. & Whalley, P. E. S., 1975. *The dictionary of butterflies and moths in colour*, xlvi, 296 pp., 405 col. figs. London.

South R., 1961. *The moths of the British Isles* (Edn 4), **2**: 379 pp., 141 pls. London.

NOLIDAE
R. J. Revell

A small family of world-wide distribution, represented in Great Britain and Ireland by two genera containing five species.

Imago (Pl.A, fig.13). Antenna with first segment dilate and a small tuft of scales at its base; haustellum present; labial palpus three-segmented, basal segment short, second long and usually porrect, third small and often very slender; maxillary palpus minute, one-segmented with a tuft of setae on inner surface distally. Forewing with characteristic tufts of raised scales at the base, middle and end of cell; veins 7 to 10 (R_5 to R_2) stalked from upper angle of cell in the British species; antemedian, postmedian and subterminal fasciae usually well marked, but stigmata absent. Hindwing usually plain with a discal spot or crescent.

Larva. Head small, partially retractable. Each body segment except the last with four pairs of verrucae – subdorsal, supraspiracular, subspiracular and subventral, the subdorsals being medially united on the prothorax; setae of varying lengths and confined to tufts on the verrucae; prolegs present on abdominal segments 4–6 and 10 only.

Pupa. Cylindrical. In the British species enclosed tightly in a tough silken cocoon which incorporates particles of the plant material to which it is attached.

Key to species (imagines) of the Nolidae

1 Forewing with antemedian fascia broad and conspicuous, enclosing a basal area distinctly darker than the rest of the forewing...................*Nola cucullatella* (p. 117)
– Forewing with antemedian fascia and basal area not as above...2
2(1) Forewing with median fascia (just proximal to the postmedian) with well-defined basally directed dentations in the dorsal half of its course..........................
....*N. confusalis* (p. 118)
– Forewing with median fascia obscure, or, if present, not prominently dentate...3
3(2) Forewing with antemedian fascia sinuate; subterminal fascia consisting of a series of conspicuous strigulae..........
....*Meganola strigula* (p. 113)
– Forewing with antemedian fascia angled, straight or obtusely curved once, but not sinuate; subterminal fascia without conspicuous strigulae...4
4(3) Forewing with antemedian fascia angled at second scale-tuft; postmedian fascia narrow, but well defined and darker than the shading adjacent to it proximally (shading not always present); ♂ antenna ciliate............
....*Nola aerugula* (p. 119)
– Forewing with antemedian fascia ill-defined or evenly curved; postmedian fascia not distinguishable by depth of colour from the shading proximal to it; ♂ antenna bipectinate.................*Meganola albula* (p. 115)

MEGANOLA Dyar

Meganola Dyar, 1898, *Jl N.Y.ent.Soc.* **6**: 43.

Only two species of this genus occur in Great Britain, neither of which occurs in Ireland. Forewing with veins 9 (R_3) and 10 (R_2) present in the British species, figure 8. Valva of male genitalia not bilobed, but bearing a ventral process (clasper) of varying shape; uncus present; aedeagus various, figure 9a, p. 114.

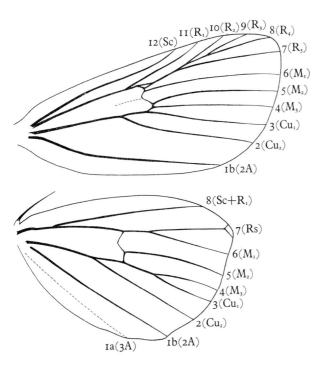

Figure 8 *Meganola albula* ([Denis & Schiffermüller]), wing venation

MEGANOLA STRIGULA ([Denis & Schiffermüller])
The Small Black Arches

Noctua strigula [Denis & Schiffermüller], 1775, *Schmett. Wien.*: 69.

Type locality: [Austria]; Vienna district.

Description of imago (Pl.6, figs 35,36)

Wingspan 18–24mm. Head with vertex white, lightly irrorate pale fuscous; frons white or buff, heavily irrorate pale and dark fuscous; antenna in male two-fifths to one-half length of costa, bipectinate almost to apex, the dorsal row of pectinations up to twice width of shaft in length, ventral row up to twice length of dorsal row, pectinations and inner surface of shaft ciliate, in female simple, minutely setose on inner surface; labial palpus short, one and one-half diameter of eye in length, white or buff, heavily irrorate pale brown and fuscous laterally. Forewing white, irrorate pale brown and fuscous, some specimens showing radial bands of ochreous suffusion, particularly between veins 1b (2A) and 2 (Cu_2); basal area with two fuscous spots, the more distal sometimes extended to form an indistinct fascia through first discal scale-tuft; antemedian fascia dark fuscous to black, minutely but distinctly dentate at second discal scale-tuft, thence sinuate to vein 1b (2A), and forming a usually well-defined lunule between this vein and dorsal margin; postmedian fascia dark fuscous to black, dentate prominently near costa and minutely dentate at veins, arcuate terminad for one-half its course, thence sinuate to dorsum; median fascia discernible in some specimens close to postmedian and approximating with its course; subterminal fascia edged distally with white, consisting mainly of dark strigulae at the veins, and forming a well-marked lunule between veins 1b (2A) and 2 (Cu_2); terminal fascia of neural dots; fringe chequered. Hindwing pale grey or pale fuscous. Genitalia, see figure 9, p. 114.

Similar species. *Nola confusalis* (Herrich-Schäffer) q.v.

Life history

Ovum. Oval, whitish and glossy. Laid on leaves of oak (*Quercus* spp.) in July.

Larva. Full-fed c.15mm long. Head blackish, glossy. Body stoutish, tapering slightly to posterior end, buff or flesh-pink; dorsal line broad, yellowish, bounded on either side by a fuscous band and interrupted on abdominal segment 3 by a fuscous blotch; similar blotches occur on abdominal segments 6 and 7 but are bisected by the dorsal line; verrucae yellowish or reddish, mostly encircled by fuscous; setae mainly pale, frequently with darker tips, the longest almost equal in length to width of body; some shorter, darker setae occur mainly on the subspiracular verrucae of the anterior half of the body, and two groups of long curved setae on abdominal segment 9 form conspicuous 'tails'.

Figure 9

Meganola strigula ([Denis & Schiffermüller])
(**a**) male genitalia (**b**) female genitalia

Nola confusalis (Herrich-Schäffer)
(**c**) male genitalia (**d**) female genitalia

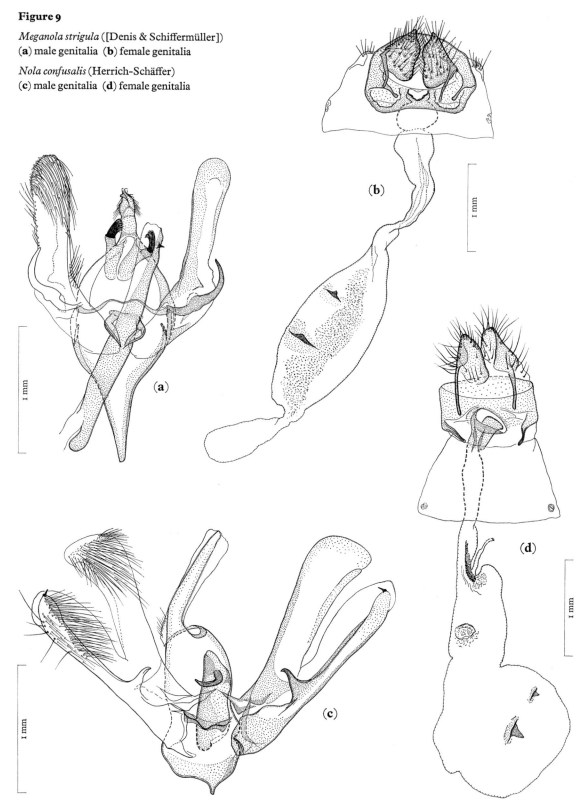

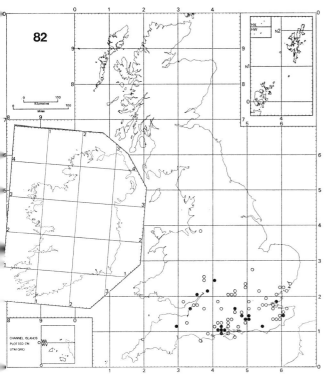

Meganola strigula

It is presumed to hatch in August, feed briefly and then make provision for overwintering as does that of *Nola cucullatella* (Linnaeus), but the details are not known. In spring it feeds on the underside of a leaf, not usually biting through to the upper surface.

Pupa. Dark brown with long reddish brown wing-cases. Pupation in June in a boat-shaped cocoon covered with particles of the bark to which it is attached.

Imago. Univoltine. Emerges in July, resting by day on an oak trunk or branch. Flies at dusk and later at night when it will visit both light and sugar.

Distribution (Map 82)

Now almost confined to oak woods in the southern and west midland counties in England; formerly more widely distributed. Abroad from southern Europe northwards to southern Fennoscandia; Asia Minor.

MEGANOLA ALBULA ([Denis & Schiffermüller])
The Kent Black Arches

Noctua albula [Denis & Schiffermüller], 1775, *Schmett. Wien.*: 69.

Type locality: [Austria]; Vienna district.

Description of imago (Pl.6, figs 42–44)

Wingspan 18–24mm. Antenna in male one-half to three-fifths length of costa, bipectinate, the pectinations ciliate and up to two and one-half width of shaft in length and subequal in the two rows, in female simple, minutely setose on inner surface; labial palpus two and one-half times diameter of eye in length, white, frequently irrorate brown laterally. Patagia and tegulae white. Forewing white, pattern fuscous or brown, inclining to ochreous brown in some specimens; basal area with a faint basal strigula and rounded costal spot containing first discal scale-tuft; antemedian fascia narrow, often very faint, curved smoothly basad and usually just missing second discal scale-tuft; postmedian fascia straightish to start with but becoming sinuate in its posterior half, broadly shaded proximally with pattern colour almost to the antemedian fascia forming a broad central band in well-marked specimens; within this band a darker but obscure median fascia runs obliquely from third discal scale-tuft to meet postmedian at the dorsum, and three white spots are prominent on costa; subterminal fascia thrice arcuate, shaded proximally with pattern colour almost to the postmedian fascia, but leaving a distinct fascia of ground colour between the two; terminal area shaded with pattern colour but leaving a white adterminal fascia and tornal blotch; neural strigulae dark brown, often conspicuous in pale specimens. Hindwing pale fuscous, suffused with very pale brown and somewhat darker towards the margin.

The species is variable in the extent of patterning on the forewing. In one form (fig.44) the pattern is reduced to a postmedian fascia, costal spots and neural strigulae. A form with completely white forewings is known.

Life history

Ovum. Spherical, whitish and glossy. Laid in July on the underside of a leaf midrib of the foodplant. Recorded foodplants are dewberry (*Rubus caesius*), raspberry (*Rubus idaeus*) and strawberry (*Fragaria* spp.), but it is probable that other members of the Rosaceae may be selected.

Larva. Full-fed c.15mm long. Head pale flesh-pink or whitish, marbled with pale brown. Body stoutish and slightly narrowed towards either end, white, flesh-pink or pinkish orange; pattern limited to small blackish subtriangular or heart-shaped markings lying between the subdorsal and supraspiracular verrucae and just anterior to them; these markings are usually obsolete on abdominal segments 1, 9 and 10, and frequently so on 2, 3 and 8;

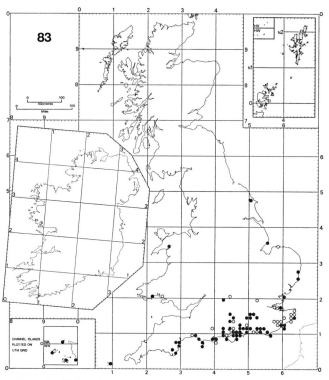

Meganola albula

verrucae more prominent than in other British Nolidae, the thoracic supraspiraculars and all subspiraculars being distinctly stalked; setae mostly short and pale, but each verruca bearing one or two which are longer and usually darker.

Hatches in the autumn, when it may feed briefly, and overwinters. Feeding begins in April and continues until June.

Pupa. Deep reddish brown, darker anteriorly; head almost black; ventral surface paler; antenna-cases margined in black. Pupation in June when a spindle-shaped cocoon with plant fibres woven into it is constructed on a grass culm or other vertical stem.

Imago. Univoltine. Emerges in late June or July, becoming active at dusk, and also later at night when it flies readily to light.

Distribution (Map 83)

Occupies a variety of coastal habitats but also woodland clearings. Mainly confined to the coast of southern Britain from the Isles of Scilly and east Devon to Suffolk in England, and Pembrokeshire in Wales, but occurring inland in Hampshire, Surrey and Berkshire. Eurasiatic; in western Europe in France and Belgium to Denmark and southern Sweden.

NOLA Leach

Nola Leach, 1815, *in* Brewster, *Edinburgh Encycl.* **9**: 135.
Roeselia Hübner, [1825], *Verz.bekannt.Schmett.*: 397.
Celama Walker, 1865, *List Specimens lepid.Insects Colln Br. Mus.* **32**: 500.

This genus is mainly Indo-Malayan and Australian; there are three species in Great Britain, one of which also occurs in Ireland. Forewing lacking veins 9 (R_3) and 10 (R_2) figure 10. In the male genitalia valva deeply bilobate, uncus absent; aedeagus often short; vesica typically with one curved cornutus, figure 9, p. 114.

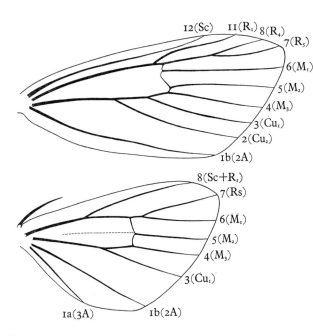

Figure 10 *Nola confusalis* (Herrich-Schäffer), wing venation

NOLA CUCULLATELLA (Linnaeus)
The Short-cloaked Moth

Phalaena (Tinea) cucullatella Linnaeus, 1758, *Syst.Nat.* (Edn 10) **1**: 537.

Type locality: [Europe].

Description of imago (Pl.6, figs 37,38)

Wingspan 15–20mm. Head with vertex white, irrorate fuscous; antenna in male approximately two-fifths length of costa, bipectinate, pectinations up to three times width of shaft in length, inner surface of shaft and pectinations long, ciliate, in female simple, inner surface minutely setose; labial palpus twice diameter of eye in length, heavily irrorate fuscous. Patagia and tegulae white, irrorate fuscous, the former sometimes with two broad bars of the same colour. Forewing white, more or less heavily suffused pale fuscous throughout; basal area heavily irrorate pale and dark fuscous, particularly in costal region; antemedian fascia blackish brown, broad at costa, but narrowing and curving evenly basad to dorsum; median fascia obscure, sinuate; postmedian fascia narrow, but usually well defined and dentate at the veins, strongly arcuate terminad from costa for one-half its length; subterminal fascia pale, sinuate, usually clear but sometimes interrupted, darker-shaded proximally, particularly at costa where shading contains two pale spots; cilia pale fuscous with paler tips. Hindwing pale fuscous. The dark basal area and antemedian fascia of the forewing form the 'short cloak' by which this species is readily recognized when at rest.

Life history

Ovum. Oval and flattened with a finely reticulate surface, whitish and glossy, upper pole darker. Laid on the underside of a leaf, usually on the midrib. Recorded foodplants are hawthorn (*Crataegus* spp.), blackthorn (*Prunus spinosa*), plum (*Prunus* spp.), apple (*Malus* spp.), pear (*Pyrus communis*) and probably other woody Rosaceae.

Larva. Full-fed *c*.15mm long. Head very dark brown, glossy. Body tapering to either end, purplish brown or reddish brown; prothoracic plate divided dorsally, dark brown and glossy; dorsal line whitish with a further whitish line close on either side of it, these lines appearing as a broad dorsal stripe to the unaided eye, being most prominent on meso- and metathorax and abdominal segments 2, 4, 6 and 8, but are interrupted by a bar of ground colour crossing them transversely on abdominal segments 1, 3, 5 and 7; verrucae pale yellowish or reddish, the subdorsals arising from dark brown, shining patches; setae mixed pale and dark, short from the subdorsal and subspiracular verrucae except for the mesothoracic subspiraculars which bear some long setae, and the subdorsal and supraspiracular of abdominal segment 9 whose long setae point posteriad forming a 'tail'; the subspiracular verrucae throughout

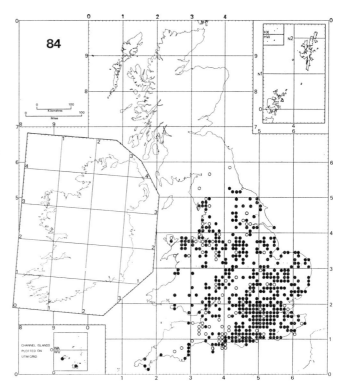

Nola cucullatella

bear some setae which are almost as long as the larval width.

Hatches in August, feeds briefly, then overwinters in a small crevice covered with a few threads of silk. In early to mid-April it begins to feed again on the developing leaves, becoming fully grown in early June. May be easily dislodged from its foodplant in May.

Pupa. Dull brown, head region slightly darker. A cocoon is spun in June on a narrow twig, often in the angle below a thin branch or thorn, whitish but rendered darker and inconspicuous by the small pieces of bark with which it is covered.

Imago. Univoltine. Flies in June and July at early dusk and later at night. Comes readily to light.

Distribution (Map 84)

Occurs in both wooded and more open localities throughout Britain from the Channel Islands to the southern Lake District and Yorkshire. Absent from Ireland. Abroad locally throughout France, Belgium to Denmark and southern Fennoscandia.

NOLA CONFUSALIS (Herrich-Schäffer)
The Least Black Arches

Roeselia confusalis Herrich-Schäffer, 1847, *Syst.Bearb. Schmett.Eur.* **2**: 164.

Type locality: Europe.

Description of imago (Pl.6, figs 45–47)

Wingspan 16–30mm. Head with vertex white, sparsely irrorate pale fuscous; antenna in male approximately one-half length of costa, bipectinate, pectinations very short and surmounted by a pencil of cilia, each pectination plus cilia up to twice width of shaft in length, in female simple, minutely setose on inner surface; labial palpus two and one-half times diameter of eye in length, white, heavily irrorate laterally and beneath with pale brown or fuscous. Patagia white, lightly irrorate pale brown or fuscous, and usually with a broad transverse bar of the same colour; tegulae similar, but with a pale brown or fuscous basal blotch. Forewing white, irrorate pale and dark fuscous, more heavily so distal to postmedian fascia; in basal area a pale ochreous brown subcostal streak extends to first discal scale-tuft; antemedian fascia dark brown, right-angled at second discal scale-tuft, thereafter curving slightly basad to dorsum; postmedian fascia dark brown, minutely dentate at veins, arcuate boldly terminad for two-thirds its length; median fascia paler than postmedian and lying close to it, bearing three well-defined basally directed dentations in dorsal half of its course; subterminal fascia white-edged distally, but indistinct and usually consisting of dots and strigulae on some veins; cilia pale buff, greyish at tips. Hindwing pale fuscous, rather whiter posteriorly in male. Genitalia, see figure 9, p. 114.

A grey form, f. *columbina* Image, occurs in Epping Forest, Essex.

Similar species. *Meganola strigula* ([Denis & Schiffermüller]) lacks the prominent dentations on the median fascia of forewing, and has the valva of the male genitalia entire, not bilobate.

Life history

Ovum. Undescribed.

Larva. Full-fed *c*.15mm long. Head pale yellowish brown or reddish brown, glossy. Body very slightly tapering to either end, yellow, marbled and streaked laterally with reddish brown and purplish brown; dorsal stripe of ground colour broad, containing three interrupted longitudinal purplish brown lines, and invaded by a conspicuous purplish brown suffusion on mesothorax, extending into metathorax, and abdominal segments 1, 3 and 7–9; verrucae prominent; setae mainly yellowish brown with some darker, mostly short, but a few exceeding the width of the larva in length arising from the thoracic supraspiracular verrucae and from most of the subspiraculars.

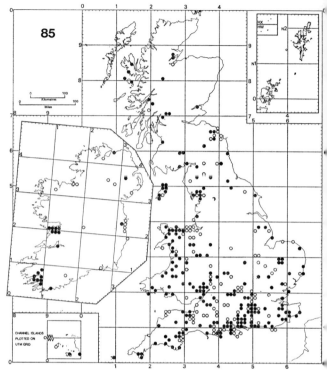

Nola confusalis

Occurs in July and August mainly on oak (*Quercus* spp.), but sometimes on beech (*Fagus sylvatica*), blackthorn (*Prunus spinosa*) and other trees, and said to feed mainly on lichens.

Pupa. Pale brown, darker dorsally, last five abdominal segments fringed laterally with delicate cilia. Overwinters from August in a rounded cocoon, broader anteriorly and resembling a small projection of the bark or paling post to which it is attached.

Imago. Univoltine. Flies in May and June, seldom occurring away from woodland. Rests by day head downwards on a tree-trunk when it greatly resembles a bird-dropping. Flies at night when it is attracted by light.

Distribution (Map 85)

Local throughout England and Wales; in Scotland very local, extending in the west to the Inner Hebrides; local in Ireland; Isle of Man; the Isles of Scilly; the Channel Islands. Abroad throughout much of Europe, as far north as southern Fennoscandia, and eastwards through Asia to Japan.

NOLA AERUGULA (Hübner)
The Scarce Black Arches

Phalaena (Bombyx) aerugula Hübner, 1793, *Samml.Vögel u.Schmett.*: 11, pl.61.

Pyralis centonalis Hübner, 1796, *Samml.eur.Schmett.* **6**: 8 [2], pl.3, fig.15.

Celama trituberculana Heslop, 1959, *Entomologist's Gaz.* **10**: 184.

Celama tuberculana sensu Edelsten, 1961, in South, *The Moths of the British Isles* (Edn 4) **2**: 36.

Type locality: Germany.

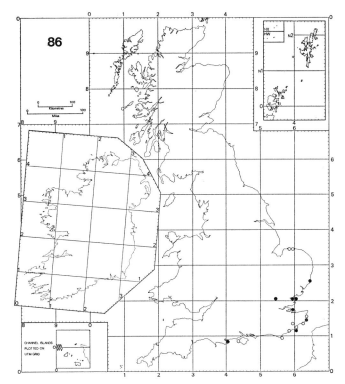

Nola aerugula

Description of imago (Pl.6, figs 39–41)
Wingspan 15–20mm. Head with vertex white, lightly irrorate pale fuscous in some specimens; antenna in male a little over half length of costa, bipectinate, pectinations very short and surmounted by a pencil of long cilia, in female simple with short setae on inner surface; labial palpus three times diameter of eye in length, irrorate fuscous, most heavily on the lateral and ventral surfaces. Patagia white, occasionally with a few pale fuscous scales; tegulae white, usually irrorate pale fuscous. Forewing white, variably irrorate with pale brown, pale fuscous and dark fuscous scales; basal area suffused with brown and grey in costal region; antemedian fascia usually well defined, brown or fuscous, right-angled at second discal scale-tuft; postmedian fascia brown or fuscous, clearly defined even in the palest specimens, right-angled close to costa with a small but distinct dentation at the angle, thence sinuate to dorsum; median fascia scarcely distinguishable in most specimens; subterminal line usually discernible, sometimes well marked and shaded proximally with fuscous; adterminal and terminal lines present, the former often dentate basad at the veins; cilia grey, pale-tipped. Hindwing white, more or less suffused with very pale fuscous, termen darker, cilia speckled with very pale fuscous.

The forewing in this species shows a good deal of variation, particularly in the degree of irroration on the whole wing, and the amount of suffusion spreading from the postmedian fascia towards the wing-base. In heavily marked specimens this suffusion may fill the whole of the area between the ante- and postmedian fasciae to form a broad central band in which the second and third discal scale-tufts show up as prominent white dots; others may have little or no irroration with the fasciae only just discernible.

Life history
Ovum. Circular and somewhat flattened with a central depression, finely but clearly ribbed and reticulate, white at first, becoming cream. Laid in August on the foodplants which include various species of *Trifolium*, *Lotus* and *Medicago*.

Larva. Full-fed *c.*15mm long. Head blackish brown, glossy. Body grey or pinkish grey; dorsal line narrow, yellow, dividing the grey or blackish velvety triangles at the anterior margin of each segment into two conspicuous wedges; a blackish, undulate, intermittent line lies ventral to the subdorsal verrucae, and a similar but more obscure line lies ventral to the subspiracular verrucae; setae pale greyish brown of medium length except for one or two arising from each subspiracular verruca, and from the supraspiracular verrucae on the pro- and mesothorax and abdominal segment 10, which equal the body width in length.

Hatches in August and feeds until September, then overwinters until the following spring, becoming full-fed in June.

Pupa. Chestnut-brown. Cocoon boat-shaped, having pieces of plant material incorporated in it, and attached to a dry stem. Tugwell (1880) records that the larva prepares much of the cocoon from the exterior; then, when the structure is nearly complete, it enters, lines the inside with silk, and finally closes the cocoon from within.

Imago. Univoltine. Occurs in late July and August. Has been observed to climb up plant stems at dusk and fly later at night when it can be attracted to light.

Distribution (Map 86)
A migrant species which may establish itself from time to time as at Deal, Kent during the 1880s. It has been recorded from the coasts of Dorset to Norfolk, including the Isle of Wight (Tugwell, 1873), but has been seen recently only at Dungeness, Kent and Orford Ness, Suffolk. There is one inland record from Parndon, Essex in 1945. Eurasiatic; in Europe from central France and Belgium to Denmark and southern Fennoscandia.

NOLA CHLAMITULALIS (Hübner)
Pyralis chlamitulalis Hübner, [1813], *Samml.eur.Schmett.* **6**: pl.25, fig.160; pl.28, fig.181.
Type locality: Europe.

A single specimen of this south European and west Asian species was found in a light-trap in Jersey on 16 July 1963 (Long, 1965).

References

Long, R., 1965. Notes on recent additions to the Macrolepidoptera of Great Britain and the Channel Islands. *Entomologist's Gaz.* **16**: 17–18.

Tugwell, W. H., 1873. *Nola centonalis* at Freshwater. *Entomologist* **6**: 317–318.

———, 1880. Life-history of *Nola centonalis*. *Ibid.* **13**: 42–45.

NOCTUIDAE
R. F. Bretherton, B. Goater and R. I. Lorimer*

The Noctuidae is one of the larger families of Lepidoptera, over 25,000 species having been described. Their distribution is world-wide, but the majority are found from the tropics northwards, a few species extending to the Arctic Circle. Kloet & Hincks (1972) include 415 species as British; these are assigned to 14 subfamilies:

Noctuinae	62	Chloephorinae	5
Hadeninae	69	Sarrothripinae	2
Cuculliinae	66	Pantheinae	3
Acronictinae	19	Plusiinae	23
Amphipyrinae	102	Catocalinae	13
Heliothinae	9	Ophiderinae	12
Acontiinae	11	Hypeninae	19

Their approximate status is summarized below. It should be borne in mind, however, that the precise number in any one category will vary from year to year as species become established or extinct as residents.

Resident	314
Immigrant, sometimes temporarily resident	9
Regular or occasional visitors	57
Imported or doubtfully British	24
Extinct	11

Most British species are medium-sized moths with fairly robust bodies and inconspicuous coloration, and are nocturnal. The largest is *Catocala fraxini* (Linnaeus) with a wingspan of *c*.100mm, and the smallest is *Hypenodes turfosalis* (Wocke) with a wingspan of *c*.15mm.

Imago (Pl.B; Pl.C). Antenna usually one-half to three-quarters length of forewing, simple, ciliate, fasciculate or bipectinate, but never thickened distally, often differing between male and female; ocelli nearly always present; eyes glabrous, lashed or hairy, providing important subfamily characters; chaetosemata absent; palpi well developed and variable, particularly long in the Hypeninae; haustellum usually well developed, but occasionally weak or absent.

* The authorship within this family is divided as follows:

R. F. Bretherton – overseas distribution for all species; migrant, adventive and extinct species in all subfamilies; Heliothinae, Acontiinae, Hypeninae.

B. Goater – Noctuinae, Amphipyrinae; keys to species for all subfamilies.

R. I. Lorimer – Hadeninae, Cuculliinae, Acronictinae, Chloephorinae, Sarrothripinae, Pantheinae, Plusiinae, Catocalinae, Ophiderinae; rearing.

Personal communications between these authors are referred to in the text by name only.

NOCTUIDAE

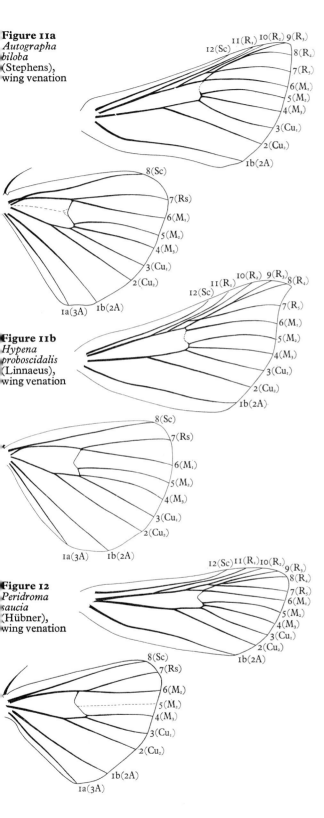

Figure 11a
Autographa biloba
(Stephens),
wing venation

Figure 11b
Hypena proboscidalis
(Linnaeus),
wing venation

Figure 12
Peridroma saucia
(Hübner),
wing venation

Tympanal organs situated laterally on posterior part of metathorax; coremata sometimes present in male, but often lying concealed until the thorax or abdomen is gently distended by injecting air. Thoracic and abdominal crests variously developed in different subfamilies. Tibiae variously adorned with spines which afford useful subfamily characters.

Wing-shape variable, though forewing usually long and narrow, resembling the blade of an oar. Forewing with vein 1c (1A) absent; 1a (3A) short, commonly anastomosing with 1b (2A) and forming a short loop near base of wing; 5 (M_2) always arising from the cell nearer to 4 (M_3) than to 6 (M_1), which is usually from, or from close to upper angle of cell; 7 (R_5) + 8 (R_4) stalked from upper angle, 9 (R_3) + 10 (R_2) stalked from upper margin of cell, 9 (R_3) anastomosing for a short distance with 8 (R_4) to form an areole; 11 (R_1) from mid-upper margin of cell. Hindwing with 1c (1A) absent; 3 (Cu_1), 4 (M_3), 6 (M_1) and 7 (Rs) usually arising from close to or from upper angle of cell; 5 (M_2) strong and either approximated to or parallel to 4 (M_3) in the quadrifine moths (figures 11a, b) or weak and obsolescent and from close to middle of discocellulars in the trifine moths (figure 12); 8 ($Sc+R_1$) connected to 7 (Rs) at a point, or anastomosing with it along part of its length. Frenulum always present, consisting, in British species, of a single strong spine in males and several weak bristles in females.

Forewing with a characteristic pattern composed of five transverse lines: subbasal which does not reach the dorsum, antemedian, postmedian, subterminal and terminal; and three discal spots, the reniform, orbicular and claviform stigmata.

Male genitalia with valva often adorned with processes, the distal region (cucullus) often separated from the main part of the valva by a narrow neck. Female genitalia with ovipositor sometimes highly specialized and equipped for probing, piercing, cutting or slitting.

The differences between the sexes of many species are very slight. In many species the abdomen is much stouter in the female and has conspicuous anal tufts in the male. With set specimens the structure of the frenulum provides a ready means of determining the sex.

Variation. The variation of British species of Noctuidae was dealt with by Tutt (1891–92) and by Turner (1926–50), but they were concerned merely with the naming of 'varieties'. Some of our species are remarkably invariable, for instance, *Naenia typica* (Linnaeus) and *Eustrotia uncula* (Clerck), while others such as *Euxoa tritici* (Linnaeus) and *Agrotis exclamationis* (Linnaeus) show seemingly endless variation which, as far as is known, has no adaptive significance. More interesting, perhaps, is the racial variation which is displayed in the Noctuidae as well as in any other family of

Lepidoptera. Kettlewell & Cadbury (1963) have made a detailed study of it in *Paradiarsia glareosa* (Esper) in the Shetlands, but the phenomenon is well known in such different species as *Euxoa cursoria* (Hufnagel), *Hadena confusa* (Hufnagel), *Aporophyla lutulenta* ([Denis & Schiffermüller]), *Cryphia muralis* (Forster) and *Luperina nickerlii* (Freyer), to name but a few. The Noctuidae and Geometridae are the two principal families in which industrial melanism has been observed. So far, little systematic work has been done on the genetics of noctuid variation. Several British taxa are of interest because they illustrate stages in speciation: the genera *Oligia* and *Amphipoea* contain groups of very closely related species, all variable, which are structurally and biologically distinct; *Aporophyla lutulenta* has, to all intents and purposes, differentiated into 'good' species *A. lutulenta* sensu stricto and *A. lueneburgensis* (Freyer), while it is a matter for debate whether the univoltine *Diarsia florida* Schmidt is really separable from the bivoltine *D. rubi* (Vieweg). Much fascinating and important work on the family remains to be done, both in the field and in the laboratory.

Ovum. Somewhat variable in shape and size relative to that of the moth, and the number laid by different species varies considerably. It is usually more or less hemispherical and ribbed, with fine transverse reticulations, but is flattened in the Acronictinae and Chloephorinae.

Larva (Pl.B; Pl.C). Usually fleshy and cylindrical, with all four pairs of prolegs and anal claspers well developed. Sparse setae arise from pinacula; exceptionally (Acronictinae, Pantheinae) the body is densely covered with secondary setae. In some subfamilies (Plusiinae, Catocalinae, Ophiderinae, Hypeninae) the number of prolegs is reduced to three or two pairs through loss of the anterior pairs, and sometimes the two anterior pairs do not develop until after the second instar.

Pupa. Usually more or less cylindrical, the thickness of the integument varying considerably; in some genera, the haustellum sheath is free distally; cremaster well developed, often possessing diagnostic characters.

Life history and behaviour

Most species are univoltine in Britain, although some are bivoltine in the south and a few may produce a third brood in favourable years. In contrast, some montane species spend two summers as larvae, and others (not all montane) tend habitually to lie over as pupae for a varying number of years. Hibernation may take place in any of the stages – ovum, larva, pupa or imago; a few species aestivate prior to oviposition. The majority of species fly during the summer months, but the first warm days of spring mark the emergence of *Orthosia* and *Cerastis* species, and late autumn is the season for *Agrochola*, *Conistra* and *Xanthia*. Some of those which emerge in autumn hibernate as adults and pair the following spring (*Conistra, Xylena, Lithophane*).

The moths are largely nocturnal though some Heliothinae, Acontiinae and Plusiinae habitually fly by day, as do species of the boreo-alpine *Anarta*. Some Catocalinae, Ophiderinae and Hypeninae are readily disturbed. Some species rest exposed on tree-trunks, palings, rocks or walls (Acronictinae, some Hadeninae, *Cucullia*), but the great majority rest on the ground, sometimes deep among grassroots, and are never seen by day, except by chance. Many species feed avidly at flowers, most particularly soon after dusk. Flowers such as those of sallow (*Salix* spp.), *Buddleia*, heather (*Calluna* and *Erica* spp.), reed (*Phragmites*), rush (*Juncus* spp.) and ivy (*Hedera*) are well-known attractants in their season, and others, such as campions (*Silene* and *Lychnis* spp.) have more specialized relationships with moths as sites of oviposition as well as sources of food. Aphis secretion, known as 'honey-dew', attracts many noctuids, and it is chiefly for this group of moths that sugaring is done. M. W. F. Tweedie (pers.comm.) has experimented with honey as an alternative to the conventional black treacle, and has found it attractive to species such as the plusias, which otherwise seem never to visit sugar. Most species are strongly attracted to light, though it is worth noting that not all will enter a trap, and many species have a characteristic and readily recognized method of approach to a light source.

The eggs may be placed on the bark, stem or leaf of the foodplant, or inserted into a bud, leaf-axil, grass-sheath or flower; eggs are only occasionally broadcast. In captivity most species can be induced to lay in a pill-box, especially if fed with dilute sugar solution. Some, *e.g. Mythimna* species, must be provided with grass-sheaths or some other special requirement and a few, notably *Dicycla oo* (Linnaeus), defy every effort to persuade them to lay. The majority of larvae feed in concealment when young, inside stems, spun leaves or seed-capsules, later becoming nocturnal and hiding during the day below or near ground level. The larvae of *Euxoa* and *Agrotis* species, known as cutworms, are almost entirely subterranean. Apart from *Parascotia fuliginaria* (Linnaeus), which feeds on fungus, and *Cryphia* species, on lichens, all feed on trees, herbaceous plants or ferns; a few Hypeninae feed on dead leaves. Many are polyphagous, others are confined to one species of foodplant. Some larvae have been described as 'cannibals' and their activities have provoked much debate. The habit of eating other larvae seems largely to be stimulated by a dearth of fresh, succulent food and may be a survival factor in early summer, when deciduous trees are defoliated. It is generally agreed that *Cosmia trapezina* (Linnaeus) and *Chilodes maritimus* (Tauscher) are carnivorous by choice, deliberately seeking out prey; opinions differ concerning most other species, but it is safe to say that fairly large numbers may be reared successfully together in close confinement, provided they are given fresh food daily.

When full-fed, most larvae leave the foodplant and move rather briskly in search of a suitable place for pupation; usually in the soil. In a few cases, their colour becomes darker at this time, but less commonly than in Sphingidae and Notodontidae. The structure of the cocoon varies greatly: often it is but a fragile cell in the soil; it may be more firmly constructed of silk, into which are woven soil particles and other debris. Many of the Acronictinae bore into rotten wood for pupation; in *Calophasia lunula* (Hufnagel), a hard capsule is woven on a paling or at the base of the foodplant, and *Parascotia fuliginaria* is exceptional in making a hammock-like cocoon suspended below fungus or bark by threads from each end. In most species, pupation occurs within a few days of the cocoon being completed, but in certain genera, *e.g. Xanthia*, the larva aestivates for two or three months, and some agrotids hibernate as larvae in the earthen cell in which they eventually pupate.

Rearing

'"Technique of Breeding" sounds very important, but I am afraid there is not much in it, apart from unfailing daily attention' (Hedges, 1949).

This quotation, the tailpiece to a paper full of useful information about some of the more 'difficult' species, sums up the basic necessity for the successful rearing of those species which complete their larval growth between early spring and leaf-fall. The majority of these can be brought through in plastic cylinders or boxes with minimal losses if ample fresh food is provided and strict attention paid to hygiene (Rivers, *MBGBI* 1: 68).

It is important first to be quite clear about one's personal object in breeding; there is almost always some distortion of the natural life cycle, as, when captive, larvae feed up more quickly than in the wild and with 'forced' stock this can so affect the date of emergence as to make it impossible for any surplus insects released to find mates. If the object is to obtain a row of bred specimens, there is no point in taking to the pupal stage vastly more than will be needed. When breeding from the egg the best course is to release surplus larvae at their place of origin while it is still possible for them to revert to the normal life cycle; by doing so, any damage caused by the removal of a gravid female has probably been offset by the restoration of as many healthy, unparasitized larvae as would have survived in the wild. In collecting wild larvae, self-discipline is particularly essential with the 'Endangered Species'; a number of them can be obtained as larvae far too easily for their own safety.

Breeding for experimental purposes involves specialized techniques because of the numbers involved and the subject is thoroughly covered by Kettlewell (1973: 319–322).

Pairing. Although members of some other families and a few noctuids will pair in very little space, some room to fly is essential for most species. For all but the largest species, such as those of *Catocala*, a 23cm × 12cm (9in × 4½in) cylinder is adequate and superior to a box of similar size because all round observation is possible. It should have a thin layer of sifted peat on the floor and some vegetation or bark for the moths to rest on. It is not wise to enclose too many moths in the same container; two of each sex, or three females to two males allows some choice, without causing the males to become bemused or to interfere with each other. For the first night or two it is unwise to feed too much; water is necessary in hot weather, but if sugar-water is always available all night, it may monopolize the attention of the moths.

If pairing is not observed it will be necessary to keep the sexes together and make provision for laying in the same container. There are several drawbacks to this; it may be impossible to identify the male and female parent, certainly the male will be useless for the cabinet and the eggs may be laid in places from whence it is difficult to remove them for examination or description.

Laying. It should be possible to discover from the text which follows the natural needs of each species and to provide suitable accommodation. The first requisite is ease of removal of the eggs for examination and, if their duration is long, for efficient storage; overwintering eggs should be kept in a refrigerator or an airy unheated outhouse. Most species will lay in a pill-box of 3cm to 5cm (1½in to 2in) diameter, but many pill-boxes now on sale are thinly covered inside with plastic; this should be removed with forceps, the bottom left loose and the inner walls scored or preferably lined with unsecured, rough paper. The top should be covered with cotton, rather than nylon, netting and, where necessary, a twig or tightly rolled ball of netting left inside for species which require crevices or terminal shoots. Only a single female should be kept in each box; it should be fed every other night, but in hot weather the cover mesh should be sprayed with water on alternate nights. There are exceptions to this general, mass-production line. The moths of some genera (*e.g. Hadena*) alight only momentarily to lay; others (*Mythimna, Archanara*) have specialized ovipositors for inserting eggs into stems or leaf-sheaths of grasses. Among the Noctuinae the females of a number of species, *e.g. Rhyacia simulans* (Hufnagel), *Noctua orbona* (Hufnagel) and *Spaelotis ravida* ([Denis & Schiffermüller]) have a maturation period of up to six weeks from the time of mating; as a general rule, they will lay quite freely if captured in late August or September but if taken before that must be given quarters large enough to inhabit for several weeks without battering themselves. It is equally important to provide a base layer, which will not be saturated by their excreta and will provide daytime cover, using finely sifted peat and a layer of dried leaves which have been previously cleared of predators such as earwigs and spiders.

Rearing larvae. Very young larvae do best in glass-topped tins or small plastic boxes. The need for care and hygiene arises as soon as the eggs hatch and at this stage the main danger is from condensation of the natural humidity of the foodplant inside their home. Paper tissues are very necessary for lining and as a resting place. The treatment of larger larvae will be influenced by the nature of their food. Most of the external feeders on leaves can be bred through without trouble in plastic boxes although those with long hair-tufts, such as the Acronictinae need more than usual space and air and are better kept in cylinders than in boxes. Flower-feeding larvae also need space and air; their food quickly becomes mouldy and releases more humidity than do most leaves. There is more than usual need for care with this kind of larva and although in extreme cases many of them will accept the foliage of their foodplant, resulting adults are small or even crippled. Seed-feeding larvae are easier to rear than the last category, but certain precautions are desirable. If they are being kept in plastic boxes or tins, there should also be a fabric bag or lining; unusual care is necessary to eliminate predators (earwigs in particular) from each batch of fresh food; the growth of the larvae (and in consequence their consumption of food) is very rapid and it is very easy to underfeed; with *Hadena* species in particular, 'foreign' stock (usually of *H. bicruris* (Hufnagel)) is often introduced with the food.

Catkin and bud-feeders. These are most conveniently reared from the egg on potted saplings kept indoors over winter; the best method of all is to induce the female to lay on the potted plant, but this cannot always be managed. What is necessary, however, is to ensure that a supply of fresh food will be regularly available, as an oak, beech or sallow small enough to pot-up is not going to feed many larvae for very long. With sallow-feeders this is easy, as the food plant can be brought forward in water; for larvae on oak, elm and beech the best course is to retard the ova by mild refrigeration. Stem-feeders as far as the Noctuidae are concerned are nearly all restricted to Gramineae. For a long time there was no accurate way of ensuring that the required species was actually being bred; one put the ova or very young larvae on to a tuft of potted grass and hoped that a few would be found some weeks later and that they would be what was needed. Haggett (1968: 67) has described a technique of individual rearing in glass tubes which requires much care and patience, but makes it possible to breed *Apamea*, *Oligia*, *Luperina* and *Amphipoea* species.

Root-feeders. These are not difficult to rear on potted foodplants, although the great quantities of wet frass necessitate frequent cleaning; many species, however, will burrow into substitute roots kept on the surface of sand and transfer themselves to fresh food when necessary. As a number of them have extremely delicate skins they should not be extracted, but encouraged gently to withdraw into ever smaller portions of the root.

Other complications arise from the nature of the life cycle. Many species will go through their first three stages between spring and autumn, overwintering as pupae; these are the easy ones. There are some, however, mainly Cuculliinae which feed up in spring and early summer but aestivate as larvae for up to ten weeks before pupating. Some of these are intolerant of disturbance during this period.

Some larvae complete their growth in autumn and overwinter before pupating. These are particularly difficult to bring through the winter. 'Leaving it to nature' usually results in massive casualties (as, indeed, probably happens in the wild) and the most successful method so far developed involves a short artificial winter in a refrigerator, followed by warmth and humidity. The needs of individual species vary and each success should be carefully noted and placed on record.

Larvae which overwinter when small and feed up during spring may be sleeved on the foodplant, kept on a potted sapling or among dried leaves in a cylinder, outdoors, against a north-facing wall. They are a good deal easier to handle than the full-fed overwintering larvae and are fully dealt with by Symes (1957) who does not, however, emphasize sufficiently how extremely difficult it is to frustrate earwigs, which, apart from their predatory habits, often use the entrance to a sleeve for their own winter quarters and are well able to force their way inside should the fastening loosen during winter.

Semi-synthetic diets. These have been successfully used for rearing some species, e.g. *Spodoptera exigua* (Hübner), *Peridroma saucia* (Hübner) (Shorey & Hale, 1965) and *Mamestra brassicae* (Linnaeus) (David & Gardiner, 1966). Formulae for some of these diets are given by Dickson (1976: 57–60) and the extensive literature on this topic has been reviewed by Singh (1977).

Pupation and storage of pupae. The majority of noctuids pupate in earth and exceptions are noted in the text; sterilized, coarsely sifted peat is the best all-round medium and should be well firmed and at least 10cm (4in) deep (more in specified cases). Although large numbers of larvae pupate between layers of cellulose wadding, there is sometimes a tendency for larvae to get down to floor level, causing an unacceptably high proportion to produce malformed pupae and in general this is a less than satisfactory substance to use, although very convenient. As pupating larvae are most intolerant of disturbance it is important that too many are not forced to go down in the same box and that they should be left undisturbed for as long as possible; it is also important to remember that peat does not stay sterilized. There is now general agreement that pupae should not be left in their pupating medium to overwinter, but should be carefully extracted and stored. Almost every breeder has his personal preference in the matter of storage and there is no generally accepted 'best method'; almost all have one thing in common, *viz.* that during the winter months the pupa should be kept in a constant temperature, preferably below 5°C (40°F) and not in actual contact with other pupae and with moderate ambient humidity. For actual emergence, there is probably nothing more efficient than the Littlewood pattern box (Littlewood, 1941), but the overriding need is for easily accessible walls with ample horizontal grooves to facilitate climbing and drying out. The question of spraying is vexed; many successful breeders do so; others, also successful, are vehemently opposed. Careful tests would be valuable and the results might well vary with different periods of the year.

Economic importance

A few noctuid species are agricultural pests; larvae of *Agrotis* species, popularly known as 'cutworms', are a constant nuisance to growers of root-crops; some Hadeninae cause serious harm to *Brassica* crops in market-gardens and to glasshouse tomatoes; *Cerapteryx graminis* (Linnaeus) is occasionally so excessively abundant in grassy uplands as to destroy tracts of pasture; *Spodoptera littoralis* (Boisduval) has been known to reach pest proportions in glasshouses, following accidental introduction. Abroad, the larvae of *S. exigua*, *Mythimna unipuncta* (Haworth), *Trichoplusia ni* (Hübner) and *Helicoverpa armigera* (Hübner) may be troublesome and are called 'army worms'. It is fortunate that some of the world's most destructive pests in warmer climates, such as *Mythimna unipuncta* and *Earias insulana* (Boisduval), are but interesting rarities here (Baker, *MBGBI* 1: 71).

Key to subfamilies (imagines) of the Noctuidae

1	Eyes hairy	2
–	Eyes glabrous	3
2(1)	Haustellum short, almost hidden by labial palpi	Pantheinae (Vol. 10)
–	Haustellum of normal length	Hadeninae (p. 196)
3(1)	Mid- and hindtibia usually spined (N.B. – only a few inconspicuous spines near apex of midtibia in *Periphanes*)	4
–	Midtibia not spined	6
4(3)	Hindwing with vein 5 (M_2) strong, approximated to 4 (M_3) at base	Catocalinae (Vol. 10)
–	Hindwing with vein 5 (M_2) arising at least one-third of the distance towards 6 (M_1) and running parallel to 4 (M_3) throughout its length, sometimes weakly developed	5
5(4)	Foretibia with an outer and sometimes also with an inner apical spur, usually stronger than other tibial spines and lying along basitarsus. Hindwing whitish with dark terminal fascia and mark at end of cell	Heliothinae (part) (Vol. 10)
–	Foretibia with apical spurs, if present, not differentiated from other tibial spines. Hindwing otherwise	Noctuinae (p. 126)
6(3)	Hindwing with vein 5 (M_2) strong, approximated to 4 (M_3) at base or connate with it	13
–	Hindwing with vein 5 (M_2) arising at least one-third of the distance towards 6 (M_1) and parallel to 4 (M_3), sometimes weakly developed or absent	7
7(6)	Eyes lashed	8
–	Eyes not lashed	9
8(7)	Labial palp porrect, length more than twice diameter of eye	Hypeninae (*Hypena crassalis*) (Vol. 10)
–	Labial palp length less than twice diameter of eye, in most species hardly projecting beyond frontal hair	Cucullinae (Vol. 10)
9(7)	Labial palp conspicuously long, either projecting above base of antenna or more than three times as long as diameter of eye. Abdomen with only a small basal crest. Slender-bodied species	Hypeninae (part) (Vol. 10)
–	Labial palp relatively shorter, if ascending above base of antenna, abdomen strongly crested. Mostly more robust species, often with abdominal crests	10
10(9)	Forewing grey or green (except *Simyra* – pale ochreous grey with black dusting forming streaks) with reniform stigma not outlined white	Acronictinae (Vol. 10)
–	Forewing ochreous, brown or fuscous; if greyish, reniform stigma partly outlined white	11

11(10)	Hindwing yellowish white with broad grey terminal fascia .... Heliothinae (*Pyrrhia*) (Vol. 10)
–	Hindwing otherwise; if with dark terminal fascia, discal area bright yellow 12
12(11)	Forewing bright green. Hindwing with veins 3 (Cu_1) and 4 (M_3) long-stalked. Wing expanse not more than 25mm .... Chloephorinae (*Earias*) (Vol. 10)
–	Forewing otherwise; or if green, hindwing with veins 3 (Cu_1) and 4 (M_3) connate and wing expanse about 40mm .... Amphipyrinae (Vol. 10)
13(6)	Forewing bright green with pale cross-lines .... Chloephorinae (part) (Vol. 10)
–	Forewing otherwise 14
14(13)	Labial palp long, porrect, third segment subequal to second. Hindwing veins 3 (Cu_1) and 4 (M_3) long-stalked, connate with 5 (M_2) .... Sarrothripinae (Vol. 10)
–	Labial palp otherwise, third segment shorter than second. Hindwing vein 5 (M_2) separated from 3 (Cu_1) and 4 (M_3) 15
15(14)	Eyes lashed. Thorax and abdomen strongly crested. Forewing of many species with metallic markings (Eyes lashed and thorax slightly crested – see *Diloba* (Notodontidae), p. 64) .... Plusiinae (Vol. 10)
–	Eyes not lashed. Thorax and abdomen usually without pronounced crests (thorax with moderate anterior crest in *Scoliopteryx*; abdomen strongly crested in *Catephia*) 16
16(15)	Face with appressed scales (with rounded prominence in *Emmelia*). Labial palp length less than twice diameter of eye. Forewing termen evenly curved; if mainly blackish, whitish areas at wing-base or on inner margin before tornus .... Acontiinae (Vol. 10)
–	Face either tufted or with scales spreading outwards in front of eye; or labial palp longer; or forewing termen excavated below apex; or wing mainly blackish with base and dorsum entirely dark .... Ophiderinae (Vol. 10)

Noctuinae

Medium-sized greyish or brownish moths. Many species are extremely variable in wing pattern both within and between populations.

Imago. Antenna variable, in male often fasciculate-ciliate or ciliate, rarely bipectinate; eyes glabrous without lashes; haustellum always well developed; labial palpi of moderate length, usually erect, occasionally directed forward. Hind-tibiae always spinose; foretibiae without spines in some genera. Hindwing with vein 5 (M_2) always absent.

Larva. Plump, soft-skinned, naked except for a few weak setae which arise from scattered pinacula on each segment; generally dull greyish or brownish with a series of oblique dark dashes subdorsally on the abdominal and especially on the posterior segments; occasionally adorned with brightly coloured longitudinal stripes. Most are nocturnal, the 'cutworms' being at least partially concealed in the soil at all times; the others ascending their foodplants at night and retiring into the soil by day. Most species overwinter in this stage and are readily 'forced' in captivity; many are those one finds most commonly in the spring, at night, feeding on the opening buds of saplings.

Pupa. Integument thin to moderately thick; cremaster usually consisting of two short, divergent spines or two parallel spines which are flanked by two pairs of much weaker, hooked bristles. Pupation usually takes place in an earthen cocoon below ground.

Behaviour. In Britain the first species of the Noctuinae to be on the wing are of the genus *Cerastis* Ochsenheimer, in late March and April, but the majority of the Noctuinae are species of high summer, flying from June to September. Most are univoltine, but a few are bivoltine. They are attracted to flowers, honey-dew and sugar, and come freely to light.

Key to species (imagines) of the Noctuinae

1	Hindwing yellow with a blackish border; cilia yellow 2
–	Hindwing not yellow 7
2(1)	Border of hindwing one-half width of wing at widest point .... *Noctua fimbriata* (p. 161)
–	Border of hindwing one-third width of wing or less ... 3
3(2)	Blackish discal spot present in hindwing 4
–	Hindwing without discal spot 5
4(3)	Sharply defined black spot near apex of forewing .... *N. orbona* (p. 159)

–	Apex of forewing without black spot, though subterminal line sometimes darkened at costa .... *N. comes* (p. 159)
5(3)	Wingspan 50mm or more; black spot near apex of forewing .... *N. pronuba* (p. 157)
–	Wingspan 45mm or less; apex of forewing without black spot 6
6(5)	Forewing purplish brown; stigmata narrowly and incompletely outlined with whitish; underside of hindwing without discal spot .... *N. janthina* (p. 162)
–	Forewing rusty brown or reddish clay-coloured; stigmata indistinct, never outlined with whitish; underside of hindwing with blackish discal spot .... *N. interjecta* (p. 163)
7(1)	Forewing richly variegated green (fades to coppery brown); an obscurely outlined pale patch beyond reniform stigma .... *Anaplectoides prasina* (p. 192)
–	Forewing without trace of green or, if green, without pale patch beyond reniform stigma 8
8(7)	Conspicuous small black spot near apex of forewing .. 9
–	Apex of forewing without black spot 12
9(8)	Blackish trapezoidal mark present between reniform and orbicular stigmata 10
–	No blackish mark between reniform and orbicular stigmata .... *Xestia baja* (p. 183)
10(9)	Orbicular stigma appearing as a pale straw-coloured lobe continuous with a costal patch of same colour, occasionally reddish-tinted .... *X. c-nigrum* (p. 179)
–	Orbicular stigma rounded or trapezoidal in shape; neither stigma nor costal patch paler than ground colour of forewing 11
11(10)	Forewing rich reddish or blackish brown, darker than hindwing .... *X. ditrapezium* (p. 180)
–	Forewing warm clay-coloured, not darker than hindwing .... *X. triangulum* (p. 181)
12(8)	Claviform stigma represented by a small round blackish dot 13
–	Claviform stigma elongate, or absent 16
13(12)	Postmedian and subterminal lines continuous, gently curving, darker than ground colour of forewing .... *Diarsia rubi/florida* (p. 176) (see text for discussion of differences).
	Postmedian and subterminal lines often broken, sinuate or angulate, never darker than ground colour of forewing 14
14(13)	Subterminal line paler than ground colour of forewing; outline of orbicular stigma indistinct; area between reniform and orbicular stigmata never darker than forewing ground colour; rusty brown (♂) or rich reddish brown with pale straw-coloured reniform stigma (♀) .... *D. dahlii* (p. 173)
–	Subterminal line rarely paler than ground colour of forewing (some northern forms of *D. mendica*); outline of orbicular stigma distinct; area between reniform and orbicular stigmata usually forming a dark brown or black patch 15
15(14)	Forewing dark purplish brown or purplish fuscous; reniform stigma paler than ground colour; subterminal line never paler than ground colour .... *D. brunnea* (p. 174)
–	Forewing pale ochreous, often reddish-tinted (southern specimens) to rich chestnut or fuscous-red, often lavender-tinted (northern specimens); reniform stigma not paler than orbicular; subterminal line in some northern forms paler than ground colour .... *D. mendica* (p. 172)
16(12)	Claviform stigma much darker than ground colour of forewing, or appearing so by being heavily outlined by darker coloration 17
–	Claviform stigma not darker than ground colour of forewing, or only narrowly outlined by darker coloration so as to be inconspicuous, or absent 33
17(16)	Claviform stigma solid, black or sometimes dark brown 18
–	Claviform stigma heavily outlined or pale-centred 23
18(17)	Forewing pale sandy-coloured, almost unmarked except for conspicuous stigmata .... *Agrotis ripae* (part) (p. 145)
–	Forewing not sandy-coloured; markings usually more numerous 19
19(18)	Black collar-mark present 20
–	Black collar-mark absent 21
20(19)	Collar-mark no more than 0.5mm at widest point; hindwing pure white (♂) or very pale grey (♀); ground colour of forewing grey (doubtfully British) .... *A. spinifera* (p. 139)
–	Collar-mark c.1mm wide at widest point; hindwing white, usually showing darker veins (♂), or light brown (♀); ground colour of forewing brown or fuscous .... *A. exclamationis* (p. 141)
21(19)	Wedge-shaped markings present in subterminal region of forewing 22
–	Wedge-shaped markings absent .... *A. trux* (part) (p. 142)
22(21)	Reniform and orbicular stigmata darker than ground colour of forewing; ochreous patch on dorsum absent .... *A. vestigialis* (part) (p. 138)
–	Reniform and orbicular stigmata paler than ground colour of forewing; ochreous patch present on dorsum (very rare immigrant) .... *Ochropleura fennica* (p. 151)

23(17) Orbicular stigma circular, pale straw-coloured or whitish .... *Agrotis trux* (part) (p. 142)
– Orbicular stigma of variable shape, if pale then not circular or no paler than ground colour of forewing .. 24
24(23) Hindwing pearly white (♂) or pearly grey (♀), with darker veins .... *A. segetum* (p. 139)
– Hindwing not pearly 25
25(24) Claviform stigma large and thick, about one-fourth length of forewing; pale streak extending between claviform and two upper stigmata; series of wedge-shaped markings in subterminal region, longest 1.5mm or more .... *A. vestigialis* (part) (p. 138)
– Claviform stigma no more than one-sixth length of forewing; pale streak between stigmata absent; wedge-shaped marks, if present, less than 1mm 26
26(25) Forewing mid-brown or fuscous brown, cross-lines and outlines of stigmata demarcated in fuscous to variable extent; region between ante- and postmedian lines unmarked except for outlines of stigmata .... *Graphiphora augur* (part) (p. 165)
– Forewing of variable colour; if mid-brown then markings more numerous and more complex, and with some at least between ante- and postmedian lines .. 27
27(26) Robust species, wingspan exceeding 40mm; antenna of ♂ bipectinate, yellow (probably in Channel Islands only) .... *Agrotis crassa* (p. 146)
– More slender species, wingspan 38mm or less; antenna of ♂ neither bipectinate nor yellow 28
28(27) Forewing sandy-coloured 29
– Forewing darker brown or fuscous 30
29(28) Underside of forewing with a small dusky suffusion around discal spot, which is kidney-shaped (figure 14b, p. 133) .... *Euxoa cursoria* (part) (p. 136)
– Underside of forewing without such a suffusion, almost unmarked; discal spot, when present, dot-like .... *Agrotis ripae* (part) (p. 145)
30(28) Reniform stigma isodiametric, rounded or square, indistinctly outlined; hindwing light brown .... *A. clavis* (p. 141)
– Reniform stigma kidney-shaped, sharply outlined; hindwing whitish, lightly suffused with fuscous 31
31(30) Usually sandy-coloured, but some northern forms darker; underside of forewing abruptly paler beyond postmedian line; discal spot kidney-shaped, surrounded by a dark suffusion (figure 14b, p. 133) .... *Euxoa cursoria* (part) (p. 136)
– Usually rich dark brown or fuscous; underside of forewing more or less concolorous beyond postmedian line; discal spot dot-like, not surrounded by dark suffusion (figure 14a, p. 133) 32

32(31) Antemedian line approaching dorsum at a right angle, and usually continuing forward through pale costal streak; imago August–September .... *E. obelisca* (p. 130)
– Antemedian line approaching dorsum at an acute angle, and usually disappearing in pale costal streak when this is present; imago July–August .... *E. tritici* (p. 132)
33(16) Claviform stigma present, not darker than ground colour 34
– Claviform stigma absent 45
34(33) Claviform stigma paler than ground colour of forewing 35
– Claviform stigma and ground colour concolorous .. 36
35(34) Forewing light green, marked with reddish .... *Ochropleura praecox* (p. 150)
– Forewing grey or blackish .... *Eurois occulta* (p. 191)
36(34) Hindwing pearly light grey with darker veins 37
– Hindwing not pearly 38
37(36) A black dash extending outwards from reniform stigma; terminal region of forewing paler than rest of wing .... *Agrotis ipsilon* (p. 143)
– Black dash absent; terminal region of forewing not paler than rest of wing .... *Peridroma saucia* (part) (p. 171)
38(36) Orbicular stigma spindle-shaped; forewing light brown (♂) or blackish (♀) .... *Agrotis puta* (p. 144)
– Orbicular stigma more or less circular in outline 39
39(38) Stigmata and cross-lines demarcated in fuscous on otherwise almost unmarked warm brown ground colour; stigmata not paler than ground colour .... *Graphiphora augur* (part) (p. 165)
– Markings more complex though sometimes less striking; if ground colour warm brown, then reniform paler than ground colour of forewing 40
40(39) Forewing rich reddish-, purplish- or blackish-brown 41
– Forewing sandy (sometimes reddish-tinted) or mid-brown, never richly reddish- or purplish-brown .. 43
41(40) Forewing blackish brown, with faintly paler stigmata and cross-lines; costa concolorous .... *Euxoa nigricans* (part) (p. 134)
– Forewing reddish- or purplish-brown; if blackish, then with pale costa 42
42(41) Costa paler than ground colour of forewing .... *Xestia agathina* (p. 188)
– Costa not paler than ground colour of forewing .... *Lycophotia porphyrea* (p. 170)
43(40) Forewing with conspicuous black antemedian fascia .... *Eugnorisma depuncta* (part) (p. 153)

–	Forewing without such fascia 44	56(53)	Small blackish dot present in lower half of reniform stigma 57
44(43)	Forewing sandy-coloured, sometimes reddish-tinted; claviform stigma shortly triangular .... *Agrotis ripae* (part) (p. 145)	–	Reniform stigma without blackish dot 58
–	Forewing dull brownish or fuscous, occasionally with a reddish tint; claviform stigma at least twice as long as broad .... *Euxoa nigricans* (part) (p. 134)	57(56)	Forewing clay-coloured, hindwing darker than forewing .... *Xestia castanea* (part) (p. 185)
45(33)	Forewing with conspicuous black antemedian fascia .... *Eugnorisma depuncta* (part) (p. 153)	–	Forewing glaucous reddish brown, hindwing paler than forewing .... *Paradiarsia sobrina* (p. 167)
–	Forewing without such fascia 46	58(56)	Forewing straw-coloured with darker brown costal region .... *Axylia putris* (part) (p. 149)
46(45)	Hindwing pearly light grey with darker veins .... *Peridroma saucia* (part) (p. 171)	–	Forewing not as above 59
–	Hindwing not pearly 47	59(58)	Orbicular stigma absent 60
47(46)	Hindwing whitish 48	–	Orbicular stigma present, though sometimes rather faint 61
–	Hindwing brownish or greyish 53	60(59)	General coloration purplish brown and straw; reniform stigma conspicuous; wedge-shaped marks in terminal region of forewing (rare immigrant) .... *Actinotia polyodon* (p. 148)
48(47)	Orbicular stigma dot-like; antenna bipectinate .... *Agrotis cinerea* ♂ (p. 137)	–	General coloration brown or blackish; reniform stigma and wedge-shaped markings absent .... *Standfussiana lucernea* (p. 154)
–	Orbicular stigma larger; antenna simple, thread-like 49	61(59)	Orbicular stigma dot-like .... *Agrotis cinerea* ♀ (p. 137)
49(48)	Colour of costa of forewing contrasting with ground colour 50	–	Orbicular stigma larger, more or less circular in outline 62
–	No contrast between colour of costa and ground colour 51	62(61)	Area between reniform and orbicular stigmata darker 63
50(49)	Costa darker than forewing ground colour; latter straw-coloured .... *Axylia putris* (part) (p. 149)	–	Area between stigmata not darker 66
–	Costa straw-coloured; forewing ground colour deep red-brown .... *Ochropleura plecta* (p. 152)	63(62)	Forewing rich blue-grey, fading to dove-grey .... *Xestia ashworthii* (p. 182)
51(49)	Wingspan 42mm or more; linear basal streak on forewing .... *Spaelotis ravida* (p. 164)	–	Forewing brown or reddish brown 64
–	Wingspan 38mm or less; basal streak absent 52	64(63)	Forewing richly patterned with red, grey and black or reddish fuscous; reniform and orbicular stigmata touching or separated by less than width of orbicular stigma .... *X. alpicola* (part) (p. 178)
52(51)	Forewing light grey or pinkish grey (often blackish in far north of Britain); black marks between reniform and orbicular stigmata and between orbicular and antemedian line; postmedian line a continuous, even curve, never represented by dots .... *Paradiarsia glareosa* (p. 169)	–	Forewing more uniformly brown or reddish brown; space between reniform and orbicular stigmata exceeding width of orbicular stigma 65
–	Forewing various shades of brown or fuscous, never grey; interstigmatal marks, if present, never black; postmedian line represented by dots or absent .... *Xestia xanthographa* (p. 187)	65(64)	Hindwing glossy, with golden tint, with distinct terminal shade and discal spot .... *Eugraphe subrosea* (part) (p. 166)
53(47)	Forewing rich reddish brown 54	–	Hindwing dull brown, lacking terminal shade and discal spot .... *Xestia rhomboidea* (p. 184)
–	Forewing not reddish, or if so, then shaded with grey, buff and/or black 56	66(62)	Forewing richly patterned with red, grey and black or reddish fuscous .... *X. alpicola* (part) (p. 178)
54(53)	Small blackish dot present in lower half of reniform stigma .... *X. castanea* (part) (p. 185)	–	Forewing lacking this combination; reddish brown, brown, grey-brown or blackish 67
–	Reniform stigma, if visible, without black dot 55	67(66)	Hindwing glossy, with golden tint, with distinct terminal shade and discal spot; forewing warm light reddish brown .... *Eugraphe subrosea* (part) (p. 166)
55(54)	Orbicular and reniform stigmata paler than ground colour of forewing .... *Cerastis leucographa* (p. 193)		
–	Orbicular and reniform stigmata invisible .... *C. rubricosa* (p. 193)		

–	Hindwing not or hardly glossy; terminal shade indistinct or absent; forewing not warm light reddish brown	68
68(67)	Forewing with conspicuous pale straw-coloured veins......*Naenia typica* (p. 189)	
–	Veins of forewing not paler than ground colour	69
69(68)	Outlines of stigmata and cross-lines conspicuously paler than ground colour of forewing (very rare immigrant)......*Mesogona acetosellae* (p. 195)	
–	Not as above	70
70(69)	Dark collar-mark present on thorax; black streak at base of forewing (very rare immigrant)......*Ochropleura flammatra* (p. 151)	
–	Collar-mark and basal streak absent	71
71(70)	Postmedian line darker than ground colour of forewing, unbroken and smoothly curved; median fascia present......*Xestia sexstrigata* (p. 187)	
–	Postmedian line not darker than ground colour of forewing, or if so then broken and jagged; median fascia absent	72
72(71)	Forewing fuscous irrorate with ochreous; hindwing dull brownish fuscous......*Rhyacia simulans* (part) (p. 156)	
–	Forewing brown or reddish brown, hindwing not fuscous	73
73(72)	Wingspan 50mm or more; ante- and postmedian lines paler than ground colour of forewing (very rare immigrant)......*R. lucipeta* (p. 157)	
–	Wingspan 45mm or less; ante- and postmedian lines, when present, darker than ground colour of forewing	74
74(73)	Ground colour of forewing smooth brown; cross-lines and stigmata darker though sometimes only faintly developed......*Graphiphora augur* (part) (p. 165)	
–	Ground colour of forewing light brown, irrorate darker; stigmata not darker than ground colour of forewing......*Rhyacia simulans* (part) (p. 156)	

EUXOA Hübner

Euxoa Hübner, [1821], *Verz.bekannt.Schmett.*: 209.
Metaxyja Hübner, [1821], *ibid.*: 223.
Exarnis Hübner, [1821], *ibid.*: 225.
Brotis Hübner, [1821], *ibid.*: 226.
Telmia Hübner, [1821], *ibid.*: 228.

A genus of world-wide distribution. Four species occur in Great Britain and Ireland.

The frons projects beyond the eyes, the projection frequently taking the form of a conical hump with a terminal, crater-like depression; antenna in male fasciculate-ciliate. All legs strongly spinose. Male genitalia with forked process on inner face of valva, formed from the clasper and sacculus (figure 13, p. 131).

Although in general the species are readily separable by eye on the characters given in the key, some individuals can only be differentiated on the structure of the genitalia.

EUXOA OBELISCA GRISEA (Tutt)
The Square-spot Dart

Noctua obelisca [Denis & Schiffermüller], 1775, *Schmett. Wien.*: 80.
Agrotis obelisca var. *grisea* Tutt, 1902, *Entomologist's Rec. J.Var.* **14**: 147.

Type locality: England; Isle of Wight.

Description of imago (Pl.7, figs 1,2)
Wingspan 35–40mm. Antenna in male notched and tufted to two-thirds, in female simple. Forewing rich chocolate-brown to purple-brown, often with a reddish tint, sometimes greyer, the ground colour more or less deeply overlaid with fuscous suffusion; reniform and orbicular stigmata, and costa, usually distinctly paler; claviform stigma darker fuscous, etched black; a black streak extends from antemedian line to orbicular stigma and continues as a black, nearly square mark between that and reniform stigma; ante- and postmedian lines usually distinctly darker than ground colour, the first extending across costal streak which, in many specimens, is distinctly paler ochreous white basad from this point; pale subterminal line usually very faint or absent. Hindwing in male whitish with terminal shade, narrow discal spot and veins fuscous, in female the whole wing is suffused fuscous. Abdomen in female a little stouter than that of male, with sclerotized ovipositor usually protruding in set specimens. Genitalia, see figure 13, p. 131.

Subsp. *grisea* differs from the nominate subspecies in being smaller, much darker and less variable. The grey form as described by Tutt is peculiar to the Isle of Wight.

Similar species. British specimens vary little and are very

Figure 13

Euxoa obelisca grisea (Tutt)
(**a**) male genitalia (**b**) female genitalia

Euxoa tritici (Linnaeus)
(**c**) male genitalia (**d**) female genitalia

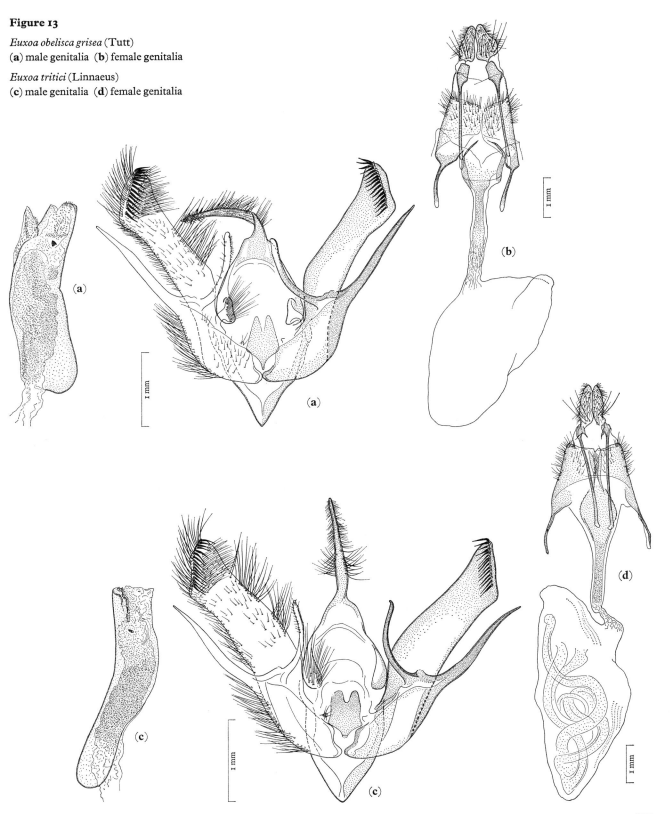

similar to *E. tritici* (Linnaeus), in which individuals occur which are virtually indistinguishable from *E. obelisca grisea*. In general, *E. obelisca grisea* has a longer, narrower forewing and the markings are simpler; in particular, the region beyond the postmedian line is almost without markings and nearly always lacks any trace of the wedge-shaped marks which are so often a feature of *E. tritici*. In *E. obelisca grisea*, the antemedian line approaches the dorsum at right angles and usually extends through the pale costal streak, whereas in *E. tritici* it meets the dorsum at an acute angle and usually disappears in the pale costal streak when this is present. The pale components of the cross-lines, often conspicuous in *E. tritici*, are rarely so in *E. obelisca grisea*. On the hindwing, dark veins are more strongly emphasized in *E. obelisca grisea*.

Life history
Ovum. Apparently undescribed.
Larva. Full-fed *c.*40mm long. Head smoky brownish, irrorate darker, the lobes darker-streaked. Prothoracic plate glossy blackish brown, pale dorsal and subdorsal lines showing plainly upon it. Body greyish brown, smoky or blackish green laterally; dorsal line pale grey, edged with blackish green; narrow subdorsal line pale greenish grey broadly edged above with a dark blackish green stripe; lateral stripe undulating, pale smoky green; subspiracular stripe broad, dirty whitish; dorsal region irrorate darker, ventral surface uniform pale smoky green; spiracles black; pinacula dark smoky green and glossy; segments rather deeply wrinkled (Buckler, 1893).

It is apparently not known whether overwintering occurs in the ovum or young larval stage. The larva is said to feed in spring and summer on species of *Helianthemum*, *Galium* and other herbaceous plants.

Pupa. Apparently undescribed.
Imago. Univoltine. Flies in late August and September, later than *E. tritici* and in localities where that species is apparently absent. Comes freely to sugar and light, and to ragwort and heather bloom. Rests concealed by day, presumably among rocks and roots.

Distribution (Map 87)
Locally common on sea-cliffs and rocky hillsides, chiefly in western Britain. Well-known localities include the Isles of Scilly, north and south coasts of Cornwall, Torquay (Devonshire) and Freshwater (Isle of Wight). It is also recorded from south Wales, west and east Scotland and, according to Donovan, the Irish coast in Cos Sligo, Donegal, Dublin, Waterford, Cork and Kerry (Baynes, 1964). Barrett (1896) also mentions the eastern Cotswolds, Gloucestershire; Herefordshire; Sutton Park near Birmingham; Brighton, Sussex; Derbyshire; Cheshire; Lancashire; Scarborough, Yorkshire and Moncrieffe Hill,

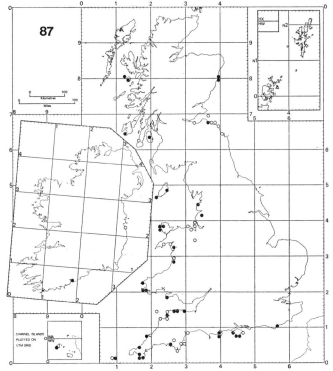
Euxoa obelisca

Perthshire, where it occurred commonly; Guernsey. Eurasiatic; ranging in western Europe from Portugal to south Norway, central Sweden and Finland, mainly in coastal or sandy localities.

EUXOA TRITICI (Linnaeus)
The White-line Dart
Phalaena (Noctua) tritici Linnaeus, 1761, *Fauna Suecica* (Edn 2): 320.
Type locality: Sweden.

Description of imago (Pl.7, figs 3–7)
Wingspan 28–40mm. Antenna in male notched and tufted to two-thirds, in female simple. Forewing with ground colour ranging from pale ochreous brown through shades of reddish brown or greyish brown to dark chocolate or blackish fuscous; a pale costal stripe is frequently present; cross-lines and stigmata are variable in development; a series of blackish wedge-shaped marks is often present in the subterminal region. Hindwing whitish with fuscous veins, narrow lunule and terminal shade; greyer in female. Genitalia, see figure 13, p. 131.

E. tritici is one of the most variable of British moths, the different forms grading into one another, so making their classification difficult. Tutt (1891–92), who specialized in the genus, named many forms which he distinguished by differences in ground colour, the degree of development of cross-lines and stigmata and the presence or absence of a pale costa, though he paid less attention to local variation, which is quite marked in the species. The usual sandhill forms tend to be rather bright brown shaded with fuscous (figs 5,6); moorland forms are blacker and strongly marked; specimens from Orkney (figs 3,4) tend to be rich, warm brown, often with a grey costal stripe. The large ochreous form, which has been confused with the continental *E. aquilina* (Hübner), has been named ab. *rhabdota* Edelsten (fig.7).

Similar species. This and related species, especially *E. cursoria* (Hufnagel), are difficult to distinguish on formal characters and they all have a comparable range of variation. Compared with *E. cursoria*, *E. tritici* has the forewing a little broader and the costa more curved; the ground colour is usually more fuscous and less sandy. On the underside, which is glossier in *E. cursoria*, the following features are helpful in determination: in *E. tritici* the forewing is more or less concolorous beyond the stigma which is dot-like or oval, not exceeding 1mm on its longer axis; the hindwing usually has a terminal shade abutting the cilia, the postmedian line is rarely present and the subterminal line always absent. In *E. cursoria* the forewing is abruptly paler beyond the postmedian line and there is a darker area usually forming a distinct cloud around the stigma which is kidney-shaped and generally exceeds 1.5mm in its longer axis; the hindwing almost always lacks a terminal shade, but the postmedian and subterminal lines are both usually represented, figure 14. The male genitalia of *E. cursoria* are asymmetrical, but symmetrical in *E. tritici* (figure 15, p.135). The differences between *E. tritici* and *E. obelisca grisea* (Tutt) have been dealt with under the latter species.

Life history

Ovum. Globular with small, flat base, very delicately striate, light in colour and glossy (Stokoe & Stovin, 1958). Deposited in loose batches.

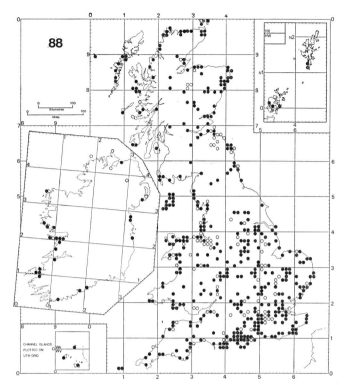

Euxoa tritici

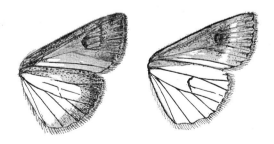

Figure 14(a) *Euxoa tritici* (Linnaeus), underside (**b**) *Euxoa cursoria* (Hufnagel), underside

Larva. Full-fed c.40mm long, plump and tapering at each end. Head rounded, shining dirty grey or sometimes dark brown and sometimes, as in *E. nigricans* (Linnaeus), with black irroration. Prothoracic plate barred or spotted with black. Body dusky grey, rather glossy, minutely irrorate blackish; dorsal lines slender, concolorous, edged with broader stripes of brownish grey clouding in which are a few black spots, one or two on each of the middle segments; subdorsal and spiracular lines dusky brownish grey; ventral surface and prolegs dirty pale grey; spiracles black. Sometimes the ground colour is dirty reddish with reddish grey stripes of deeper shade. It closely resembles that of *Agrotis vestigialis* (Hufnagel) (Buckler, 1893).

The method of hibernation is unknown. In spring it feeds nocturnally until June on a variety of small herbs such as species of *Cerastium*, *Stellaria*, *Spergula* and *Galium*, lying in the soil during the day. On the continent it is reported to be a pest of cultivated buckwheat (*Fagopyrum esculentum*) (Barrett, 1896).

Pupa. Yellow-brown, glossy, smooth, with thin integument; anal segment rounded with two short, parallel cremastral spines. Subterranean.

Imago. Univoltine. Flies in July and August, but, in north-west Surrey, R. F. Bretherton finds that the large ab. *rhabdota* occurs, not commonly, in gardens and woods at this season and that the small dark heathland form emerges later, with a peak in the last week of August and the first week of September. The two forms seem to be distinct biological races. Mainly nocturnal, visiting flowers of ragwort, heather and marram and coming freely to sugar and light. Occasionally flies by day in hot weather, and is sometimes found resting on flower-heads, though more usually it hides amongst tangles of overhanging roots on sandhills, *etc.*, from whence it may be raked.

Distribution (Map 88)
Widely distributed throughout the British Isles, chiefly on coastal sandhills but also on heaths and moors or even in gardens. Apparently absent from the sandhills of Unst, though common in south Shetland, where *E. cursoria* is evidently absent. Eurasiatic; in western Europe it occurs from Portugal to south Norway, eastern Sweden and Finland.

EUXOA NIGRICANS (Linnaeus)
The Garden Dart
Phalaena (Noctua) nigricans Linnaeus, 1761, *Fauna Suecica* (Edn 2): 322.
Type locality: Sweden.

Description of imago (Pl.7, figs 8–10)
Wingspan 32–40mm. Antenna in male notched and tufted to two-thirds, in female simple. Forewing blackish fuscous to ochreous brown, grey-brown or reddish brown; reniform and orbicular stigmata usually distinct, paler than ground colour and often united by a blackish bar; claviform stigma concolorous, narrowly edged with dark fuscous, not discernible in dark specimens; cross-lines consist of dark and pale components, often obsolete, but represented in darker specimens by series of light ochreous dots, and in pale ones by indistinct darker lines; a narrow median fascia often present. Hindwing whitish, clouded with smoky brownish grey, especially towards margin; veins and crescent-shaped discal spot darker. Sexes differ as in other members of the genus, but ovipositor is even more conspicuous. Genitalia, see figure 15, p. 135.

Though rather an obscurely marked species it is, nevertheless, subject to considerable variation. The typical dark form (fig.8) predominates in northern British populations, the darkest being stated to occur on the mosses of Cheshire and Lancashire. Southern specimens are of more variable colour, and usually paler.

Similar species. Some forms can be confused with *E. tritici* (Linnaeus) but that is a neater, more sharply marked insect. The presence of the claviform stigma, as well as the relatively longer and narrower forewing, should preclude

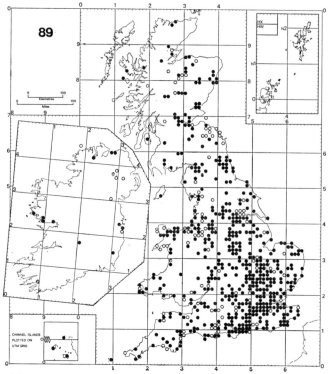

Euxoa nigricans

confusion with *Xestia xanthographa* ([Denis & Schiffermüller]).

Life history
Ovum. Globular, with small flat base, very delicately striate, reticulate and glossy. In captivity laid in loose batches on clover and other plants (Stokoe & Stovin, 1958).

Larva. Full-fed *c.*40mm long, smooth and cylindrical. Head light brown, irrorate blackish. Body pale ochreous dorsally, inclined to greenish laterally; lines greenish grey, edged with black. The presence of a double whitish stripe low down along the sides and conspicuous black pinacula distinguish it from *E. tritici* (Buckler, 1893).

The stage in which overwintering takes place is uncertain, but in spring, from March onwards, it is sometimes very destructive to herbaceous plants, particularly to species of *Trifolium*, *Plantago* and Umbelliferae. Buckler (*loc.cit.*) records that a field of four hectares (ten acres), sown in autumn with clover and wheat, had the whole of the clover eaten by the larvae during the following spring, after which they moved to the hedgerows, destroying *Heracleum* and all other herbs, but leaving the wheat and wild grasses.

Pupa. Light brown with thin, glossy integument; cremaster with two short, slightly divergent spines. In a subterranean earthen cocoon.

Figure 15

Euxoa nigricans (Linnaeus)
(**a**) male genitalia (**b**) female genitalia

Euxoa cursoria (Linnaeus)
(**c**) male genitalia (**d**) female genitalia

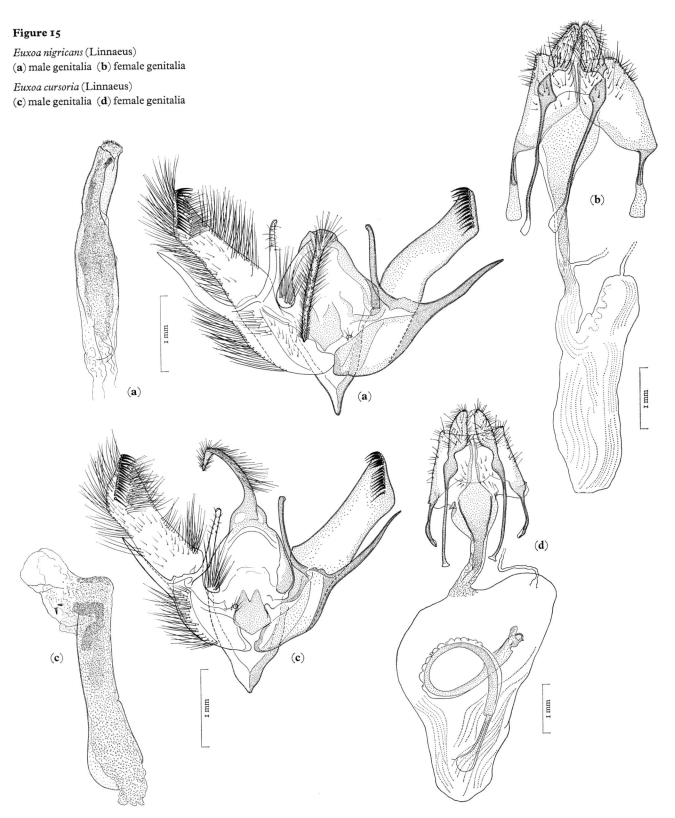

Imago. Univoltine. Flies in August. Strongly attracted to flowers of ragwort, burdock, tansy and *Buddleia* at night, and is also sometimes common at sugar. At light its appearance is rather unpredictable, but is apparently due in some way to the siting of the lamp.

Distribution (Map 89)
A lowland species of rich farmland, allotments and gardens, marshes, river valleys, fens, salt-marshes and open downland. It is widely distributed in lowland England, though evidently more common in the eastern counties. It is local in the lowlands and valleys in Scotland, and is widespread in Ireland; the Channel Islands. Eurasiatic; in Europe it occurs from Portugal to central Scandinavia and Finland.

EUXOA CURSORIA (Hufnagel)
The Coast Dart
Phalaena cursoria Hufnagel, 1766, *Berlin.Mag.* **3**: 416.
Type locality: Germany; Berlin.

Description of imago (Pl.7, figs 11–18)
Wingspan 34–38mm. Antenna in male notched and tufted to two-thirds, in female simple. Forewing ground colour varies from whitish ochreous through shades of greyish or ochreous brown to rich mahogany brown; markings vary greatly in intensity; stigmata always outlined, but in some specimens reniform and orbicular stigmata are conspicuously paler than ground colour, in others hardly discernible; a dark cloud may be present in the dorsal half of the reniform stigma; claviform stigma varies from being barely visible to intensely blackish brown; cross-lines may be well developed or absent; sometimes a dark terminal shade is a conspicuous feature of an otherwise almost unmarked forewing. Hindwing whitish, suffused basad from the cilia to a greater or lesser extent with sandy grey; a very narrow darker discal spot and darker veins; cilia whitish. Genitalia, see figure 15, p. 135.

The species is extremely variable. In the southern part of its range in Britain, sandy-coloured or whitish forms predominate, while those from the north are smaller and darker, the darkest occurring in isolated colonies at Norwick and Burrafirth, on Unst. The most heavily marked forms are ab. *distincta* Tutt (fig.14) and ab. *sagitta* Hübner (fig.17) which are more common northwards. In Shetland, a form resembling ab. *sagitta*, with ground colour rich mahogany red and with whitish costa and reniform and orbicular stigmata, is not uncommon. Some beautiful forms have the veins strikingly paler than the ground colour, and in others the costa is pale.

Similar species. Darker specimens are easily confused with *E. tritici* (Linnaeus) *q.v.* Sandy-coloured specimens can be

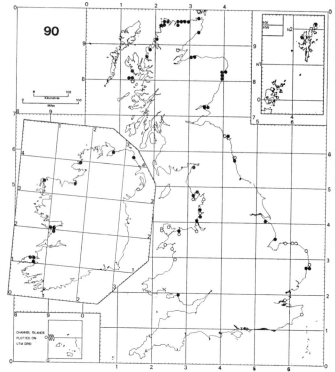

Euxoa cursoria

mistaken for *Agrotis ripae* (Hübner), but in that species the reniform stigma is less than one-fifth the width of the wing, whereas in *Euxoa cursoria* it is about one-third of its width. The hindwing of *Agrotis ripae* lacks a discal spot and terminal shade, and the underside of the forewing is almost unmarked, lacking the dusky cloud around the stigma which, when present, is dot-like and not kidney-shaped.

Life history
Ovum. Rather small and somewhat oval-globular with a small flattened base, delicately ribbed and reticulate; pinkish buff. In captivity laid in clusters of 40–50 (Stokoe & Stovin, 1958).

Larva. Full-fed *c.*42mm long, plump and tapering at each end; segmental divisions and subdividing wrinkles very well defined; in early instars long and slender for a noctuine, smooth-skinned and glossy. Head and prothoracic plate pale brownish buff. Body colour varies according to age; in early instars it is a vivid glaucous green dorsally and bluish or greenish grey laterally with narrow greyish or whitish longitudinal lines, brown pinacula and black spiracles; in later instars it becomes pale ochreous brown, the colour change beginning at the posterior segments and gradually spreading forward, the pale lines persist, the pinacula become increasingly conspicuous (Buckler, 1893).

It is not certain how the insect spends the winter, but it

has been supposed that the larva overwinters when young. It feeds nocturnally during spring and summer until as late as July on sandhill plants such as sea-sandwort (*Honkenya peploides*), *Viola* spp., early hair-grass (*Aira praecox*) and sand-couch (*Agropyron junceiforme*), hiding in sand by day.

Pupa. Moderately stout, pale golden brown, smooth and glossy, with thin integument; cremastral spines a little shorter, thicker and more divergent than in *E. tritici*. Rather deep in the sand in a compact, ovoid cocoon.

Imago. Univoltine. Flies from late July to September and occurs much more commonly in certain years than in others. Comes readily to flowers of marram and ragwort, as well as heathers, also comes freely to light. It has been taken in daytime by raking overhanging tangles of marram roots, when it drops inert to the ground.

Distribution (Map 90)

It is confined to coastal dunes where it is commonest some little distance behind the foredunes, as opposed to *Agrotis ripae* which prefers the foreshore. Colonies occur on many, perhaps most, extensive tracts of sandhills northwards from Suffolk and Cheshire; in Scotland it is especially common on the west and north coasts, and extends to the Shetlands, though it has not been found recently in the Orkneys and is apparently absent from seemingly suitable ground at Quendale, south Shetland. One specimen, possibly an immigrant, was recorded from Dover, Kent on 6 August 1957 (Chalmers-Hunt, 1962–70). It is widespread on the Irish coasts but apparently absent from the south-east (Baynes, 1964). Northern Eurasiatic, halophile; in France it is known with certainty only from the estuaries of the R. Somme and R. Canche; in the Netherlands, north Germany and Denmark it is widespread on dunes and heaths; it is very local on coasts in south Norway and Sweden, reaching about 65°N in Finland.

AGROTIS Ochsenheimer

Agrotis Ochsenheimer, 1816, *Schmett.Eur.* **4**: 66.
Scotia Hübner, [1821], *Verz.bekannt.Schmett.*: 226.
Agronoma Hübner, [1821], *ibid.*: 227.
Georyx Hübner, [1821], *ibid.*: 227.
Psammophila Stephens, 1850, *List Specimens Br.Anim. Colln Br.Mus.* **5**: 72.

A genus containing numerous species widespread in the Holarctic and other regions. Eight are resident in Britain, one is a regular immigrant, one is resident only in the Channel Islands and one other is doubtfully on the British list.

The frons projects as in *Euxoa*, but has rather long vertical ridges; antenna in male variable, bipectinate, tapering to a filiform tip. Distinguished from *Euxoa* by the structure of the male genitalia: on the inner face of the valva, the single clasper is elongate and the sacculus shortly rounded, so that there is only a single spine-like process.

AGROTIS CINEREA ([Denis & Schiffermüller])
The Light Feathered Rustic

Noctua cinerea [Denis & Schiffermüller], 1775, *Schmett. Wien.*: 80.
Bombyx denticulatus Haworth, 1803, *Lepid.Br.*: 133.
Type locality: [Austria]; Vienna district.

Description of imago (Pl.7, figs 19–24)

Wingspan 33–40mm, males larger than females. Antenna in male strongly bipectinate to about three-quarters with thick, finely ciliate teeth, tip filiform, in female simple. Patagia with black collar-mark near base. Forewing from light brown or greyish to almost black; markings extremely variable: constant features are the small reniform stigma, hardly one-sixth width of wing, the orbicular stigma dot-like or obsolete, and the claviform stigma absent; cross-lines dentate, darker than ground colour, strongly or weakly represented, or obsolete; a dark median fascia, running through the reniform stigma, is often present. Hindwing whitish, suffused fuscous, with small dark discal spot and darker veins; cilia glossy, whitish. Female with narrower forewing and darker hindwing. Though most authors have depicted the female as being much darker than the male, A. J. Wightman amassed a long series of matched pairs varying from pale whitish grey to nearly black, now in BMNH.

Variation lies chiefly in the ground colour, definition of the cross-lines and degree of development of the median fascia, which may occasionally completely fill the area between the ante- and postmedian lines. According to Wightman (pers.comm.), the small, light-coloured form

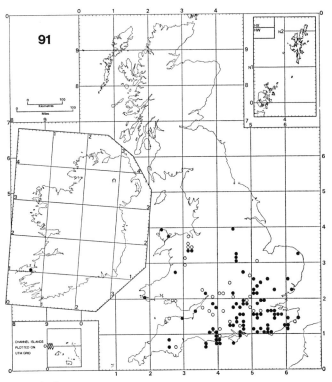

Agrotis cinerea

freely to light and both sexes have been taken at sugar, but females are comparatively seldom seen. Wightman (pers. comm.) reported that once in Sussex, after an extremely hot day, females were found in abundance after nightfall resting on short grass-stems on the downs. Barrett (*loc.cit.*) states that females have occasionally been taken in hot sunshine, running or tumbling over the grass; otherwise they are hardly ever seen by day.

Distribution (Map 91)

Local on chalk and limestone hills and downs in southern England, on shingle at Dungeness, Kent, on old spoil heaps in the Forest of Dean, Gloucestershire, on the mountains of north Wales and on Anglesey. Very rare in Ireland, having been reported, mostly as single specimens, only from Cos Tyrone, Antrim and Kerry (Baynes, 1964; 1970). Palaearctic; in Europe it is widely distributed on calcareous and sandy soils from Spain to Denmark; it occurs locally on the coasts of Sweden, but only just reaches Norway and Finland.

described as subsp. *tephrina* Staudinger from the southern chalk-hills merely falls within the range of normal variation.

Life history

Ovum. Hemispherical with about 30 ribs, some shorter than others, extending to a circular raised ring at the top, minutely reticulate; very pale pinkish grey when laid, gradually becoming darker. In captivity laid in small clusters, hatching in about 12 days (Buckler, 1893).

Larva. Full-fed *c.*37mm long, rather plump. Head glossy black, mandibles greenish. Prothoracic plate jet-black with fine, pale median sulcus. Body blackish green with a thread-like, pale, darker-edged dorsal line; lateral lines rather indistinct; ventral surface paler greenish; segments each with a number of blackish pinacula, largest anteriorly, each with a fine seta; spiracles large, black (Buckler, *loc.cit.*).

Feeds on wild thyme (*Thymus drucei*) and perhaps other small herbaceous plants, though it is recorded as having refused sheep's-fescue (*Festuca ovina*) (Buckler, *loc.cit.*). Feeds through the summer and autumn, overwintering in the soil when full-grown.

Pupa. Short and stout, light brown with very conspicuous spiracles (Barrett, 1896). Pupation takes place in the spring; subterranean.

Imago. Univoltine. Flies in May and June. Males come very

AGROTIS VESTIGIALIS (Hufnagel)
The Archer's Dart

Phalaena vestigialis Hufnagel, 1766, *Berlin.Mag.* **3**: 422.
Noctua valligera [Denis & Schiffermüller], 1775, *Schmett. Wien.*: 80.

Type locality: Germany; Berlin.

Description of imago (Pl.7, figs 25–30)

Wingspan 30–40mm. Antenna in male strongly bipectinate tapering to filiform tip, in female simple. Patagia brown or fuscous with narrow, pale margins; tegulae paler, usually whitish or light brown with darker dorsal margins. Fore wing varying from whitish to deep brown with darker clouding; cross-lines indistinct but stigmata, especially claviform, and subterminal wedge-shaped marks conspicuous. Hindwing whitish with dark veins and discal spot, often with dark terminal band also; in female more uniformly brownish fuscous with whitish cilia.

The overall pattern of markings is fairly constant, but in series the species shows considerable population differences. The 'typical' sandhill form (figs 25,26) is whitish brown with dark fuscous-brown markings; series from the Hebrides and Orkneys are smaller and paler, those from Cornwall very bright light brown, and from Braunton Burrows, Devon duller light brown with a more conspicuous pale medial longitudinal streak. Specimens from Studland, Dorset (figs 27,28) are silvery grey, often with a distinct vinous tint, and those from the New Forest, Hampshire and the Surrey heaths similar but darker. Browner specimens occur inland in the Breckland (figs 29,30) and the dullest and brownest of all are found near Oxford and Birmingham.

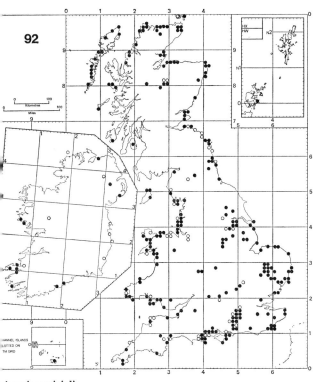

Agrotis vestigialis

Life history

Ovum. Apparently undescribed.

Larva. Full-fed *c.*40mm long, moderately plump, smooth and glossy. Head light brown, lobes edged by a darker band or irroration. Prothoracic plate brownish, darker-barred. Body pale olive-grey tinged with brownish or purplish dorsally; dorsal and subdorsal lines rather obscure; ventral surface very pale grey; spiracles and pinacula black.

Feeds nocturnally on bedstraws (*Galium* spp.) and stitchwort (*Stellaria* spp.), grasses and probably other small herbaceous plants; in captivity it accepts lettuce (*Lactuca sativa*) and dandelion (*Taraxacum officinale*). By day it rests below the surface of the soil. Feeds from August throughout the winter, becoming dormant only in the coldest weather (Barrett, 1896).

Pupa. Pale reddish brown with rather thin integument; cremaster with two very short, divergent spines. Subterranean.

Imago. Univoltine. Flies from late June to September, usually at night but occasionally by day. Visits in abundance flowers such as ragwort, heathers and marram, and also sugar. Both sexes come freely to light, even to a hand-lamp. Generally rests concealed, though it sometimes spends the day on ragwort inflorescences.

Distribution (Map 92)

Predominantly coastal on sandhills but local inland on light, sandy soils and heaths. It occurs locally all round the coasts of Great Britain and Ireland as far as Orkney and the Outer Hebrides, and in the Channel Islands. Eurasiatic; in Europe it is widespread from Portugal to Denmark, more local in south Norway, Sweden and Finland, but has been recorded within the Arctic Circle.

AGROTIS SPINIFERA (Hübner)
Gregson's Dart

Noctua spinifera Hübner, [1808], *Samml.eur.Schmett.* **4**: pl.83.

Type locality: Europe.

Very doubtfully British. The only record is of one, or possibly two, specimens said to have been caught in the Isle of Man, flying in the afternoon sunshine in August 1869 with *A. vestigialis* (Hufnagel), and originally thought to be aberrations of that species. It is questionable whether these specimens were the same as those later identified by Doubleday and others as *A. spinifera*. The species is subtropical and Mediterranean; recorded in western Europe from Portugal, south Spain and south-east France. The imago is figured in South (1961, **1**: pl.46, fig.2).

AGROTIS SEGETUM ([Denis & Schiffermüller])
The Turnip Moth

Noctua segetum [Denis & Schiffermüller], 1775, *Schmett. Wien.*: 81.

Type locality: [Austria]; Vienna district.

Description of imago (Pl.7, figs 31–36)

Wingspan 32–42mm. Antenna in male strongly bipectinate, tapering to filiform tip, in female simple. Patagia sometimes narrowly dark-edged. Forewing varies from pale whitish grey through shades of brown to black; lighter forms usually finely irrorate with darker scales; stigmata outlined dark brown or blackish, clouded within, the orbicular usually annulate; claviform stigma usually not exceeding 2mm in length, often obsolescent in darker specimens; cross-lines variably developed: in darkest specimens, subterminal line often represented by a series of pale dots (ab. *subatratus* Haworth, fig.34); females often darker than males, though not always so. Hindwing in male pearly white with darker veins, in female pearly light grey, more heavily suffused in darker specimens; fringe whitish, often with a fine dark line running through it.

Similar species. Only two other related species, *A. ipsilon* (Hufnagel) and *Peridroma saucia* (Hübner), have pearly hindwings. Both are usually larger than *Agrotis segetum*;

A. ipsilon has a black dash extending from the reniform stigma to the postmedian line; in *Peridroma saucia* the forewing is more ample, the stigmata, especially the orbicular, are larger, and a discal spot is present in the hindwing.

Life history

Ovum. Globular; milk-white when laid, becoming light cream after about 24 hours; ribbed and reticulate with two concentric rings round micropyle. Laid in small irregular clusters, hatching in about a fortnight. Oviposition is inhibited at temperatures above 25°C (77°F), by relative humidities below 60 per cent and by keeping females in continuous darkness (Singh & Kevan, 1956).

Larva. Full-fed *c.*47mm long, the typical cutworm, rather plump, smooth and glossy, tapering abruptly at each end. Head rather large, light brown with two darker bars. Prothoracic plate pale and glossy. Body greyish white or pale grey tinged with purple or pink, ventrally faintly tinted with yellowish brown; pinacula grey, each bearing a fine seta; spiracles black.

It is more or less completely subterranean. Barrett (1896) wrote that probably no species was more destructive in cultivated fields and gardens. Besides boring into the roots of turnips, beet, swedes and carrots, it also destroys cabbages, mustard and many other herbaceous plants, entering roots, gnawing off young shoots at ground level and eating seedlings. The majority, hatching in summer, feed up slowly in the winter and become full-fed in spring, but some attain full growth quickly and produce moths in autumn; apparently it is not known how the progeny of these moths pass the winter.

Pupa. Length to *c.*20mm; light reddish brown, wing-cases paler and abdominal segments darker anteriorly; integument rather thin, glossy; wing-cases smooth; abdominal segments lightly and finely punctate anteriorly; cremaster with two slightly divergent, tapering spines set on a short, conical projection. Subterranean in a small earthen cocoon.

Imago. Mainly univoltine. Flies in May and June, but there is sometimes a small second generation in September and October. Seldom seen by day but at night it comes to flowers, sugar and light.

Distribution (Map 93)

Common to abundant throughout England and Ireland, but scarcer in Scotland except in the south and east; present in the Channel Islands. Eurasiatic; it is found throughout western Europe, reaching about 64°N.

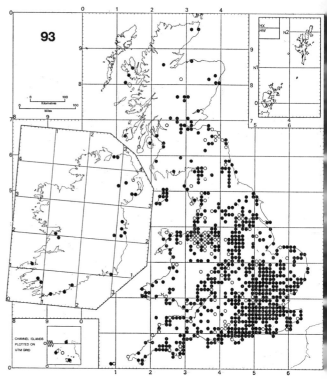

Agrotis segetum

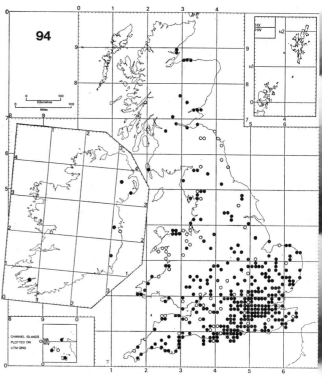

Agrotis clavis

AGROTIS CLAVIS (Hufnagel)
The Heart and Club

Phalaena clavis Hufnagel, 1766, *Berlin.Mag.* **3**: 426.
Noctua corticea [Denis & Schiffermüller], 1775, *Schmett. Wien.*: 81.

Type locality: Germany; Berlin.

Description of imago (Pl.7, figs 37–43)

Wingspan 35–40mm. Antenna in male strongly bipectinate, tapering to filiform tip, in female simple. Forewing varying from pale stone-colour through shades of brown and fuscous to black, the pale ground colour being often so heavily irrorate with dark fuscous as to appear almost black; stigmata almost always present; reniform often dark, indistinct and large, about one-third width of wing; orbicular variable in shape and size; claviform nearly always short and pale-centred; cross-lines usually well defined though sometimes very indistinct or absent. Hindwing light brownish fuscous, hardly darker in female, with narrow discal spot and slightly darker veins. Wings of female relatively longer than those of male.

Both sexes are equally variable. A striking form has pale veins (ab. *venosa* Tutt, fig.43). Others have the median area of the forewing much darker (fig.42), or this condition is reversed with the basal and terminal regions blackened. Several of these forms have been named but they tend to run into one another and are difficult to classify.

Life history

Ovum. Likened by Buckler (1893) to an orange in shape, with underside flattened and having a central boss, containing the micropyle, in a small depression at top; surface finely ribbed and reticulate; straw-coloured when laid, becoming flesh-coloured with darker markings.

Larva. Full-fed *c.*35mm long, of even girth, stout and rugose. Head with anterior margins of lobes broadly streaked blackish. Prothoracic plate hardly darker than ground colour. Body brownish grey, finely irrorate all over, and with rather indistinct longitudinal lines; pinacula large, blackish, each with a short seta; spiracles rather small, black. Differs from *A. segetum* ([Denis & Schiffermüller]) in rugosity and lack of contrast in colour of dorsal and lateral surfaces (Buckler, *loc.cit.*).

Polyphagous, recorded feeding on species of *Chenopodium*, *Polygonum*, *Trifolium*, *Rumex*, *Verbascum* and wild carrot (*Daucus carota*). At first it feeds on the leaves, making holes in them, but when *c.*10mm long it goes underground and eats the roots. Feeds from August to November and overwinters full-grown in the cell in which it will pupate in the spring.

Pupa. Dark brown, integument rather thick; abdominal segments punctate; cremaster with two short divergent spines set on a prominent papilla. Subterranean.

Imago. Univoltine. Flies in late June and July. Visits flowers including lime, privet and red valerian, and comes freely to sugar and to light where males usually predominate. Rarely encountered by day.

Distribution (Map 94)

Locally abundant on coastal sandhills, less so inland though sometimes frequent on chalk-hills and clays of the London Basin. It is found chiefly in the southern counties of England, but extends north to Sutherland and to northern Ireland; it is apparently absent from much of the Midlands. It occurs in the Channel Islands. Eurasiatic; in Europe occurring from Spain to the Arctic, but mainly in the west.

AGROTIS EXCLAMATIONIS (Linnaeus)
The Heart and Dart

Phalaena (Noctua) exclamationis Linnaeus, 1758, *Syst.Nat.* (Edn 10) **1**: 515.

Type locality: Europe.

Description of imago (Pl.7, figs 44–52)

Wingspan 35–44mm. Antenna in male shortly bipectinate, in female simple. Patagia with a conspicuous dark brown or fuscous collar-mark. Forewing varying from pale whitish brown through shades of grey-brown to mahogany or dark fuscous, smooth or finely irrorate with darker coloration; stigmata usually well defined, the claviform being the most prominent and the orbicular most variable; cross-lines variable in development and sometimes absent. Hindwing in male white with veins fuscous, occasionally with darkened border, in female brownish fuscous; a minute, dark, dot-like discal spot, placed well forward in the cell, is often present.

Both sexes are equally variable. Sometimes the costa is reddish (ab. *costata* Tutt) or dark brown. Sometimes all the stigmata are united (ab. *juncta* Tutt, fig.51), and occasionally they form a blackish patch from which lineolae extend to the postmedian line (ab. *lineolatus* Tutt, fig.52). Some specimens have the cross-lines very strongly represented (ab. *catenata* Wize, fig.48). Details of variation are discussed by Goater (1969).

Similar species. The presence of the dark mark on the patagia distinguishes *A. exclamationis* from the occasional female *A. clavis* (Hufnagel) which resembles it.

Life history

Ovum. Hemispherical, ribbed with finer reticulations; whitish when laid, becoming dull pinkish. In captivity loosely attached to sides of container, soon falling free, hatching in about 12 days.

Larva. Full-fed *c.*38mm long; a typical cutworm. Head brownish drab, darkest at mouth, with blackish brown streak on each lobe. Prothoracic plate dark brown, glossy, bisected by pale continuation of dorsal line. Body warm brown above level of spiracles, greyish white below; each segment bearing an ovoid or pyriform dorsal blotch through which passes a faint, pale, thread-like dorsal line; subdorsal lines also faint; pinacula dark brown, the posterior-dorsal pair on each segment being the largest; spiracles particularly large, enabling the larva to be distinguished at a glance from *A. segetum* ([Denis & Schiffermüller]) (Buckler, 1893).

In captivity feeds on a variety of herbaceous foliage, including species of *Rumex*, *Chenopodium* and *Plantago*; it also accepts slices of carrot and turnip. Barrett (1896) records it doing considerable damage to linen spread out to bleach during manufacture. Feeds from July onwards: some larvae pupate in late summer to produce a small second brood in September, but the majority attain full growth in the autumn and hibernate in an earthen cocoon in which they eventually pupate.

Pupa. Reddish brown; spiracles large; cremaster consisting of two short, divergent spines. Pupation occurs in the spring in the cocoon in which the larva overwintered.

Imago. Mainly univoltine. Flies from mid-May to late July with a peak in late June; the occasional examples seen in September are generally smaller than those of the summer generation. In parts of southern Europe, a second emergence is usual. Comes to flowers such as red valerian and *Buddleia*, to honey-dew, to sugar in great abundance, and likewise to light. Rests concealed by day.

Distribution (Map 95)
Found in a wide range of habitats but commonest, perhaps, in cultivated country. It is abundant through most of the British Isles except Scotland, where it decreases rapidly northwards. Eurasiatic; in Europe widespread to central Scandinavia and Finland.

AGROTIS TRUX LUNIGERA Stephens
The Crescent Dart

Noctua trux Hübner, [1824], *Samml.eur.Schmett.* **4**: pls 155,163.
Agrotis lunigera Stephens, 1829, *Ill.Br.Ent.* (Haust.) **2**: 113.
Type locality: Ireland; Cork.

Description of imago (Pl.7, figs 53,54)
Wingspan 35–42mm. Sexually dimorphic. *Male* (fig.53). Antenna strongly bipectinate, tapering to filiform tip. Forewing pale greyish ochreous, clouded and irrorate dark fuscous, occasionally darker; claviform stigma black, short and thick, solid; orbicular stigma rounded, outlined in black

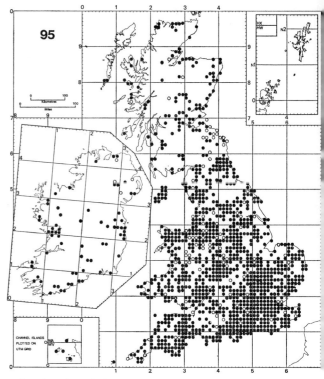

Agrotis exclamationis

and variously shaded within, but usually paler than ground colour and often a conspicuous and distinctive feature; reniform stigma rather large, constant in shape, outlined with dark fuscous; usual cross-lines present, together with narrow median fascia passing through reniform stigma. Hindwing shining white, sparsely irrorate with fuscous, veins darker. *Female* (fig.54). Antenna simple. Forewing dark blackish brown or nearly black; stigmata as in male but pale orbicular even more conspicuous; a pale patch often present in space between reniform stigma and postmedian line; cross-lines inconspicuous. Hindwing more heavily suffused fuscous, with narrow fuscous discal spot and darker veins; cilia whitish.

Series from different localities present visible differences: specimens from the Isle of Wight are particularly large and pale, with sexual dimorphism more pronounced than, for example, in those from Portland, Dorset. The nominate subspecies is very different and, unlike the British subspecies, extremely variable; while occasional specimens occur on the Continent which are hard to distinguish from subsp. *lunigera*, the various continental forms apparently never occur in British populations. Accordingly, many authors have treated subsp. *lunigera* as a distinct species but Boursin (1964) considers it be only a form of *A. trux*.

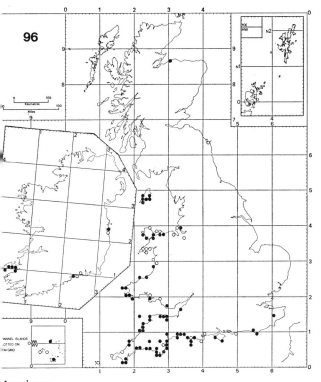
Agrotis trux

Life history
Ovum. Small, nearly spherical, slightly flattened beneath; dirty white when laid, developing a pinkish spot near micropyle and a transverse zone of the same colour. Oviposition in the wild has not been observed; in captivity eggs were crammed into a honeyed sponge where they were not easily detected (Browne *in* Barrett, 1896). They hatch in about 12 days (Buckler, 1893).

Larva. Full-fed *c.*42mm long, plump and wrinkled. Head mottled brownish, with dark spot on each side. Body dirty brown dorsally, greyer on sides, each segment with several conspicuous black pinacula which eventually change to small, brown annular markings. In earlier instars fine dorsal and subdorsal lines are present, and faint dorsal chevrons, but these markings disappear at maturity, the remains of the dorsal line persisting as a fine pale line through the prothoracic plate.

In the wild is said to feed on herbaceous plants on and near sea-cliffs; in captivity it has been reared on dandelion (*Taraxacum officinale*), plantain (*Plantago* spp.), knotgrass (*Polygonum aviculare*) and also sliced carrot. Usually becomes full-fed by November, passing the winter in this state, eating little, and pupates in the spring; however, in captivity larvae may be induced to feed up quickly and produce moths in the autumn.

Pupa. Undescribed. Large larvae, unearthed in early November in Cornwall among rock sea-spurrey (*Spergularia rupicola*), did not pupate until February.

Imago. Flies in July and August, occasionally in September. Comes freely to light and sugar, especially when this is applied to thistle and knapweed heads, and can also be netted flying along cliff-edges at night.

Distribution (Map 96)
Chiefly an inhabitant of sea-cliffs of varying geology, but sometimes found a few hundred metres inland in valleys running up from the coast. In Britain it is locally common from the Isle of Wight westwards to north Wales; in Scotland it has recently been recorded from Findhorn, Morayshire; Isle of Man. In Ireland it occurs on the east and south-west coasts in Cos Louth, Dublin, Waterford, Cork and Kerry (Baynes, 1964); the Channel Islands. Mediterranean-Asiatic; the nominate subspecies extends up the Atlantic coasts of Portugal, Spain and France at least as far as Brittany and also occurs inland in the Pyrenees.

AGROTIS IPSILON (Hufnagel)
The Dark Sword Grass

Phalaena ipsilon Hufnagel, 1766, *Berlin.Mag.* **3**: 416.
Noctua suffusa [Denis & Schiffermüller], 1775, *Schmett. Wien.*: 80.
Agrotis ypsilon misspelling.
Type locality: Germany; Berlin.

Description of imago (Pl.8, figs 14,15)
Wingspan 40–55mm. Antenna in male strongly bipectinate, in female simple. Forewing dark purple-brown to pale straw; claviform stigma distinct; reniform stigma followed by a strong black 'dart' pointing outwards, with two smaller 'darts' facing inwards near margin. Hindwing translucent, apart from brown suffusion on veins and near tornal angle and inner margin.

Ab. *albescens* Clark has ground colour of forewing almost white. Lesser variation in ground colour is largely seasonal, moths taken in spring being usually paler and having the markings less darkened than those in autumn.

Similar species. *Peridroma saucia* (Hübner) has antenna simple in male and lacks 'darts' on forewing. *Agrotis segetum* ([Denis & Schiffermüller]) is also without 'darts' and male hindwing is pure white with very slight darker suffusion.

Life history
Ovum. Very small, conical, flattened above and below and slightly ribbed; yellow, later brownish grey. In captivity laid apparently at random on foliage or muslin; eggs laid on 9 July hatched 1 August.

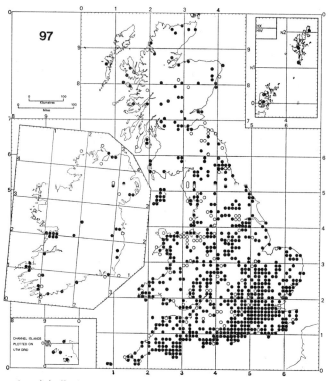

Agrotis ipsilon

Larva. Full-fed c.37mm long; in early instars purplish black and greasy-looking, but in its last two instars becoming rich purplish brown above and yellow or green beneath, with two white spots on the head, a necklace of four black spots on the first two segments, and pairs on the remaining segments; last instar figured by Buckler (1893), but not there described.

Polyphagous on herbaceous plants, but seldom seen in the wild, as it feeds mainly nocturnally and at ground level or below. Larvae which hatched on 1 August were kept indoors in closed tins and fed mainly on dandelion (*Taraxacum officinale*); they were full-fed about 1 September (Bretherton, 1969). See also Hellins (1868), Barrett (1896, **3**: 283) and Cockayne (1936a).

Pupa. Light red-brown, with two short setae on the pointed cremaster. Subterranean, in a very slight cocoon.

Imago. A regular annual immigrant, but varying greatly in numbers; counts at a light-trap in Surrey from 1964 to 1978 ranged from 275 in 1966 to one in 1972. Has been seen in every month of the year except January and February; occurs in most years in small numbers in May, June and July, and is most numerous from mid-August to October. Moths which arrive early in the year can probably produce at least one generation in Britain, with a natural life-cycle of about three months, varying with the seasonal temperature; but most late autumn females are immature and there is no clear evidence of overwintering in Britain in any stage. Reared moths, emerging from 25 to 30 September, and captured moths, which were caged out of doors, died in the first severe frost.

Distribution (Map 97)

Occurs throughout the British Isles to Shetland. Cosmopolitan. Its limits of residence in Europe are not known, but migrants are common as far north as Denmark and southern Fennoscandia, occasional in Norway to 65°N, and frequent in Iceland. It is believed to migrate from north Africa to Europe in the spring and to return in the autumn (Williams, 1958: 142). In Egypt, where the larvae are a pest of cotton and wheat, the moths are winter immigrants, breeding in several generations from September to April, but then disappearing suddenly, at the time of their greatest abundance, while still sexually immature.

AGROTIS PUTA (Hübner)
The Shuttle-shaped Dart
Noctua puta Hübner, [1803], *Samml.eur.Schmett.* **4**: pl.52.
Bombyx radius Haworth, 1803, *Lepid.Br.*: 133.
Type locality: Europe.

Description of imago (Pl.8, figs 1–4)

Wingspan 30–32mm. *Male.* Antenna pectinate, with rather short, ciliate teeth, tapering to filiform tip. Forewing light brown with blackish basal patch; reniform stigma also more or less completely black, as is area between reniform stigma and costa; orbicular stigma usually narrowly oval, horizontal, pale with narrow dark outline and central streak; claviform stigma short, pale-centred; cross-lines usually indistinct, but occasionally well defined. Hindwing whitish with darker veins and moderately distinct discal spot. *Female.* Antenna simple. Forewing blackish with pale ochreous orbicular stigma and more or less distinct pale lines above and below it. Hindwing grey with whitish cilia.

Subsp. *insula* Richardson
Agrotis puta insula Richardson, 1958, *Entomologist's Gaz.* **9**: 129.

(Pl.8, figs 3,4). Forewing in male more ash-coloured with stronger markings, in female hardly darker than that of male.

Life history

Ovum. Globular with small, flat base, strongly ribbed from rather wide micropyle; dark in colour, with darker markings (Stokoe & Stovin, 1958). Laid in captivity in large batches on foodplant or sides of container, hatching in seven days.

Larva. Full-fed c.32mm long, plump, with small head and

nal segment and intersegmental divisions markedly contricted. Head light brown, glossy, lobes and mouth marked with darker brown. Body dull dirty brown with darker mottling dorsally and in spiracular region; dorsal line narrow, brown, broadening towards posterior of each segment and narrowly edged with darker brown; subdorsal lines dark brown; subspiracular lines faint, whitish grey; pinacula very dark brown; spiracles black.

Polyphagous on herbaceous plants, including species of *Rumex*, *Taraxacum*, *Polygonum* and *Lactuca*. Those obtained in autumn are known to overwinter at least until March, but it is uncertain whether the spring generation of moths arises from these or from those which have pupated in the autumn.

Pupa. Light brown, integument rather thin; abdominal segments minutely punctate anteriorly; cremaster with two rather widely set, divergent, short spines. Subterranean.

Imago. Flies in May and early June, in July and August, and often again in late September and October. While the summer generation is undoubtedly the progeny of the spring one, it is still uncertain whether the other flights constitute distinct generations or result from variation in the rate of larval development. Rests concealed by day, and at night visits flowers, including those of lime and *Buddleia*, honey-dew and sugar; it also comes commonly to light.

Distribution (Map 98)

Inhabits pastures, cultivated ground, marshes and woodland. It is common in southern England, decreasing northwards; it is present in the Channel Islands. It has not been reliably recorded from Ireland. Subsp. *insula* is known only from the Isles of Scilly. Mediterranean-Asiatic; in western Europe it extends northwards through Portugal, Spain and France to western Belgium and south Germany, but is local and rare in the last two regions.

AGROTIS RIPAE (Hübner)
The Sand Dart

Noctua ripae Hübner, [1823], *Samml.eur.Schmett.* 4: pl.151.
Type locality: Europe.

Description of imago (Pl.8, figs 7–13)

Wingspan 32–42mm. Antenna in male acutely dentate to near tip, in female simple. Forewing typically sand-coloured, variously shaded, with small but distinct stigmata; crosslines represented by series of dark dots; wedge-shaped marks in subterminal area are frequently present; veins are sometimes beautifully demarcated in white. Hindwing in male white, veins often slightly darkened, in female pale fuscous with darker veins and trace of discal spot.

Local races from different parts of the country are recognizable in series; every population, however, shows

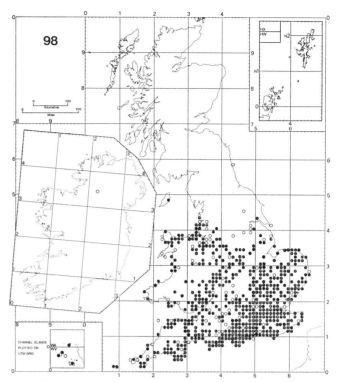

Agrotis puta

considerable variation within itself. Very white forms with the markings indistinct (ab. *obotritica* Schmidt, fig.7) or well developed (ab. *weissenbornii* Freyer, fig.11) occur in Hampshire and Dorset. South Devon specimens are usually reddish and referable to ab. *desillii* Pierret (fig.10), though the specimen illustrated came from Wittering, Sussex. On the Lincolnshire coast a grey form (ab. *grisea* Tutt) is found, while the reddish brown ab. *brunnea* Tutt (fig.12) occurs farther north, in Yorkshire. A very striking form, pale with strong blackish markings, occurs at Rosslare in south-east Ireland (fig.13) and also in Cornwall.

Similar species. *Euxoa cursoria* (Hufnagel) differs in having a longer, narrower forewing, and the space between the reniform stigma and the dorsum is only slightly greater than the depth of the reniform stigma itself, whereas in *Agrotis ripae* it is about twice that depth. The hindwing of *Euxoa cursoria* is bordered and usually has a distinct discal spot; in *Agrotis ripae* it is not bordered and the discal spot is usually absent. Confusion is possible with *Luperina nickerlii gueneei* Doubleday but that has a chequered fringe on the forewing, a white-edged reniform stigma and clearer, pale, dark-edged cross-lines.

Life history

Ovum. Rather large for size of moth, globular with small,

flat base, delicately striate, dark and glossy (Stokoe & Stovin, 1958).

Larva. Full-fed *c.*40mm long, cylindrical. Head light brown, glossy. Prothoracic and anal plates dull yellowish. Body dull ochreous or greenish, obscurely marked; spiracles large, black; legs and prolegs tipped dark brown.

Feeds from August until late in the autumn on a variety of dune plants, but especially on prickly saltwort (*Salsola kali*), sea-rocket (*Cakile maritima*) and species of *Atriplex*, feeding at night and hiding in the sand by day, from whence it can be raked with the fingers, often in large numbers, in the autumn. When so disturbed, it is extremely sluggish and appears to be dead. It can be reared in large flowerpots or biscuit-tins of sea-sand covered with glass, and fed on slices of carrot. Overwinters deep in the sand. The containers should be kept out of doors and should be left undisturbed.

Pupa. Very pale brown with thin integument; wing-cases extremely finely sculptured; two cremastral spines widely divergent, short and slender. Pupation takes place in the sand in late spring.

Imago. Univoltine. Flies from mid-June to mid-July, earlier than *Euxoa cursoria*. Can be found at night resting on sand or feeding at the flowers of marram and lyme-grass, and comes plentifully to sugar and light.

Distribution (Map 99)

It is strictly coastal, occurring on sand-dunes, especially on those just above high-water-mark. It is found on all suitable coasts of England and Wales, east Scotland, Jersey, Alderney and the Isle of Man; it also occurs on the east and south coasts of Ireland. Eurasiatic, halophile; in western Europe, local on sandy coasts from Portugal to Denmark and the Baltic, just reaching the extreme south of Sweden.

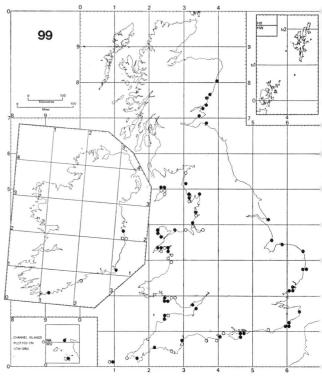

Agrotis ripae

AGROTIS CRASSA (Hübner)
The Great Dart

Noctua crassa Hübner, [1803], *Samml.eur.Schmett.* 4: pl.32
Type locality: Europe.

Description of imago (Pl.8, fig.16)

Wingspan 40–50mm. Antenna in male strongly pectinate, in female simple. Forewing ground colour varying from yellowish white to brown with reniform, orbicular and claviform stigmata darker but sometimes obscure; reniform stigma large and squarish; antemedian fascia double, with sharp outward extension at the dorsum; submarginal fascia outwardly dentate, with inward facing 'darts', often obscured. Hindwing in male pure white, in female with much dark suffusion. Abdomen with conspicuous lateral tufts, in male white, in female dark.

Similar species. *A. segetum* ([Denis & Schiffermüller]) is usually smaller, has the reniform stigma less square and the abdominal tufts small and inconspicuous; *A. trux lunigera* Stephens has the orbicular stigma conspicuously white; *Peridroma saucia* (Hübner) has the antenna in male fasciculate-ciliate and the hindwing translucent.

Life history
Early stages unknown in Britain.
Larva. Head small, shining brown with lighter central lines. Body glossy, dirty brown with obscure darker lines on the dorsum and clearer lines on the sides; long grey-white bristles on the pinacula. Overwinters, eating roots of grasses and other plants. Has been reported as a pest of vines in Alsace. Figured in Spuler (1910, **4**: Nachtrag pl.2, fig.33).
Pupa. Notably short and stout; light red-brown; cremaster hollowed with two short points (Forster & Wohlfahrt, 1963, **4**: 16, 17).
Imago. According to Barrett (1896, **3**: 287) the only British mainland specimen is one received by S. Stevens among a lot of common species taken at Dover – probably an accidental importation. It could, however, be overlooked in the field among *Agrotis segetum*. Flies in August.

Distribution
In the Channel Islands there have been occasional captures from 1874 onwards, most recently in Jersey in 1961 and 1962, and in Alderney in August 1974 (Long, 1965 and pers. comm.): it may be immigrant or possibly precariously established. Eurasiatic; the species occurs in Morocco, Portugal and Spain; it is widespread in south and central France, but is not mentioned in recent lists for Brittany and Le Havre; very rare in south-east Belgium.

AGROTIS DEPRIVATA Walker
Agrotis deprivata Walker, 1857, *List lepid.Insects Colln Br. Mus.* **11**: 739.
Euxoa bilitura Hampson, 1903, *Cat.Lepid.Phalaenae Br. Mus.* **4**: 290, *nec* Guénee (1852).
Type locality: Chile; Valdivia.

A female of this species, which is widespread in Chile and Argentina, was exhibited on behalf of Mr. I. G. Sims at a meeting of the British Entomological and Natural History Society on 8 September 1977 (Emmet, 1978). The specimen was reared from a larva found by Mr. Sims on the floor of a warehouse in Spitalfields Market, London on 24 June 1977, which pupated almost immediately in soil without a cocoon; the moth emerged on 19 July 1977. The forewing is blackish grey, with the orbicular and reniform stigmata paler, joined by a black bar; the hindwing, including cilia, is white. This larva had probably been imported with onions believed to have come from Chile.

FELTIA Walker
Feltia Walker, 1856, *List lepid.Insects Colln Br.Mus.* **9**: 165, 202.

FELTIA SUBGOTHICA (Haworth)
Noctua subgothica Haworth, 1809, *Lepid.Br.*: 224.
Type locality: England.

Very doubtfully British. The insect described by Haworth in 1809, and said by him 'to have been taken once recently', was regarded by Tutt (1891, **2**: 46, 51) as a form of *Euxoa tritici* (Linnaeus). Stephens (1829) figured under the name *subgothica* a specimen said to have been taken with some others near Barnstaple, north Devon. This specimen is now in BMNH (RCK coll.). It is undoubtedly a North American species, which was later known there as *Agrotis jaculifera* Guenée. Stephens also mentioned captures near London and in Norfolk, but there is no subsequent record in Britain. Barrett (1896, **3**: 351, pl.132), who also figured the species, suggested that the Barnstaple specimens might have been introduced with rough timber, which was then frequently imported.

FELTIA SUBTERRANEA (Fabricius)
Noctua subterranea Fabricius, 1794, *Ent.syst.* **3** (2): 70.
Agrotis annexa Treitschke, 1825, *Schmett.Eur.* **5**: 154.
Type locality: America.

Probably introduced, possibly also misidentified. The species was treated as British by Stephens (1829), under the name *Agrotis annexa* Treitschke, on the strength of three specimens: 'one taken nearly 30 years ago and later destroyed; one in June 1817, near West Ham; and one near Cork in June 1826, now in the cabinet of Mr. Stone'. The species is North American. Tutt (1891, **2**: 8) considered that the insect referred to by Stephens was a small variety of *A. ipsilon* (Hufnagel). There is no later record.

ACTINOTIA Hübner

Actinotia Hübner, [1821], *Verz.bekannt.Schmett.*: 244.
Chloantha Boisduval, Rambur & Graslin, 1836, *Colln icon.hist.Chenilles Eur.*: Noctuélides text & pl.22, figs 5,6.

Four species occur in the Palaearctic region, one of which occasionally reaches Britain as a rare immigrant.

Frons smooth; antenna in male ciliate; palp porrect, hairy below. Mid- and hindtibia spined; foretibia fringed with hair only. Wings with termen subcrenulate. Forewing marked with longitudinal dark streaks between the veins; cross-lines more or less obsolete; series of wedge-shaped marks arising from subterminal line.

Larvae, where known, always on *Hypericum* spp., hiding by day under leaves at ground level. Bivoltine, moths in April to May and August.

ACTINOTIA POLYODON (Clerck)
The Purple Cloud

[*Phalaena*] *polyodon* Clerck, 1759, *Icones Insect.rar.* 1: pl.11, 2.
Phalaena (Noctua) perspicillaris Linnaeus, 1761, *Fauna Suecica* (Edn 2): 317.

Type locality: [Sweden].

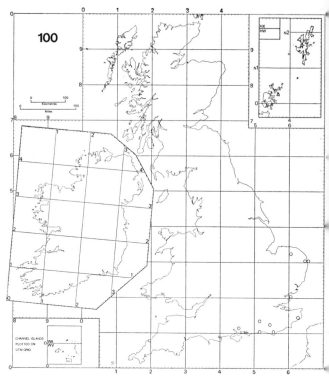

Actinotia polyodon

Description of imago (Pl.8, fig.17)
Wingspan 31–36mm. Forewing yellow-brown, with purplish costal suffusion; two whitish bars from base to reniform stigma, which is large and ringed with white; submarginal fascia with four long teeth extending into the cilia. Hindwing dull white, outwardly darker.

Life history
The early stages have not been recorded in Britain.

Ovum. Hemispherical, with many strongly raised ribs, of which only about one-third reach the micropyle.

Larva. Head small, light brown and glossy. Body red-brown, spotted with black, darker on dorsum; dorsal and lateral lines yellow, slender, and obscure, with black oblique streaks between them and broad yellow stripes below; pinacula black, in dark spots above the stripes (Forster & Wohlfahrt, 1965, 4: 123; figured by Spuler, 1910, 2: pl.28, fig.3 and Hoffmeyer, 1962). Feeds nocturnally on flowers and seeds of St. John's-wort (*Hypericum* spp.), usually in July and August but occasionally as a second generation later.

Pupa. Short and stout, sharply contracted towards the cremaster, which has two thin hooks and several setae. Usually overwinters.

Imago. It is a very scarce immigrant to south-east England, probably from northern France, Belgium and the Netherlands. There are less than a dozen records, about half at sugar, one in a spider's web, others at light, only the first generation being known here. Abroad, frequently found at flowers, and sometimes on tree-trunks by day, mainly in woods on light, sandy soil.

Distribution (Map 100)
The British records are as follows – Hampshire: one reported in 1855, but caught earlier. Sussex: Worthing, 15 May 1919; Hailsham, a fresh male 30 May 1954; Tilgate Forest, a battered male 5 June 1954. Kent: Reinden Wood, Folkestone, 4 June 1892. Essex: Bradwell-on-Sea, 27 May 1954, 16 May 1960. Norfolk: Yarmouth, June 1839; Gorleston lightship, 1938. The only record from Ireland, in 1891, is unconfirmed. The species is Eurasiatic; in western Europe it occurs in Spain and is widespread in France; in Belgium, in the south of the Netherlands and Denmark it is local and usually scarce; it reaches south Norway, central Sweden and Finland.

AXYLIA Hübner
Axylia Hübner, [1821], *Verz.bekannt.Schmett.*: 242.

A single species in the Palaearctic region. A genus of somewhat uncertain affinity and placed by some authors in Heliothinae. Habits and larval characters are essentially noctuine (Beck, 1960), although the male genitalia are unlike any others of the subfamily.

AXYLIA PUTRIS (Linnaeus)
The Flame

Phalaena (Noctua) putris Linnaeus, 1761, *Fauna Suecica* (Edn 2): 315.

Type locality: Sweden.

Description of imago (Pl.8, fig.18)

Wingspan 30–38mm. Antenna in male finely ciliate, in female simple. Forewing pale straw-coloured, costal region above stigmata dark brownish fuscous; a streak of similar colour extends from reniform stigma to fringe; reniform stigma bluish grey, more or less filled in with brownish fuscous; orbicular stigma small, circular, straw-coloured, similarly darkened; claviform stigma dot-like or absent; antemedian line strongly zigzagged, sometimes hardly discernible; postmedian line a series of black dots; a dark, irregular spot in tornal region; fringe chequered. Hindwing light brownish fuscous, veins and discal spot a little darker; a series of small, semilunular dots adjacent to fringe, which is chequered, most strongly anteriorly; underside with distinct postmedian line.

A remarkably invariable species but the straw ground colour of the forewing is sometimes tinged with reddish or brown.

Similar species. *A. putris* could only be confused with *Mesapamea secalis* (Linnaeus) ab. *oculea* Haworth which has a broader forewing with a more arched costa and a pale instead of dark reniform stigma, lacks the dark streak extending from the reniform stigma to the hind margin and rests with its wings flat instead of pleated.

Life history

Ovum. Hemispherical with flattened base, ribbed and reticulate; pale creamy yellow developing dark clouding round micropyle and in a broken median zone. Laid in large batches on the underside of leaves of the foodplant, hatching in five days.

Larva. Full-fed *c*.35mm long, body stout, tapering anteriorly from first abdominal segment and truncate posteriorly; abdominal segments 1, 2 and 8 noticeably swollen. Head brown, glossy with darker markings. Body greyish brown, mottled and irrorate with blackish; a small yellowish dorsal spot on each segment, more conspicuous on

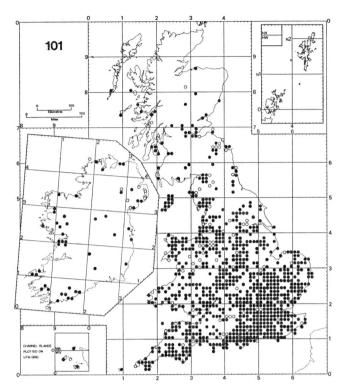

Axylia putris

abdominal segments 1 and 2; thorax with dark mediodorsal line; abdomen with series of greenish grey, yellow-edged, cuspidate dorsolateral marks; a yellowish transverse band on abdominal segment 8; spiracular stripe pale; spiracles rather inconspicuous, pale, peritreme black.

Feeds at night from July to October on herbaceous plants including species of *Rumex*, *Polygonum*, *Lamium*, *Taraxacum*, *Chenopodium*, *Lactuca*, hedge-bedstraw (*Galium mollugo*), and hound's-tongue (*Cynoglossum officinale*).

Pupa. Short, stout, rounded; reddish brown, greyer dorsally, not very glossy; wing-cases minutely sculptured; cremaster with two widely divergent short spines. Subterranean from October to June in a brittle, earthen cocoon.

Imago. Mainly univoltine. Flies in June and July, sometimes with a few specimens from late August to October. Comes to flowers (Barrett, 1897), sugar and light.

Distribution (Map 101)

Inhabits cultivated land, gardens, lanes and borders of woods and is common to very common throughout lowland Britain and Ireland; the Channel Islands. Eurasiatic; in western Europe widespread from Portugal to the south of Norway, Sweden and Finland.

OCHROPLEURA Hübner

Ochropleura Hübner, [1821], *Verz.bekannt.Schmett.*: 223.
Hapalia Hübner, [1821], *ibid.*: 220, *nec* Hübner, 1818.
Ogygia Hübner, [1821], *ibid.*: 225.
Actebia Stephens, 1829, *Nom.Br.Insects*: 41.

Numerous species in the Holarctic region. There are two species which are resident in the British Isles and two which are very rare immigrants.

Frons smooth; antenna in male rather finely fasciculate-ciliate. All legs with spines; tarsi with three rows of spines. Male genitalia strongly spinose at tip of valva, clasper rather weak; aedeagus with a strong spine.

OCHROPLEURA PRAECOX (Linnaeus)
The Portland Moth

Phalaena (Noctua) praecox Linnaeus, 1758, *Syst.Nat.* (Edn 10) **1**: 517.

Type locality: [Europe].

Description of imago (Pl.8, fig.19)

Wingspan 40–46mm. Antenna fasciculate-ciliate in male, slightly more coarsely than in female. Thorax pale grey-green; patagia a little paler and frons distinctly so. Forewing pale greyish green, very finely irrorate with fuscous; stigmata large, whitish, the reniform and orbicular containing darker greenish or brownish clouds; cross-lines narrow and irregular, fuscous with pale edges; a distinct reddish subterminal band; fringe chequered. Hindwing light brown, glossy, with narrow, paler terminal band separated from whitish fringe by a fine dark line; discal spot faint, narrow. Abdomen light brown.

Varies little except in minor detail.

Life history

Ovum. Apparently undescribed.

Larva. Full-fed *c.*48mm long, plump, cylindrical with well marked segmental divisions. Head light brown with blackish V-shaped mark. Body dark or light grey; dorsal line formed from a series of whitish spindle-shaped marks, edged with black, on each segment; on each side is a broad brown or reddish dorsolateral stripe, irrorate above with fuscous and separated from a narrower white lateral stripe by a blackish line; spiracular stripe broad, of various shades of bluish grey; spiracles pale with broad, dark rim; ventral surface pale grey.

Feeds nocturnally from September until June on creeping willow (dwarf sallow) (*Salix repens*) when that is present, but also on other sand-dune plants such as species of *Lotus, Stellaria, Cerastium* and *Artemisia*. Rests by day in the sand, raising a small hillock at the end of the trail from its feeding-place. Will thrive in captivity on narrow-leaved species of *Salix*.

Pupa. Dull greyish brown, finely and sparsely speckled with black; abdominal segments 4–7 with a fine black line in posterior third; spiracles dark brown; cremaster rugose with fine longitudinal wrinkles and two terminal hooks (Nordström *et al.*, 1935–41). Subterranean.

Imago. Univoltine. Flies in late August and September. Extremely docile. Sometimes it may be shaken from overhanging roots of marram by day, when it rolls down the sandy bank with wings clasped tightly round its body, somewhat resembling a dead marram sheath. After dark it soon settles down to feed at flowers such as ragwort, marram and heather, and when it comes to light it does so by fluttering along the ground in a series of short hops.

Distribution (Map 102)

Occurs chiefly on coastal sandhills where there are slacks containing well-established vegetation including creeping willow. It is found on sandy coasts of England and Ireland and is commonest, perhaps, on the coasts of north Devon, Cheshire, Lancashire, Cumbria and Ireland. It occurs inland in north Nottinghamshire and north Lincolnshire. In Scotland it occurs locally on the coast, and inland in

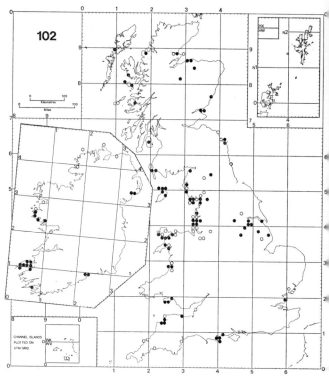

Ochropleura praecox

Strathspey, Inverness-shire where it is believed to breed on river shingle. It was recorded from Jersey before 1948 but not since. Eurasiatic; in Europe it is less restricted to coasts, occurring from Spain to central Scandinavia and Finland.

OCHROPLEURA FENNICA (Tauscher)
Eversmann's Rustic

Noctua fennica Tauscher, 1806, *Mém.Soc.Nat.Moscou* **1**: 210.

Type locality: Russia; Petropol, Capta Menfe.

Description of imago (Pl.8, fig.20)

Wingspan 38–44mm. Forewing narrow; in male dark brown near costa with broad band of pinkish orange beneath, broadening from base to wing-tip, reniform and orbicular stigmata conspicuously pale; in female forewing mottled purple-grey near costa, with area beneath yellowish white. Hindwing in male dull white, in female darkened outwardly.

Life history

The early stages have not been recorded in Britain.

Larva. Described by Spuler (1908, **1**: 153) as slender and cylindrical, speckled laterally with grey; a thin white subdorsal line, below which the sides are deep black; above the legs a rather broader irregular white stripe, heavily irrorate brown and black; the small head, very large prothoracic plate and anal plate are black. Said to be polyphagous on herbaceous plants, including willowherb (*Epilobium* spp.).

Imago. Probably immigrant from north-east Europe. Only four British specimens are known – one caught in Burney Wood, Chesterfield, Derbyshire in August 1850, and later acquired by Allis of York (Doubleday, 1850); the second, a worn female found in his light-trap at Shepperton, Middlesex by L. A. Durden in August 1972, probably on the night of either 1/2 or 4/5 (Durden, 1974 and *in litt.*), now in BMNH; the third in an m.v. light-trap at Mapperley, Nottingham, 9 August 1972 (Marchant, 1978); the fourth, a male, in a light-trap near Barthol Chapel, Aberdeen, on the night of 20 August 1977, collected by C. Marsden (Marsden & Young, 1978), this specimen is now in the Royal Scottish Museum. The 1972 specimens may have been associated with an immigration of the Finnish form of *Syngrapha interrogationis* (Linnaeus) which was recorded in small numbers in Lincolnshire, Norfolk and Orkney, and in much greater numbers in the Netherlands and Denmark, between 28 July and 1 August; that in 1977 probably accompanied a large immigration of *Eurois occulta* (Linnaeus) recorded in Aberdeenshire and elsewhere.

Distribution

Euro-Siberian and North American; in western Europe the moth is sporadically frequent in migration years in Finland; there are scattered records in Sweden, one in Norway and five in Denmark in 1968.

OCHROPLEURA FLAMMATRA ([Denis & Schiffermüller])
The Black Collar

Noctua flammatra [Denis & Schiffermüller], 1775, *Schmett. Wien.*: 80.

Type locality: [Austria]; Vienna district.

Description of imago (Pl.8, fig.21)

Wingspan 42–52mm. Thorax with prominent collar, in male black, in female suffused grey. Forewing pinkish grey; two lozenge-shaped marks and one quadrate spot between base and reniform stigma, and a faint dark pre-apical mark on costa. Hindwing greyish white with slight dark suffusion terminally and on the veins.

Similar species. It might be overlooked among pale examples of *Xestia c-nigrum* (Linnaeus) and *X. triangulum* (Hufnagel), from which it is distinguished by the conspicuous collar, slighter pre-apical spot and paler hindwing.

Life history

The early stages have not been recorded in Britain.

Larva. Head golden brown with two black curved stripes. Prothoracic plate darker brown with light central lines. Body light golden grey to green, with clear lateral stripes dark-edged above (Forster & Wohlfahrt, 1963, **4**: 21). Overwinters. In France is said to feed on species of *Fragaria* and *Taraxacum* (Lhomme, 1923–35, **1**: 158–178).

Imago. It was probably an occasional immigrant to England in the past. Barrett (1897, **4**: 49) cited three examples of whose authenticity he was satisfied; two caught at Freshwater, Isle of Wight, in 1859 and 1876, and one at Cromer lighthouse, Norfolk in 1875. Two of these are now in BMNH. No later record is known. The moth is said to fly from May to September (Lhomme, *loc.cit.*).

Distribution

Eurasiatic; in Europe it occurs in Portugal, Spain, southeast and central France, south Germany, and eastwards in the Balkans; it seems to be rare in most places and little known.

OCHROPLEURA PLECTA (Linnaeus)
The Flame Shoulder

Phalaena (Noctua) plecta Linnaeus, 1761, *Fauna Suecica* (Edn 2): 321.

Type locality: Sweden.

Description of imago (Pl.8, figs 5,6)

Wingspan 28–34mm. Antenna ciliate, more coarsely in male than in female. Thorax rich purplish brown or chestnut-brown; patagia a little paler. Forewing glossy purplish brown or chestnut-brown, fading to warm reddish brown; a characteristic broad, pale straw-coloured costal streak extending three-quarters length of wing from base, edged dorsally by a diffuse black line which extends through reniform and orbicular stigmata; these are straw-coloured, more or less filled with greyish brown clouding; orbicular stigma varies from circular to spindle-shaped; reniform stigma constant; claviform stigma absent; an indistinct, pale subterminal shade is sometimes present. Hindwing shining white with pale, reddish brown cilia and occasionally a few elongate dots on veins representing postmedian line.

Variation is slight but occasionally the costal stripe is heavily suffused with reddish (ab. *rubricosta* Fuchs, fig.6).

Similar species. Unlikely to be confounded with any other British species, but *O. leucogaster* (Freyer) occurs in Brittany and might just reach Britain. It is a little larger, has a white patch on the anterior abdominal segments and the black mark on the forewing extends farther beyond the reniform stigma; the hindwing is even purer white than in *O. plecta*.

Life history

Ovum. Hemispherical with flattened base, upper half strongly ribbed, less so below, with very delicate reticulations; bright yellow when laid. In captivity laid in loose batches or singly, hatching in eight days.

Larva. Full-fed *c.*35mm long, stout, slightly attenuated anteriorly. Head small, rounded, glossy brown, shaded and reticulate with darker coloration. Body in early instars greyish brown with very conspicuous, broad, white spiracular line; in later instars yellowish brown irrorate with paler colour, sometimes reddish or greenish; dorsal and subdorsal lines inconspicuous, thread-like and interrupted, whitish, shaded with grey on each side; spiracular stripe very broad and distinct, yellowish ochreous or whitish, upper edge bounded by a dark brown line through spiracles; pinacula distinct, black; spiracles whitish, surrounded by dark brown.

Feeds nocturnally during the summer and is polyphagous on herbaceous plants including species of *Plantago*, *Rumex* and *Galium*, groundsel (*Senecio vulgaris*), cultivated lettuce and beet.

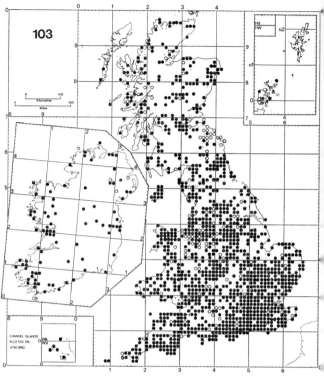

Ochropleura plecta

Pupa. Dark purplish brown, broadly spindle-shaped, tapering to a point posteriorly; wing-cases minutely and abundantly sculptured with fine lines; abdominal segments punctate, 4–6 with smooth pale rim posteriorly, resembling hoops of a barrel; cremaster with two close-set parallel hooked spines and two very fine hooked bristles offset on each side. On or just below the surface of the soil; the winter is probably passed in this stage.

Imago. Has two peaks of emergence, in May and early June, and second half of August, but occurs from April all through summer, straggling into autumn. Comes very freely to light, sugar and ragwort flowers. It is a wild flier and the author has witnessed at least three instances when it has entered the ear of a fellow entomologist who was working at light, disappearing inside and causing acute discomfort!

Distribution (Map 103)

Inhabits open woodland, farmland, gardens, meadows and marshes. Common in suitable habitats throughout the British Isles from the Channel Islands to Orkney, and the Hebrides. Holarctic; its range includes North Africa and Europe to central Scandinavia and Finland.

EUGNORISMA Boursin
Eugnorisma Boursin, 1946, *Revue fr.Lépidopt.* **10**: 188.

The genus includes numerous species occurring in the Palaearctic region, chiefly in central Asia and the Middle East. Of these, only one is found in Britain.

Distinguished by the characteristic shape of the valva, and the spiny armature of the aedeagus in the male genitalia.

EUGNORISMA DEPUNCTA (Linnaeus)
The Plain Clay

Phalaena (Noctua) depuncta Linnaeus, 1761, *Fauna Suecica* (Edn 2): 321.

Type locality: Sweden.

Description of imago (Pl.9, fig.30)
Wingspan 36–44mm. Antenna in male setose, in female simple. Forewing light brownish ochreous, sometimes reddish-tinted, finely irrorate with fuscous; reniform and orbicular stigmata narrowly outlined ochreous and clouded within with fuscous; basal line consists of two conspicuous black spots separated by pale vein 11 (R_1); antemedian line consists of an oblique black band, widest at level of orbicular stigma which it touches, divided by the pale veins 2 (Cu_2) and 11 (R_1) and usually failing before dorsum; postmedian line narrow and often reduced to dots, light brown or fuscous; wavy brown subterminal line broadest in apical region; area between postmedian and subterminal lines distinctly paler and containing a row of elongate blackish dots on veins; a faint median fascia passes between stigmata where it often forms a blackish patch. Hindwing glossy light brown, sometimes showing faint discal spot and even fainter postmedian line; cilia pale straw-coloured, lightly tinted pinkish. Abdomen in female distinctly shorter and stouter than in male.

The species varies little except in tone of ground coloration. Continental specimens are rather larger and distinctly greyer.

Life history
Ovum. Globular, base flattened; cream-coloured when laid, developing reddish spot in micropylar region and narrow reddish belt. In captivity eggs have been laid in masses on and through muslin, hatching in about three weeks.

Larva. Full-fed *c.*47mm long, plump. Head narrower than first thoracic segment, glossy ochreous with blackish dash on each side. Body from dorsum to spiracular region greyish ochreous, the colour of old limestone, finely irrorate with dark grey; dorsal line pale, thread-like and broken; a darker lozenge-shaped mark on each segment, from the posterior edge of which black wedge-shaped marks extend forward like an inverted V, more conspicuous on abdominal

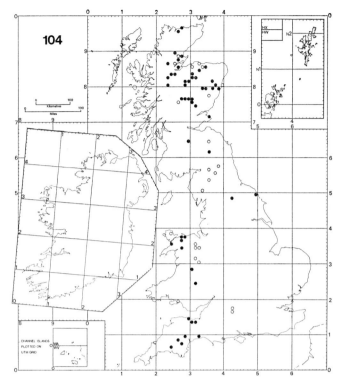

Eugnorisma depuncta

segments 6–8; colour of dorsal region gradually intensifies towards a broad, pale sulphur-yellow spiracular stripe; ventral surface, including legs and prolegs, abruptly greyish ochreous; spiracles whitish, each surrounded by a narrow black peritreme; in early instars the dorsal region is rich mahogany or purplish brown, contrasting even more strongly with the sulphur-yellow spiracular stripe.

Immediately on hatching, young larvae overwinter gregariously in dry, hollow grass-stems. Kept in a refrigerator, they will come out and start feeding in the new year, as soon as they are brought into the warmth. Once this has happened, they will feed up quickly on dock (*Rumex* spp.) without trouble, though mould is a danger just before diapause is broken. According to Barrett (1897) the young larvae can be swept in the wild as early as January. Recorded foodplants include species of *Rumex*, *Primula* and *Lamium*; the larva feeds until May in the wild.

Pupa. *C.*15mm long, reddish brown, glossy; wing-cases smooth; abdominal segments sparsely punctate anteriorly; cremaster a short, flattened, parallel-sided projection, lightly sclerotized except at tip and bearing two slightly convergent, straight spines and, laterally, two pairs of very fine hooked bristles. Pupation takes place in the soil in May.

Imago. Univoltine. Flies from July to early September. Visits flowers of ragwort, also sugar and light.

Distribution (Map 104)

Inhabits deciduous woodland. It is locally common in suitable situations in central and east Scotland, extending southwards into the northern counties of England and north Wales, becoming progressively less frequent. It has been found as far south as Savernake Forest, Wiltshire and Devon, but is very rare in these localities for reasons that are not understood. Euro-Siberian; in western Europe local and often rare in hills and mountains from the Pyrenees through southern and central France to central Germany; it is absent from Belgium, the Netherlands and north-west Germany, but occurs in Denmark and southern parts of Norway, Sweden and Finland, where it is mainly coastal.

STANDFUSSIANA Boursin
Standfussiana Boursin, 1946, *Revue fr.Lépidopt.* **10**: 190.

About 12 species in the Palaearctic region, mostly in mountainous places. There is only one British representative.

Frons smooth; antenna in male coarsely ciliate. Distinguished from related genera by the structure of the male genitalia, the cucullus being long and narrow, the costal region of the valva expanded into a flattened plate and the aedeagus broad, scobinate and without cornuti.

STANDFUSSIANA LUCERNEA (Linnaeus)
The Northern Rustic

Phalaena (Noctua) lucernea Linnaeus, 1758, *Syst.Nat.* (Edn 10) **1**: 510.

Type locality: Europe.

Description of imago (Pl.9, figs 1,2)

Wingspan 36–46mm, more variation in size occurring in specimens from different populations, rather than within them. Antenna in male more coarsely ciliate than in female. Forewing from light ochreous grey to nearly black; reniform and orbicular stigmata present but often hardly discernible, claviform stigma absent; ante- and postmedian lines edged with ochreous, varying in intensity: on either pale or dark specimens, darker or paler components of the lines may be the more conspicuous; a narrow, dark central fascia often present. Hindwing heavily suffused fuscous, usually having a broad but indistinctly outlined dark border in outer third of wing, contrasting with whitish cilia. Underside conspicuously banded on both fore- and hindwing.

This elegant but indistinctly-marked species is much given to local variation. The largest and darkest specimens come from Shetland, where they are almost blue-black when fresh. Orkney forms are smaller and more strongly marked, and fit the description of ab. *renigera* Stephens (fig.2), which also occurs round much of Scotland including the Hebrides. Dark forms also occur in parts of Ireland. Coffee-coloured forms occur on the Great Orme, the mountains of north Wales and the Lake District; those from Portland and the Isle of Wight are similar except that they are larger on average and have a tendency for the terminal area of the forewing to be darkened.

Similar species. Rhyacia simulans (Hufnagel) could be confused with *Standfussiana lucernea*, but it has a much larger and more conspicuous orbicular stigma, and often a dark interstigmatal patch; the cross-lines, which lack the pale component, are represented by a series of dark dots; the markings on the underside are less contrasting. *Spaelotis ravida* ([Denis & Schiffermüller]) is distinguished by the presence of a basal dash on its forewing and by the paler hindwing.

Life history

Ovum. Flattened-hemispherical with slightly elevated micropyle and about 48 longitudinal ribs; light reddish grey (Nordström et al., 1935–41: 96). Eggs laid in captivity were stuffed into a pad of honeyed cotton-wool and hatched in 18 days.

Larva. Full-fed c.45mm long, stout with each segment thickened and rather wrinkled laterally. Head dark fuscous with two small, pale spots. Body dusky olive-brown or olive-green, mottled and irrorate with small blackish streaks; subdorsal region of each segment with a faint, pale olive-green spot, below which there is an oblique shading of blackish green; spiracles inconspicuous, pale with dark peritreme; ventral surface pale olive-green; thoracic legs pale brown, darker-tipped.

Wild larvae feed up in spring and may be found at night stretched out on rocks near their foodplants which include harebell (*Campanula rotundifolia*), biting stonecrop (*Sedum acre*), species of *Primula* and *Saxifraga*, sheep's-fescue (*Festuca ovina*) and probably other grasses. Apparently it is a difficult species to rear from the egg; the larvae die about half-grown from an unknown cause.

Pupa. Very smooth and shining; wing-cases tumid and highly transparent, but encasements of other appendages hardly perceptibly raised except for those of the palpi which form a small knob-like projection; anal segment rather swollen, rounded and terminating in a blunt point from which arise two parallel, slender spikes.

Imago. Univoltine. Flies from July to September, later in the north. It has a spectacular flight on sunny afternoons from about 15.00 to 16.00 hours, when it dashes about at great speed along cliffs and over scree. At night, however, it flies gently, when its banded underside is conspicuous by lamplight. It comes to flowers of red valerian, heather, wood-sage and ragwort as well as to sugar and light.

Distribution (Map 105)

It inhabits mountains, ravines, scree, quarries and rocky places by the sea. It is chiefly western and northern in Britain, though it also occurs locally on rocky coasts from the Isle of Wight to Cornwall. It is recorded from all coastal counties of Ireland except Cos Louth, Wicklow and Limerick (Baynes, 1964). Probably Eurasiatic, though the eastern limits of its range are still uncertain; in western Europe it is found as several subspecies in the Sierra Nevada and Pyrenees, the Massif Central of France, the Alps where it reaches a height of 3,000m (10,000ft), and the mountains of south-west Norway; it also occurs on the coasts of Sweden and south Finland.

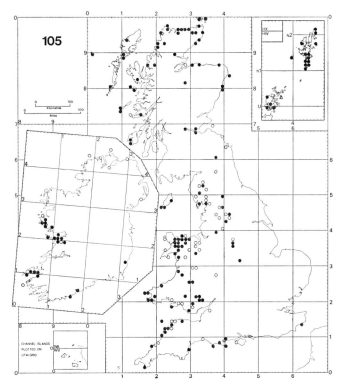

Standfussiana lucernea

RHYACIA Hübner
Rhyacia Hübner, [1821], *Verz.bekannt.Schmett.*: 210.
Epipsilia Hübner, [1821], *ibid.*: 210.

A Holarctic genus. One species is resident in the British Isles, and a second has occurred on a single occasion.

Frons smooth and domed, in side view hardly extending beyond eyes; antenna in male ciliate.

RHYACIA SIMULANS (Hufnagel)
The Dotted Rustic

Phalaena simulans Hufnagel, 1766, *Berlin.Mag.* **3**: 396.
Noctua pyrophila [Denis & Schiffermüller], 1775, *Schmett. Wien.*: 71.

Type locality: Germany; Berlin.

Description of imago (Pl.9, figs 3–5)
Wingspan 45–60mm. Antenna in male shortly ciliate, in female simple. Forewing colour varies greatly according to locality, from pale yellowish brown (fig.3, Portland), biscuit-colour (fig.4, Cotswolds and north Berkshire), dark brown (Inverness-shire), to nearly black in f. *suffusa* Tutt (fig.5, Hebrides, Orkney, Shetland); stigmata faintly outlined darker; transverse lines faint, the antemedian consisting of darker dots and the subterminal shortly dentate. Hindwing brown, darker on veins; cilia whitish.

Similar species. *R. lucipeta* ([Denis & Schiffermüller]), a very scarce immigrant, is larger, lighter-coloured, and has pale stigmata strongly outlined darker.

Life history
Ovum. Undescribed.
Larva. Full-fed *c*.43mm long. Head pale brown, glossy. Body thick-set, pale buff, darkening in later instars; eight lozenge-shaped black lateral markings with diagonal pale lines beneath joining a whitish spiracular stripe; anal plate ochreous, freckled darker.

The early stages are hardly known in the wild, and in captivity the species is difficult to rear. Eggs laid in late September hatched in 16 days; the larvae fed exclusively on grasses, but herbaceous plants such as dock and dandelion were needed in the later instars. They went down into peat for pupation at the end of November, but died later (R. I. Lorimer; Haggett, 1968). In the wild larvae probably overwinter and complete their growth in spring, as has been reported in Germany.
Pupa. Undescribed.
Imago. In England moths emerge and fly in late June and early July, then aestivate and reappear in late August and September, varying greatly in number from one year to another. Especially attracted to blossoms of red valerian

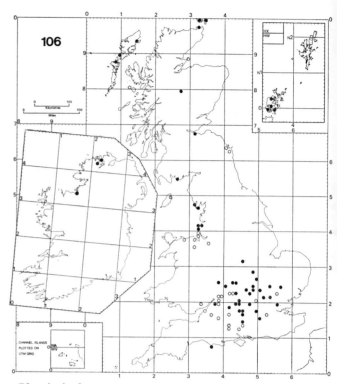

Rhyacia simulans

and *Buddleia*. In Orkney moths fly freely from mid-July to early August, and a fertile female has been taken in early September.

Distribution (Map 106)
Discontinuous. In central southern England, from Dorset to north Berkshire, Oxfordshire and Warwickshire, local on calcareous soils, but found singly more widely in years of abundance; in north Wales and the Scottish Highlands, near mountain scree; in the Hebrides and Orkney, mainly on moorland and pasture. In Ireland there are records from Co. Sligo, and examples of f. *suffusa* were taken in mid-August 1969 in Co. Donegal (Baynes, 1964; 1970). Eurasiatic; in western Europe local and seldom common, from Portugal to south Norway and farther north in Sweden and Finland. In the Alps it reaches 2,000m (6,500ft).

NOTE. See *Entomologist's Rec. J. Var.* **90**: 324 for records in Kent and Essex in 1978.

RHYACIA LUCIPETA ([Denis & Schiffermüller])
The Southern Rustic

Noctua lucipeta [Denis & Schiffermüller], 1775, *Schmett. Wien.*: 71.

Type locality: [Austria]; Vienna district.

Description of imago (Pl.8, fig.22)

Wingspan 57mm. Forewing dull greenish grey, fading when worn; orbicular and reniform stigmata conspicuously white-ringed; four broken white fasciae. Hindwing dull grey.

Similar species. *Standfussiana lucernea* (Linnaeus) is smaller, darker, without greenish tinge on forewing, and lacks pale-ringed stigmata, as does *Rhyacia simulans* (Hufnagel) *q.v.*

Life history

The early stages are unknown in Britain but have been described in Germany. Eggs are laid on many herbaceous plants, among which the larvae overwinter, completing their growth in May (Bergmann, 1954, **4**).

Larva. Head golden brown without markings. Body grey-green, with obscure paler dorsal, and yellowish lateral stripes; four black pinacula on each segment (Forster & Wohlfahrt, 1963, **4**: 28; figured by Spuler, 1910, **4**: Nachtrag pl.2, fig.25).

Imago. Probably a very occasional immigrant. The only British capture was made by Wightman (1969) in a light-trap at Pulborough, Sussex, after 02.00 hours on 15 July 1968. This specimen, a rather worn female which lacks the greenish tinge, is now in BMNH. July 1968 was a month of great migratory activity, during which more than a dozen scarce immigrant species were recorded. The beginning of this activity coincided with a fall of wind-borne red dust over much of southern Britain on the night of 30 June/1 July. Arrivals about that date were traced meteorologically from north-west Spain or Morocco (French & Hurst, 1969); but those caught later in the month may have come from elsewhere.

Distribution

Palaearctic; occurring in western Europe mainly on limestone in south Spain, and in France in the Pyrenees, Auvergne and Alps, reaching the Ardennes in Belgium and Limburg in the Netherlands; it is also known in west and central Germany, where it is scarce.

NOCTUA Linnaeus

Noctua Linnaeus, 1758, *Syst.Nat.* (Edn 10) **1**: 508.
Triphaena Ochsenheimer, 1816, *Schmett.Eur.* **4**: 69.
Lampra Hübner, [1821], *Verz.bekannt.Schmett.*: 221.
Euschesis Hübner, [1821], *ibid.*: 221.

A small Holarctic genus, six species of which are found in Britain.

Frons smooth; antenna in male ciliate or setose; palpus hairy with short terminal segment. Forewing long and narrow. Hindwing yellow with blackish border.

NOTE. The sexes of the species in this genus are very similar; examination of the frenulum (easily seen with a lens) provides a ready means of determination.

NOCTUA PRONUBA (Linnaeus)
The Large Yellow Underwing

Phalaena (Noctua) pronuba Linnaeus, 1758, *Syst.Nat.* (Edn 10) **1**: 512.

Type locality: [Sweden].

Description of imago (Pl.8, figs 25–30)

Wingspan 50–60mm. Sexually dimorphic, a fact ignored by most previous authors. Size and general shape of the two sexes extremely similar, and both sexes polymorphic but dark specimens are males and pale ones females. Antenna finely setose in both sexes. Forewing ground colour variable (see below); orbicular and reniform stigmata large, rounded; claviform stigma absent; cross-lines weakly represented; a conspicuous black double spot at junction of subterminal line with costa. Hindwing orange-yellow with a black border about one-sixth width of wing; discal spot nearly always absent; cilia orange-yellow.

The greyish Linnaean form is evidently rare in Britain, and the usual male forms are: ab. *ochreabrunnea* Tutt (fig.25), which has the forewing marbled warm brown and ochreous, the ochreous colouring being especially prominent on the costa; ab. *distinctacaerulescens* Tutt, which is similar but has the ochreous colouring replaced by glaucous blue-grey; ab. *brunnea* Tutt, which is deep reddish brown or liver-coloured; and ab. *innuba* Tutt (fig.26), which is blackish brown. The commonest female forms are: ab. *ochrea* Tutt (fig.27), which is ochreous, freckled with brownish scales and with darker brown reniform stigma; ab. *caerulescens* Tutt (fig.28), similar but with ground colour slaty grey; and ab. *rufa* Tutt (fig.29), which is light reddish ochreous with the reniform stigma often darkened and the orbicular stigma often paler than the ground colour. Variation in the hindwing is rare but probably overlooked; the orange-yellow may be replaced by lemon or creamy white or suffused blackish to a greater or lesser extent (ab. *postnigra* Turner, fig.30 and ab. *suffusa* Cockayne). The border may be broader or narrower, or even absent; rarely a narrow discal spot is present (ab. *hoegi* Herrich-Schäffer).

Similar species. N. *orbona* (Hufnagel) is the only other similar species with a black mark at the apex of the forewing, but it is much smaller and has a large discal spot on the hindwing.

Life history

Ovum. Globular, ribbed and reticulate; pale creamy white when laid. More than 1,000 eggs are laid in large, compact masses on the underside of leaves of the foodplant or in grass inflorescences. Apparently there is an obligate pre-oviposition period of at least a month (Singh & Kevan, 1956). Eggs hatch in about a month (Scorer, 1913).

Larva. Full-fed *c*.50mm long, robust, slightly tapered anteriorly, of cutworm form and habit. Head light brown with darker markings. Prothoracic plate conspicuous, blackish in darker specimens but brown in green examples, with a narrow, pale median sulcus. Body brown, ochreous or even bright pea-green; dorsal line narrow, pale; subdorsal lines a little wider, marked above by a series of black dashes on abdominal segments; spiracles black, surrounded by dark clouding; ventral surface below spiracles dull, dingy brown.

Feeds through the winter and may be found on almost any mild night on a wide range of wild and cultivated herbs and grasses. Has more the habits of a cutworm than other members of its genus, spending much time below ground.

Pupa. C.25mm long, chestnut-brown, smooth and glossy; abdominal segments minutely punctate anteriorly; cremaster with two rather bulbous-based, parallel tapering spines and two pairs of minute bristles, set on a smooth, conical base. Pupation occurs in soil in late spring.

Imago. Univoltine, with a prolonged period of emergence, the peak of abundance occurring in late August. It is one of the commonest and greediest visitors to sugar and may fill a light-trap to the detriment of other species. Also common at valerian and *Buddleia* and other flowers. It roosts by day on the ground, in hay, dead leaves or brushwood and is readily disturbed, dashing off wildly and displaying the bright hindwings (flash coloration); nevertheless it is often spotted and pursued, with success, by birds.

Distribution (Map 107)

Though commonest in lowland habitats, it is also quite frequent on moors. It occurs throughout the British Isles, including the Channel Islands and outlying islands. Palaearctic; its range including north Africa and the whole of western Europe to Iceland, central Scandinavia and Finland.

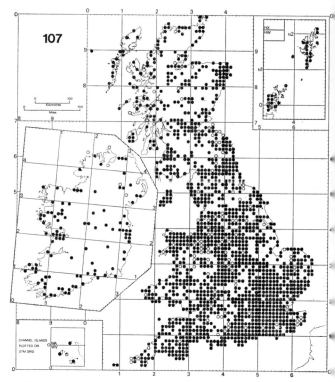

Noctua pronuba

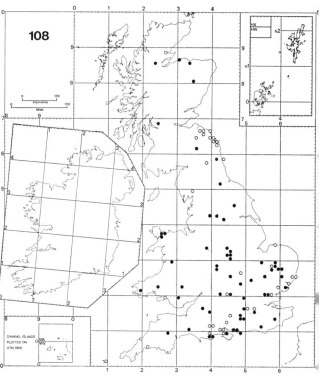

Noctua orbona

NOCTUA ORBONA (Hufnagel)
The Lunar Yellow Underwing

Phalaena orbona Hufnagel, 1766, *Berlin.Mag.* **3**: 304.
Noctua subsequa [Denis & Schiffermüller], 1775, *Schmett. Wien.*: 79.

Type locality: Germany; Berlin.

Description of imago (Pl.8, fig.23)
Wingspan 38–45mm. Antenna very finely setose in both sexes. Forewing with ground colour varying from greyish ochreous to reddish brown; reniform stigma more or less conspicuous according to the amount of darker clouding within; orbicular stigma pale-outlined but only occasionally darker within than ground colour; a conspicuous black, wedge-shaped spot at junction of subterminal line with costa. Hindwing with large and conspicuous discal spot. Less variable than *N. comes* (Hübner).

Similar species. *N. orbona* is distinguished from *N. comes* by its narrower forewing and straighter costa, and by the presence of a black mark towards the apex. *N. pronuba* (Linnaeus) is larger and lacks the discal spot on the hindwing. Boursin (1963a,b) separated *N. interposita* (Hübner) from *N. orbona*, the clearest distinctions being in the structure of the genitalia; superficially, *N. interposita* is rather larger and has broader wings, a large reniform stigma, a white costal end to the postmedian line and a broader band on the hindwing. It has been recorded from Spain and France and could occur in Britain.

Life history
Ovum. Hemispherical, somewhat flattened above, glistening, with 30 blunt ribs and faint reticulations; dirty white when laid, becoming pale grey with greyish brown blotch round micropyle and zone of irregular blotches round middle (Buckler, 1893).

Larva. Full-fed *c.*38mm long, stout, cylindrical, tapering slightly from first abdominal segment to head. Head light brown, lobes reticulate and edged with blackish stripe. Body greyish brown or pale drab; a conspicuous pale dorsal stripe narrows anteriorly and is slightly constricted between each segment; subdorsal stripe nearly as wide, fainter anteriorly, with a series of oblong blackish marks above, becoming larger posteriorly; spiracular stripe broad, greyish ochreous, pale above and below, its upper limit defined by a thread-like brown line through spiracles; spiracles white, rimmed with black, and enclosed in larger dark blotches (Buckler, *loc.cit.*).

Occurs from August or September to April, feeding through the winter and maturing before its congeners. Feeds on grasses such as cock's-foot (*Dactylis glomerata*), common couch (*Agropyron repens*) and reed canary-grass (*Phalaris arundinacea*) and on other herbaceous plants such as cowslip (*Primula veris*), creeping cinquefoil (*Potentilla reptans*) and *Ranunculus* spp. Feeds chiefly at night but often rests on grasses on mild, especially humid, days throughout winter, when it may be taken by sweeping.

Pupa. Smaller and neater than that of *N. comes q.v.*; cremaster with two nearly straight parallel spines set so close as to appear one, flanked by two pairs of minute bristles. Subterranean, in a weak, friable cocoon.

Imago. Univoltine, occurring from late June to August or September, but apparently aestivating prior to oviposition. Comes to sugar and light.

Distribution (Map 108)
In the south of its range it appears to occur mainly in open, wooded country on lighter soils including both sand and chalk; in the north it is found on moorland and sand-dunes. It is decreasing everywhere and is now absent from many places where it was once plentiful. It is probably now seen most often in the Breck district of East Anglia. The New Forest used to be a well-known locality and it was quite widespread in Scotland. Records from Shetland and Ireland require confirmation. Western Palaearctic; in western Europe it extends from south Spain and Portugal to Denmark and south Scandinavia, but is generally local and uncommon.

NOCTUA COMES (Hübner)
The Lesser Yellow Underwing

Noctua comes Hübner, [1813], *Samml.eur.Schmett.* **4**: pl.111.
Triphaena orbona sensu Pierce, 1909, *Genitalia Noct.Br. Islands*: 47.

Type locality: Europe.

Description of imago (Pl.8, figs 35–40)
Wingspan 38–48mm, southern specimens being larger than those from Scotland. Antenna shortly ciliate in both sexes. Forewing varies from pale clay towards greyish, brownish or reddish clay; reniform and orbicular stigmata, usually hardly darker than ground colour, narrowly outlined pale yellowish; a darker shade in dorsal half of reniform stigma is frequent and sometimes both stigmata are much darker than ground colour; cross-lines usually very faint, ante- and postmedian lines often being represented by series of blackish dots. Hindwing orange-yellow with irregular black subterminal band and black discal spot.

A wide range of variation occurs both within and between populations. In parts of north-east Scotland and the Western Isles it is strongly polymorphic, the forewing varying from black with stigmata and cross-lines indicated in reddish brown (ab. *nigrescens* Tutt, fig.39), through rich mahogany, variously marked with black and yellow (fig.40) to dark or lighter red, grey or clay-coloured; extreme examples of the red form, which are commoner in Scotland than elsewhere, are referable to ab. *rufa* Tutt (fig.38). Forms with dark stigmata are also more common in Scotland. Some forms are mottled, with a whitish postmedian line and a pale mark at the apex of the wing distad of the subterminal line. Specimens with very prominent, dentate cross-lines (ab. *sagittifer* Cockayne, fig.36) are frequent in the Isles of Scilly but rare elsewhere. The hindwing is less variable. In darker Scottish specimens the yellow may be more or less obliterated by black scaling. Sometimes it is replaced by pale lemon-yellow or brown. The width of the band varies slightly and the discal spot may be reduced or, occasionally, absent (ab. *connuba* Hübner).

Similar species. *N. comes* is broader-winged than *N. orbona* (Hufnagel), has the costa more arched and lacks the black mark in the apex of forewing; however, it should be noted that the junction of the subterminal line with the costa is often darkened and this could be mistaken for an apical black spot.

Life history
Ovum. Hemispherical, strongly ribbed, reticulate; pale cream-coloured when laid, becoming dark brown with darker clouding round micropyle. In captivity laid in irregular batches firmly attached to sides of container, hatching in ten to 11 days (Scorer, 1913).

Larva. Full-fed c.45–50mm long, robust, fleshy. Head light brown, banded and reticulate. Body light ochreous grey to light brown, often with greenish tint, marbled ochreous; dorsal line thread-like, pale and distinct only in thoracic region; abdominal segments 6–8 with conspicuous, dark brown, dorsolateral wedge-shaped marks edged below with pale ochreous; a transverse bar of similar colour behind wedges on segment 8; spiracles white, surrounded by rather faint, dark blotches; subspiracular line broad and pale; ventral surface slightly paler than dorsum, flecked blackish; a small black spot on each proleg.

Feeds nocturnally from September to April. Polyphagous, its foodplants including hawthorn (*Crataegus* spp.), birch (*Betula* spp.), sallow (*Salix* spp.), heather (*Calluna vulgaris*), primrose (*Primula vulgaris*), foxglove (*Digitalis purpurea*) and *Rumex* spp. According to R. I. Lorimer, very young larvae found in January and February are always on evergreen species of *Helianthemum*, *Vinca*, heather *etc.* and die if transferred to dock or other herbaceous plants, but the larger ones, which may be found commonly at night in March and April, feed on these without trouble. Larvae obtained *ab ovis* may be readily forced at normal indoor temperatures.

Pupa. 17–19mm long, dark chestnut-brown, glossy; wing-cases smooth; abdominal segments strongly punctate anteriorly; cremastral spines parallel, hooked outwards

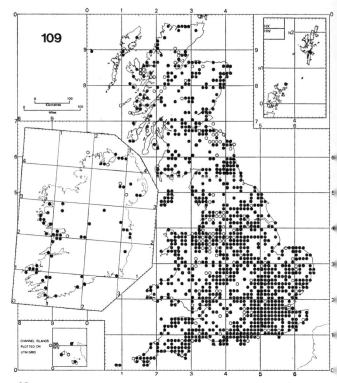

Noctua comes

at tip, flanked by two pairs of extremely fine, shorter bristles. Subterranean, in an earthen cocoon.

Imago. Univoltine, flying from July to September. Rests concealed by day and is not often seen. On the sandhills at Findhorn, Moray it flies with great rapidity at early dusk and is extremely difficult to net; this habit has not been noted in southern England. Later at night, it is common at flowers of burdock, tansy, heather and *Buddleia*, as well as at sugar and light.

Distribution (Map 109)
It inhabits all kinds of terrain and occurs plentifully throughout the British Isles to Shetland. Western Palaearctic; occurring from the Canary Isles, north Africa and Asia Minor through southern and central Europe to southern Scandinavia, being generally more widespread and abundant than *N. orbona*.

NOCTUA FIMBRIATA (Schreber)
The Broad-bordered Yellow Underwing

Phalaena fimbriata Schreber, 1759, *Nov.Spec.Ins.*: 13.
Phalaena (Noctua) fimbria Linnaeus, 1767, *Syst.Nat.* (Edn 12) **1** (2): 842.

Type locality: Germany; Halne.

Description of imago (Pl.8, figs 31,32)

Wingspan 50–58mm. Sexually dimorphic; as in *N. pronuba* (Linnaeus), dark forms are male and pale ones female. Antenna in male shortly ciliate, in female simple. Forewing ground colour variable (see below); orbicular and reniform stigmata concolorous, narrowly outlined whitish; basal line extending from costa halfway to dorsum; antemedian line oblique, darker than ground colour; postmedian line geniculate, paler than ground colour; subterminal line pale, crenulate with a blackish mark at junction with costa. Hindwing deep orange-yellow with intense black border one-half width of wing; cilia orange-yellow.

The species varies extensively throughout its range. The commonest form of the male is ab. *brunnea* Tutt, in which the forewing is reddish brown and clearly marked; similarly marked and almost as common is ab. *solani* Fabricius, in which the reddish brown is replaced by olive-green. The darkest male form is ab. *nigrescens* Busse, which has the hindwing more or less suffused fuscous. The typical female is pale ochreous; other female forms include ab. *virescens* Tutt, in which the forewing is tinted light greenish, and ab. *rufa* Tutt, in which the colour is light reddish brown. Some of the less common forms, such as the mahogany-coloured ab. *brunnea*, may belong to either sex. The brilliantly coloured hindwing seldom varies.

Life history

Ovum. Dome-shaped, strongly ribbed and delicately reticulate between ribs; micropyle papilliform; apple-green when laid, becoming paler but turning grey-green immediately before hatching. Laid in early autumn in neat batches, hatching in about ten days.

Larva. Full-fed *c.*55mm long, very robust. Head pale brown, irrorate darker. Body ochreous brown, sometimes reddish-tinted, irrorate with blackish; dorsal line fine, spiracular line broad, both pale; abdominal segments 8 and 9 have pale, transverse bands, anterior to which are pairs of rather obscure, dark, wedge-shaped marks; spiracles large, white, behind each a conspicuous black spot. In early instars the dorsum carries a diamond pattern; the large, white spiracles are already characteristic, but the black spots posterior to them are absent.

Feeds at first on birch (*Betula* spp.), sallow (*Salix* spp.) and several other species of tree, remaining on the tree until the approach of winter, when it descends to the

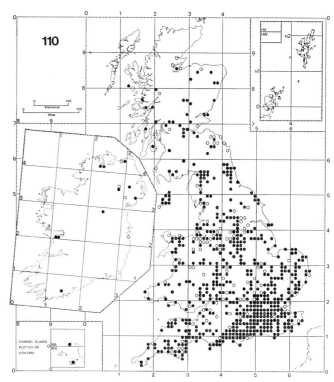

Noctua fimbriata

ground and overwinters (Newman, 1869). In spring it feeds on the opening buds of young sycamores (*Acer pseudoplatanus*) or on the autumn foodplants and is full-grown in early May.

Pupa. Rich chestnut-coloured, very glossy; abdominal segments punctate anteriorly; cremaster with two parallel, down-curved spines and two pairs of very fine, hook-tipped lateral bristles. Subterranean.

Imago. Univoltine, appearing in July, then aestivating and reappearing in late August, with stragglers lasting until late September. Varies in abundance from year to year. On the rare occasions when it is disturbed by day, its habits are similar to those of *N. pronuba*. By night it often visits sugar and light.

Distribution (Map 110)

It frequents extensive deciduous woodland and parkland where it is widespread throughout mainland Britain and Ireland; Isle of Man; the Channel Islands. Western Palaearctic; in western Europe it occurs from Portugal to southern Scandinavia and Finland.

NOCTUA JANTHINA [Denis & Schiffermüller]
The Lesser Broad-bordered Yellow Underwing

Noctua janthina [Denis & Schiffermüller], 1775, *Schmett. Wien.*: 78.

Type locality: [Austria]; Vienna district.

Description of imago (Pl.8, figs 33,34)

Wingspan 34–44mm. Antenna in both sexes very finely setose. Thorax with conspicuous pale patagia. Forewing rich purplish brown, clouded with blue-grey and reddish purple; stigmata sometimes indistinctly outlined whitish, forming partial '80' mark; subterminal line often defined in same colour. Hindwing bright orange-yellow, with suffused blackish basal patch; irregular subterminal band black; cilia orange-yellow; discal spot absent both on upper and under surfaces.

This species varies little. Ab. *rufa* Tutt (fig.34) has the forewing brighter red with mauve suffusion. Very occasionally the hindwing has the bright yellow ground colour replaced by a much paler shade.

Life history

Ovum. Dome-shaped, with numerous ribs; whitish when laid, developing darker clouding. Hatches in 13 days (R. I. Lorimer).

Larva. Full-fed *c.*40mm long, plump, tapering slightly anteriorly. Head brown, tinged with light or dark greyish or ochreous, often with a faint greenish tint, and finely reticulate with darker colour. Body ochreous brown with a greenish tint; dorsal line thread-like, white, passing through a series of rather faint, dark V-shaped marks, represented on abdominal segments 7 and 8 by shorter, intensely black marks; ochreous transverse dash across posterior of segment 8; spiracular stripe broad, pale brownish grey or brownish ochreous, with dark olive-grey wavy stripe above; spiracles whitish; ventral surface ochreous, reticulate with grey and dotted with black; anal comb and prolegs blackish.

Feeds from August or September to April or May. In autumn it feeds on species of *Rumex*, *Primula* and bramble (*Rubus fruticosus* agg.) and, according to Barrett (1897), on species of *Arum*, *Atriplex*, *Matricaria* and *Stellaria*. In spring it is to be found at night on species of *Salix*, *Crataegus*, *Ulmus* and *Corylus avellana* as well as on herbaceous plants. Larvae obtained *ab ovis* are easily forced.

Pupa. Rather stout, chestnut-brown, very glossy; similar to that of *N. interjecta* (Hübner) *q.v.*, but cremastral spines more slender. Subterranean, in a fragile cocoon.

Imago. Univoltine, flying in late July and August. Rests concealed by day but may occasionally be seen flying vigorously along hedgerows towards sunset on very hot days. After dark it comes freely to light and to flowers of tansy, ragwort, *Buddleia* and *Hebe*, but is less frequent at sugar.

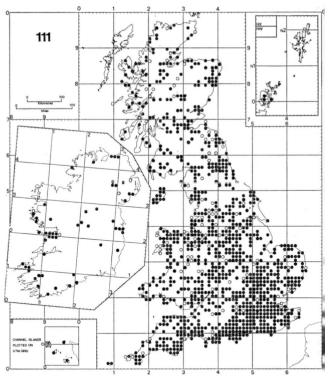

Noctua janthina

Distribution (Map 111)

Inhabits deciduous woodland, hedgerows and gardens. It occurs throughout the British Isles from the Channel Islands to Orkney and is often abundant in the south. Western Palaearctic; in Europe it is found from the south to Denmark, just reaching west Norway and the extreme south of Sweden and Finland.

NOCTUA INTERJECTA CALIGINOSA (Schawerda)
The Least Yellow Underwing

Noctua interjecta Hübner, [1803], *Samml.eur.Schmett.* **4**: pl.23.

Agrotis interjecta caliginosa Schawerda, 1919, *Verh.zool.-bot.Ges.Wien* **69**: (115).

Type locality: Austria; Mährisch-Ostrau.

Description of imago (Pl.8, fig.24)

Wingspan 31–36mm, female a little smaller and shorter-winged. Antenna in male shortly fasciculate, in female finely setose. Forewing reddish brown or warm clay-coloured; stigmata very obscure except for a small dark shade sometimes present in dorsal half of reniform; crosslines faint and delicate; a line of minute black dots between postmedian and subterminal lines; a darker shade usually occurs on basal side of subterminal line and a narrow median fascia is common, though often faint. Hindwing deep, rich yellow with moderately broad, black border and rays of blackish scales extending across wing from base, and small discal spot; underside with a distinct blackish discal spot.

Varies little. The nominate subspecies differs from subsp. *caliginosa* in the paler coloration without reddish tint of the forewing, and the paler yellow hindwing, which has a narrower border without dark rays and less dark suffusion on the underside. Ab. *clara* Lempke (1962: 194) from the Netherlands and Britain, closely resembles the nominate subspecies in having a narrowly banded hindwing, less black on the forewing and a similar underside, but the ground colour of the hindwing resembles that of subsp. *caliginosa*.

Life history

Ovum. Hemispherical; ribbing heavy but obtuse; reticulations also obtuse but very light; distinct depression around micropyle; whitish when laid, but darkening quickly around micropyle and median zone. In captivity laid in irregular heaps and batches on edges of container rather than on flat surfaces, hatching in 11 days (R. I. Lorimer).

Larva. Full-fed *c.*38mm long, moderately plump, broadest at abdominal segment 8, tapering anteriorly. Head pale ochreous with darker lines. Body yellowish drab or pale yellow-brown, dorsal region streaked with fine, darker longitudinal irroration; dorsal line whitish; subdorsal stripe pale purplish brown or brownish; broad lateral stripe of same colour, edged whitish; ventral surface yellowish grey; a curved black streak or a pair of dots on each proleg.

Feeds from September to May on grasses and other herbaceous plants, including common mallow (*Malva sylvestris*), primrose (*Primula vulgaris*) and species of *Rumex*. In spring it also eats the opening buds of *Salix* spp. Apparently the larva is seldom found.

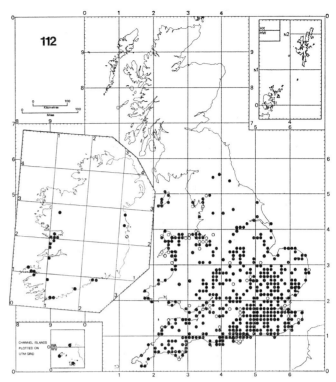

Noctua interjecta

Pupa. Rather stout, reddish brown, very glossy; wing-cases and segmental divisions well defined, former with fine radiating striae; posterior extremity bluntly rounded, cremastral spines parallel, curved at tips, with two pairs of extremely fine bristles adjacent; very similar to that of *N. janthina* [Denis & Schiffermüller], but cremastral spines a little more robust. Subterranean.

Imago. Univoltine, flying in July and August. Rests concealed most of the day but often indulges in a rapid flight along hedgerows in late afternoon. Flies again at night when it visits flowers such as ragwort, lavender and *Juncus*. Comes frequently to light but seldom to sugar.

Distribution (Map 112)

Inhabits open country, lanes and hedgerows. It is widespread, but local and seldom common, in southern and midland England, Wales, Ireland as far north as Co. Antrim; Isle of Man; Isles of Scilly; the Channel Islands. Western Palaearctic; the nominate subspecies is Mediterranean; subsp. *caliginosa* is central and north European, occurring in Austria, west and north-west Germany, the Netherlands, Belgium and the north of France.

SPAELOTIS Boisduval

Spaelotis Boisduval, 1840, *Genera Index meth.eur.Lepid.*: 106.

Largely Holarctic; only one species occurs in Britain, but closely allied species occur in the Alps and in Fennoscandia.

Antenna in male ciliate; palpi hairy on upper surface. Abdomen of female with a pair of deep hair-covered concavities on the sides of the penultimate segment. Male genitalia with valva long, with extended ampulla.

SPAELOTIS RAVIDA ([Denis & Schiffermüller])
The Stout Dart

Noctua ravida [Denis & Schiffermüller], 1775, *Schmett. Wien.*: 80.
Noctua obscura Brahm, 1790, *Insekten Kal.Samml.Oek.* **2**: 191.
Type locality: [Austria]; Vienna district.

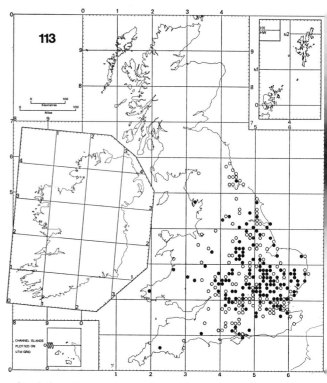

Spaelotis ravida

Description of imago (Pl.9, figs 10,11)
Wingspan 42–50mm. Forewing brown, when fresh with strong rosy tinge, sometimes covering much of area below costa; antemedian, postmedian and subterminal fasciae, paler; orbicular stigma distinct, flattened, and enclosed by a dark streak from the antemedian fascia to the rather obscure reniform stigma. Hindwing pale whitish, darker greyish marginal suffusion slight in male, more extensive in female.

Variation usually slight. In ab. *abducta* Esper fasciae and stigmata are outlined in white; and in ab. *confluens* Cockayne the orbicular stigma is elongated to join the reniform.

Similar species. Graphiphora augur (Fabricius) has no streak connecting the stigmata and the hindwing is wholly dark brown; *Standfussiana lucernea* (Linnaeus) *q.v.* also lacks this streak, has barely visible stigmata, and much darker brown hindwing; *Euxoa nigricans* (Linnaeus) is smaller and darker, with hindwing more suffused in both sexes.

Life history
The early stages in the wild are little known in Britain, and in captivity probably only one moth has been reared *ab ovo* in recent times (Haggett *in litt.*).
Ovum. Slightly flattened, grey-black, with whitish ridges and ring of white dots round base. Laid in September in large heaps covered with scales from the female abdomen. Hatches in December in captivity. Eggs are difficult to obtain in captivity; many darken but fail to hatch.
Larva. Described (1865) and figured (1893, **5**: pl.73, fig.3) by Buckler as having three colour-forms in the last instar, greyish white, dull green, and brown, all showing paler dorsal lines and a series of oblique yellowish subdorsal marks. Head grey, mottled and streaked with brown. The brown form is also figured by Hoffmeyer (1962) and Haggett (1960a; 1963) gives a fuller description.

Feeds slowly through winter on young leaves, and later on roots, of sow-thistle (*Sonchus* spp.), dandelion (*Taraxacum officinale* agg.) and dock (*Rumex* spp.), but dies if brought indoors, and in the open there is also heavy mortality before pupation in April or May.
Pupa. Undescribed.
Imago. Moths emerge in late June and early July, but soon aestivate, often gregariously in sheds and under loose bark; they reappear in September when the females are sexually mature. The moths frequent mainly the edges of marshes and meadows and come, usually sparingly, to sugar and light. There are great cyclical variations in abundance and temporary extensions of range; on the east coast the species is possibly reinforced by immigration (Bretherton, 1957).

Distribution (Map 113)
In England mainly eastern and midland, sometimes appearing more widely; not reliably recorded this century in Scotland; unknown in Ireland. Eurasiatic; in western Europe in Portugal and Spain, widespread in France and Belgium but very local and variable in numbers in the Netherlands; in Denmark only in east Jutland and the islands; very local in Norway, but common in Sweden and Finland, where it reaches 64°N.

GRAPHIPHORA Ochsenheimer

Graphiphora Ochsenheimer, 1816, *Schmett.Eur.* **4**: 68.

A small number of species in the Holarctic region, only one of which is European.

Frons smooth; antenna in male ciliate. All tibiae spinose; midleg with three rows of spines. In male genitalia, the valva tapers to a narrow point, and the clasper is elongate and strongly elbowed.

GRAPHIPHORA AUGUR (Fabricius)
The Double Dart

Noctua augur Fabricius, 1775, *Syst.Ent.*: 604.

Type locality: Germany.

Description of imago (Pl.9, figs 6,7)
Wingspan 38–48mm. Antenna ciliate. Forewing broad, glossy fuscous-brown; the three stigmata are outlined with black to a greater or lesser extent; cross-lines scalloped but sometimes reduced to a series of blackish dots. Hindwing brownish fuscous with slightly darker veins and faint discal spot.

The usual form in the south of England (fig.6) is lighter brown, often reddish-tinted, and larger than Scottish specimens (fig.7).

Similar species. *Rhyacia simulans* (Hufnagel) has a narrower forewing and never has black outlines to the stigmata; *Spaelotis ravida* ([Denis & Schiffermüller]) also has a narrower forewing with a fine, black streak at the base, no claviform stigma and a much paler hindwing.

Life history
Ovum. Rounded and button-like, strongly ribbed from raised micropyle to rim on upper surface; pale whitish when laid. Deposited in large numbers in orderly array, not quite touching one another, on the foodplant (Stokoe & Stovin, 1958).

Larva. Full-fed *c.*47mm long, plump, tapering anteriorly from about abdominal segment 3. Head small, pointed, glossy light brown, reticulate with dark brown and with a broad blackish dash on each lobe. Body reddish brown; dorsally a series of rather indistinct V-shaped marks, and on each segment two pairs of small but conspicuous white points, anterior pair closer together; a narrow, transverse yellowish band on abdominal segment 8, anterior to which is a conspicuous black double mark; spiracular line irregular and dark; spiracles white; subspiracular stripe paler, reticulate with grey; ventral surface dull purplish brown with greenish tinge. The ground colour is somewhat variable but the pairs of white dots are characteristic.

The young larva is stated to go into hibernation early. In spring it is one of the commoner species to be found feeding

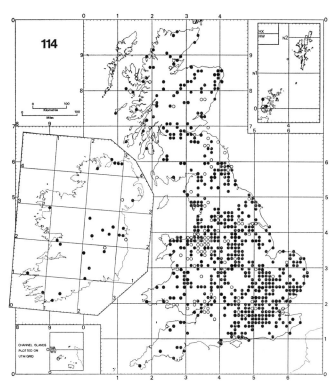

Graphiphora augur

nocturnally on the opening buds of various trees and shrubs including hawthorn (*Crataegus* spp.), blackthorn (*Prunus spinosa*), birch (*Betula* spp.) and sallow (*Salix* spp.), as well as dock (*Rumex* spp.).

Pupa. Rather stout, rich mahogany-coloured; antenna-cases well marked and sculptured in ridges; palp-cases prominent; wing-cases with edges raised and densely and minutely sculptured in fine channels; dorsal surface and abdomen glossy, abdominal segments minutely punctate; cremaster triangular, heavily sclerotized, with two divergent spines, tips of which are slightly hooked. In a brittle cocoon just below the surface of the soil.

Imago. Univoltine. Flies in June and July. Attracted to flowers of tansy (Barrett, 1896), to honey-dew, sugar and light.

Distribution (Map 114)
It inhabits deciduous woodland, parkland, fens and marshes, occurring rather locally throughout the British Isles; rare or absent in some tracts of apparently suitable country although common elsewhere in the district. Holarctic; in Europe its range extends from south-west France and north Italy northwards almost to the Arctic Circle, though it is apparently confined to mountainous regions in Spain and south-east France.

EUGRAPHE Hübner

Eugraphe Hübner, [1821], *Verz.bekannt.Schmett.*: 224.
Coenophila Stephens, 1850, *List Specimens Br.Anim.Colln Br.Mus.* 5: 74.
Ammogrotis Staudinger, 1895, *Dt.ent.Z.Iris* 8: 358.

A number of species in the Palaearctic region, one of which occurs in Britain.

Frons domed, but smooth and lightly sclerotized; antenna in male (of *E. subrosea* (Stephens)) strongly bipectinate. Midtarsus with four rows of spines, the outer row consisting of only a few spines.

EUGRAPHE SUBROSEA (Stephens)
The Rosy Marsh Moth

Graphiphora subrosea Stephens, 1829, *Ill.Br.Ent.* (Haust.) 2: 200.

Type locality: England; Whittlesea.

Description of imago (Pl.9, figs 8,9)
Wingspan 36–42mm. Sexes rather similar, forewing in female narrower. Antenna in male strongly pectinate, in female simple. Forewing rosy ochreous or pale rosy grey, finely irrorate with reddish brown or rosy fuscous; glaucous suffusion sometimes present, especially along costal region; reniform and orbicular stigmata concolorous with ground colour, but appearing paler against surrounding darker clouding; reddish fuscous interstigmatal bar sometimes present, extending to antemedian line; claviform stigma absent or represented by a very fine streak, dorsal to which is a small, pale cloud; scalloped cross-lines moderately distinct, each commencing as a dark costal mark; subterminal line with darker clouding basad. Hindwing ample, shining ochreous gold with fuscous clouding towards margin, darker veins and rounded discal spot.

The extinct fenland form was evidently rather larger and redder than the Welsh specimens; Meyrick (1928) gives the wingspan as 38–42mm, whereas a Welsh series measured 36–40mm. The ground colour in bred insects varies considerably, though the glaucous-grey hue seldom disappears entirely. Variation occurs, too, in the intensity of darker clouding, development of cross-lines and of the interstigmatal patch.

Similar species. The strongly pectinate antenna of the male and the rosy grey coloration of the forewing should preclude confusion with any other species, though in lamplight *Diarsia mendica* (Fabricius) or even worn *D. brunnea* ([Denis & Schiffermüller]) might be mistaken for it; each of these has a distinct, dot-like, black claviform stigma.

Life history
Ovum. Almost hemispherical with about 20 longitudinal ribs. Laid singly or a few at a time (Nordström *et al.*, 1935–41). Hatches in about ten days (de Worms, 1968).
Larva. Full-fed *c.*40mm long, plump and smooth, tapering sharply at each extremity. Head pale grey-brown, reticulate and with two widely-spaced, tapering vertical dark stripes. Body pinkish, yellow-tinted, with fine blackish irroration dorsally which is intensified towards subdorsal stripes, there forming conspicuous black clouds on each segment; clear, unbroken, pale primrose dorsal stripe; dorsolateral stripe a little wider, of similar colour and contrasting strongly with black clouds described above; between dorsolateral stripes and spiracles extend two stripes, the upper rusty and the lower purplish, edged along spiracles with black; spiracular stripe very broad, cream-coloured with pink clouding and edged below by broken line of very dark, purplish brown; ventral surface warm flesh-pink; legs and prolegs pale.

In Wales, larvae have been found from April to June or even July; they probably grow at different rates during winter and thus reach maturity over an extended period. They feed nocturnally in spring on the unopened buds of bog-myrtle (*Myrica gale*), where the young shoots grow up through sphagnum. In Fenland likewise they used to be taken on bog-myrtle and Gardiner (1968) indicates that it is unlikely that in the wild any were ever found on any other foodplant in this country. Though bog-rosemary (*Andromeda polifolia*) is recorded as an alternative foodplant on the Continent and also occurs in the Welsh localities, larvae have never there been reported on it. Other continental foodplants given are bog-bilberry (*Vaccinium uliginosum*), Labrador-tea (*Ledum groenlandicum*) and heather (*Calluna vulgaris*). The larva drops into the sphagnum in response to any touch or vibration. In captivity it feeds readily on crack-willow (*Salix fragilis*) and other narrow-leaved species of *Salix*, preferring ripe female catkins.

Pupa. *C.*18mm long, cylindrical, tapering rather sharply posteriorly; dark chestnut-brown, hardly glossy except on wing-cases; abdominal segments rather heavily sclerotized, granular and strongly punctate anteriorly; cremaster consisting of two outwardly curved, tapering, horn-like spines set on slender, tapering tip of abdomen and flanked on each side by a more slender, hooked spine. Pupation takes place in a loose, silken cocoon spun almost vertically amongst sphagnum or bog debris, but in captivity it occurs successfully in dry cellulose wadding. From a pupa formed on 5 June, the adult emerged on 25 July; emergences occur chiefly in the evening between 18.00 and 22.00 hours.

Imago. Univoltine. Flies in late July and August. Comes freely to light.

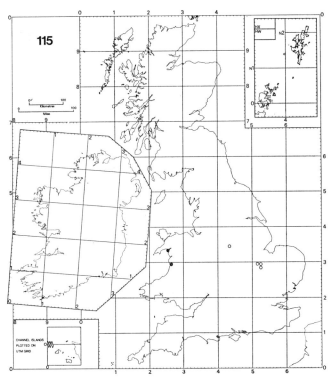

Eugraphe subrosea

Conservation. This species is known from very few localities in Britain and it is desirable that collecting of all stages should be kept to a minimum.

Distribution (Map 115)

Inhabits the wetter parts of acid bog which, in its Welsh locality, is periodically burnt. The species was discovered, probably in 1828, by Weaver in Yaxley Fen, Huntingdonshire, and occurred quite commonly there and in the adjacent localities of Whittlesea Mere and Holme Fen up to c.1850, when the draining of the fens was completed. It was not seen again until 9 August 1965, when a specimen was taken in north Wales (Revell, 1965). Moths were subsequently discovered in quantity in Borth Bog, Cardiganshire in August 1967 and larvae in the following spring. The species continues to occur there, shifting its ground, and has since been reported from a third locality in Wales. Eurasiatic or possibly Holarctic (Boursin, 1964); in western Europe it occurs from Finland to Denmark, north Germany and Austria. One was taken at light in the Netherlands in 1958, and the species was discovered in the Massif Central, France, in Monts du Forez, in August 1968. The continental form is greyer and less red and has been referred to subsp. *subcaerulea* Staudinger, but red forms similar to the nominate subspecies are found in Denmark.

PARADIARSIA McDunnough

Paradiarsia McDunnough, [1929], *Bull.Dep.Mines, Can.* 55: 48.

A few species widespread in the Holarctic region, two of which occur in Britain.

Frons smooth; antenna in male ciliate or fasciculate. All tibiae with spines; tarsi each with four rows of spines. Differs from other genera mainly in the structure of the male genitalia.

PARADIARSIA SOBRINA (Duponchel)
The Cousin German

Noctua sobrina Duponchel, [1843], *in* Godart & Duponchel, *Hist.nat.Lépid.Fr.* Suppl. 4: 224.

Type locality: France; Pyrénées-Orientales.

Description of imago (Pl.9, fig.18)

Wingspan 34–39mm. Antenna in male shortly ciliate, in female finely setose. Forewing purplish brown, more or less tinged with an evanescent glaucous bloom; markings obscure; orbicular stigma large, concolorous with ground colour but often with faint clouding in centre; reniform stigma with a more conspicuous blackish patch in dorsal half; claviform stigma absent; cross-lines faint, a row of minute black dots often marking outer edge of postmedian line; subterminal line pale, wavy; narrow median fascia often present. Hindwing shining brownish grey with darker veins; fringe paler, tinged pinkish with faint dark central line. Abdomen in female more pointed and lacking scale-tufts of male.

This species varies little; the individual figured is more red than the majority of specimens. The most glaucous form is referable to ab. *mista* Freyer and dull, dark grey specimens to ab. *suffusa* Tutt.

Similar species. Wild-caught specimens are often in less than perfect condition, and can be confused with several other species. Some forms of *Xestia xanthographa* ([Denis & Schiffermüller]) have the dark reniform spot and postmedian row of dots but the forewing is less ample and the glaucous tinge absent; the hindwing is paler and has a more sharply defined border. Some moorland forms of *Diarsia mendica* (Fabricius) are very glaucous, but the markings are more clearly defined and a small, black claviform dot is present. Male *D. dahlii* (Hübner) has a more rust-coloured forewing with clearer markings, and in this species, too, the black claviform dot is present.

Life history

Ovum. Globular, with small micropylar depression and about 30 ribs; bright pale yellow when laid, becoming deep flesh-coloured (Buckler, 1893). Eggs are said to be laid loosely, hatching within about a fortnight.

Larva. Full-fed *c.*30mm long, plump, cylindrical, tapering slightly at extremities. Head glossy brownish ochreous with a black dot towards the front of each lobe. Body of soft, velvety texture, reddish or red-brown, slightly irrorate grey; dorsal line narrow, pale, slightly wider in middle of each segment and merging into a characteristic pale, ochreous spot towards posterior of each segment; each segment also has a pair of small ochreous dorsolateral spots, equally characteristic; spiracular stripe pale, dirty pinkish brown; ventral surface a little darker.

Hibernates when small and does not seem to be encountered in springtime until the end of May or first week in June. Feeds nocturnally on heather (*Calluna vulgaris*), bilberry (*Vaccinium myrtillus*) and low saplings of birch (*Betula* spp.).

Pupa. Yellowish brown, lustrous; dorsal surfaces of abdominal segments 5–7 punctate anteriorly; cremaster inconspicuous with two closely set, short, stiff, terminal bristles (Nordström *et al.*, 1935–41). Subterranean, in peaty soil.

Imago. Univoltine. Flies in July and August. Attracted to heather bloom, sugar and light.

Distribution (Map 116)
Very local in those birch woods in Scotland where all the foodplants are present from *c.*200m (600ft) upwards. The temporary nature of such conditions, and the build-up of parasites, seem to be responsible for the insect's changing its ground every few years. The best-known localities are Rannoch, Perthshire and Aviemore, Inverness-shire, but it is evidently widespread in suitable places as far north as Ross and Sutherland. Eurasiatic; in western Europe it occurs as a montane species in the Pyrenees, Alps, Ardennes and the high moors of central Germany; it is very rare at low levels in the Netherlands but widespread in Denmark and occurs northwards through Fennoscandia to the Arctic.

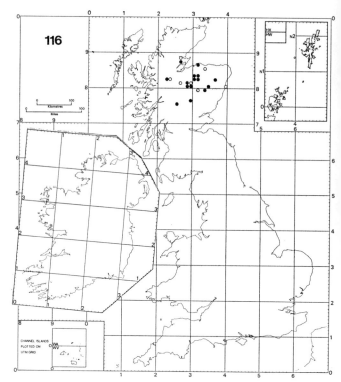

Paradiarsia sobrina

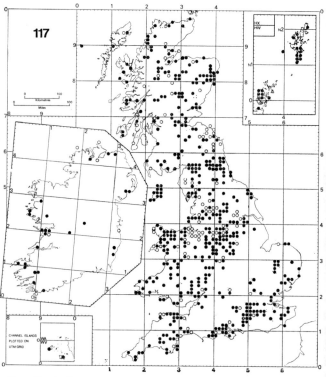

Paradiarsia glareosa

PARADIARSIA GLAREOSA (Esper)
The Autumnal Rustic

Phalaena (Noctua) glareosa Esper, 1788, *Schmett.* **4**: 387, pl.128, fig.3.

Type locality: Europe.

Description of imago (Pl.9, figs 12–15)

Wingspan 32–38mm. Sexes weakly dimorphic. Antenna in male ciliate, in female simple. Forewing narrow with very straight costa, light grey, very finely speckled with black scales, and with conspicuous black markings between the reniform and orbicular stigmata and between the orbicular stigma and the antemedian line; antemedian line with a smaller black spot above and below the latter mark; two smaller spots on basal line, one on costa and one on median vein; postmedian line nearly straight, curving basad towards costa, narrow, pale, darker-edged on each side. Hindwing in male nearly white with darker veins, in female suffused light brownish grey.

Variation occurs both between and within populations. Specimens with the forewing strongly suffused with pink are most frequent in south-west Britain and are referable to ab. *rosea* Tutt. Scottish specimens (fig.14) are in general deeper grey. The Shetland population is dimorphic, one form being little darker than examples from the Scottish Highlands, the other, ab. *edda* Staudinger (fig. 15), being nearly black with the outlines of the stigmata and cross-lines pale buffish grey (Kettlewell & Gibson, 1973). Forms closely approaching ab. *edda* have recently been taken in Orkney and Caithness, some of the intermediates from the former locality being of a peculiar coffee-colour, tinted with grey. The hindwing is considerably suffused with fuscous in the darker specimens of both sexes.

Similar species. The nominate subspecies is unlikely to be confused with any other species, but some very dark *Diarsia mendica thulei* (Staudinger) from Shetland bear some resemblance to *Paradiarsia glareosa* ab. *edda*. The latter can be recognized by its more fuscous, less brown, thorax and abdomen, narrower wings (though the forewing of subsp. *thulei* is narrower than that of the nominate subspecies), straighter cross-lines on the forewing, and the suffusion on the hindwing which is fuscous rather than brownish.

Life history

Ovum. Hemispherical, prominently ribbed, rather glossy; whitish, strongly marked with darker coloration on ribs on upper half. Hatches in autumn about five weeks after being laid.

Larva. Full-fed *c.*33mm long, cylindrical. Head pale brown, lightly banded and reticulate with darker coloration. Body in early instars purplish brown dorsally with a broad bright yellow or whitish yellow spiracular stripe; in later instars yellowish brown; dorsal and subdorsal lines narrow, yellowish, edged each side by a darker, shaded line; region between subdorsal line and spiracles umber-brown; spiracular stripe yellowish; ventral surface pale brown; pinacula small, blackish.

Hibernates when small. In early spring it may be found at night resting on grass-stems in a characteristic posture, the anterior segments being raised in the form of a question mark. Later it may be beaten from scrub birch or found at night on heather and herbaceous plants. Polyphagous; its foodplants include bluebell (*Endymion non-scriptus*), heather (*Calluna vulgaris*), bell-heather (*Erica cinerea*) and species of *Rumex*, *Plantago*, *Galium*, *Crepis* and *Salix*; it also accepts grasses such as sheep's-fescue (*Festuca ovina*) and, in captivity, annual meadow-grass (*Poa annua*) and cock's-foot (*Dactylis glomerata*).

Pupa. Reddish brown, hardly glossy; wing-cases rugose, marked with fine longitudinal and transverse striae; abdominal segments strongly punctate anteriorly; cremaster a blackish, flat, very rugose lobe bearing a pair of nearly parallel spines. Subterranean, in a fragile cocoon.

Imago. Univoltine. Flies from mid-August to mid-September. At night it comes frequently to heather bloom and light, but rarely to sugar. Rests concealed and is seldom seen by day. Nevertheless, visual predation by common gulls (*Larus canus*) and golden plover (*Pluvialis apricaria*) by day in Shetland is evidently a causal factor in the existence of the *edda* cline; the blackish form constitutes 97 per cent of the population in Unst, where the soil is dark and peaty, but only about 2 per cent in South Mainland where the soil is paler and less organic (Kettlewell, 1961; Kettlewell & Cadbury, 1963).

Distribution (Map 117)

It occurs on sandy or sometimes gravelly soils at the edges of woodland and, in the north, also on peaty moorland. It is local throughout the British Isles but more common in the north; it also is found in Jersey and Guernsey. Atlantico-Mediterranean, being apparently confined to Europe, with a curiously disjunct distribution extending from Portugal, north Spain and Italy through south and central France, Switzerland, west Germany, Belgium and the Netherlands to Denmark and southern Fennoscandia.

LYCOPHOTIA Hübner

Lycophotia Hübner, [1821], Verz.bekannt.Schmett.: 215.
Scotophila Stephens, 1829, Nom.Br.Insects: 41.

There are three European species in this genus, one of which occurs in the British Isles.

Frons smooth; antenna in male ciliate. Foreleg only weakly spined. Male genitalia very characteristic; valva bifurcate at tip; clasper reduced to a fold; sacculus extends far into cucullus and is rounded and scobinate; uncus narrow and elongate; aedeagus scobinate at orifice.

LYCOPHOTIA PORPHYREA ([Denis & Schiffermüller])
The True Lover's Knot

Noctua porphyrea [Denis & Schiffermüller], 1775, Schmett. Wien.: 83.
Agrotis strigula Thunberg, 1788, D.D.Mus.Nat.Ac.Upsal.: 72.
Phalaena (Noctua) varia Villers, 1789, Linn.Ent. 2: 276.
Type locality: [Austria]; Vienna district.

Description of imago (Pl.9, figs 16,17)
Wingspan 24–34mm. Antenna in male serrate, in female simple, finely ciliate in both sexes. Forewing with ground colour varying from pale reddish brown with considerable admixture of whitish grey scaling (fig.16), through shades of purplish brown; reniform and orbicular stigmata usually whitish grey, reniform variously clouded within, orbicular very small, circular, pyriform, or occasionally absent; claviform stigma usually present, dark-outlined; cross-lines whitish grey, black-edged, antemedian strongly angled; several narrow, whitish streaks in terminal region. Hindwing brownish fuscous with paler cilia, narrow discal spot and postmedian line sometimes just discernible.

Variation lies mainly in the ground colour of the forewing rather than in the general pattern, which is remarkably constant. Specimens from Shetland (ab. *suffusa* Tutt, fig.17), are larger on average and rich purplish grey with the pattern more or less obscured, and are more variable in this respect than those from southern England.

Similar species. In most forms of *Xestia agathina* (Duponchel) the costa is paler than the rest of the forewing, the orbicular stigma nearly as large as the reniform, the cross-lines dark, not whitish, and a series of wedge-shaped marks is present in the subterminal region; the hindwing is paler and more distinctly banded.

Life history
Ovum. Globular, base slightly flattened, with numerous delicate ribs; pearly white when laid.
Larva. Full-fed *c.*27mm long, cylindrical, hardly tapering. Head glossy ochreous brown, reticulate and having two vertical dark marks. Body reddish brown; narrow cream dorsal line interrupted anteriorly on abdominal segments and edged with dark brown lunules; subdorsal line ochreous, narrow and failing anteriorly; spiracular stripe ochreous with pink markings and indistinct brownish marks above; spiracles black; ventral surface pale reddish brown.

Feeds nocturnally on heather (*Calluna vulgaris*) and bell-heather (*Erica cinerea*) from August to May, hibernating when already well grown. Rests by day amongst leaf-litter under the foodplant, to which it bears a close resemblance.
Pupa. Cylindrical, abdominal segments short and rapidly tapering, dark red-brown, very glossy; cremaster a narrowly triangular projection surmounted by two parallel spines with down-curved tips. In a slight cocoon on or just below the surface of the soil.
Imago. Univoltine. Flies from June to August with a peak in July. Though it sometimes flies in the afternoon, it is mainly nocturnal, when it visits flowers of heather. Strongly attracted even to weak light but has rarely been recorded at sugar.

Distribution (Map 118)
The most common noctuid of heathland throughout the British Isles, including the Channel Islands. Even small, isolated patches of the foodplants in woods or gardens will support a small colony. Eurasiatic; occurring from northwest Spain to Fennoscandia, reaching the Arctic Circle; it is a common species of moorland but is sometimes found in localities where heather is not present.

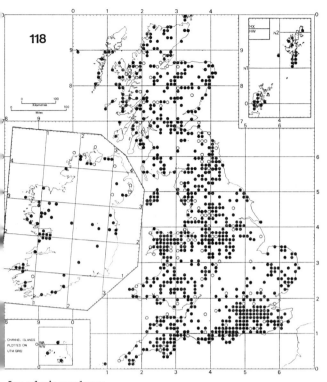

Lycophotia porphyrea

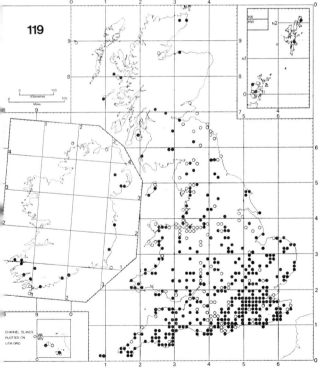

Peridroma saucia

PERIDROMA Hübner
Peridroma Hübner, [1821], *Verz.bekannt.Schmett.*: 227.

There are ten Neotropical and Hawaiian species, one of which is cosmopolitan and frequently reaches the British Isles as an immigrant.

Frons naked; antenna in male fasciculate-ciliate. All tibiae with spines. Valva in male genitalia with conspicuous corona.

PERIDROMA SAUCIA (Hübner)
The Pearly Underwing

Noctua saucia Hübner, [1808], *Samml.eur.Schmett.* **4**: pl.81.
Peridroma porphyrea sensu Edelsten, 1939, *in* South, *Moths Br.Isles* **1**: 138.

Type locality: Europe.

Description of imago (Pl.9, figs 36–38)
Wingspan 45–56mm. Antenna in male fasciculate-ciliate, in female simple. Thorax with longitudinal crest, often silvery grey. Forewing in the typical form (fig.36) light brown with markings darker; reniform stigma black and distinct, other stigmata and fasciae usually indistinct. Hindwing translucent, except for dark suffusion on veins, on inner margin and towards the tornus, rather more pronounced in female.

Variation in colour and pattern of forewing very great. In f. *nigricosta* Tutt (fig.38), the costal half is dark with the dorsal half pale; in the very common f. *margaritosa* Haworth (fig.37) the whole wing is fairly uniformly dark reddish brown with whitish suffusion near the base. Many other forms have been named.

Life history
Ovum. Small, flattish and button-shaped, strongly ribbed and reticulate, light yellow and glossy.

Larva. Full-fed *c.*42mm long, plump and cylindrical. Head dark brown with light central stripe. Body in early instars pale with conspicuous brown spiracular line, later becoming purplish above and whitish beneath; a strong undulating stripe above the spiracles; nine pairs of lateral white spots; anal plate white with two large and two small black spots.

Its habits, foodplants and life cycle in the wild are little known in Britain. In captivity it can be reared on many herbaceous plants; if forced at 21°C (70°F) the span from egg to moth is about three months. Barrett (1896, **3**: 369) records that from eggs obtained in October larvae fed outdoors until most of them died after frosts in January; but two survived and produced moths in March and April. He also mentions a pupa which was dug up in April, from which a moth emerged on 1 May. There is strong presumption that early immigrant moths can produce a later generation in the wild in Britain; but winter survival in any stage appears to be exceptional, and there is no clear evidence to support the suggestion that the species is a permanent resident in sheltered nooks on the south coast.

Pupa. Red-brown; cremaster with two points. Subterranean, in an oval earthen cocoon.

Imago. Annually immigrant, but with large variations in abundance: light-trap records for 13 years in Surrey show catches ranging from 66 moths in 1966 to none in 1971 and 1975. These variations often, but not always, coincide with those of *Agrotis ipsilon* (Hufnagel), but *Peridroma saucia* is much less common and more south-western in its distribution. Has been reported in every month from May to November, but is most numerous in late September and early October. Comes readily to flowers, sugar and light, hiding by day among foliage, usually at ground level.

Distribution (Map 119)

Occurs throughout the British Isles, but most numerous in the south-west; rare inland in Scotland and Ireland. Cosmopolitan; in western Europe it is possibly resident in the Mediterranean zone; in northern Europe it is only a moderately common immigrant to the Netherlands and Denmark and is hardly known in Norway and Sweden, though in some years it reaches the Faroes and Iceland.

DIARSIA Hübner
Diarsia Hübner, [1821], *Verz.bekannt.Schmett.*: 222.

A large genus of world-wide distribution. There are five species recognized in Britain.

Frons smooth; antenna in male ciliate. Spines on foretibia only moderately developed. Forewing of all species with a small black dot representing claviform stigma. Distinguished from related genera by structure of male genitalia.

DIARSIA MENDICA (Fabricius)
The Ingrailed Clay
Noctua mendica Fabricius, 1775, *Syst.Ent.*: 611.
Noctua festiva [Denis & Schiffermüller], 1775, *Schmett. Wien.*: 314.
Phalaena (Noctua) primulae Esper, 1788, *Schmett.* **4**: 136.
Type locality: Germany.

Description of imago (Pl.9, figs 19–28)

Wingspan 28–36mm, male larger and ampler-winged than female, Scottish specimens tend to be smaller. Antenna in male regularly notched, each notch bearing a bristle and densely clothed with fine cilia, in female very finely ciliate. Forewing in south Britain ranging from pale straw to bright chestnut (figs 19–22), in Scotland darker, often with a delicate glaucous tint (figs 23,24); reniform and orbicular stigmata paler than ground colour, containing darker clouding; a small black quadrate mark between antemedian line and orbicular stigma, and a similar larger mark between the orbicular and reniform stigmata, both sometimes paler brown or obsolete; claviform stigma represented by a black dot or, occasionally, a short streak; cross-lines weak; a narrow median shade which passes through reniform stigma sometimes present. Hindwing pale grey-brown with golden sheen, darker in northern specimens; discal spot and postmedian line faint or absent; cilia light golden, often with a pinkish tint.

Subsp. *thulei* (Staudinger)

Agrotis festiva var. *thulei* Staudinger, 1891, *Dt.ent.Z.Iris* **4**: 266.

(Pl.9, figs 25–28). Forewing darker than in nominate subspecies, usually dark chestnut-brown to black-brown. Hindwing heavily suffused fuscous. Subterminal line often conspicuously pale.

Subsp. *orkneyensis* (Bytinski-Salz)

Rhyacia festiva var. *orkneyensis* Bytinski-Salz, 1939, *Entomologist's Rec.J.Var.* **51**: 31.

Forewing a little brighter than in subsp. *thulei*, being described as 'Prussian red' to 'cameo brown', otherwise differing little.

All populations and both sexes vary considerably in ground colour, clarity of markings and the development of the black quadrate marks on the forewing.

Similar species. Certain specimens from the northern isles can be mistaken for *D. rubi* (Vieweg), but the forewing of that species has a more evenly curved, dark-coloured postmedian line and a dark terminal shade. The male of *D. dahlii* (Hübner) has the forewing with the costa more arched and the subterminal line clear, pale and thread-like; the hindwing is darker. *D. brunnea* ([Denis & Schiffermüller]), compared with southern *D. mendica*, has the forewing darker, purple-glossed and with the reniform stigma nearly always conspicuously paler than the ground colour. The underside of southern *D. mendica* is paler than that of either *D. brunnea* or *D. dahlii* and the postmedian line, when present, is never so clearly and sharply defined as in those species.

Life history

Ovum. Hemispherical with rather wide base, ribbed and reticulate; glossy yellowish white when laid, becoming greyish and developing pale brown clouding in micropylar area and in a ring near base. Laid in captivity in batches of up to 100 firmly attached to sides of container, hatching in about two weeks.

Larva. Full-fed *c.*32mm long, cylindrical, plump and flaccid. Head small and retractile, light brown with two vertical dark streaks. Prothoracic plate brownish with three very indistinct lines. Body varying in colour from light to dark reddish brown or olive-green, delicately reticulate with darker coloration; a series of dark but indistinct lozenge-shaped dorsal marks on abdominal segments 2–8, interrupted by a slender, ochreous dorsal line; subdorsal line clearer posteriorly, ochreous, surmounted on abdominal segments 3–8 by black, wedge-shaped marks which increase in size posteriorly; a transverse pale band on abdominal segment 8; spiracular stripe paler than ground colour; spiracles black, ringed ochreous, behind each an oblique, dark lateral blotch; ventral surface greyish ochreous. Larvae from the north of Britain are often more richly coloured.

Hibernates and reaches maturity in the spring, when it feeds on primrose (*Primula vulgaris*), bramble (*Rubus fruticosus* agg.), hawthorn (*Crataegus* spp.), sallow (*Salix* spp.), bilberry (*Vaccinium myrtillus*), birch (*Betula* spp.) and heather (*Calluna vulgaris*). In captivity it can be reared *ab ovo* on dock (*Rumex* spp.) or knotgrass (*Polygonum aviculare* agg.) and can be induced to feed up before Christmas to produce the adult from November to January.

Pupa. Red-brown, very glossy; wing-cases finely sculptured; abdominal segments strongly but finely punctate anteriorly; cremaster with two rather thick spines with recurved tips,

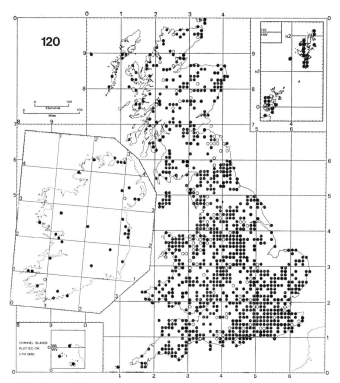

Diarsia mendica

flanked by two similar pairs of much finer structures. Subterranean.

Imago. Univoltine. Flies in June and July. Comes to honeydew and flowers, as well as to sugar and light. Mainly nocturnal, it is sometimes seen flying in sunshine at high altitudes.

Distribution (Map 120)

Inhabits deciduous woodland in the south, and moorland as well as woodland in the north and west. It is common throughout the British Isles. Represented in Orkney by subsp. *orkneyensis* and in Shetland by subsp. *thulei*. Holarctic; in western Europe widespread from the Pyrenees to North Cape, but from central France southwards it is mainly confined to high ground.

DIARSIA DAHLII (Hübner)
The Barred Chestnut
Noctua dahlii Hübner [1813], *Samml.eur.Schmett.* **4**: pl.99.
Type locality: Europe.

Description of imago (Pl.9, figs 31,32)

Wingspan 32–42mm. The most strongly sexually dimorphic *Diarsia* species, though each sex varies comparatively little within itself. *Male* (fig.31). Antenna coarsely notched and fasciculate. Forewing very broad, costa strongly arched; reddish brown or bistre-brown, mottled and shaded with purplish brown; reniform and orbicular stigmata inconspicuous though large, centre of reniform usually paler than ground colour; claviform stigma represented by the small black dot; cross-lines ill-defined with exception of a pale, wavy subterminal line which lies in a broad, purplish brown band occupying most of the area beyond the postmedian line. Hindwing greyish brown with golden sheen, almost unmarked except for faint lunule; cilia paler, with pinkish gold sheen. Underside irrorate with red-brown scales and with strongly developed wavy postmedian line on both fore- and hindwing. *Female* (fig.32). Antenna setose. Forewing narrower, though costa still arched; deep purplish brown with glaucous bloom, usually with conspicuous pale brown reniform stigma; other markings indistinct, save for pale, wavy subterminal line. Hindwing without golden sheen. Underside as in male.

Specimens from southern England are usually more ochreous brown in males, reddish in females; those from Scotland and Ireland are more richly marked, the males redder and the females purplish brown or blackish. Ab. *perfusca* Kane has the forewing dark sepia with a clear whitish reniform stigma.

Similar species. Some forms of *D. mendica* (Fabricius) may be mistaken for male *D. dahlii*, but have the costa of the forewing less strongly arched, the hindwing paler and usually with a faint postmedian band and a differently marked underside. The female can be confused with *D. brunnea* ([Denis & Schiffermüller]) *q.v.*

Life history
Ovum. Hemispherical, flattened and slightly concave beneath, ribbed and finely reticulate, drab with brown central zone (Buckler, 1893). In captivity eggs are laid in small batches, each separated from the other, hatching in about 12 days.
Larva. Full-fed *c.*34mm long, tapering slightly at each end. Head pale brown. Prothoracic plate dark brown, velvety, margined anteriorly with even darker brown. Body varies from whitish ochreous through greyish ochreous, ochreous yellow or cinnamon-brown to rich orange-brown and mahogany; a constant feature is the strong contrast between colour of dorsum and rest of body; dorsum delicately freckled darker, each segment having an ochreous brown diamond mark which obliterates dorsal line except at anterior edge of segment; subdorsal lines narrow, rather paler than ground colour; laterally grey or brownish grey, space between subdorsal line and spiracles very thickly freckled dark greyish brown; spiracular stripe pale greyish, hardly distinguishable from ventral surface; pinacula black, narrowly circled pale ochreous; spiracles black.

October to May; feeding on dock (*Rumex* spp.) and plantain (*Plantago* spp.) before overwintering, and young shoots of sallow (*Salix* spp.) (Barrett, 1897), hawthorn (*Crataegus* spp.) and bramble (*Rubus fruticosus* agg.) (Stokoe & Stovin, 1958) in spring when they are said to be hard to find. Larvae *ab ovis* may be forced readily in captivity.
Pupa. Rich reddish brown, wing-cases paler; glossy with fine sculpturing on wing-cases; abdominal segments finely punctate anteriorly; cremaster with two rather thick, closely set, parallel spines with recurved tips, flanked by two pairs of much finer bristles of similar shape. Pupation takes place in the ground in a loose cocoon.
Imago. Univoltine. Flies in August and September. Comes freely to flowers of heather, ragwort and wood-sage but only occasionally to sugar; both sexes come to light. Rests concealed and is seldom if ever seen by day.

Distribution (Map 121)
Inhabits heaths, moors and woodland on acid soils. In Britain it occurs chiefly in the Midlands and north, where it is locally fairly common to common; it is very local in southern England and Wales. In Ireland it is locally abundant. Eurasiatic, essentially northern; in Europe it is local and rare in north and east France, Belgium, the Netherlands and west Germany; fairly widespread in Denmark and southern Norway; generally common in Sweden and the southern half of Finland.

DIARSIA BRUNNEA ([Denis & Schiffermüller])
The Purple Clay
Noctua brunnea [Denis & Schiffermüller], 1775, *Schmett. Wien.*: 83.
Type locality: [Austria]; Vienna district.

Description of imago (Pl.9, fig.29)
Wingspan 36–45mm. Antenna in male slightly more coarsely ciliate than in female. Forewing broader and more arched in male; deep purple-brown varying in intensity, sometimes with glaucous bloom; reniform stigma pale straw-coloured, usually conspicuously paler than ground colour, but containing darker clouding which occasionally obliterates it; orbicular stigma concolorous with ground colour, sometimes with very narrow pale outline; claviform stigma represented by a blackish dot; a quadrate, black mark usually present between reniform and orbicular stigmata, and a mark of similar colour extends for a greater or lesser distance basad from orbicular stigma; cross-lines rather faint, subterminal wavy and separating darker purplish brown band between it and postmedian line from clearer purple terminal region. Hindwing grey-brown with golden sheen, fringe paler with pinkish gold

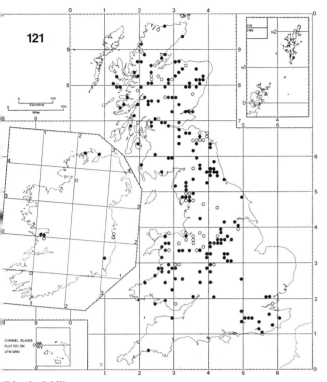

Diarsia dahlii

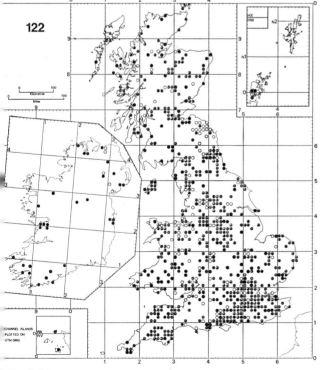

Diarsia brunnea

sheen and separated from rest of wing by fine, broken dark line; discal spot large and moderately conspicuous, postmedian line absent. Abdomen in male with conspicuous anal tuft.

All British populations seem to vary in shade of forewing, but specimens from Scotland and Ireland are often darkest, with reniform stigma either strikingly pale, or obscured.

Similar species. Distinguished from female *D. dahlii* (Hübner) by the darker wavy band between the postmedian and subterminal lines, the absence of the pale subterminal line which is characteristic of that species, and the blacker quadrate mark between the reniform and orbicular stigmata. The underside is shinier and less freckled than in *D. dahlii* but has, as in that species, a well-developed postmedian line; the discal spot is, on average, more strongly developed than in *D. dahlii*.

Life history

Ovum. Hemispherical, with small, flat base; strongly ribbed; creamy yellow when laid, becoming darker brownish grey and with even darker clouding in region of micropyle. Hatches in eight days (R. I. Lorimer).

Larva. Full-fed *c.*32mm long, plump, soft-textured, gradually tapering anteriorly from middle of body. Head brown. Body rather variable in colour, usually pale velvety chocolate-brown with yellowish tint; finely reticulate and marked with narrow, oblique, pale dashes in dorsolateral region; conspicuous yellow transverse mark on abdominal segment 8, edged anteriorly with dark brown.

Occurs from August until April or May. Before hibernation it is found on many herbaceous plants, but especially on species of *Rumex*. In spring it feeds at night on bramble (*Rubus fruticosus* agg.), sallow (*Salix* spp.), birch (*Betula* spp.), great wood-rush (*Luzula sylvatica*), bracken (*Pteridium aquilinum*), bilberry (*Vaccinium myrtillus*), etc. Larvae obtained *ab ovis* can easily be forced.

Pupa. Rather elongate, slender, deep dark red, extremely glossy; wing-cases finely sculptured and outlined dark grey; incisions of segments darkened and minutely punctate; cremaster consists of two rather long, nearly parallel spines, tips recurved, flanked by two pairs of fine bristles of similar shape. Subterranean.

Imago. Univoltine. Flies in June and July. Rests concealed by day, but by night comes freely to sugar and light.

Distribution (Map 122)

Common in deciduous woodland throughout most of the British Isles to the Orkneys; in the Channel Islands it is recorded from Jersey and Alderney. Holarctic; in western Europe it is generally common from Portugal to central Fennoscandia.

DIARSIA RUBI (Vieweg)
The Small Square Spot

Noctua rubi Vieweg, 1790, *Tabl.Verz.Brand.Schmett.* **2**: 57.

Phalaena (Noctua) bella Borkhausen, 1792, *Naturgesch.eur. Schmett.* **4**: 605.

Type locality: Germany; Berlin.

Description of imago (Pl.9, figs 33–35)

Wingspan 30–38mm, males a little larger than females, with more ample wings. Antenna in male fasciculate, in female finely ciliate. Forewing darker or lighter reddish brown, often with greyish tint, veins fuscous; reniform and orbicular stigmata concolorous with ground colour or a little paler, often narrowly outlined pale straw-colour with a darker shade between; claviform stigma represented by a blackish dot, sometimes extended as a fine streak; ante- and postmedian lines fine, dark, the latter evenly curved and pale-edged basally; subterminal line sinuate, terminal shade beyond it rather darker than rest of forewing. Hindwing ample, light greyish brown with slightly darker terminal shade and pale brown, pinkish-tinged cilia; veins and discal spot fuscous, fairly conspicuous. Abdomen in male with conspicuous anal tuft, in female tapering sharply posteriorly.

The summer brood tends to be smaller and darker than the spring brood (figs 33,34). Specimens from Shetland (fig.35) are very large, dull-coloured insects and, like those from elsewhere in northern Britain, are univoltine and may properly belong to *D. florida* (Schmidt) *q.v.* Otherwise variation is slight, but some specimens from Orkney are a uniform or mottled rich chestnut-brown with cross-lines a little darker.

Similar species. Can only be confused with *D. florida* and, in northern Britain, with small, dark *D. mendica* (Fabricius). In these, however, the postmedian line is pale on a dark ground and more curved, whereas in *D. rubi* its outer edge is blackish, giving the appearance of a darker cross-line which is nearly straight, though curving basad towards the costa; usually the dark veins contrast with the paler ground colour; the hindwing is more ample than that of *D. mendica*. The male genitalia of *D. rubi* have the clasper elbowed at a wide base and a large antler-shaped ampulla; in *D. mendica* the clasper is evenly curved and the ampulla is a slender arm, bulbous at its base (Pierce, 1909).

Life history

Ovum. Almost spherical with flattened base, strongly ribbed and reticulate above; pale yellowish when laid, developing darker clouding in region of micropyle; hatching within ten days.

Larva. Full-fed *c.*30mm long, stout, tapering anteriorly. Head brown with paler reticulation. Body pale brown to dark slaty brown, freckled darker; a series of narrow, oval brown marks along dorsum, through which passes a very narrow, pale dorsal line; a series of dark, rather indistinct, dorsolateral marks; spiracular stripe broad and pale, separated from darker region above by fine blackish line through black spiracles; ventral region light brown.

Larvae from eggs laid in the autumn occur from September to May and those of the other generation in June and July. They feed on heather (*Calluna vulgaris*) and a variety of herbaceous plants, including dandelion (*Taraxacum officinale* agg.) and species of *Rumex*, *Stellaria* and *Plantago*; overwintered larvae apparently do not ascend in spring, like their congeners, to feed on the unopened buds of woody plants. Larvae obtained *ab ovis* in the autumn, both from the southern and northern populations, are easily forced.

Pupa. Reddish brown, smooth and glossy; cremaster consisting of two short, nearly parallel spines with recurved tips, on either side of which are three very weak, flexible bristles with hooked or curled tips. Subterranean, in an earthen cocoon.

Imago. Bivoltine in the south, flying in May and June and again in August and September. Univoltine in Scotland and the Scottish Isles, flying in July and August, such specimens being possibly referable to *D. florida*. Rarely seen by day but by night it comes freely to flowers such as bramble, ragwort, *Buddleia* and heather, and those of marsh grasses and rushes. Comes readily to light and sugar.

Distribution (Maps 123,124)

It occurs most commonly in marshy places and damp woodland, though it is also found in pasture and cultivated land. It is common throughout the British Isles to Shetland; in Ireland it is particularly abundant in the second generation (Baynes, 1964). Eurasiatic; in north-west Europe widespread from north-west Spain to Fennoscandia, reaching sparsely beyond the Arctic Circle.

DIARSIA FLORIDA (Schmidt)
The Fen Square Spot

Noctua florida Schmidt, 1859, *Stettin.ent.Ztg* **20**: 46.

Type locality: not stated.

Description of imago (Pl.9, figs 39,40)

Wingspan 30–38mm. This species, if species it is, was introduced to the British list by Cockayne (1950b). It has always been confused with *D. rubi* (Vieweg) and no constant diagnostic character has been revealed for separating it. It is stated to be larger, paler, brighter in colour and more strongly marked than *D. rubi*, and to be univoltine, flying in July between the broods of *D. rubi*.

Status. Urbahn (1970) reopened the question of the status

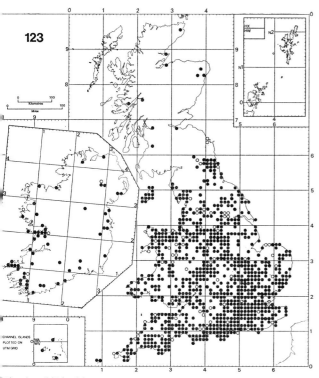

Diarsia rubi bivoltine

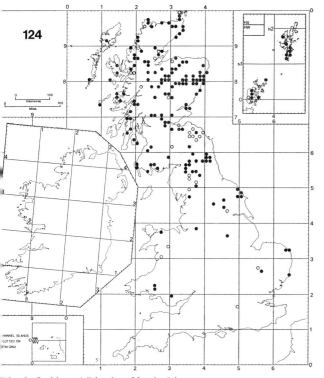

Diarsia florida and *Diarsia rubi* univoltine

of *D. florida* and refuted the differences in early stages and imago claimed by previous workers. Moreover, despite the fact that in his breeding experiments Urbahn induced *D. rubi* to produce three, four or even more generations a year and could never get *D. florida* to produce more than two, he found that the two insects would pair readily and give rise to fertile hybrids (*florida* female × *rubi* male). There remains the fact that in parts of England such as Fenland, the Lake District and the classic locality, Askham Bog, Yorkshire, single-brooded *D. florida* flies in July between the broods of *D. rubi* and appears to be biologically isolated from them. Also, according to Goodson (1951), *D. florida* flies much later in the night than *D. rubi*. In the north, however, *rubi*-like populations occur which are univoltine, the adults flying in July and August. Cockayne (*loc.cit.*) stated that the rare recessive forms ab. *flava* Walker (fig.40) and ab. *ochracea* Walker belong to *D. florida* and have never been found in *D. rubi*.

The author accepts Urbahn's conclusions that *D. florida* is in the process of separation from *D. rubi*: in England and Ireland, the concept of *D. florida* may be used to distinguish the univoltine taxon which flies between the broods of *D. rubi*, but in Scotland, no such distinction exists and the Scottish population is kept under *D. rubi* for the time being. The situation in north England and the Scottish border counties is anomalous and requires elucidation.

Distribution (Map 124)

In the parts of Britain where bivoltine *D. rubi* occurs *D. florida* has been reported from East Anglia, Yorkshire, Cumbria, Northumberland and Wales. On the Continent, it has been reported from north France, the Netherlands, north Germany, Denmark, south Sweden and Finland.

XESTIA Hübner

Xestia Hübner, 1818, *Zuträge Samml.exot.Schmett.* **1**: 16.
Amathes Hübner, [1821], *Verz.bekannt.Schmett.*: 222.
Megasema Hübner, [1821], *ibid.*: 222.
Lytaea Stephens, 1829, *Ill.Br.Ent.* (Haust.) **2**: 107.
Segetia Stephens, 1829, *ibid.* **2**: 153.

A large genus of world-wide distribution. There are 11 species in Great Britain and nine in Ireland.

Very similar to *Diarsia* but distinguished by the male genitalia, the weaker spines on the foreleg (except in *X. alpicola* (Zetterstedt)), smoother thoracic setae and more strongly developed thoracic crests.

XESTIA ALPICOLA ALPINA (Humphreys & Westwood)
The Northern Dart

Hadena alpicola Zetterstedt, 1839, *Ins.Lapp.*: 938.
Hadena hyperborea Zetterstedt, 1839, *ibid.*: 938.
Agrotis alpina Humphreys & Westwood, 1843, *Br.Moths Transform.* **1**: 118.

Type locality: Scotland; Cairn Gowr, Perthshire.

Description of imago (Pl.10, figs 1-4)

Wingspan 34-40mm, female with less ample wings than male. Antenna in male bipectinate, in female simple. Forewing soft fawn-grey or dove-grey, finely irrorate with fuscous and clouded with rose-red; superimposed are longitudinal rich reddish brown or black markings; reniform and orbicular stigmata clearly defined, grey, variously clouded with red; dorsal rim of reniform stigma often forms a narrow pale crescent, and a pale orbicular stigma frequently contrasts with rest of wing; fine, scalloped cross-lines usually present, and dark wedge-shaped marks in subterminal region also usually occur. Hindwing grey-brown, glossy, with trace of discal spot; cilia pale.

Both sexes are equally variable and local populations have their own characteristics. The nominate subspecies, which occurs in Scandinavia but not in Britain, lacks the rich red coloration and strong markings of subsp. *alpina*, and is altogether greyer. The longitudinal markings are reddish brown in the form from Rannoch, Perthshire (figs 1,2) but black in specimens from Aviemore, Inverness-shire (figs 3,4). The most richly coloured forms are said to come from Shetland. Specimens from recently discovered colonies in the north of England and Ireland resemble those from Perthshire, but are still less brightly coloured.

Life history

Ovum. Flattened-hemispherical with flat base; strongly ribbed and reticulate, slightly glossy; pale yellow when laid, becoming dirty whitish or pale straw-coloured, very strongly marked with purplish pink round micropyle and in an irregular girdle. In captivity eggs laid in a large batch hatched in nine days.

Larva. Full-fed *c*.35mm long. In penultimate instar body varying from velvety purplish brown to purplish pink, with narrow, pale dorsal and subdorsal lines and black dashes which are most strongly developed posteriorly; in its last instar 'its skin is like that of an elephant, leathery and rough with wrinkles' (Buckler, 1893), and less strongly patterned, being deep brownish red with black irroration; ventral surface reddish, irrorate more sparsely with paler red or flesh-colour.

It takes two years to complete its growth, hatching in July and becoming full-fed in the May of its second year of feeding. Natural foodplants are crowberry (*Empetrum* spp.) and, probably, heather (*Calluna vulgaris*). Larvae hatching in captivity, when offered a choice of bell-heather (*Erica cinerea*), hawthorn (*Crataegus monogyna*) and silver birch (*Betula pendula*), started feeding on the last-named, eating holes in the leaves, but died after one month.

Pupa. Light brown, hardly glossy; wing-cases sculptured with many small, circular depressions; cremaster with four strong, divergent hook-tipped spines set on a conical, wrinkled, heavily sclerotized papilla. Pupation occurs in the last week of May without a cocoon on the surface of peat under lichen. Pupae have been obtained in 'even' years by scraping the surface-covering of lichens from damp channels between clumps of the larval foodplants.

Imago. Univoltine. Flies from mid-June to the middle of August. In the populations occurring in the Scottish mountains, which have been closely studied, the species has a biennial rhythm in which it is much more common in the 'even' years, but there is evidence that in the north of England it is more common in 'odd' years. Occasionally it is found at rest on rocks or scrambling about over sparse vegetation at high altitudes. It is rarely taken at mercury vapour light at lower elevations in Scotland but more often by this method in the north of England. Nineteenth-century entomologists recorded that it did not come to sugar and they used to take it in flight, often in appalling conditions, on mountains at night.

Distribution (Map 125)

Though essentially a species of high mountain-tops, it occurs near sea-level in Shetland, though it does not seem to have been recorded there for many years. Known formerly on the Scottish mainland mostly from the mountains south of Loch Rannoch, Perthshire, it was discovered in 1942 near Aviemore, Inverness-shire (Russell & de Worms, 1944) and subsequently in other parts of the Highlands. It was first recorded from England in 1950 when one was taken at 800m (2,500ft) on Skiddaw in the Lake District (Vallins, 1951); subsequently one was taken at light at

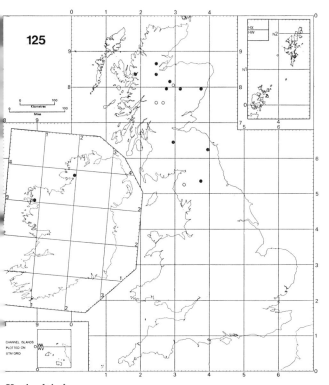

Xestia alpicola

Moor House National Nature Reserve in the Pennines at 450m (1,500ft) in 1963 (Redway & Heath, 1973) and confirmed as established there in 1973 (Withers, 1974). In 1975 a single male was found at 850m (2,600ft) near the summit of the Cheviot, Northumberland (Young, 1976). It also occurs in Ireland, having been reported from Co. Mayo in 1972 (Redway & Heath, *loc.cit.*) and Co. Donegal in 1973 (Redway, 1973). Eurasiatic, boreo-alpine; subsp. *alpina* is endemic in Britain; *X. alpicola* (Zetterstedt) occurs as subsp. *carnica* Hering and subsp. *riffelensis* Oberthür in different parts of the Alps; the nominate subspecies is widespread in Fennoscandia to the Arctic and has been recorded on three occasions from Denmark.

XESTIA C-NIGRUM (Linnaeus)
The Setaceous Hebrew Character

Phalaena (*Noctua*) *c-nigrum* Linnaeus, 1758, *Syst.Nat.* (Edn 10) **1**: 516.

Type locality: Europe.

Description of imago (Pl.10, fig.5)

Wingspan 35–45mm. Antenna finely ciliate. Forewing from greyish ochreous to reddish ochreous, heavily suffused with fuscous; reniform stigma concolorous, clouded with reddish fuscous; orbicular stigma creamy, creamy grey or reddish ochreous, continuous with patch of same colour which spreads along costa towards area above reniform stigma; a conspicuous black mark between stigmata and adjacent to pale costal patch extending basad to antemedian line; claviform stigma small, dark-tipped, or absent; crosslines fine, evenly curved; a darker shade between subterminal line and margin; a subapical black spot. Hindwing whitish, suffused fuscous with darker veins.

Variation occurs in the depth of fuscous suffusion and the amount of reddish admixture in the ground colour. The orbicular stigma is rarely more or less circular in outline and is very occasionally absent (1970, *Proc.Trans. Br.ent.nat.Hist.Soc.* **3**: pl.11, fig.9). Albino specimens have been reported.

Similar species. Easily distinguished from *X. triangulum* (Hufnagel) and *X. ditrapezium* ([Denis & Schiffermüller]) by the heavy fuscous suffusion and pale costal patch on the forewing, and the much paler hindwing; *Ochropleura flammatra* ([Denis & Schiffermüller]) *q.v.*

Life history

Ovum. Hemispherical, with numerous prominent ribs radiating to rather broad base; finely reticulate; creamy white when laid, glossy. According to Singh & Kevan (1956) there is a pre-oviposition period of three to four days, or even less, and the average number of eggs laid is less than 100. They are scattered, usually singly, on the foodplant or surrounding soil, and hatch in eight to nine days (R. I. Lorimer).

Larva. Full-fed *c*.37mm long, tapering slightly anteriorly and sharply attenuate behind abdominal segment 8. Head ochreous, streaked dark brown. Body in early instars bright green, without markings; in later instars olive-brown or pale brownish grey; dorsal line extremely faint, dark-edged; a series of narrowly wedge-shaped, black subdorsal streaks on abdominal segments enlarging posteriorly; a narrow, pale transverse band on abdominal segment 8; spiracles white with black peritreme; broad spiracular stripe yellowish or greenish yellow; ventral surface pale, greyish brown.

The larva is widely polyphagous and has been reported feeding on common chickweed (*Stellaria media*), creeping willow (*Salix repens*), bilberry (*Vaccinium myrtillus*), white dead-nettle (*Lamium album*), groundsel (*Senecio vulgaris*) and species of *Epilobium*, *Verbascum*, *Plantago* and *Arctium*. Larvae obtained *ab ovis* in the autumn may readily be forced in captivity on dock (*Rumex* spp.).

Pupa. Chestnut-brown, hardly glossy; wing-cases paler and very finely sculptured; abdominal segments punctate anteriorly, with fine, transverse dorsal lines; cremaster with two tapering, flexible, slightly convergent, hooked spines flanked by two very minute bristles of similar shape.

Imago. In the laboratory, up to six generations have been obtained in one year (Singh & Kevan, 1956), but in the wild the situation is rather obscure. Most moths occur in autumn, but there is a small emergence in May and June. It is apparently unknown how the latter individuals spend the winter and it seems unlikely that the huge numbers in autumn can be their progeny. The moth comes freely to many species of flowers including ragwort, *Buddleia*, valerian and ivy and also to honey-dew, sugar and light.

Distribution (Map 126)

Inhabits woodland, marshes, cultivated ground and heathland. It occurs throughout the British Isles, preferring lowland situations; it is usually abundant in the south, but rare in outlying islands. Its numbers are evidently subject to long-term fluctuations, and in autumn native stock is reinforced by immigration. Holarctic; occurring throughout western Europe from Portugal to central Scandinavia and Finland.

XESTIA DITRAPEZIUM ([Denis & Schiffermüller])
The Triple-spotted Clay

Noctua ditrapezium [Denis & Schiffermüller], 1775, *Schmett.Wien.*: 312.

Type locality: [Austria]; Vienna district.

Description of imago (Pl.10, fig.6)

Wingspan 39–46mm. Sexes very similar, but females often a little darker. Antenna in male finely notched and ciliate, in female ciliate. Forewing rich reddish brown to purplish brown, suffused fuscous, often with coppery tint; reniform and orbicular stigmata concolorous with ground colour, with greater or lesser dark suffusion; black mark between stigmata extends basad to antemedian line; costa anterior to this a little paler than ground colour; claviform stigma absent or represented by a narrow black streak or faint blur; black mark at base of wing bisected by light ochreous basal line; ante- and postmedian lines pale-edged; subterminal line with darker shade basally; a conspicuous black mark before apex of wing, near junction of subterminal line with costa. Hindwing whitish, suffused greybrown, with faint golden sheen; veins and narrow discal spot darker; cilia pinkish white, glossy.

Little variation occurs, but, in series, differences can be detected between specimens from different localities, some being darker or more mottled than others.

Similar species. Pattern very similar to that of *X. triangulum* (Hufnagel), but the colour of the forewing is a more intense, redder brown, more heavily suffused with fuscous, and the hindwing is relatively paler. The underside of *X. ditra-*

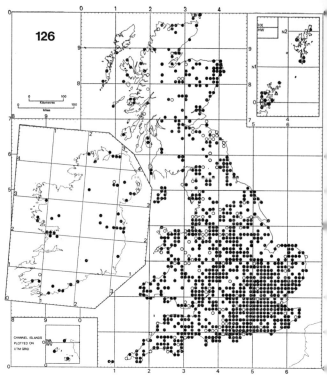

Xestia c-nigrum

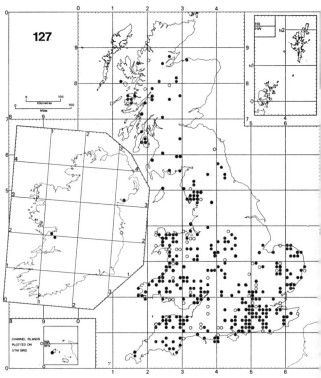

Xestia ditrapezium

pezium is very much more glossy; the forewing has a darker fuscous suffusion which extends to the subterminal line, and the veins are conspicuously paler; in *X. triangulum* the suffusion is less intense and tends to stop at the postmedian line and the veins are not conspicuously paler. On the underside of the hindwing, in *X. ditrapezium* dark irroration is usually confined to costal region and the discal spot is dot-like or slightly tailed posteriorly, being usually *c*.1mm in diameter; in *X. triangulum* the irroration is more evenly spread across the wing and the discal spot is V-shaped or triangular, *c*.2.5mm in diameter.

Life history

Ovum. Hemispherical, with small micropylar depression, numerous ribs and fine reticulations; whitish when laid, darkening to dull dirty grey before hatching.

Larva. Full-fed *c*.40mm long, cylindrical, plump and soft-textured. Head yellowish brown, each lobe with dark brown vertical stripe. Body ochreous brown to purplish brown, each segment with a lozenge-shaped, dark brown dorsal blotch; dorsal line thread-like, grey; a series of black subdorsal dashes becoming larger posteriorly to abdominal segment 8, on which they are triangular and united by a transverse black band edged posteriorly with yellow; a series of dark, oblique marks along sides; spiracles yellowish white, conspicuous; ventral surface from spiracles paler ochreous brown or purplish brown. Some forms have fairly conspicuous whitish subdorsal lines below the black markings.

In autumn it feeds on primrose (*Primula vulgaris*), dandelion (*Taraxacum officinale* agg.) and species of *Rumex* and *Stellaria*, and hibernates when small. In spring it feeds nocturnally on the opening buds of sallow (*Salix* spp.), dogwood (*Swida sanguinea*), birch (*Betula* spp.), hazel (*Corylus avellana*) and bramble (*Rubus fruticosus* agg.). Larvae obtained *ab ovis* in autumn are easily forced.

Pupa. Rather blunt anteriorly, abdominal segments rapidly tapering, bright red-brown, smooth and shining; cremaster with two projecting spines lying closely side by side (Barrett, 1897). Subterranean, in a flimsy cocoon.

Imago. Univoltine. Flies in July. Rests concealed by day but after dark it visits the flowers of wood-sage, ragwort and rush, and comes commonly to light and sugar.

Distribution (Map 127)

Occurs in damp, deciduous woodland and parkland, apparently preferring loamy or clay soils. It is found locally from Kent and East Anglia to Devon, in Wales and in central and west Scotland. Rarely recorded in Ireland. Holarctic; in Europe mainly montane in Portugal and Spain, south France and Italy, extending northwards to the Netherlands and the Danish islands.

XESTIA TRIANGULUM (Hufnagel)
The Double Square-spot
Phalaena triangulum Hufnagel, 1766, *Berlin.Mag.* **3**: 306.
Type locality: Germany; Berlin.

Description of imago (Pl.10, fig.7)
Wingspan 36–46mm. Antenna in male finely notched, ciliate, in female ciliate. Forewing pattern closely resembles that of *X. ditrapezium* ([Denis & Schiffermüller]), but ground colour is pinkish grey or ochreous grey, lightly mottled with darker shades. Hindwing greyish brown, discal spot and veins slightly darker; cilia pale pinkish ochreous. Markings very constant.

Similar species. X. ditrapezium; *Ochropleura flammatra* ([Denis & Schiffermüller]) *q.v.*

Life history

Ovum. Hemispherical with slightly hollowed, flattened base, numerous ribs and fine reticulations; cream-coloured when laid, glossy, becoming pale grey with darker micropyle prior to hatching; larger and slightly more cone-shaped than that of *X. ditrapezium*. Hatches in seven days (R. I. Lorimer).

Larva. Full-fed *c*.42mm long, plump, tapering anteriorly and somewhat attenuate posterior to abdominal segment 8. Head light brown reticulate with darker brown, and with a vertical blackish stripe on each side. Body ochreous to ochreous brown, sometimes grey-tinted, delicately reticulate blackish; dorsal line slender, pale; a series of black, dorsolateral marks becoming more pronounced posteriorly, those on abdominal segments 7 and 8 very conspicuous; posterior to last pair is a transverse, straw-coloured band; subdorsal lines faint, broken, absent anteriorly; spiracles white with black peritreme; ventral surface below them dingy ochreous grey.

Feeds from August to May, hibernating when small. In spring it is often one of the commonest larvae feeding nocturnally on bramble (*Rubus fruticosus* agg.) and species of *Rumex* and *Primula*; it also attacks the opening buds of sallow (*Salix* spp.), birch (*Betula* spp.), hawthorn (*Crataegus* spp.), blackthorn (*Prunus spinosa*) and hazel (*Corylus avellana*). Larvae obtained *ab ovis* are easily forced.

Pupa. Bright reddish brown, glossy; wing-cases finely sculptured; abdominal segments minutely punctate anteriorly; cremaster with two strong, almost parallel spines, tips recurved; on either side two very fine, hooked bristles. Subterranean, in a brittle, earthen cocoon.

Imago. Univoltine. Flies in June and July. Rests concealed by day and at night comes to lime-blossom, honey-dew, sugar and light.

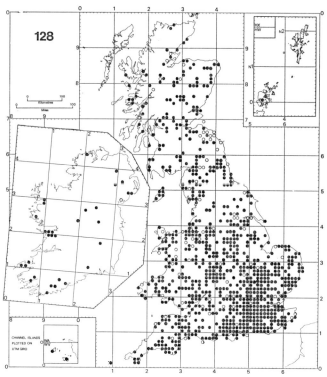
Xestia triangulum

Distribution (Map 128)

Inhabits wooded country. It is very much more common and widespread than the preceding species and occurs over much of Britain and Ireland, though it is apparently less common in south-west England; the Channel Islands. Eurasiatic; in western Europe its range extends from Spain and north Italy to central Scandinavia and Finland.

XESTIA ASHWORTHII (Doubleday)
Ashworth's Rustic

Agrotis ashworthii Doubleday, 1855, *Zoologist* **13**: 4749.
Type locality: Wales; Llangollen, Denbighshire (Clwyd).

Description of imago (Pl.10, figs 8,9)

Wingspan 35–40mm. Sexes very similar. Antenna in male finely notched, ciliate, in female ciliate. Forewing bluish grey, quickly fading to slate or dove-grey, variable in amount of blackish irroration which in some specimens almost obscures the grey ground colour, and in development of dark clouding; reniform and orbicular stigmata rather small, concolorous with ground colour, though sometimes rendered conspicuous by darker clouds around them; claviform stigma usually absent but sometimes represented by a small, dark cloud; cross-lines blackish, antemedian strongly angled basad at *c*.1.5mm from dorsum, postmedian sharply dentate throughout its length; subterminal line sinuate, darker-shaded basally and developing into a black blotch at costa; a shaded median fascia usually passes between reniform and orbicular stigmata and may occasionally fill whole area between ante- and postmedian lines. Hindwing light greyish brown with indistinct discal spot and whitish cilia.

Occasionally specimens occur which resemble subsp. *candelarum* (Staudinger), having the forewing light ashy grey with faint markings.

Life history

Ovum. Hemispherical, rather strongly ribbed; creamy white, developing dark micropylar spot and girdle.

Larva. Full-fed *c*.40mm long, plump, flaccid, tapering slightly anteriorly. Head bright chestnut. Body very dark grey-black, often with greenish tint, each segment with conspicuous, squarish, black lateral spot; thoracic legs pale chestnut.

Occurs from September to May, and overwinters when small. Polyphagous, especially in limestone districts; recorded foodplants include thyme (*Thymus drucei*), common rock-rose (*Helianthemum chamaecistus*), goldenrod (*Solidago virgaurea*), foxglove (*Digitalis purpurea*), harebell (*Campanula rotundifolia*), heather (*Calluna vulgaris*) and species of *Hieracium* and *Galium*. Seems to favour steep, rocky ground and scree where its foodplants grow in small, isolated patches. Though it feeds most actively at night, on sunny spring days it basks on its foodplant or a nearby rock, feeding intermittently and retiring when the weather turns cold. Larvae obtained *ab ovis* can be forced and those found in spring feed well on willows (*Salix* spp.), enjoying ripening female catkins.

Pupa. Bright red-brown, glossy; wing-cases finely sculptured with longitudinal striae; cremaster consisting of two

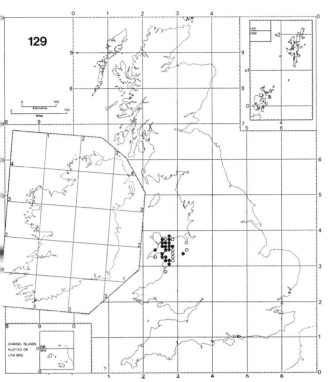

Xestia ashworthii

forcipulate spines set on a ventrally grooved papilla and flanked by two extremely minute, short bristles. Subterranean or under moss, in a flimsy cocoon; the pupal period lasts about one month.

Imago. Univoltine. Flies from mid-June to August. Rests by day among loose rocks and in hot weather may take flight when disturbed. At night comes to light and to sugar, especially when this is smeared on foliage.

Distribution (Map 129)

Occurs in the mountains of north Wales, on slate and limestone. Well-known localities include Llangollen, Denbighshire; Penmaenmawr, Sychnant Pass and Snowdon, Caernarvonshire; Cader Idris, Merionethshire.

The species is represented in the mountains of France, Switzerland, Germany and Austria by subsp. *candelarum* and in south Norway, south Sweden and Finland, where it is mainly coastal, by subsp. *jotunensis* (Schöy).

XESTIA BAJA ([Denis & Schiffermüller])
The Dotted Clay

Noctua baja [Denis & Schiffermüller], 1775, *Schmett.Wien.*: 77.

Type locality: [Austria]; Vienna district.

Description of imago (Pl.10, fig.10)

Wingspan 38–44mm. Antenna in male finely notched and ciliate, in female ciliate. Forewing with costa arched and apex rather pointed; light reddish brown, less commonly greyish or ochreous, sometimes with strong glaucous bloom; reniform and orbicular stigmata concolorous with ground colour, narrowly outlined ochreous, otherwise inconspicuous; dorsal half of reniform stigma sometimes clouded with fuscous; claviform stigma very faint or absent; cross-lines fine, outer edge of postmedian with row of minute black dots; another minute black dot nearly always present between basal and antemedian lines, and two or three larger, contiguous dots together form a conspicuous black subapical spot. Hindwing greyish brown, with indistinctly darker veins and discal spot; cilia shining, pinkish ochreous. Abdomen in male with conspicuous anal tufts.

There is considerable variation in the ground colour of the forewing; a darker cloud often occurs between the stigmata extending to the costa and a narrow median fascia is sometimes present.

Life history

Ovum. Hemispherical with mamillate projection at micropyle, having about 40 conspicuous, radiating ribs and fine reticulations; creamy white when laid. Hatches from eighth day (R. I. Lorimer).

Larva. Full-fed *c.*40mm long, large and stout for size of moth, rather flaccid. Head light brown, glossy. Body variable in colour from ochreous brown to reddish brown; dorsal and subdorsal lines yellowish, narrow, interrupted and inconspicuous except on thorax; oblique dark brown marks, edged ventrally with yellowish, in region between dorsal and subdorsal lines making a series of posteriorly directed Vs or sometimes Xs, the two posterior pairs being larger and more conspicuous; laterally mottled and variegated with greyish brown; spiracles black; ventral surface light brown.

Occurs from September to May. In autumn it feeds on primrose (*Primula vulgaris*) and dock (*Rumex* spp.). After overwintering it may be found commonly at night on bogmyrtle (*Myrica gale*), blackthorn (*Prunus spinosa*), bramble (*Rubus fruticosus* agg.), sallow (*Salix* spp.), hawthorn (*Crataegus* spp.) and saplings of birch (*Betula* spp.). Larvae obtained *ab ovis* may easily be forced and will accept slices of carrot.

Pupa. Chestnut-brown, glossy; wing-cases with fine, radiating, sculptured lines; abdominal segments punctate

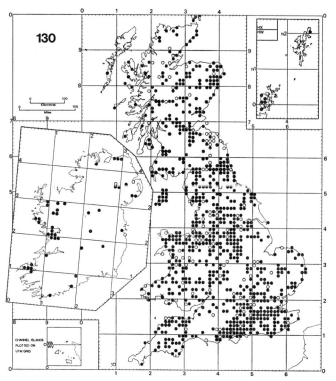

Xestia baja

anteriorly; cremaster a narrow, bilobed, rugose, heavily sclerotized projection bearing two parallel or slightly convergent, downward-curved spines with hooked tips. Subterranean, in a flimsy cocoon.

Imago. Univoltine. Flies in late July and August. Comes to blossoms of ragwort and tansy (Barrett, 1897), sugar and light.

Distribution (Map 130)
Occurs in woodland, especially on sandy soil, and on heathland. It is common throughout most of the British Isles to Shetland. Eurasiatic; in western Europe common from Portugal to the Arctic Circle.

XESTIA RHOMBOIDEA (Esper)
The Square-spotted Clay

Phalaena (Noctua) rhomboidea, Esper, 1790, *Schmett.* **4**: 485.
Noctua stigmatica Hübner, [1813], *Samml.eur.Schmett.* **4**: pl.100.
Type locality: Europe.

Description of imago (Pl.10, fig.11)
Wingspan 37–44mm. Sexes very similar; abdomen in female rather slender but more pointed than in male. Antenna in male very slightly more notched and ciliate than in female. Forewing shorter and broader than that of congeners, dark purplish brown or fuscous brown; reniform and orbicular stigmata concolorous with ground colour, inconspicuous by contrast with black quadrate mark between them, and with rhomboidal mark, almost as large, between orbicular stigma and antemedian line; separated from this mark by median vein is a third, rather shadowy dark mark, representing claviform stigma; basal line marked by small black spots; postmedian line gently curved, a little paler than ground colour; subterminal line with dark shade basally. Hindwing fuscous brown with faint purplish gloss in fresh specimens; veins slightly darker; discal spot very faint.

Similar species. Distinguished from *X. ditrapezium* ([Denis & Schiffermüller]) by its broader wing-shape, duller colour and the absence of a subapical spot on the forewing.

Life history
Ovum. Flattened-spherical, with slightly elevated micropyle and about 28 strong longitudinal ribs and fine transverse reticulations; yellowish-white, developing a brownish micropylar spot and irregular transverse band. Laid singly or in small groups (Nordström *et al.*, 1935–41).

Larva. Full-fed *c.*38mm long, plump, flaccid, slightly tapering anteriorly. Head very glossy reddish brown, reticulate with darker brown and with a blackish blotch in middle of each lobe. Body umber-brown, sometimes with an olive or reddish tinge, between subdorsal lines dull ochreous on posterior segments; dorsal line thread-like, white, distinct only on thoracic segments; subdorsal lines also thread-like, and in region between a series of oblique, dark grey or brownish dashes which form posteriorly directed V-shaped marks along dorsum, most prominent on abdominal segments 7 and 8; spiracles dark greyish brown, distinct; spiracular stripe broad, ochreous, white-edged above, especially on thorax; ventral surface pale greenish brown; two black dots above each proleg and two on prolegs themselves (Barrett, 1897).

Feeds from September to May, hibernating when small. Seldom, if ever, found in the wild and its natural foodplants are probably unknown, but it is stated to eat dandelion

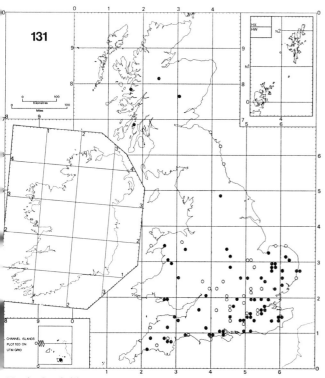

Xestia rhomboidea

(*Taraxacum officinale* agg.), common chickweed (*Stellaria media*), dock (*Rumex* spp.), plantain (*Plantago* spp.) and sallow (*Salix* spp.). Easily forced in confinement, when it will accept sliced carrot and potato. When full-grown it lies in the soil for several weeks before pupating, a point which should be remembered when captive larvae have gone down.

Pupa. Light reddish brown, integument rather thin, lacking gloss; wing-cases rather strongly sculptured with fine lines; abdominal segments minutely punctate anteriorly; cremaster with four flexuous spines set side by side, central ones somewhat stouter.

Imago. Univoltine. Flies in August. Comes to flowers of burdock, wood-sage, rosebay and ragwort, and also to sugar and light.

Distribution (Map 131)

An insect of deciduous woodland, preferably where the undergrowth is sparse, chiefly on chalk and gravel but also occurring on heavy clay soils. It is very local and rather elusive. In the south of England it occurs from Kent and the eastern counties westwards to Devon; it is now very rare in the New Forest, Hampshire, one of its former strongholds; it is perhaps most regularly encountered in the beech woods of the Chilterns (Berkshire, Oxfordshire and Buckinghamshire); abundant in Breckland, 1978. Older records from Scotland are widely scattered up to Morayshire but recently it has been found rarely and only on the west coast as far north as Arisaig, Inverness-shire. Eurasiatic; in western Europe it occurs in hilly country from the Pyrenees through France to the Ardennes, becoming commoner northwards; in west Germany it is widespread, though not common, in damp woodland; though not recorded from the Netherlands, it is common in Denmark and the Baltic islands and extends sparsely to south Norway and Sweden.

XESTIA CASTANEA (Esper)
The Neglected, or Grey Rustic

Phalaena (*Noctua*) *castanea* Esper, 1796, *Schmett.* **4** (Abs.2): 27.

Noctua neglecta Hübner, [1803], *Samml.eur.Schmett.* **4**: pl.34.

Type locality: Europe.

Description of imago (Pl.10, figs 12–15)

Wingspan 36–42mm. Antenna in male coarsely serrate, each serration bearing short, stiff bristles, in female shortly ciliate. Forewing deep mahogany-red; reniform and orbicular stigmata inconspicuous, concolorous with ground colour, outlined ochreous; reniform stigma with dark dot in dorsal half; cross-lines very faint, postmedian represented by a row of minute black dots; subterminal line more conspicuous, fine, sinuous, ochreous. Hindwing fuscous with golden sheen; fringe ochreous with fine red line.

The typical form (fig.12) prevails in the north of Britain. In the south the common form is ab. *neglecta* Hübner (fig.15), in which the forewing is ochreous or putty-coloured with a faint pinkish flush, more especially on the fringe; the markings are only slightly more distinct than in the typical form, with the dark dot in the reniform stigma still the most outstanding feature. Most colonies in the south consist almost entirely of ab. *neglecta*, varying little among themselves, but in some, red forms, usually paler than those from the north, occur fairly frequently. A Sussex colony, now believed to be extinct, produced rich yellow-ochre forms; in Shropshire and Surrey yellow forms occur (fig.13). In the Highlands two forms appear to predominate, typical *X. castanea* and a duller, paler, deep reddish grey form (fig.14). Specimens from Orkney also vary considerably, some being extremely deep rich red, others deep, dull, purplish grey.

Similar species. In Orkney *X. castanea* is easily confused, in the moth-trap, with *X. xanthographa* ([Denis & Schiffermüller]), the more slender outline of the former species and the dark dot in the reniform stigma being useful distinguishing characters when the dark hindwing cannot be seen (R. I. Lorimer).

Life history

Ovum. Hemispherical, glossy, rather strongly ribbed and striate; creamy yellow when laid. Hatches in 21 days (R. I. Lorimer).

Larva. Full-fed c.37mm long, plump and firm-textured. Head glossy reddish brown with a dark brown streak on the inner face of each lobe. Dimorphic. Body either predominantly green or mid-brown, minutely irrorate with greyish; spiracles white, black-rimmed, conspicuous; spiracular stripe broad, conspicuous, whitish, with fine dark line through spiracles; ventral surface a little paler than upper surface.

Occurs from September to May, hibernating when small. Feeds nocturnally chiefly on heather (*Calluna vulgaris*), but also on bell-heather (*Erica cinerea*) and cross-leaved heather (*E. tetralix*). In captivity it takes readily to grey sallow (*Salix cinerea*).

Pupa. Light brown, glossy; wing-cases smooth, very glossy; abdominal segments minutely punctate anteriorly; cremaster consisting of six broad-based, flexuous spines, slightly convergent with hooked tips, outer two pairs more slender. Subterranean.

Imago. Univoltine. Flies in August and September. May sometimes be found by day at rest among the lower stems of long heather, but more usually at night feeding on the flowers which have a great attraction for it. Comes freely to sugar and light.

Distribution (Map 132)

It occurs on heaths and moorland, and in heathy woodland. It is locally common throughout Britain to the Orkneys, but is sometimes absent from apparently suitable ground, as in east Suffolk. Recent records from Ireland suggest that it has been overlooked there. Eurasiatic; in Europe mainly western and always among heather, its range extending from Portugal to southern Fennoscandia.

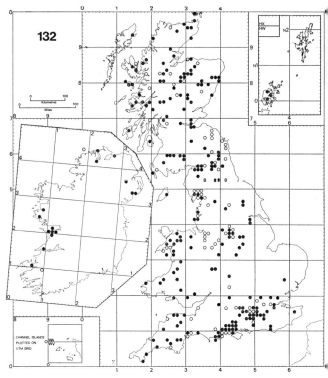

Xestia castanea

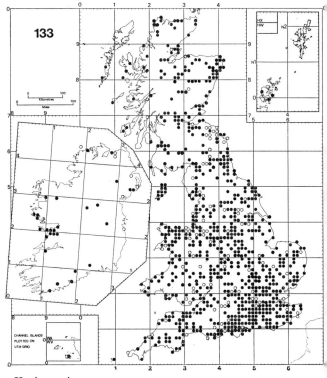

Xestia sextrigata

XESTIA SEXSTRIGATA (Haworth)
The Six-striped Rustic

Noctua sexstrigata Haworth, 1809, *Lepid.Br.*: 228.
Noctua umbrosa Hübner, [1813], *Samml.eur.Schmett.* **4**: pl.97.

Type locality: [Britain].

Description of imago (Pl.10, fig.21)

Wingspan 36–38mm, male on average larger and with broader wings than female. The female has the abdomen slender and more pointed than the male. Antenna in male notched and finely pectinate, in female finely ciliate. Forewing light greyish brown, veins darker; reniform and orbicular stigmata concolorous with ground colour, finely outlined dark brown; a minute claviform dot below orbicular stigma; basal, ante- and postmedian lines fine, dark brown; subterminal line narrow, shaded; central fascia rather narrow, passing just basad of reniform stigma; a conspicuous, dark, finely-pencilled marginal line before fringe forms sixth cross-line. Hindwing light ochreous brown, glossy, with darker veins; in male, often with darker but faint postmedian line and border; in female, more uniformly darker, unmarked.

Life history

Ovum. Hemispherical, delicately ribbed and reticulate, yellowish white when laid.

Larva. Full-fed *c.*33mm long. Virtually indistinguishable from that of *X. xanthographa* ([Denis & Schiffermüller]) *q.v.* Body yellowish to greyish brown or brown.

Its habits are obscure owing to confusion with the next species. Believed to start feeding in the autumn and to overwinter. Stated to feed on bramble (*Rubus fruticosus* agg.) and to be polyphagous on herbaceous plants, water figwort (*Scrophularia auriculata*) and bluebell (*Endymion nonscriptus*) being mentioned specifically. According to Buckler (1893), larvae refused to eat grass, but were reared successfully on dock (*Rumex* spp.), ribwort plantain (*Plantago lanceolata*), hedge-bedstraw (*Galium mollugo*) and garden strawberry (*Fragaria ananassa*), completing their growth on greater periwinkle (*Vinca major*).

Pupa. Moderately stout and smooth, dark brown and rather shining (Buckler, *loc.cit.*). Subterranean, in a weak cocoon.

Imago. Univoltine. Flies in July and August. Comes freely to ragwort bloom, honey-dew, sugar and light.

Distribution (Map 133)

Occurs mainly in damp places such as meadows, marshes, fens, boggy hollows and damp woodlands, but sometimes in other types of lowland habitat such as gardens, hedgerows and downs. It occurs commonly throughout most of the British Isles, including Jersey, the Outer Hebrides and Orkney. Eurasiatic; in western Europe it ranges from Portugal and north Spain through central and northern France, becoming common northwards through Belgium, the Netherlands and Denmark, to south Norway, central Sweden and Finland.

XESTIA XANTHOGRAPHA ([Denis & Schiffermüller])
The Square-spot Rustic

Noctua xanthographa [Denis & Schiffermüller], 1775, *Schmett.Wien.*: 83.

Type locality: [Austria]; Vienna district.

Description of imago (Pl.10, figs 16–20)

Wingspan 32–40mm. Sexes very similar. Antenna in male densely clothed with short, stiff bristles, in female sparsely ciliate. Forewing usually some shade of ochreous tinged with red or grey, or light or mid-brown; reniform stigma squarish, concave on outer face, orbicular stigma more or less circular, both usually, though sometimes only reniform, paler than ground colour; dark interstigmatal shade sometimes present; cross-lines usually faint, often broken; a row of small dots beyond postmedian line; subterminal line usually paler than ground colour with darker shade basad, though variable in expression. Hindwing whitish, veins grey-brown; discal spot very faint or absent; darker border usually distinct, of variable width.

All populations show considerable minor variation in ground colour and development of markings. In northern specimens the forewing is often rich red-brown or fuscous and the hindwing of the female is usually much darker than that of the male. Some Scottish island forms are so dark as to be truly melanic and dark specimens are appearing in the London area. Albinos have been reported.

Similar species. Paradiarsia sobrina (Duponchel) *q.v.*; *Euxoa nigricans* (Linnaeus) has the forewing narrower and a conspicuous claviform stigma. The pale reniform stigma of *Xestia xanthographa* should preclude confusion with any other species.

Life history

Ovum. Hemispherical, prominently ribbed and delicately reticulate; creamy white when laid, glossy, developing dark blotch round micropyle and belt of similar colour. Hatches in about three weeks (Scorer, 1913).

Larva. Full-fed *c.*35mm long, cylindrical, plump, firm-textured. Head glossy, light brown with dark stripe down each lobe. Body pale brown or mid-brown; dorsal line narrow, unbroken, pale, edged with dark brown clouding; subdorsal lines similar, edged above in abdominal segments by narrow, black marks which increase in size and emphasis to abdominal segment 7, then much fainter on 8; yellowish stripe above spiracles; spiracles small, black; ventral surface pale purplish brown.

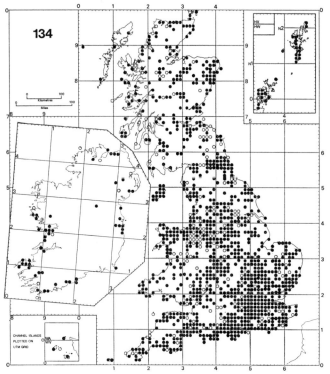

Xestia xanthographa

Reaches maturity in March or April, feeding nocturnally through the winter chiefly on Gramineae but also on species of dock (*Rumex*) and plantain (*Plantago*), primrose (*Primula vulgaris*) and, according to Barrett (1897), in the spring on young shoots of sallow (*Salix* spp.), oak (*Quercus* spp.) and other trees.

Pupa. Light brown, slightly glossy; wing-cases rather strongly sculptured; abdominal segments punctate anteriorly; cremaster shortly conical, with two rather long, slightly divergent, downward-pointing spines. Subterranean cocoon is formed in March or April, but pupation does not occur for about six weeks.

Imago. Univoltine. Flies in August and September. May sometimes be seen flying in multitudes low over the ground in grassy places soon after dusk, and later visiting flowers such as ragwort, tansy, *Buddleia*, burdock and heather. Comes readily to sugar and light.

Distribution (Map 134)

Abundant throughout Great Britain and Ireland, but less common in the outlying islands. Eurasiatic; it occurs throughout western Europe to southern Scandinavia and Finland.

XESTIA AGATHINA (Duponchel)
The Heath Rustic

Noctua agathina Duponchel, 1827, *in* Godart & Duponchel, *Hist.nat.Lépid.Fr.* 4: 361.

Type locality: France; Montpellier.

Description of imago (Pl.10, figs 22–26)

Wingspan 28–36mm, southern specimens are largest. Antenna in male fasciculate, in female finely ciliate. Forewing pale rosy brown or purplish pink, irrorate with darker scales; a broad-based, pale costal streak to level of reniform stigma; reniform and orbicular stigmata small, equally pale, orbicular very oblique; interstigmatal patch black, continued between orbicular stigma and antemedian line as a black wedge; claviform stigma small, concolorous with ground colour but with distinct narrow black outline; cross-lines dentate, not very conspicuous; three or four narrow, wedge-shaped black marks extend basad from subterminal line near apex of wing. Hindwing light greyish brown, very glossy, with slightly darker discal spot and postmedian and subterminal shades.

Minor variation occurs in all populations. In north Wales and parts of south and west England, the bright red ab. *rosea* Tutt (fig.24) is frequent. The blackish ab. *scopariae* Millière (fig.25) is found in Cheshire, Yorkshire and other parts of north England. Specimens from the Scottish Highlands approach subsp. *hebridicola* (Staudinger).

Subsp. *hebridicola* (Staudinger)

Agrotis agathina hebridicola Staudinger, *in* Staudinger & Rebel, 1901, *Cat.Lep.pal.Faunengeb.* (1): 138.

(Pl.10, fig.26). A little smaller on average and forewing with ground colour ashy grey; otherwise as nominate subspecies.

Similar species. *Lycophotia porphyrea* ([Denis & Schiffermüller]) is smaller, has more distinct, pale cross-lines, and lacks the pale costal streak contrasting with the black interstigmatal patch.

Life history

Ovum. Broadly conical in shape, rounded into flattened base and truncate at tip; about 30–35 very pronounced ribs, diminishing in number upwards, with transverse reticulations. Hatches in about three weeks (Scorer, 1913).

Larva. Full-fed c.30mm long, slender, hardly tapered. Polymorphic. Head pale brown, lobes outlined blackish. Body bright green, brown or greyish brown, darker dorsally and shaded blackish laterally; dorsal line narrow, whitish; subdorsal line broad, white, margined above by a series of short, black, longitudinal streaks, below with reddish; spiracular stripe broad, yellowish, marked on each segment with a spot of pale purplish, greenish, orange or red; spiracles and pinacula black; ventral surface paler.

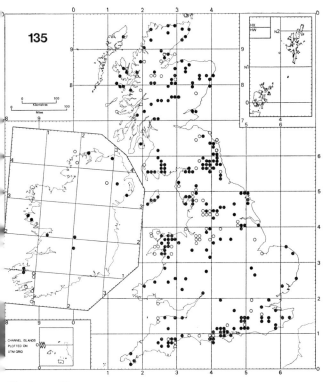

Xestia agathina

Feeds nocturnally from September until late May or June on heather (*Calluna vulgaris*). Generally considered to be difficult to rear, the exact requirements of the larva being unknown, especially towards the time of pupation.

Pupa. Small and delicate, light brown, glossy; wing-cases smooth; abdominal segments weakly punctate anteriorly; cremaster with two parallel, needle-like short spines on a shortly conical, rugose base. Pupates in a silken cocoon spun low down among the stems of the foodplant (Stokoe & Stovin, 1958).

Imago. Univoltine. Flies in September. Rests by day among the heather. Flies over the heather at dusk and again late at night, feeding meanwhile at flowers of its foodplant.

Distribution (Map 135)

Occurs on heathland and moorland, and is said to prefer long heather growing among well-spaced trees. It is probably to be found on all large tracts of heathland in Britain to Orkney, though numbers fluctuate; it is stated by Baynes (1964) to be scarce, though widespread, in Ireland. It is represented in the Hebrides by subsp. *hebridicola*. Atlantico-Mediterranean; in western Europe it occurs in Portugal and Spain, is widespread in France and Belgium and local in the Netherlands and southern Denmark, extending into western Germany.

NAENIA Stephens

Naenia Stephens, 1827, *in* Anonymous, *Retrospective Rev.* (2) **1**: 243.

There are two species in the Palaearctic region, one of which occurs in the British Isles.

Frons smooth; antenna in male setose-ciliate; palpus upturned, with naked terminal segment. Foreleg without spines. Forewing broad.

NAENIA TYPICA (Linnaeus)
The Gothic

Phalaena (*Noctua*) *typica* Linnaeus, 1758, *Syst.Nat.* (Edn 10) **1**: 518.

Type locality: [Europe].

Description of imago (Pl.10, fig.27)

Wingspan 36–46mm, female usually considerably larger and more robust than male. Antenna setose-ciliate, more finely in female. Forewing whitish brown clouded with dark fuscous; veins, stigmata and cross-lines demarcated in whitish brown; claviform stigma indistinct or absent; reniform stigma containing pale line. Hindwing fuscous; fringe whitish with narrow dark line at base. Varies little.

Similar species. *Heliophobus reticulata* (Goeze) has the forewing narrower, with the ground colour brighter brown and a series of wedge-shaped marks in subterminal region; its hindwing is pale with a distinct discal spot and dark border. *Tholera decimalis* (Poda) has the antenna strongly bipectinate in the male and also has a narrower forewing with wedge-shaped marks in subterminal region, and a bordered hindwing. Both species have hairy eyes.

Life history

Ovum. Flattened dome-shaped, ribbed and reticulate, whitish and glistening. Laid in clusters on foliage of fruit-trees and other trees, usually on upper surface, hatching in ten to 14 days (Newman, 1869).

Larva. Full-fed *c.*48mm long, plump, tapering anteriorly, abdominal segment 8 thicker, sharply attenuate posteriorly. Head greyish ochreous with brown markings. Body pale greyish ochreous, irrorate brown; a series of indistinct, double, pale lateral markings on abdominal segments; two pairs of oblique black marks on segments 7 and 8, posterior pair larger; spiracular line wavy, pale, edged blackish above; spiracles whitish; pinacula conspicuous, whitish; ventral surface light greyish brown with greenish tint. Markings strongest when the larva is about half-grown.

Hatches in late July and is at first gregarious, feeding in large groups on the underside of leaves; it hibernates and resumes feeding in the spring, when it is solitary. Polyphagous; feeding on many herbaceous plants including dandelion (*Taraxacum officinale* agg.) and species of *Rumex* and *Sonchus*, as well as on shrubs and trees such as sallow (*Salix* spp.), crab-apple (*Malus sylvestris*) and blackthorn (*Prunus spinosa*).

Pupa. Robust, dark chestnut-brown, glossy; wing-cases smooth; abdominal segments densely punctate anteriorly; cremaster with two tapering, convergent, downward-curving spines flanked by two pairs of lateral bristles with hooked tips. Pupation normally takes place in May, underground in a well-formed but brittle cocoon of earth and silk.

Imago. Univoltine. Flies in June and July. Rests concealed by day, usually on the ground but sometimes inside buildings. Has been recorded feeding commonly at privet bloom (Millward, 1907). Comes freely to sugar soon after dusk, but seldom to light, though it may often be seen in the vicinity of a powerful lamp, flitting amongst the herbage.

Distribution (Map 136)

Occurs in gardens, allotments and waste places and along weedy banks and river margins. The species appears to have decreased in numbers in the last half-century, though it is still widespread in much of lowland Britain and Ireland; it was recorded from Guernsey and Jersey prior to 1940. Eurasiatic; in western Europe it is found from Spain, through western and northern France to southern Scandinavia and Finland.

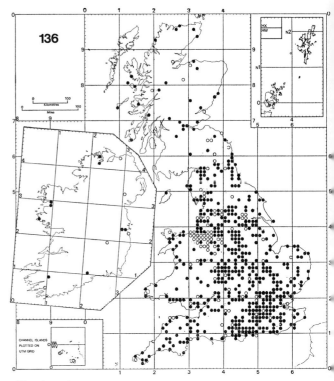

Naenia typica

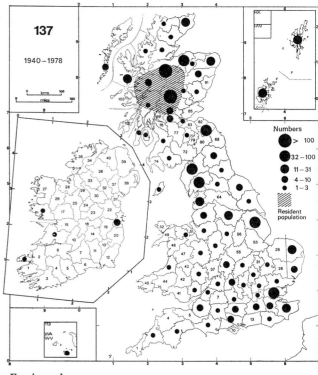

Eurois occulta

EUROIS Hübner
Eurois Hübner, [1821], *Verz.bekannt.Schmett.*: 218.

A small Holarctic genus with circumpolar distribution. One species is resident in Scotland and is also frequently immigrant to various parts of the British Isles.

Very similar to *Anaplectoides* McDunnough *q.v.*, from which it differs in its tibiae, which have four rows of spines, and in the structure of the male genitalia.

EUROIS OCCULTA (Linnaeus)
The Great Brocade

Phalaena (Noctua) occulta Linnaeus, 1758, *Syst.Nat.* (Edn 10) **1**: 514.

Type locality: Europe.

Description of imago (Pl.10, figs 28–30)
Wingspan 52–64mm. Antenna in male finely ciliate, in female simple. Forewing with orbicular stigma usually white and flattened; reniform stigma outlined in white and markedly excavated distally; subbasal, postmedian and subterminal fasciae dentate and whitish grey, but often obscured. Hindwing dark brown; cilia white.

Variation in ground colour of forewing, thorax and tegulae is very great. In the resident populations of the central Scottish Highlands (which may merit subspecific status but have not yet been so described) it is usually almost black (fig.30), resembling f. *passetii* Thierry-Mieg; but in some specimens it is dark grey with much lighter suffusion. Most immigrants are a still lighter, steel-grey, and have the thorax and tegulae also pale (fig.28); in Shetland, specimens with pale ground colour but strong dark markings are frequent (fig.29), possibly reflecting a different origin.

Similar species. *Polia nebulosa* (Hufnagel), especially in its melanic forms; but this has the orbicular stigma rounded, the reniform stigma less excavated, and hindwing paler with a dark marginal line before the cilia.

Life history
Ovum. Globular, depressed on the summit and flattened beneath, ribbed; pale yellow, becoming pinkish and later leaden. Laid in small clusters at random, hatching after about one month.

Larva. Full-fed *c*.60mm long, one of the most beautiful noctuid larvae. Body in early instars greenish white, with black spots; later purplish brown, with broken white lines along the dorsum and wedge-shaped spots above a conspicuous, pink-edged, white spiracular line; yellow ventrally (figured by Buckler, 1895, **6**: pl.92, fig.2; Spuler, 1910, **4**: pl.24, fig.11; Hoffmeyer, 1962).

The main food is bog myrtle (*Myrica gale*) growing on peaty ground, though many herbaceous plants and also birch (*Betula* spp.) and sallow (*Salix* spp.) are eaten, especially in later instars. In the wild it overwinters early and resumes feeding in April and May, when it may be found after dark. In confinement diapause can be prevented by moderate, even warmth and moths obtained the same year, though it responds unevenly to such treatment and casualties may be high.

Pupa. *C*.25mm long, stout, blackish purple; cremaster with two points. In moss on the surface of the soil, without a cocoon.

Imago. The resident populations are found close to growths of bog myrtle, especially where these are bordered by trees. The moths usually emerge in July and are mostly worn by early August. They come freely to sugar, to light less often and usually late at night, but are seldom found at rest by day. The pale forms, which are presumed immigrants, are reported in most years widely and sometimes in considerable numbers over the Scottish mainland and islands. These overlap geographically with the residents, especially in Inverness-shire; but they appear usually rather later, with a peak period about the end of August and extending into September, and there is at present no evidence either of establishment or of interbreeding with residents. In England immigrant forms have twice been bred from larvae found in the wild, and often from eggs of captured females; but there is no indication of permanent colonization (Bretherton, 1972).

Distribution (Map 137)
Resident probably only in the central and western Scottish Highlands, where it was discovered about 1850. In England, a moth was caught by Curtis at Dover, Kent, before 1830. Presumed immigrants to a total of over 500 in the period 1945 to 1978 have been recorded mostly in north Scotland and the Scottish islands, the whole of north England, and down the east coast to Essex and Kent, with scattered records inland and as far west as Cornwall, Isle of Man, and Co. Clare in Ireland. Large immigrations occurred since 1945 in 1948, 1954, 1955, 1960, 1964 and 1977 with a few records in most other years, the range varying considerably between the good years. Meteorological evidence suggests that Norway is the main source, and that variations in the dispersal in the British Isles are determined by wind direction; but the association of *Eurois occulta* with other immigrant species in some years makes it probable that some come from Denmark, the Netherlands, or north Germany. The species is Holarctic, including Greenland, Canada and Iceland. In northern Europe it is widespread both at low altitudes and in the mountains; from Belgium through eastern France it is mainly montane, reaching 2,000m (6,500ft) in the Alps. It is rare in the Massif Central and unknown in the Pyrenees.

ANAPLECTOIDES McDunnough

Anaplectoides McDunnough, [1929], *Bull.Dep.Mines, Can.* **55**: 65.

Several species in the Holarctic region, one of which is represented in Europe and in the British Isles.

Frons smooth; antenna in male setose-ciliate; palpus hairy, directed forward, terminal segment not elongate. Foreleg hairy with, or (as in *A. prasina* ([Denis & Schiffermüller])), without spines; tarsi with three rows of spines.

ANAPLECTOIDES PRASINA ([Denis & Schiffermüller])
The Green Arches

Noctua prasina [Denis & Schiffermüller], 1775, *Schmett. Wien.*: 82.

Noctua herbida [Denis & Schiffermüller], 1775, *ibid.*: 313.

Type locality: Germany; Saxony.

Description of imago (Pl.10, figs 31–33)

Wingspan 43–53mm. Antenna setose-ciliate. Forewing typically bright pale green, more or less strongly marbled with black, coppery brown and white; reniform stigma large; orbicular stigma prominent, often paler than ground colour; claviform stigma short and blunt, black-edged, frequently faint or absent; a clear, whitish or sage-green patch usually present distal to reniform stigma; cross-lines usually conspicuous. Hindwing shining dark fuscous, with conspicuous discal spot and whitish cilia. Underside rather strongly marked, hindwing light fuscous, densely irrorate with blackish scales, with prominent small, round discal spot, wavy postmedian and, sometimes, subterminal line.

There is considerable variation in the detail of the marbled markings; sometimes there is a distinct dark central shade between the ante- and postmedian lines. Ab. *demuthi* Richardson (fig.33) is a melanic form which has been bred from Cannock Chase, Staffordshire. The beautiful green ground colour is unstable, quickly fading to coppery brown.

Life history

Ovum. Hemispherical with numerous fine ribs; creamy white when laid. In captivity laid in large, irregular batches on the sides of the container, hatching in seven days.

Larva. Full-fed *c.*50mm long, plump, fleshy, thickened posteriorly. Head glossy light brown, reticulate, with blackish dash in middle of each lobe. Body dark grey-brown or mid-brown, irrorate with darker grey or black, with an inconspicuous series of rhomboidal dorsal markings, sometimes reduced to faint dots; spiracular stripe abruptly paler, dirty whitish, irrorate dark grey; spiracles white, black-rimmed, each in a triangular blackish blotch; thoracic legs pale reddish brown, prolegs light brown.

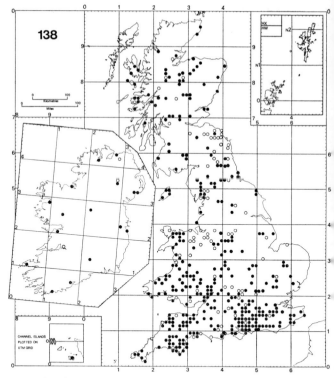

Anaplectoides prasina

Overwinters and completes its growth in spring, feeding on species of *Polygonum*, *Rumex* and *Rubus*, and on the opening buds of sallow (*Salix* spp.) and bilberry (*Vaccinium myrtillus*). In captivity it is easily forced and will eat cabbage and slices of carrot as well as *Rumex*.

Pupa. *C.*20mm long, reddish brown, glossy, robust; wing-cases finely sculptured; cremaster with two downwardly directed, slender spines, hooked outwards at tips, set on rugose, conical base. Subterranean in a brittle earthen cocoon woven with a little silk.

Imago. Univoltine. Flies from about 20 June onwards into July. By day it rests concealed or, occasionally, on tree-trunks. After dark it comes freely to sugar and light, when it is remarkably docile.

Distribution (Map 138)

Inhabits deciduous woodland. It is locally common throughout Britain except in the Midlands, northern England and parts of Scotland; it is widely distributed in Ireland and has been recorded from Jersey. Holarctic; in western Europe it is widespread from Spain and northern Italy to southern Norway, central Sweden and Finland.

CERASTIS Ochsenheimer
Cerastis Ochsenheimer, 1816, *Schmett.Eur.* **4**: 84.
Cerastia Stephens, 1850, *List Specimens Br.Anim.Colln Br. Mus.* **5**: 79.
Gypsitea Tams, 1939, *Entomologist* **72**: 135.

Six species occur in the Palaearctic region, two of which are found in Britain.
  Frons smooth. Thorax and abdomen hairy. Foretibia without spines.
  Imago flies in spring; larvae feed in the summer months and winter is passed as a pupa.

CERASTIS RUBRICOSA ([Denis & Schiffermüller])
The Red Chestnut
Noctua rubricosa [Denis & Schiffermüller], 1775, *Schmett. Wien.*: 77.

Type locality: [Austria]; Vienna district.

Description of imago (Pl.10, figs 35–37)
Wingspan 32–38mm. Antenna in male shortly bipectinate, in female simple. Forewing bright reddish brown, often with a distinct glaucous hue and almost unmarked except for some dark marks on costa which are vestiges of the cross-lines, and three or four tiny straw-coloured costal dots near apex; subterminal line faint, pale and wavy. Hindwing pale straw-coloured, suffused reddish fuscous, glossy; discal spot faintly indicated; fringe pale straw-coloured, containing a reddish line.
  Variation is mainly geographical. The form described is that which is found in the south of England (fig.35) and Ireland. Scottish specimens are less red, darker and more glaucous; similar forms occur in montane localities in Wales and the Scottish borders. Two forms found in the north of Britain are ab. *mucida* Esper (fig.36), which has the forewing coloured dull purplish slate, and ab. *pilicornis* Brahm, which is slaty grey, entirely lacking reddish coloration. Ab. *mista* Hübner has a strong glaucous suffusion on the normal reddish ground colour, and the rather rare ab. *pallida* Tutt (fig.37) is very much paler than the typical form.
Similar species. C. *leucographa* ([Denis & Schiffermüller]) *q.v.*

Life history
Ovum. Hemispherical with small, flat base; ribbed and reticulate; whitish and glistening, developing dark cloud round micropyle and in a more or less broken central girdle. Laid in batches, hatching in 12 days.
Larva. Full-fed more than 40mm long and large for size of moth, plump, flaccid, tapering from abdominal segment 8 towards head and abruptly truncate posteriorly. Head small, dull brown, reticulate darker brown. Body rich purplish chocolate, irrorate with grey; dorsal line ochreous grey, thread-like and broken, complete only along thorax; a series of inconspicuous, forwardly directed V-shaped dorsal dark markings; a broad, yellowish subdorsal dash on each segment, lying in a dark brown patch edged above by a black, triangular mark containing a white dot; spiracles greyish, black-rimmed; spiracular stripe broad, paler than ground colour; ventral surface pale brown.
  Feeds in May and June on groundsel (*Senecio vulgaris*), species of *Rumex* and *Galium*, other herbaceous plants and sallow (*Salix* spp.). Feeds only at night but remains on the foodplant by day (Barrett, 1899).
Pupa. Very stout, glossy mahogany-red; wing-cases rugosely sculptured; cremaster with two closely converging, tapering spines with outwardly hooked tips, flanked by two pairs of very fine hooked bristles. Subterranean, in a hard brittle earthen cocoon. The larva shrinks considerably before pupation; pupa from June to March or April.
Imago. Univoltine. The moth emerges in the first warm days of March and April. It is a regular visitor at sallow-bloom with *Orthosia* spp. and comes boisterously to light.

Distribution (Map 139)
It occurs chiefly in and around deciduous woodland in the south and also on boggy moorland in the north. It is usually common throughout the British Isles as far north as Orkney, and was recorded from Guernsey prior to 1940. Eurasiatic; in Europe it is rare in the Mediterranean region, widespread and common northwards to Denmark, eastern Sweden and Finland, but local in Norway.

CERASTIS LEUCOGRAPHA ([Denis & Schiffermüller])
The White-marked
Noctua leucographa [Denis & Schiffermüller], 1775, *Schmett.Wien.*: 83.

Type locality: Germany; Saxony.

Description of imago (Pl.10, figs 38–40)
Wingspan 35–39mm. Antenna in male strongly bipectinate, in female serrate-setose. Forewing matt brownish red or purplish red clouded with fuscous; reniform and orbicular stigmata usually conspicuously pale, yellowish grey; cross-lines indistinct; postmedian line a curved row of dots, pale towards tip of wing and black towards base; subterminal line sinuous, pale; paler shade usually present between postmedian and subterminal lines, and darker shade between subterminal line and margin of wing, emphasizing paler cilia. Hindwing grey-brown with golden sheen; veins and narrow discal spot darker; fringe straw-coloured, containing a reddish grey line.
  Considerable minor variation occurs in the clarity of markings, the amount of clouding within the stigmata and,

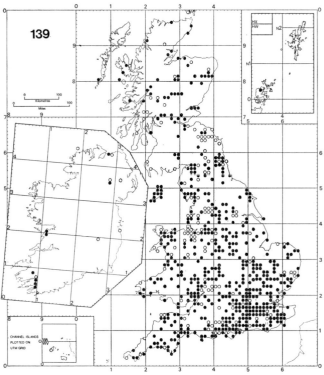

Cerastis rubricosa

to some extent, in their shape. The ground colour varies from bright ochreous red (ab. *rufa* Haworth) to intense blackish red (ab. *suffusa* Tutt, fig.40).

Similar species. Differs from *C. rubricosa* ([Denis & Schiffermüller]) as follows: antenna in male more strongly pectinate; forewing duller and more strongly marked, particularly the pale stigmata; hindwing paler.

Life history

Ovum. Globular and slightly cone-shaped with small, flat base, finely ribbed and reticulate; pale straw-coloured when laid, becoming flesh-coloured with brown micropylar spot and broken central girdle. In captivity usually laid on the muslin covering of the container or on tufts of thread-ends provided for the purpose.

Larva. Full-fed *c*.37mm long, posterior segments slightly thickened. Head light brown. Dimorphic. Body either bright pea-green or pale purplish brown, mottled and irrorate with irregular whitish dots and lines; dorsal line narrow, whitish, edged deeper green or brown; subdorsal line a series of oblique, undulating, green or brown dashes; spiracular line darker green or brown; spiracles white, dark-rimmed; ventral surface and legs uniform pale green or brown.

The larva has seldom if ever been found in the wild. In captivity it is easily reared on sallow (*Salix* spp.) or plantain (*Plantago* spp.). Other foodplants recorded are dock (*Rumex* spp.), chickweed (*Stellaria* spp.), bilberry (*Vaccinium myrtillus*) and oak (*Quercus* spp.). Feeds in May and June.

Pupa. More delicate than that of *C. rubricosa*, chestnut-brown; wing-cases rugose; cremaster with two divergent, outwardly hooked spines set on a conical base; adjacent are two pairs of fine, thread-like bristles. Subterranean. The winter is passed in this stage.

Imago. Univoltine. Seldom emerges before the end of March, a little later than most of the other early spring noctuids. Comes freely to sallow-bloom and light.

Distribution (Map 140)

It occurs in rather open deciduous woodland, apparently requiring the presence of both oak and sallow. It is extremely local, being sometimes confined to one wood in a district and absent from many seemingly suitable localities. It occurs sporadically in south-east England, Northamptonshire, Herefordshire, mid-Wales, the Lake District and Yorkshire, with a few colonies outside these main centres. It is unknown in Ireland and the Channel Islands. Eurasiatic; in western Europe it is scattered and very local through central and northern France, the Netherlands and western Germany; it is common in Denmark, scattered in Finland, only occasional in the extreme south of Sweden and absent from Norway. This discontinuity parallels that in England, but on a larger scale and with a similar lack of explanation.

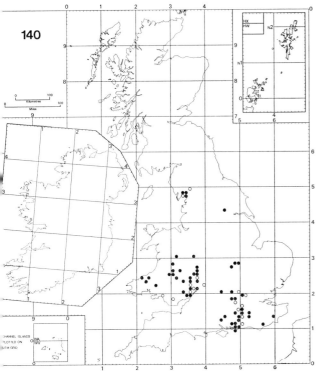

Cerastis leucographa

MESOGONA Boisduval

Mesogona Boisduval, 1840, *Genera Index meth.eur.Lepid.*: 1944.

A very distinct genus, with two Palaearctic species, one of which has once been found in Britain.

Frons naked; vertex hairy; antenna in male ciliate. Foreleg densely hairy without spines; midtarsus with three rows of spines.

MESOGONA ACETOSELLAE ([Denis & Schiffermüller])
The Pale Stigma

Noctua acetosellae [Denis & Schiffermüller], 1775, *Schmett. Wien.*: 84.

Type locality: [Austria]; Vienna district.

Description of imago (Pl.10, fig.34)

Wingspan 36–46mm. Forewing light pinkish brown, stigmata large and outlined paler; subbasal and postmedian fasciae pale brown, darker-edged, subbasal curved outwards and postmedian curved inwards from costa to dorsum. Hindwing brownish, with darker terminal shade.

Life history

Early stages unknown in Britain.

Ovum. Hemispherical with strong ribs, clear reddish yellow with white flecks.

Larva. Head and prothoracic plate brown. Body dark flesh-coloured with black streaks and slender faint, dorsal stripes; pinacula light-coloured (figured by Spuler, 1910, 4: pl.31, fig.4). In Germany it hatches in the spring and is polyphagous on scrub oaks (*Quercus* spp.) and other small trees in May and June.

Pupa. Red; in the soil in a slight cocoon (Forster & Wohlfahrt, 1964, 4: 63).

Imago. Probably an occasional immigrant. The only recorded British specimen is a female (figured) which was taken by Thomas Salvage at sugar in his garden at Arlington, Sussex on 26 October 1895 (Adkin, 1895) at a time of much immigrant activity.

Distribution

Eurasiatic; in western Europe it occurs in Spain, and is widespread in France northwards to Normandy and the Paris region; supposedly extinct in Belgium.

Hadeninae

A large subfamily of the Noctuidae of world-wide distribution with 69 resident species in Great Britain and Ireland. The eyes are hairy, a character shared by only one other noctuid subfamily, the Pantheinae (with one resident species) which has vein 5 (M_2) in the hindwing clearly present.

Imago. Antenna of normal noctuid length, variable in character; haustellum well developed except in *Tholera decimalis* (Poda), in which it is reduced. Foreleg without tibial spine, except in *Mamestra brassicae* (Linnaeus). Forewing on average broader than in most noctuid subfamilies; venation as in Noctuinae (figure 12, p. 121). Only two species show clear sexual differences in the forewing; in a number of others the female hindwing is darker, or there are antennal or abdominal differences. In some species of *Mythimna* Ochsenheimer the abdomen in the male has a ventral tuft of dark hairs.

Ovum. In most species flattened-spherical to hemispherical with fine ribbing and finer reticulation, dimpled overall, or finely sculptured. In a rather high proportion of species (about 60 per cent) the eggs are laid in large batches, broadcast by three species and in the remainder laid on the foodplant. With the exception of three species, which overwinter in this stage, they all hatch within two weeks of laying.

Larva. Variable in length relative to maximum girth and in the comparative size of the head, these characters being directly related to feeding habits. Thoracic legs and eight abdominal prolegs present; pinacula present, with short bristles; no species with long hair tufts. Except *Eriopygodes imbecilla* (Fabricius), the larvae of all resident as well as several immigrant species are figured either in Buckler (1891; 1893; 1895; 1899) or Haggett (1957; 1960b; 1963; 1968).

With one exception, all feed above ground on the leaves, flowers or seeds of vascular plants.

Pupa. In most species long and cylindrical, although some have stouter thoracic segments; in one genus the haustellum is free posteriorly; cremaster well developed, usually showing specific characters. Pupation takes place at or below ground level, or in a hollow stem; in a cocoon which is in most cases rather flimsy.

Life history. Most species are univoltine, some bivoltine in the southern part of Britain. No species spends more than a single season as a larva, but several may lie over in the pupa.

Economic importance. Several species may damage vegetable crops and one causes irregular and sporadic damage to upland pasture; a few species appear to have changed their normal feeding habits in the past two decades and may attack young conifers. Several of our rarer visitors which do not achieve the necessary breeding rate here are serious pests in warmer climates.

Key to species (imagines) of the Hadeninae

NOTE. Some species key out in more than one couplet.

1	Forewing blackish with white bar in tornal region .... *Egira conspicillaris* ab. *melaleuca* (p. 244)	
–	Forewing not blackish or if so then lacking white tornal bar	2
2(1)	Cross-lines on forewing absent or represented by dots	
–	Cross-lines present	2?
3(2)	Postmedian line represented by two or more dots	
–	Postmedian line absent (subterminal line may consist of dots)	1?
4(3)	Forewing with longitudinal striations	
–	Forewing without longitudinal striations	1?
5(4)	Postmedian line represented by a series of dots	6
–	Postmedian line represented by two dots only	?
6(5)	Dots distinctly elongate; costa strongly arched, broad costal streak distinctly paler than ground colour of forewing .... *Senta flammea* (p. 273)	
–	Dots round; costa nearly straight; costal area no paler than ground colour of forewing	
7(6)	Hindwing light fuscous, with a discal spot .... *Mythimna obsoleta* (p. 269)	
–	Hindwing white with darker veins, without a discal spot	
8(7)	Forewing whitish ochreous with heavy wedges of brownish fuscous shading medially and in terminal region; postmedian line of dots curving sharply inwards on to costa .... *M. putrescens* (p. 271)	
–	Forewing light yellowish brown; terminal and median shading hardly darker than ground colour and not forming wedges; postmedian line of dots meeting costa at right angles .... *M. loreyi* (p. 272)	
9(5)	Ground colour of forewing usually smooth clay, café-au-lait or reddish clay-coloured with very faint striations; hindwing usually with conspicuously darker veins; salt-marshes in south and east England .... *M. favicolor* (p. 269) (See text for differences from *M. pallens*)	
–	Ground colour of forewing whitish ochreous, ochreous or reddish-tinted, finely striated especially along median vein and in terminal region; veins of hindwing hardly darker than ground colour; in various habitats	1?

10(9)	Hindwing with postmedian line of dots *M. straminea* (p. 262)
–	Hindwing without dots ... 11
11(10)	Hindwing uniformly light fuscous with paler cilia *M. impura* (p. 263)
–	Hindwing whitish, often with some faint fuscous shading towards termen, but with cilia hardly paler than ground colour *M. pallens* (p. 264)
12(4)	Reniform stigma containing a small white fleck ... 13
–	Reniform stigma without a white fleck, or absent ... 14
13(12)	Oblique dark streak present in apex of forewing *M. unipuncta* (p. 268)
–	Apical streak absent *M. ferrago* (p. 258)
14(12)	Reniform stigma present, darker than ground colour *Orthosia cruda* (p. 246)
–	Reniform stigma absent *Mythimna favicolor* (p. 265)
15(3)	Forewing without longitudinal streaks ... 16
–	Forewing with longitudinal streaks ... 18
16(15)	Claviform stigma present; hindwing white with fuscous veins *Egira conspicillaris* (p. 244)
–	Claviform stigma absent; hindwing fuscous ... 17
17(16)	Forewing greyish; fringe with contrasting median line *Orthosia populeti* (p. 249)
–	Forewing some shade of ochreous or reddish brown, never grey; fringe without contrasting central line *O. munda* (p. 254)
18(15)	Orbicular and claviform stigmata present, sometimes very faint ... 19
–	Orbicular and claviform stigmata absent ... 20
19(18)	Orbicular and claviform stigmata paler than ground colour of forewing; hindwing with broad fuscous terminal shade *Cerapteryx graminis* (p. 238)
–	Orbicular and claviform stigmata concolorous, usually inconspicuous; hindwing white with fuscous veins *Egira conspicillaris* (p. 244)
20(18)	Forewing with conspicuous white, hooked bar in median region; hindwing fringe with dark central line *Mythimna l-album* (p. 267)
–	Forewing lacking white bar; hindwing fringe without dark line ... 21
21(20)	Hindwing white *M. litoralis* (p. 266)
–	Hindwing fuscous ... 22
22(21)	Forewing with black basal streak, not pinkish-tinged *M. comma* (p. 270)
–	Forewing without black basal streak, pinkish-tinged *M. pudorina* (p. 261)
23(2)	Claviform stigma absent ... 24
–	Claviform stigma present ... 50

24(23)	Reniform stigma white, or containing a white dot, or outlined in white ... 25
–	Reniform stigma lacking any white ... 34
25(24)	Reniform stigma containing a white dot ... 26
–	Reniform stigma predominantly white or white-outlined, but lacking a dot ... 29
26(25)	Reniform stigma linear; wingspan 44mm or more *M. turca* (p. 256)
–	Reniform stigma bilobed or kidney-shaped; wingspan 38mm or less ... 27
27(26)	Reniform dot oval; antemedian line elbowed *M. conigera* (p. 257)
–	Reniform dot round; antemedian line not elbowed ... 28
28(27)	Reniform stigma clearly outlined; dark line along termen of forewing at base of fringe *Eriopygodes imbecilla* (p. 237)
–	Outline of costal lobe of reniform stigma obscure; no dark line along termen *Mythimna albipuncta* (p. 259)
29(25)	Forewing with white in stigmata only ... 30
–	Forewing with white elsewhere than in stigmata ... 32
30(29)	Hindwing yellow with black border *Anarta cordigera* (p. 202)
–	Hindwing not yellow ... 31
31(30)	Forewing blue-black; reniform stigma conspicuously white *Melanchra persicariae* (p. 216)
–	Forewing variegated red, ochreous and grey; reniform and orbicular stigmata both whitish *Panolis flammea* (p. 242)
32(29)	White median fascia reaching dorsum *Hadena compta* (p. 232)
–	White median fascia not reaching dorsum ... 33
33(32)	Orbicular stigma usually entirely white; white or pale patches in apex and tornal regions of forewing *H. confusa* (p. 233)
–	Orbicular stigma dark-centred; apical and tornal regions of forewing no paler than surrounding ground colour *H. albimacula* (p. 235)
34(24)	Orbicular stigma contained in a rectangular mark, which may be broken into two parts, one lying between the orbicular and reniform stigmata, the other between the orbicular stigma and antemedian line ... 35
–	No such mark containing orbicular stigma ... 36
35(34)	Rectangular mark black; an abundant and widespread form *Orthosia gothica* (p. 255)
–	Rectangular mark brownish, sometimes almost concolorous with ground colour; chiefly a northern form *O. gothica* ab. *gothicina* (p. 255)
36(34)	Fringe of forewing chequered (sometimes best seen on underside) ... 37
–	Fringe of forewing not chequered ... 41

37(36)	Hindwing yellow with broad black border......... *Anarta myrtilli* (p. 200)	49(48)	Subterminal line of forewing irregularly undulate reaching dorsum at a right angle.............*Orthosia incerta** (p. 253)
–	Hindwing not yellow.............38	–	Subterminal line nearly straight, reaching dorsum a an oblique angle.............*O. opima* (p. 248)
38(37)	Hindwing white with broad black border and large discal spot.............*A. melanopa* (p. 203)	50(23)	Forewing with black basal streak.............5
–	Hindwing light brownish or fuscous with crescent-shaped discal spot.............39	–	Forewing without basal streak.............5
39(38)	Orbicular and reniform stigmata conspicuously outlined.............40	51(50)	Area between postmedian and subterminal lines o forewing greyish or whitish, paler than ground colou of forewing.............5
–	Outlines of stigmata blurred, or stigmata absent.............*Hadena caesia* (p. 236)	–	Area between postmedian and subterminal line brown, not paler than ground colour of forewing.......5
40(39)	Ground colour of forewing pale whitish buff; subterminal line not widening to form a tornal blotch.............*H. irregularis* (p. 231)	52(51)	Area below reniform and orbicular stigmata pale than ground colour of forewing.............*Lacanobia contigua* (p. 217
–	Ground colour of forewing rich rusty red or red-brown, sometimes mauve-tinted; yellowish subterminal line widening to form a rounded tornal blotch.............*Ceramica pisi* (p. 224)	–	Area below reniform and orbicular stigmata darke than ground colour.............*L. w-latinum* (p. 218
41(36)	Forewing straw-coloured to reddish clay; reniform stigma containing a small black dot; hindwing glossy white with darker veins.............*Mythimna vitellina* (p. 260)	53(51)	Forewing purplish brown; claviform stigma connected to postmedian line by a black bar.............*L. thalassina* (p. 219
–	Forewing greyish, brown, reddish or blackish; reniform stigma without black dot; hindwing darker, or if white, then with lunule and postmedian line.......42	–	Forewing light brown to mid-brown; claviform stigma without black bar.............*L. suasa* (p. 220
42(41)	Forewing with row of dots or small chevrons along termen adjacent to cilia.............43	54(50)	Forewing intense blue-black; reniform stigma whit with yellowish centre.............*Melanchra persicariae* (p. 216
–	Terminal row of dots or chevrons lacking.............46	–	Not as above.............5
43(42)	Hindwing with postmedian band of elongate dots.............*Orthosia gracilis* (p. 250)	55(54)	Subterminal line distinctly toothed, often wit blackish or dark-coloured wedge-shaped mark arising from it basally.............5
–	Postmedian dots absent, or if present, not elongate.............44	–	Subterminal line nearly straight or slightly undulate never toothed and never having wedge-shaped mark arising from it; or occasionally absent.............8
44(43)	Hindwing white, with postmedian and subterminal lines, cilia pinkish.............*O. miniosa* (p. 247)	56(55)	Veins of forewing conspicuously pale right acros wing.............5
–	Hindwing fuscous or brownish.............45	–	Veins of forewing not paler than ground colour, or i so, then only in median area of wing.............5
45(44)	Subterminal line on forewing nearly straight, conspicuously pale; wingspan 32mm or more.............*O. stabilis* (p. 251)	57(56)	Ante- and postmedian lines of forewing as pale a veins; ♂ antenna not bipectinate.............*Heliophobus reticulata* (p. 213
–	Subterminal line wavy, not conspicuously pale; wingspan 32mm or less.............*O. cruda* (p. 246)	–	Ante- and postmedian lines absent, or not as pale a veins; ♂ antenna bipectinate.............*Tholera decimalis* (p. 240
46(42)	Two or more black dots present in subterminal line of forewing.............47	58(56)	Tegulae unicolorous.............5
–	Subterminal black dots absent.............48	–	Tegulae with bands of darker or lighter colour.............6
47(46)	Forewing grey, fringe with contrasting median line.............*O. populeti* (p. 249)	59(58)	Basal region of forewing sooty; subterminal regio ochreous; reniform stigma linked to dorsum by a almost straight double line; hindwing white. (Ver rare accidental introduction).............*Brithys crini* (p. 24
–	Forewing brownish or reddish clay, fringe unicolorous.............*O. munda** (p. 254)	–	Not as above.............6
48(46)	Orbicular stigma absent.............*Eriopygodes imbecilla* (p. 237)		
–	Orbicular stigma present, though sometimes faint.......49		

* *O. munda* ab. *immaculata* and pale forms of *O. incerta* are easi confused. For antennal and other differences, see text.

60(59) Forewing reddish brown 61
— Forewing light grey-brown, brown or blackish, with clearly defined markings 63

61(60) Subterminal line distinctly paler than forewing ground colour; wingspan 42mm or less 62
— Subterminal line not paler than ground colour of forewing; wingspan 47mm or more .... *Polia bombycina* (p. 207)

62(61) Subterminal line pale yellow, enlarging to form a tornal blotch of same colour; not forming a W near its mid-point .... *Ceramica pisi* (p. 224)
— Subterminal line white, not enlarging at tornus, but forming a distinct W near the mid-point .... *Lacanobia oleracea* (p. 221)

63(60) Forewing blackish fuscous; reniform and orbicular stigmata clearly outlined with ochreous .... *Tholera cespitis* (p. 240)
— Forewing ochreous grey, ochreous brown or ochreous fuscous; stigmata not pale-outlined 64

64(63) Subterminal line broken into a series of pale dots; two small white dots in heel of reniform stigma .... *Sideridis albicolon* (p. 212)
— Subterminal line complete, strongly toothed; white reniform dots absent .... *Discestra trifolii* (p. 204)

65(58) Postmedian region of forewing whitish or light greyish 66
— Postmedian region of forewing grey, brown or yellowish 69

66(65) Forewing grey, tinted with rosy red and mauve; subterminal line thickened at tornus .... *Polia hepatica* (p. 208)
— Forewing not rosy- or mauve-tinted; subterminal line not thickened at tornus 67

67(66) Forewing with blackish central fascia; wingspan 35mm or less .... *Hecatera bicolorata* (p. 225)
— Forewing without central fascia; wingspan 38mm or more 68

68(67) Claviform stigma blackish; wedge-shaped marks present in terminal region of forewing; ♂ antenna bipectinate .... *Pachetra sagittigera* (p. 210)
— Claviform stigma not or hardly darker than ground colour of forewing; wedge-shaped marks absent; ♂ antenna not bipectinate .... *Polia nebulosa* (p. 209)

69(65) Black bar (occasionally merely outlined) linking ante- and postmedian lines through or just below claviform stigma 70
— No such junction between ante- and postmedian lines 72

70(69) Forewing dark ashy grey to ashy fuscous, often with bluish or mauvish tint; reniform stigma whitish with a dark central cloud; other stigmata also usually outlined whitish .... *Papestra biren* (p. 223)
— Forewing brown or brownish ochreous; stigmata without white, concolorous with ground colour 71

71(70) Subterminal region of forewing paler than median and terminal regions; veins in median region paler than surrounding areas of wing; pale mark below reniform and orbicular stigmata usually 2- or 3-lobed .... *Hada nana* (p. 206)
— Subterminal region of forewing no paler than surrounding areas of forewing, often darker than terminal region; veins not paler than adjacent areas of wing; pale mark below stigmata, when present, an obscure rounded blotch .... *Hadena luteago* (p. 231)

72(69) Forewing dark fuscous with obscurely paler stigmata and cilia, or black with white cilia and white flecks in reniform and orbicular stigmata; tegulae pale-centred .... *Polia nebulosa*, melanic forms (p. 209)
— Not as above 73

73(72) Forewing ashy fuscous, the three stigmata usually conspicuously paler than ground colour .... *Papestra biren* (p. 223)
— Forewing ochreous or brown, claviform stigma never paler than ground colour 74

74(73) Anterior tibia bearing a spine which lies nearly parallel to first tarsal segment .... *Mamestra brassicae* (p. 214)
— Anterior tibia without a spine 75

75(74) Forewing whitish, ochreous or ochreous buff 76
— Forewing light brown to brownish fuscous 77

76(75) Dark chequers in fringe of forewing preceded by blackish or dark-coloured chevrons (sometimes faint in pale specimens) .... *Hadena perplexa* (p. 229)
— Dark chequers in fringe of forewing not preceded by chevrons, giving them a square rather than rounded appearance .... *H. irregularis* (p. 231)

77(75) Orbicular stigma paler than reniform; hindwing whitish with a broad terminal shade, but lacking any trace of postmedian line .... *Discestra trifolii* (p. 204)
— Orbicular stigma not paler than reniform; hindwing light brown or brownish fuscous, if paler then with at least a trace of postmedian line 78

78(77) Subterminal line inclined outwards at tornus 79
— Subterminal line not inclined at tornus, meeting dorsum at an oblique angle and nearly parallel to termen .... *Hadena bicruris* (p. 235)

79(78) Forewing violet-tinged; orbicular stigma obliquely elongate and usually touching reniform stigma dorsally; postmedian line curved....*H. rivularis* (p. 228)
- Forewing not violet-tinged; orbicular stigma nearly circular and never touching reniform stigma; postmedian line toothed....*H. perplexa capsophila* (p. 229)

80(55) Postmedian line represented by a row of dots, or absent....81
- Postmedian line complete....82

81(80) Postmedian line of hindwing represented by elongate dots....*Orthosia gracilis* (p. 250)
- Postmedian line of hindwing absent or represented by round dots....*O. stabilis* (p. 251)

82(80) Hindwing white or whitish, without a distinct terminal shade....83
- Hindwing brownish or fuscous; if whitish, then with a distinct terminal shade....84

83(82) Reniform stigma containing a minute black dot; antemedian line sharply elbowed; hindwing without postmedian or terminal lines....*Mythimna vitellina* (p. 260)
- Reniform stigma without black dot; antemedian line not elbowed; hindwing with terminal and usually also postmedian lines....*Orthosia miniosa* (p. 247)

84(82) Postmedian line strongly toothed; subterminal line absent or represented by a broken series of orange dots....*Hecatera dysodea* (p. 227)
- Postmedian line not or very slightly toothed; subterminal line well developed, continuous and nearly straight....*Orthosia opima* (p. 248)

ANARTA Ochsenheimer
Anarta Ochsenheimer, 1816, *Schmett.Eur.* **4**: 90.
Charelia Sodoffsky, 1837, *Bull.Soc.Nat.Moscou* **1837** (6): 88.

Small, diurnal noctuids, all inhabiting acid soils and of mainly boreo-alpine distribution. Of a number of Holarctic species, three occur in Great Britain, only one in Ireland; all easily recognizable.

Imago. Wingspan 24–32mm. Forewing short, compact, cryptically coloured. Hindwing strongly contrasting. The sexes can be differentiated by small differences in the antenna.

Ovum. Almost globular. Laid singly on the foodplant.

Larva. Cylindrical, slightly attenuate at extremities. All three species feed by preference on Ericaceae.

Pupa. Colour variable, thoracic and abdominal segments 1–4 stout, thence rapidly tapering. In a cocoon on the surface of the soil. Overwinters in this stage.

ANARTA MYRTILLI (Linnaeus)
The Beautiful Yellow Underwing
Phalaena (Noctua) myrtilli Linnaeus, 1761, *Fauna Suecica* (Edn 2): 311.
Type locality: Sweden.

Description of imago (Pl.11, figs 1,2)
Wingspan 24–28mm. Antenna in male shortly, but densely, ciliate to tip, in female very shortly and sparsely ciliate. Forewing from dull brownish grey in the typical form (fig.2) to bright purplish red-brown (f. *rufescens* Tutt, fig.1); subbasal and antemedian lines white in costal half, ochreous in dorsal half; postmedian and subterminal lines white; random white streaks on the veins and an almost unbroken white subdorsal streak give a reticulate appearance; stigmata obscure, the centre of the wing being dominated by a white arrow-head mark directed to the tornus; fringe chequered brown and white. Hindwing dark yellow with black terminal shade, broad at the termen, narrow at costa and dorsum. Abdomen dark brown, each segment with yellowish posterior margin.

There seems to be no geographical significance in the distribution of the two forms; f. *rufescens* is usually the commoner, but both occur throughout the range of the species in Britain.

Similar species. Worn specimens of the typical form have been confused with *A. cordigera* (Thunberg) in which the reniform stigma is white and in which there are no pale bands on the abdomen.

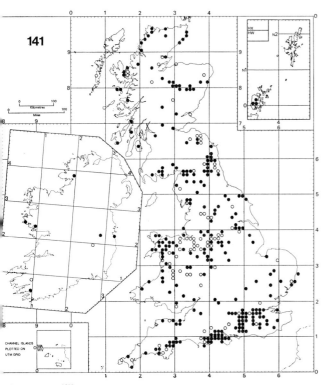

Anarta myrtilli

Life history

Ovum. Flattened-spherical, ribbed and reticulate, micropyle surrounded by a small depression, orange-yellow. Laid singly or in pairs on leaves of the foodplant, hatching in about 12 days.

Larva. Full-fed *c.*30mm long, rather stout. Head small, greenish brown with darker dots. Body green, usually bright; dorsal line pale yellow, broken in centre of each segment, wider on abdominal than on thoracic segments; subdorsal line of disconnected, slightly curved, dashes pale yellow; supra- and subspiracular lines pale and obscure; spiracles whitish, outlined black.

May be found from April to October, most commonly in July and August. Feeds on the terminal shoots of heather (*Calluna vulgaris*) and bell-heather (*Erica cinerea*), mostly by day, although it may be found by night; it is often heavily parasitized. White willow (*Salix alba*) is recorded as an alternative foodplant in captivity.

Pupa. Thoracic segments stout, brownish green, eyes and antennae conspicuous; abdominal segments yellowish brown, quickly tapering; cremaster abruptly squared, with three pairs of apical and several pairs of subapical bristles, all short and divergent. On or just below the surface of the soil, in a tough cocoon of silk mixed with earth and leaf litter.

Imago. Univoltine in the north where it flies in June. The life-cycle in southern England is obscure, as it appears from late April to August, being most abundant in June and July. Flies in sunshine rapidly and low over heather and feeds at moorland flowers; when the sun is obscured flight ceases and the moth may be found at rest on heather, well camouflaged by its cryptic coloration. A night flight is recorded occasionally (Bretherton, 1974).

Distribution (Map 141)

On acid moorland throughout Great Britain and Ireland to the Outer Hebrides and Orkney, but not Shetland. Abroad Atlantico-Mediterranean; in western Europe on heaths and mountains from the Iberian Peninsula to the Arctic Circle, reaching 2,000m (6,500ft) in the Alps.

ANARTA CORDIGERA (Thunberg)
The Small Dark Yellow Underwing

Noctua cordigera Thunberg, 1788, *D.D.Mus.Nat.Ac.Upsal.*: 72.

Type locality: Sweden.

Description of imago (Pl.11, figs 3–5)
Wingspan 24–28mm. Antenna in male densely but shortly ciliate, in female sparsely and very shortly ciliate. Forewing grey-black, variably dusted silvery in basal and subterminal areas; reniform stigma large, white, with small central blackish mark; other stigmata visible only as paler suffusion; median fascia broad, enclosing reniform stigma and more intensely black than basal and subterminal areas; underside pattern a straw-coloured disc with margins and basal area sooty brown. Hindwing bright pale yellow with black border at costa and termen.

Variation affecting mainly the amount and degree of silvery dusting in basal and subterminal areas may be considerable in both, confined to either, or completely absent, as in ab. *suffusa* Tutt (fig.5).

Similar species. A. *myrtilli* (Linnaeus), typical form, worn specimens only, *q.v.*

Life history
Ovum. Almost spherical, with light ribbing and reticulation; creamy white (Hellins *in* Buckler, 1895); Tonge's photograph, reproduced in Stokoe & Stovin (1958), suggests heavy reticulation and considerable darkening. Laid singly on the foodplant, hatching in about 12 days.

Larva. Full-fed *c.*30mm long, slender and of velvety texture. Head dark purplish brown. Body varying from purplish brown to reddish brown; dorsal and subdorsal lines white; subspiracular stripe wide, paler than ground colour, with darker freckling; spiracles black, those on the prothorax most conspicuous; dull brown ventrally. There is also a blackish form with all markings obscured.

Feeds nocturnally on bearberry (*Arctostaphylos uva-ursi*) in June and July, hiding low down by day. In captivity it has been successfully reared on *Vaccinium* spp. and on strawberry-tree (*Arbutus unedo*), feeding up in about five weeks.

Pupa. Short and stout, glossy red-brown; abdominal segments 5–7 have keeled dorsal ridges bearing a row of hooked spines; cremaster reduced, with two short, ventrally curved apical spines. On the surface of the soil in a long silken cocoon, along which the pupa is able to move. For detailed description and enlarged photograph see Hedges (1954), who does not mention the cremastral spines.

Imago. Univoltine. Flies in sunshine from mid-May to mid-June. Feeds at flowers and like *A. melanopa* (Thunberg) will come to bait made of small heaps of *Arctostaphylos* flowers, often feeding actually inside the pile. From late afternoon to sunset, single specimens and pairs *in copula* may be found sitting on rocks. The ecology of this species is obscure; although the foodplant is local, the moth is even more so and seems to have definite limits of vertical distribution, being usually found at altitudes of 200–650m (600–2,000ft).

Distribution (Map 142)
Very local in Perthshire, Inverness-shire, Aberdeenshire and Angus. In the original description of ab. *aethiops* Hofmann, 1893 (ab. *suffusa*) there is a reference to specimens of this form from Shetland, but there is no recent confirmation. Abroad Holarctic; in Europe at low levels in Scandinavia and Denmark, but wholly montane farther south, reaching the Eifel Mountains, French Alps and Pyrenees.

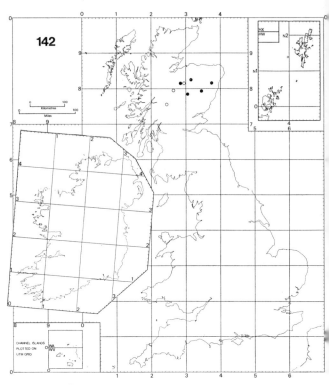

Anarta cordigera

ANARTA MELANOPA (Thunberg)
The Broad-bordered White Underwing

Noctua melanopa Thunberg, 1791, *Ins. Suecica* (2): 42.
Type locality: Sweden.

Description of imago (Pl.11, figs 6–8)
Wingspan 26–32mm. Antenna in male shortly ciliate, in female sparsely and very shortly ciliate. Forewing dark brown with variable ochreous suffusion; subbasal and antemedian lines of darker, unconnected spots and lunules; subterminal line darker brown, complete; orbicular and claviform stigmata small, blackish; reniform stigma usually conspicuous, blackish, with pale centre and, occasionally, pale outline. Hindwing whitish, suffused brownish grey at base and dorsum; discal spot large and lunular, blackish; terminal fascia broad, blackish.

There is considerable variation in the colour and degree of contrast in the forewing markings; specimens from Shetland are mainly ochreous-tinted (ab. *wistromi* Lampa (fig.8)). The hindwing varies less, but the basal suffusion may extend into and beyond the median area, and in the rare ab. *rupestralis* Hübner (fig.7) to the terminal fascia.

Life history
Ovum. Almost spherical, finely, but deeply ribbed and faintly reticulate; glossy light orange-yellow at first, becoming pink. Laid singly or in pairs on leaves of the foodplant, hatching in about 12 days.

Larva. Full-fed *c.*37mm long, slender and of velvety texture. Head brown. Body purplish pink; dorsal line rather obscure, ochreous brown, lined black; subdorsal line yellowish white, lined black; the whole dorsal area covered by a pattern of red-brown triangles and wedges; subspiracular stripe pale reddish brown; spiracles whitish, ringed black.

Feeds by night in July and August on crowberry (*Empetrum nigrum*), bilberry (*Vaccinium myrtillus*), cowberry (*Vaccinium vitis-idaea*) and bearberry (*Arctostaphylos uva-ursi*). An extremely wide range of plants is recorded as having been accepted in captivity and may indicate an adaptability necessary for survival after unseasonable snow at high altitudes; among them are knotgrass (*Polygonum aviculare*), hairy wood-rush (*Luzula pilosa*), goat willow (*Salix caprea*) and strawberry-tree (*Arbutus unedo*).

Pupa. Thoracic segments stout, black, smooth and very glossy; abdominal segments tapering rapidly to a short, rugose, blunt-tipped cremaster. Among moss, on the surface of the soil.

Imago. Univoltine. Occurs in late May and June. Flies with great speed, generally in bright sunshine, but on cloudy days, if warm, feeds at flowers of bearberry, cowberry and bilberry.

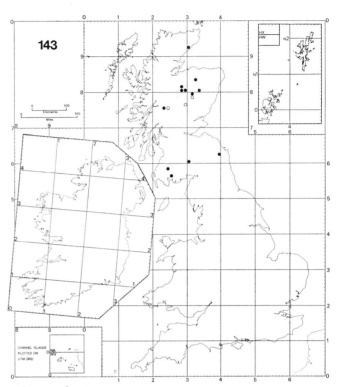

Anarta melanopa

Distribution (Map 143)
Although much less local than *A. cordigera* (Thunberg) the species is restricted to altitudes of above 650m (2,000ft) on the mainland, but occurs much lower down, almost to sea level in the Hebrides and Shetland. The statement in Barrett (1900), frequently repeated elsewhere, that this moth requires the presence of *Rhacomitrium lanuginosum* for survival is not invariably correct, although no doubt the moss occurs in many of its haunts (B. Goater). England, Northumberland; Scotland on the higher mountains from Galloway to Sutherland and, according to old specimens in BMNH, from unspecified localities at lower altitudes in the Hebrides and Shetland. Abroad Holarctic, boreo-alpine; in western Europe, the Alps and Apennines, Scandinavia and north Finland.

DISCESTRA Hampson

Discestra Hampson, 1905, *Cat.Lepid.Phalaenae Br.Mus.* **5**: vii, 14.

There are several Palaearctic species, of which only one occurs in Great Britain and Ireland. The genus is distinguished by the strongly projecting frons, without sclerotized plate below. Thorax crested. Forelegs with hooked apical spines. Abdomen crested.

DISCESTRA TRIFOLII (Hufnagel)
The Nutmeg

Phalaena trifolii Hufnagel, 1766, *Berlin.Mag.* **3**: 398.
Noctua chenopodii [Denis & Schiffermüller], 1775, *Schmett. Wien.*: 82.

Type locality: Germany; Berlin.

Description of imago (Pl.**11**, figs 9–12)
Wingspan 23–39mm (see Pierce, C. W. (1971) for note on extreme variation of size). Antenna in male shortly and densely ciliate, in female very shortly and sparsely ciliate. Forewing from pale to blackish brown, sometimes tinted ochreous (ab. *saucia* Esper, fig.11) with darker pencilling; that of female narrower and with more pointed apex than in male; subbasal, ante- and postmedian lines obscurely paler, subterminal line yellowish white with two teeth to the termen; stigmata extremely variable, reniform, especially the dorsal lobe, usually dark, orbicular of ground colour, outlined paler and claviform outlined darker or completely dark. Hindwing very pale fuscous, veins brownish and conspicuous; discal spot elongate, brownish; terminal fascia wide, ochreous brown to grey-brown.
Similar species. *Mamestra brassicae* (Linnaeus) is usually larger, with a curved spine on tibia of foreleg; forewing teeth of subterminal line do not reach termen; veins and terminal fascia of hindwing much less contrasting. *Sideridis albicolon* (Hübner) has the reniform stigma marked with white dots on the distal margin.

Life history
Ovum. Hemispherical, rather heavily ribbed, with deep pitting between the ribs; pale greenish white; apart from darker rings at micropyle and equator, there is little colour change before hatching. Laid singly on leaves or stems of the foodplant, sometimes hatching in only five days.
Larva. Full-fed *c.*45mm long, rather stout, but capable of great elongation of anterior segments. Head pale brown. Body variable from bright green, through brown to purplish green, completely brown forms predominating; dorsal line dark, complete, broken or vestigial; very often a row of broad, black subdorsal dashes; supra- and subspiracular lines variable, the latter usually broad and pale, often pink-tinged; spiracles white, thickly outlined black.
The larva is very intolerant of light and feeds by night, mainly on Chenopodiaceae such as orache (*Atriplex* spp.) and goosefoot (*Chenopodium* spp.), but onion (*Allium cepa*), knotgrass (*Polygonum aviculare*) and yellow alison (*Alyssum saxatile*) have also been recorded. Styles (1960) mentions the species as a pest on young conifers in forest nurseries.
Pupa. Slender; thoracic segments yellow-brown to orange-brown, wing-cases darker, often tinged green; abdominal segments evenly tapering, smooth with an anterior band of punctation on each; cremaster reduced and squared, with two small lateral spines. Subterranean in a frail cocoon.
Imago. Bivoltine in the south, flying May to June and August to September, univoltine from the Midlands northward, flying in late June and July. Sometimes feeds at flowers such as ragwort and knapweed by day, but is mainly nocturnal and comes to flowers, sugar and light.

Distribution (Map 144)
This species seems to prefer gravels, light or sandy soils, coastal areas and river valleys; it is liable to population explosions which may be caused by the rapid increase of the foodplants on newly cleared ground, or possibly to local migration. Common and well distributed in south-east England and East Anglia; more local from Hampshire to the west and north and becoming less common from the Midlands northwards; local in Scotland, where it may be reinforced by immigration; has temporarily established itself in Orkney; rare, possibly only immigrant, in Ireland; the Isles of Scilly; the Channel Islands. Holarctic, with local races in North America; in western Europe widespread from Portugal to Denmark, south Scandinavia and Finland.

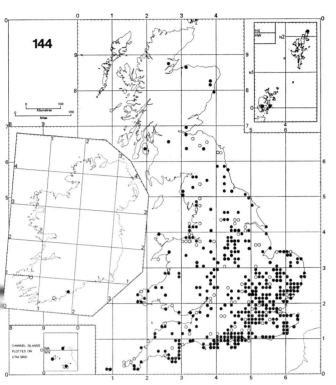

Discestra trifolii

LACINIPOLIA McDunnough
Lacinipolia McDunnough, 1937, *Can.Ent.* **69**: 43.

Of this very large genus in the Americas, two species have been recorded in Britain, probably accidentally introduced.

LACINIPOLIA RENIGERA (Stephens)
Celaena renigera Stephens, 1829, *Ill.Br.Ent.* (Haust.) **3**: 16.
Type locality: England; near London.

This was described and figured as a species new both to Britain and to science on the basis of three specimens, one in Stephens' own cabinet, obtained from the Marsham collection, the other two in the collection of Mr Stone, believed by Stephens to have been taken near London many years before. No later record is known. The species is North American, and these examples, if really taken in Britain, were probably accidentally introduced. This species must not be confused with *Ochropleura renigera* (Hübner), a European alpine species, which was erroneously described as British by Stephens through misidentification of specimens later found to belong to a form of *Standfussiana lucernea* (Linnaeus) (Noctuinae) *q.v.*

LACINIPOLIA LAUDABILIS (Guenée)
Hecatera laudabilis Guenée, 1852, *in* Boisduval & Guenée, *Hist.nat.Insectes* (Lépid.) **6**: 30.
Type locality: America.

A specimen of this North American species was captured by C. M. Jones at the corner of Valentia and Cable Roads, Hoylake, Cheshire, in the forenoon of 31 July 1936. The locality is near the maritime approaches to the Mersey. It was supposed that it had been accidentally imported or had escaped from an incoming ship (Smith, 1947; 1948).

HADA Billberg

Hada Billberg, 1820, *Enum.Ins.Mus.Blbg*: 86.

This genus has numerous Holarctic species, of which only one is found in Great Britain and Ireland. Its members are rather similar to *Polia* Ochsenheimer, but *Hada* species are much smaller. There are also differences in the male genitalia, particularly the deeply bifurcate juxta and the strongly developed ampulla.

HADA NANA (Hufnagel)
The Shears

Phalaena nana Hufnagel, 1766, *Berlin.Mag.* **3**: 398.
Noctua dentina [Denis & Schiffermüller], 1775, *Schmett. Wien.*: 82.

Type locality: Germany; Berlin.

Description of imago (Pl.11, figs 13–16)

Wingspan 28–38mm. Antenna in male densely but shortly ciliate, in female very shortly and sparsely ciliate. Forewing in female longer and narrower than in male; ochreous white to greyish ochreous; basal streak short, black, usually with an ochreous blotch immediately above; median fascia about twice as wide at costa as at dorsum and enclosing all stigmata; reniform and orbicular stigmata of ground colour, outlined darker; claviform stigma paler, surrounded by a triangular blackish mark which is often forked distally. Fresh specimens have an orange-ochreous patch at base of dorsum; it soon fades. Hindwing in male greyish brown to ochreous brown with darker terminal shade, in female usually darker basally, the terminal shade less contrasting; in both sexes cilia whitish.

Specimens from the shingle beaches of southern England are often very pale (ab. *leucostigma* Haworth, fig.16); ochreous specimens (ab. *ochrea* Tutt, fig.15) occur in most southern and midland counties and the species tends to become progressively darker from south to north, although even in Shetland intermediate as well as very dark forms occur.

Life history

Ovum. Low-conical, heavily ribbed with light reticulation; whitish at first, darkening gradually at micropyle and equator and finally becoming pinkish brown. Laid singly or in small batches on the foodplant, hatching in about ten days.

Larva. Full-fed *c.*35mm long. Head grey-brown. Body dark greenish brown; dorsal and subdorsal lines fine and white, both broken; dorsal area covered with obscure brownish diamonds and more clearly defined greyish subdorsal blotches; spiracular line grey; spiracles black; ventral surface and legs pale greenish brown.

Nocturnal in habit and feeds on a variety of plants, mainly Compositae and probably largely on the roots. Recorded foodplants include dandelion (*Taraxacum officinale*), hawk's-beard (*Crepis* spp.), hawkweed (*Hieracium* spp.), knotgrass (*Polygonum aviculare*) and common chickweed (*Stellaria media*). The main brood is fully grown in late July or August.

Pupa. Atypical of the subfamily in shape, with heavily armoured head and long wing-cases, mahogany-red, darker dorsally; abdominal segments 5–8 with posteriorly pointing lateral spines; cremaster with two strong apical spines. In a slight cocoon among surface litter (see Haggett (1954) for fuller details).

Imago. Flies from late May to early July, with a small second brood in the south. May often be found at rest by day on fence-posts and tree-trunks and will feed at flowers in full sunshine, but the main period of activity is late dusk when it feeds at flowers of campion and wood-sage; later it comes to sugar and light.

Distribution (Map 145)

This species seems to be equally at home on the coast, moorland or in woodland with, perhaps, a preference for light soils. Common throughout Great Britain to Shetland and the whole of Ireland. Abroad Eurasiatic; in Spain and France especially montane, northwards to the Arctic.

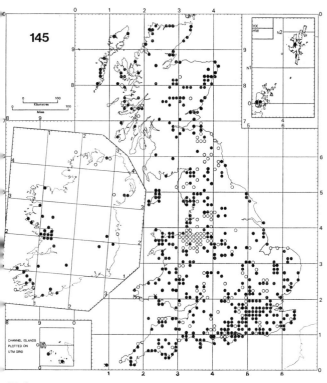

Hada nana

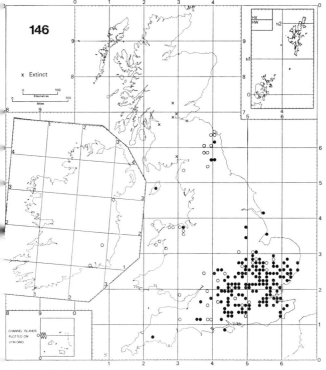

Polia bombycina

POLIA Ochsenheimer

Polia Ochsenheimer, 1816, *Schmett.Eur.* **4**: 73.
Aplecta Guenée, 1838, *Annls Soc.ent.Fr.* **7**: 217.

A small genus containing large species in which the forewing is exceptionally long in relation to its breadth. There are three British species, each filling a distinct ecological niche.

Imago. Wingspan 43–58mm. Antenna in male fasciculate or ciliate. Thorax and abdomen with crests.

Ovum. On the foodplant, in large batches.

Larva. Cylindrical, usually some shade of brown with darker diamond-shaped dorsal pattern. On herbaceous plants in autumn, on deciduous trees and shrubs in spring.

Pupa. Large and well proportioned, widest at posterior end of wing-cases, thence evenly and gradually tapering; cremaster with good specific characters. Below ground in a cocoon.

POLIA BOMBYCINA (Hufnagel)
The Pale Shining Brown

Phalaena bombycina Hufnagel, 1766, *Berlin.Mag.* **3**: 410.
Noctua advena [Denis & Schiffermüller], 1775, *Schmett. Wien.*: 77.
Noctua nitens Haworth, 1809, *Lepid.Br.*: 188.
Type locality: Germany; Berlin.

Description of imago (Pl.11, figs 17,18)

Wingspan 47–52mm. Antenna in male shortly fasciculate, in female sparsely ciliate. Forewing glossy, pinkish brown, sometimes with blue-grey or grey suffusion; median fascia only slightly darker, up to four times as broad at costa as at dorsum; stigmata variable in prominence, reniform usually most distinct, slightly paler than ground colour, outlined slightly darker; orbicular and claviform usually more obscure, almost touching and forming a rough figure '8'; subterminal line usually a series of disconnected blackish dots and dashes, the largest immediately above the tornus, but occasionally complete or nearly so. Hindwing pale brown, with slightly darker terminal shade.

Variation affects mainly the ground colour, the degree of grey or grey-blue suffusion and the prominence of the stigmata and subterminal line. In ab. *unicolor* Tutt (fig.18) all markings are obscure.

Life history

Ovum. Low dome-shaped, irregularly ribbed and weakly reticulate, the micropyle slightly raised; reddish violet. Laid in large, neat batches, hatching in about ten days (Hawkins, 1954).

Larva. Full-fed *c.*55mm long. Head pale brown. Body pale brown with a dorsal pattern of obscure, slightly darker brown diamonds, which extend laterally to give a rather barred appearance; spiracles golden brown, outlined black; ventrally pale yellowish, sometimes greenish brown. Buckler (1895) described and figured an unusual grey-green form with sage-green dorsal diamonds.

The life history of the wild larvae is obscure. In captivity they feed up quickly if kept at a temperature of 18°–20°C (64°–68°F) on sow-thistle (*Sonchus* spp.), dandelion (*Taraxacum officinale*), cultivated lettuce (*Lactuca* spp.) or knotgrass (*Polygonum aviculare*) and will produce moths in late autumn. Almost certainly the larva hibernates in the wild, but it is not known whether, like its congeners, it completes its growth on the leaf-buds and leaves of deciduous trees and shrubs, or on herbaceous plants, such as the Compositae which it favours in captivity.

Pupa. Elongate, reddish brown, with slightly darker wing-cases; thoracic segments finely sculptured; abdominal segments clearly separated, the anterior part of each lightly punctate; cremaster blackish, striate and laterally flattened with two lateral, two submedian and two central apices, the outer with fine and the two central with strong bristles.

Imago. Univoltine. Flies from mid-June. Comes freely to light and to sugar, and prefers flowers such as campion, wood-sage, martagon lily and viper's bugloss.

Distribution (Map 146)

Although the species is most common on light or calcareous soils, it tends to wander and may migrate locally, as several at a time may be taken at fairly long intervals at one trap site. Discontinuous; locally common south of the Thames and in the Fens and East Anglia; much less common, although widely distributed, in the Midlands; Northumberland; north Wales; widely distributed in southern Scotland until mid-nineteenth century, since when there are no definite records. There are few Irish records – one each from Cos Louth, Carlow and Waterford (Baynes, 1964). Eurasiatic; in western Europe from Spain to central Scandinavia and Finland.

POLIA HEPATICA (Clerck)
The Silvery Arches

[*Phalaena*] *hepatica* Clerck, 1759, *Icones Insect.rar.* **1**: pl.8, fig.3.
Phalaena (*Noctua*) *tincta* Brahm, 1791, *Ins.Kal.* **2** (1): 393.
Type locality: [Sweden].

Description of imago (Pl.11, figs 19,20)

Wingspan 43–52mm. Antenna in male densely and shortly fasciculate, in female sparsely ciliate. Forewing silvery blue-grey; ante- and postmedian lines rather indefinite, greyish white; subterminal line blackish brown, incomplete and most clearly defined at the extremities and in the middle, with considerable thickening just before the tornus; upper stigmata enclosed by a grey-brown subcostal shading, orbicular pale greyish, reniform variably suffused brown; claviform stigma a clear greyish brown wedge, all three outlined blackish brown. Hindwing fuscous, with obscure lunular discal spot and broad, brownish fuscous terminal shade.

The species varies less than its congeners, but in Scottish specimens (fig.20) the ground colour is more strongly bluish-tinged and the subcostal shading is purplish brown; throughout the range there is minor variation in the degree of contrast between the ground colour and the lines and stigmata.

Life history

Ovum. Hemispherical, strongly ribbed and reticulate; pale brownish violet. Laid in large batches, hatching in about ten days.

Larva. Full-fed *c.*50mm long. Head pale brown. Prothoracic plate light grey-brown. Body various shades of red-brown, with subdued dorsal pattern of darker diamonds; spiracles blackish and rather clear; ventrally slightly paler brown than dorsally and laterally. Where this species and *P. nebulosa* (Hufnagel) are found together, the larva of *P. hepatica* is usually of a more reddish shade of brown with less clearly defined dorsal diamonds and a weak lateral pattern, the spiracles not being obscured by extensions of the dorsal diamonds.

The young larva feeds mainly on herbaceous plants such as dock (*Rumex* spp.) and dandelion (*Taraxacum officinale*), although it may be beaten by day from birch or sallow in early autumn, when it is *c.*20mm long. After hibernation it feeds mainly on the buds and young leaves of birch (*Betula* spp.), rough-leaved willow (*Salix* spp.), hawthorn (*Crataegus* spp.) and bog-myrtle (*Myrica gale*), ascending the trees or bushes very rapidly soon after dusk. Pupates in May.

Pupa. Elongate; thoracic segments, wing-cases and appendages reddish brown and finely sculptured; abdominal segments more glossy with deep intersegmental divisions; cremaster triangular, black and heavily striate, with several basal bristles and two apical spikes with curved tips. In a slight cocoon. The moth emerges in two to three weeks.

Imago. Univoltine. Flies in June and July and is often found at rest by day on tree-trunks or fences. Feeds at flowers after dusk and comes to sugar and light.

Distribution (Map 147)

This species is apparently restricted to lightly or moderately wooded country on acid soils (*cf. Lacanobia contigua* ([Denis & Schiffermüller])). The distribution poses problems such as the moth's comparative abundance locally on

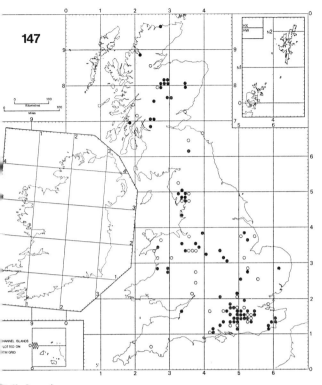

147

Polia hepatica

the tornus; stigmata outlined with black, orbicular and reniform more prominent in typical and melanic forms, the short, thick claviform in pale forms. Hindwing pale fuscous with a faint discal spot, the colour of the base merging gradually into a slightly darker terminal shade.

Variation is considerable and largely geographical. In western England, most of Scotland and Ireland the pale ab. *pallida* Tutt (fig.24) predominates; the melanic forms ab. *robsoni* Collins (heterozygote) and ab. *thompsoni* Arkle (homozygote) (fig.25) were first recorded about 1890 from industrial Cheshire, Lancashire and Yorkshire, but seem now to have been replaced completely by the more generally distributed ab. *plumbosa* Mansbridge (fig.26). Kettlewell (1973: 272–273) discusses this in more detail. Intermediate forms closer to the typical form (fig.23) occur over the rest of Britain and are in general paler in the extreme south and north of the Midlands, whence they merge with ab. *pallida*, but darker from the London area to the Midlands.

Similar species. *Eurois occulta* (Linnaeus) (Noctuinae) – only the pale grey, presumed immigrant form, *q.v.*

Life history

Ovum. Hemispherical, ribbed and lightly reticulate; pale, slightly bluish green at first, becoming grey-green after about a week and darkening rapidly three or four days later, immediately before hatching. Laid on the pre-hibernation foodplants in large, orderly batches.

Larva. Full-fed *c.*55mm long. Head dark, golden brown. Body variable, from glaucous grey to dark brown, usually ochreous brown to brown; rather paler dorsally; dorsal line very pale, traversing a series of very much darker dorsal diamonds which extend laterally almost to touch a row of equally dark, oblique supraspiracular streaks; spiracles black; ventrally pale greyish brown.

Feeds on herbaceous plants before and immediately after overwintering, but climbs to feed on the buds and leaves of deciduous trees and shrubs, such as birch (*Betula* spp.), rough-leaved willow (*Salix* spp.), sycamore (*Acer pseudoplatanus*) and bramble (*Rubus fruticosus* agg.) as soon as they are available. Full-fed in May.

Pupa. Elongate; glossy reddish brown with darker, lightly sculptured wing-cases; abdominal segments tapering evenly to segment 8, which tapers sharply and has two fine dorsolateral bristles before a heavily grained, blackish cremaster with several finer bristles and a pair of slightly bulbous-tipped apical spines. In a subterranean cocoon, in which the larva has spent well over a week before ecdysis.

Imago. Univoltine. Emerges from early June to mid-July and usually hides by day, although sometimes found at rest on tree-trunks or fences. Feeds at flowers, such as campion and wood-sage soon after dusk, and comes to sugar in numbers and to light.

the Surrey heaths, but relative scarcity on similar ground in Hampshire and Dorset. Occasionally single specimens are taken in light-traps well away from the normal habitat. Locally common in southern England, more locally distributed in northern England, Wales and Scotland, but apparently absent from the western and northern islands and Ireland. Eurasiatic; in western Europe, mainly montane in France, Spain and Belgium, at lower levels to central Scandinavia and Finland.

POLIA NEBULOSA (Hufnagel)
The Grey Arches

Phalaena nebulosa Hufnagel, 1766, *Berlin.Mag.* **3**: 418.

Type locality: Germany; Berlin.

Description of imago (Pl.**11**, figs 23–26)

Wingspan 46–58mm. Antenna shortly ciliate, in male densely, in female sparsely. In female costa of forewing more curved subapically than male. Forewing of various shades of grey, from almost white to blackish, with darker subbasal, median and subterminal pencillings; clarity of lines very variable and greatest in the medium grey colour-range; subterminal line usually a series of separate blackish wedges, with a more conspicuous blackish streak before

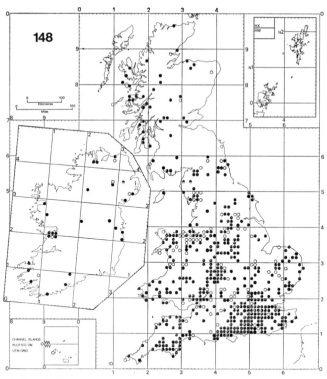

Polia nebulosa

PACHETRA Guenée
Pachetra Guenée, 1841, *Annls Soc.ent.Fr.* **10**: 241.

Contains a single, widely distributed Palaearctic species, which is very local in England.

Similar to *Polia* Ochsenheimer, from which it can be separated by the shorter, more ample forewing and the strongly bipectinate antenna of the male.

PACHETRA SAGITTIGERA BRITANNICA Turner
The Feathered Ear

Phalaena sagittigera Hufnagel, 1766, *Berlin.Mag.* **3**: 410.

Noctua leucophaea [Denis & Schiffermüller], 1775, *Schmett. Wien.*: 82.

Bombyx fulminea Fabricius, 1777, *Gen.Ins.*: 282.

Pachetra sagittigera britannica Turner, 1933, *Entomologist's Rec.J.Var.* **45**: 284.

Type locality: Britain.

Description of imago (Pl.11, figs 28,29)

Wingspan 38–50mm. Antenna in male strongly bipectinate, in female ciliate. Forewing whitish, suffused to a varying degree with very pale brown; main veins whitish; subbasal, ante- and postmedian lines blackish, edged white, the wing thus appearing reticulate; subterminal line of separate blackish, inward-pointing wedges distad of a wide, ill-defined pale fuscous band; reniform and orbicular stigmata usually greyish brown, outlined whitish; claviform stigma dark grey-brown, most conspicuous on paler specimens. Hindwing in male whitish, in female pale fuscous; in both sexes a lunular discal spot and sometimes a very pale greyish terminal shade.

The British subspecies, for which Turner used Barrett's (1897: 147) description, is separated by the ground colour, which is consistently more whitish than in other races and therefore shows more contrast in the markings. Three British specimens in BMNH resemble the darker continental typical form.

Life history

Ovum. Hemispherical, strongly ribbed and more lightly reticulate; pale straw-colour at first, becoming purplish brown and hatching in about 12 days. Apparently laid in large batches around grass stems (Stokoe & Stovin, 1958: pl.55, fig.3); captive females lay freely in chip-boxes.

Larva. Full-fed c.40mm long, rather stout and of velvety texture. Head pale golden brown. Body brownish ochreous; markings variable: the most lightly marked form has a pale dorsal line, parallel rows of subdorsal dots, several fine, brownish lateral lines and is pale ventrally; at the other extreme the subdorsal dots are replaced by thick,

Distribution (Map 148)

Although well distributed, and occurring in most inland habitats in southern Britain, this species is most common in well-wooded country and is more or less confined to woods in the northern part of its range. Much more local in Scotland, where it is most common in the west; recorded from the Isles of Scilly, several islands of the Inner Hebrides and from the greater part of Ireland; the Channel Islands. Eurasiatic; throughout western Europe to about 62°N.

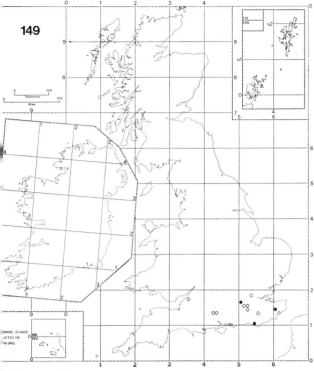

Pachetra sagittigera

disconnected dashes and there is sometimes a broad, dark supraspiracular stripe, followed by an equal stripe of the ground colour containing a double row of black dots.

The larva is green for the first four instars and feeds on grasses, preferring annual and wood meadow-grass (*Poa annua* and *P. nemoralis*). Probably feeds through the winter in mild weather, but is very difficult to overwinter in captivity Numbers have been found clinging to grass-stems at night in January and February. Full-fed in April.

Pupa. Reddish brown; thoracic segments sculptured; abdominal segments with dorsal and subdorsal hollows; cremaster black and rugose, with two parallel apical spines. On the surface of the soil, in a fragile cocoon.

Imago. Univoltine. Emerges towards the end of May and in June. Feeds at flowers, especially those of privet, and comes to sugar, but has not often been taken at light. The ecology of this species is obscure; it occurs only on chalk and limestone but is extremely local and may be almost extinct in Britain.

Conservation. Of the 'endangered' species in this subfamily, *P. sagittigera* is the most greatly at risk and should not be collected at all unless new and vigorous colonies are discovered.

Distribution (Map 149)

There are old records from the Bristol area; during the present century there have been a number of records along the North Downs from Folkestone to Dorking, including several formerly strong colonies; elsewhere there are isolated records from Wiltshire, Hampshire and the Chilterns in Buckinghamshire. Eurasiatic; in western Europe the nominate subspecies ranges from Spain to south Sweden, mainly on calcareous soils.

SIDERIDIS Hübner
Sideridis Hübner, [1821], *Verz.bekannt.Schmett.*: 232.

Of a number of Holarctic species only one is found in the British Isles.

Frons simple, rounded, with a sclerotized plate below. Anterior tarsus with claw-like apical spines.

SIDERIDIS ALBICOLON (Hübner)
The White Colon

Noctua albicolon Hübner, [1813], *Samml.eur.Schmett.* **4**: pl.117, figs 542,543.
Type locality: Europe.

Description of imago (Pl.11, figs 21,22)
Wingspan 38–44mm. Antenna ciliate, in female more shortly. Forewing usually pale grey-brown, but may be ochreous tinged, darker grey or even blackish grey; cross-lines ill-defined, ante- and postmedian slightly paler than ground colour, subterminal usually a series of whitish dots; stigmata equally obscure, reniform indicated by a pair of vertically opposed whitish dots on distal margin of the dorsal half. Hindwing very pale fuscous; lunular discal spot, terminal fascia and veins pale brownish fuscous; cilia white.

There is a considerable amount of local variation; dark, almost blackish forms occur on the heaths of east Hampshire, but the very dark forms from north-east Scotland mentioned in earlier works are elusive and only one dark grey, somewhat ochreous-suffused specimen in BMNH labelled 'Moray', has been traced.

Similar species. Mamestra brassicae (Linnaeus): foreleg with a slightly curved, tapering spur on the tibia; forewing usually with a clearly outlined reniform stigma; often a clear, complete whitish subterminal line; hindwing with a fine dark line in the whitish fringe. *Apamea remissa* (Hübner) ab. *obscura* Haworth (Amphipyrinae): eyes glabrous; usually a complete, irregular whitish subterminal line; hindwing uniformly brownish fuscous. *A. furva* ([Denis & Schiffermüller]) (Amphipyrinae): eyes glabrous; subterminal line whitish, complete; reniform stigma usually outlined whitish. *Discestra trifolii* (Hufnagel) *q.v.*

Life history
Ovum. Almost spherical, ribbed and more lightly reticulate; whitish at first, gradually darkening at micropyle and equator. Laid singly on the foodplant, hatching in about 12 days.

Larva. Full-fed *c.*50mm long; dorsal surface tapers sharply from abdominal segment 8. Head and prothoracic plate yellowish brown. Body usually bluish green, but sometimes grey-green or blue-grey, darker dorsally; dorsal line paler than ground colour, edged darker; spiracles white, ringed black; ventral surface soft, pale green.

Nocturnal in habit; a large number of foodplants have been recorded, including goosefoot (*Chenopodium* spp.), orache (*Atriplex* spp.), common restharrow (*Ononis repens*), dandelion (*Taraxacum officinale*), sea-bindweed (*Calystegia soldanella*), knotgrass (*Polygonum* spp.), dock (*Rumex* spp.), sea-rocket (*Cakile maritima*) and common chickweed (*Stellaria media*) with, in some cases, a preference for the flowers. The variety of this list suggests that the larva is in fact polyphagous. Full-fed from late July.

Pupa. In sand; Huggins (1957) notes that a depth of at least 25cm (10in) is necessary for successful pupation in captivity.

Imago. Univoltine, partially bivoltine in the south. Flies from late May and through June with a small second brood in August in the south. Inhabits coastal sand-dunes and occurs locally inland in sandy areas, hiding by day under dune crests and among marram and lyme-grass from which it may be disturbed by raking or beating. At night it feeds at the flower-heads of grasses and comes to sugared grass-heads and light.

Distribution (Map 150)
Coastal dunes of England, Wales and eastern Scotland to Moray. There are also strong inland colonies in the Breckland area of Norfolk and on the heaths of Surrey and east Hampshire; both banks of the lower Thames valley west of London have produced a fair number of records. Local and rather uncommon in Ireland, where it has been recorded mainly from the north and south-west coasts. The Channel Islands. Eurasiatic; in western Europe local and mainly coastal from south-west France to Denmark, only just reaching south Norway and Sweden.

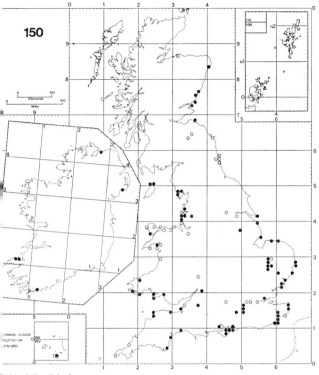

Sideridis albicolon

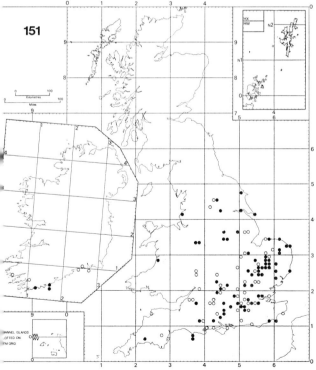

Heliophobus reticulata

HELIOPHOBUS Boisduval

Heliophobus Boisduval, 1828, *Eur.Lepid.Index method.*: 69.

There are two Palaearctic species, of which one is found in the British Isles, where it is represented by two subspecies.

Frons flat. Dorsal and ventral surface of thorax long and densely haired. Femur long-haired. Abdomen with long and dense lateral fringes. Male genitalia with sacculus and juxta very elongate.

HELIOPHOBUS RETICULATA (Goeze)
The Bordered Gothic

Noctua (Phalaena) reticulata Goeze, 1781, *Ent.Beyträge* **3** (3): 254.
Noctua calcatrippae Vieweg, 1790, *Tabl.Verz.Brand.Schmett.* **2**: 71.
Phalaena (Noctua) saponariae Borkhausen, 1792, *Naturgesch.eur.Schmett.* **4**: 370.

Type locality: [Europe].

Description of imago (Pl.11, figs 32–34)
Wingspan 35–40mm. Antenna in male shortly ciliate, in female simple.

Subsp. *marginosa* (Haworth) (figs 32,33).
Noctua marginosa Haworth, 1809, *Lepid.Br.*: 195.

Forewing pale brown; lines yellowish white, sometimes tinged with pink, subbasal broken and vestigial, antemedian concave, postmedian sharply angled before costa, subterminal broad and very clear; main veins clear, whitish; the markings of the basal part are less clear cut than those beyond the orbicular stigma. Hindwing whitish basally, with pale greyish lunular discal spot and terminal fascia.

Subsp. *hibernica* Cockayne (fig.34)
Heliophobus saponariae hibernica Cockayne, 1944, *Entomologist's Rec.J.Var.* **56**: 55.

Ground colour of forewing pale purplish brown; otherwise as in subsp. *marginosa*.

There is little variation within each subspecies.

Similar species. Tholera decimalis (Poda) in which the male has strongly bipectinate antenna; forewing darker brown; lines, other than subterminal, inconspicuous; flies late August to September. *Naenia typica* (Linnaeus) (Noctuinae) q.v.

Life history
Ovum. Dome-shaped, ribbed and reticulate; glossy, yellowish white at first, gradually darkening at micropyle and equator before hatching. Usually laid in rows on the foodplant, hatching in about 14 days.
Larva. Full-fed *c.*45mm long, rather stout and tapering little at extremities. Head pale brown. Body putty-coloured,

tinged greenish or reddish; dorsal line whitish, finely edged grey-brown; a series of fine, grey-brown subdorsal and lateral lines; spiracles small, whitish, ringed greyish; ventrally slightly paler than ground colour.

The larva has seldom been successfully reared in captivity; Lees (1954) records only moderate success on knotgrass (*Polygonum aviculare*) and bladder campion (*Silene inflata*) and among somewhat speculative suggestions are the unripe seeds of various Caryophyllaceae, 'various low plants' or 'grasses'. It has been established, however, that it is full-fed in August and that it pupates below ground level.

Pupa. Glossy red-brown; thoracic segments (except metathorax) sculptured; metathorax and abdominal segments smooth, the latter with anterior band of punctation; cremaster small, with four fine subapical bristles and two robust, close-set apical spines.

Imago. Univoltine. The moth flies in June and early July and feeds at flowers such as red valerian, campion, viper's bugloss and wood-sage, and comes to sugar and light.

Distribution (Map 151)

Subsp. *marginosa*: locally common in the counties south of the Thames, more generally common in East Anglia, but local and becoming more scarce in the Midland counties and northward to Yorkshire, with a preference for calcareous soils. Chalmers-Hunt (1970) quotes an old record for the Isle of Man.

Subsp. *hibernica*: Ireland, so far recorded only in the extreme south from Co. Wexford to Co. Kerry. Eurasiatic; the nominate subspecies ranges through western Europe from Spain to Denmark, south Norway, Sweden and Finland to about 64°N.

MAMESTRA Ochsenheimer

Mamestra Ochsenheimer, 1816, *Schmett.Eur.* **4**: 76.
Barathra Hübner, [1821], *Verz.bekannt.Schmett.*: 218.

Mamestra contains only a few species, one of which occurs in the Palaearctic region.

Frons flat; palpus porrect, third segment hairy and very short. Thorax and abdomen with crests. Anterior tibia with a stout, curved spine.

MAMESTRA BRASSICAE (Linnaeus)
The Cabbage Moth

Phalaena (*Noctua*) *brassicae* Linnaeus, 1758, *Syst.Nat.* (Edn 10) **1**: 516.
Type locality: not stated.

Description of imago (Pl.11, figs 30,31)

Wingspan 34–50mm. Antenna shortly ciliate, in male densely, in female sparsely. Forewing grey-brown to blackish brown with variable reddish brown scaling; subbasal, antemedian and postmedian lines inconspicuous, slightly paler than ground colour, finely edged darker; subterminal line, when present (see below), whitish to yellowish white with two angular projections which do not reach the termen; stigmata outlined black, reniform with whitish distal margin and less clearly defined proximal margin. Hindwing fuscous; discal spot greyish fuscous; terminal shade brownish fuscous; fringe white with a fine greyish central line. A slightly curved, apically pointed spur on the tibia of the foreleg (figure 16) distinguishes *M. brassicae* from all similar species.

The subterminal line varies; in most specimens from southern Britain it is complete and clear, but in most specimens from the north it is absent or vestigial (fig.31). A small form, apparently recurrent, occurs in the Dingle peninsula, Co. Kerry (Huggins, 1964; 1967).

Figure 16 *Mamestra brassicae* (Linnaeus), foreleg

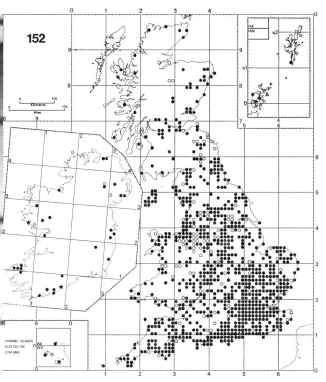

Mamestra brassicae

Similar species. Sideridis albicolon (Hübner) *q.v.* and *Discestra trifolii* (Hufnagel) *q.v.*

Life history

Ovum. Relatively small, hemispherical, ribbed and reticulate; whitish at first, darkening gradually to purplish brown. Laid on almost any foliage in large regular batches, hatching in about eight days.

Larva. Full-fed *c.*50mm long, elongate, with a slight dorsal hump on abdominal segment 8. Head light brown. Body from fairly bright green, through brownish green and greenish brown to almost black; dorsal line fine, black; subdorsal line of blackish bars; spiracular line broad, pale green or pale ochreous; spiracles white; yellowish green ventrally. Pale green with yellowish intersegmental bands when young, becoming gradually darker with each moult; when full-grown the majority are brownish green or blackish green.

Nocturnal in habit and completely polyphagous, although particularly addicted to *Brassica* crops, on which it may be a serious pest as it spoils by boring and fouling more than it actually eats.

Pupa. Elongate, reddish brown; wing- and limb-cases finely sculptured; abdominal segments evenly tapering, darker brown and smooth, with a finely pitted anterior band on each; segment 8 sharply excavated to a narrowly conical cremaster with two short, apically hooked spines. In earth in a flimsy cocoon.

Imago. Univoltine. May be encountered in any month of the year including, exceptionally, December and January; most common in June and July and from late August to the end of September and there is some evidence that it may overwinter in the north, as there are a few indoor records in midwinter. The species seems to have no ecological preferences, although market-gardening has provided a very congenial environment.

Distribution (Map 152)

Common throughout Great Britain and Ireland although rather less so in the north than in the south. Eurasiatic, probably Holarctic; all Europe to central Scandinavia and Finland.

MELANCHRA Hübner
Melanchra Hübner, [1820], *Verz.bekannt.Schmett.*: 207.

Contains only one species in the British Isles.

Very similar to *Mamestra* Ochsenheimer, but lacking the tibial spine on the foreleg. The male genitalia are unusual, with an extremely large, diamond-shaped gnathos, unlike any other British noctuid.

MELANCHRA PERSICARIAE (Linnaeus)
The Dot

Phalaena (*Noctua*) *persicariae* Linnaeus, 1761, *Fauna Suecica* (Edn 2): 319.
Type locality: Sweden.

Description of imago (Pl.11, fig.35)
Wingspan 38–50mm. Antenna in male setose-ciliate, in female sparsely ciliate. Forewing bluish black; all lines obscure except subterminal, which usually consists of a series of ochreous dots, but may vary from obsolescent to, extremely rarely, an unbroken line; reniform stigma white; a brownish ochreous central line most developed at the costal extremity and roughly parallel to the distal margin; other stigmata usually obscure although the orbicular may be lightly outlined pale ochreous. Hindwing whitish basally; discal spot greyish fuscous, varying from an obsolescent to a prominent lunule; terminal fascia greyish fuscous, variable in extent and frequently reaching the median area.

There is no local variation, but individual specimens vary in the degree of ochreous suffusion. In ab. *unicolor* Staudinger the reniform stigma is obscured by the ground colour and at the other extreme in ab. *ochrorenis* Kardakoff the reniform is golden yellow and the underside heavily suffused ochreous; these aberrations are extremely rare in Britain, but intermediate forms are not uncommon.

Life history
Ovum. Hemispherical, with rather flattened base, ribbed and lightly reticulate; whitish green at first, slowly darkening to pinkish brown. Laid singly or in rather untidy masses on the foodplant, hatching in about eight days.

Larva. Full-fed *c*.44mm long, of velvety texture. Head pale brown. Prothoracic plate greenish brown, traversed by whitish dorsal and subdorsal lines. Body varying from pale grey-green, through shades of green and brown to purplish brown; dorsal line pale; subdorsal line fine, greyish; darker, backward-pointing chevrons on abdominal segments 1 and 2 giving illusion of a hump; less pronounced chevrons on segments 3–7; pronounced dorsal hump on segment 8; spiracles white, ringed black; a series of dark, oblique bars from immediately above the spiracles, where they touch the lateral extremity of the dorsal chevrons, to the base of the prolegs; legs and prolegs yellow-brown. In the early instars green with a broad, pale lateral stripe; the humped appearance is gradually acquired in the last two instars.

Polyphagous, finding much to suit it in the average garden, and feeding also by day or night in particular on nettle (*Urtica* spp.) and low bushes of elder (*Sambucus* spp.) or rough-leaved willow (*Salix* spp.); Styles (1960) records it as a nuisance on larch (*Larix decidua*) in forest plantations. Feeds rather slowly, well into autumn, and is full-fed in September or October.

Pupa. Rather stout, glossy red-brown; head and thorax with dorsal sculpturing; abdominal segments evenly tapering, lightly sculptured, each with an anterior band of punctation; segment 8 abruptly rounded ventrally to a blackish cremaster with two short, divergent, slightly hooked apical spines. Below ground, in a cocoon.

Imago. Univoltine, with a long period of emergence from late June to August. Not often seen by day, but feeds at flowers and sugar at night and comes to light. The ecology of this species is rather obscure because, throughout its range in Britain, it seems to have adapted to life among or on the fringes of human habitation, where it is now commoner than in open country.

Distribution (Map 153)
Very common from the south coast of England to the Midlands, progressively less so in northern England, except north Lancashire and south Westmorland (VC69), where it occurs regularly. Local but well distributed in Wales and Ireland and has been recorded from the Isle of Man, and recently from a few localities in southern Scotland; the Isles of Scilly; the Channel Islands. Eurasiatic, probably Holarctic; through western Europe to Denmark, south Sweden and Finland, but not known in Norway.

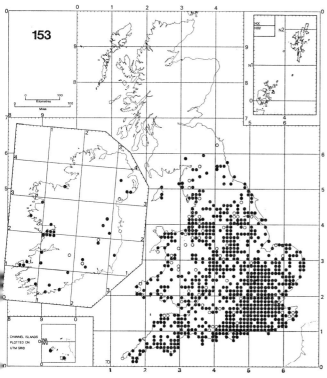

Melanchra persicariae

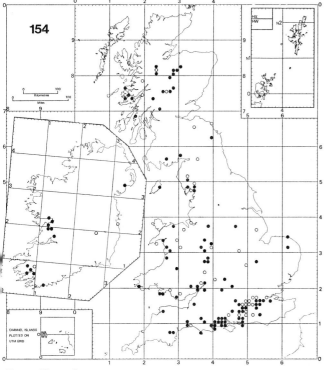

Lacanobia contigua

LACANOBIA Billberg

Lacanobia Billberg, 1820, *Enum.Ins.Mus.Blbg*: 87.
Barathra Hübner, [1821], *Verz.bekannt.Schmett.*: 207.
Peucephila Hampson, 1909, *Trans.ent.Soc.Lond.* **1909**: 461.

A rather diverse genus with numerous Palaearctic and Holarctic species, of which five occur in Great Britain and four also in Ireland; the status of a sixth species is doubtful.

Frons smooth; antenna in male ciliate. Thorax and abdomen with more or less strong crests. Forewing with subterminal line strongly dentate.

LACANOBIA CONTIGUA ([Denis & Schiffermüller])
The Beautiful Brocade

Noctua contigua [Denis & Schiffermüller], 1775, *Schmett. Wien.*: 82.

Type locality: [Austria]; Vienna district.

Description of imago (Pl.11, fig.27)

Wingspan 36–42mm. Antenna shortly ciliate, in male densely, in female sparsely. Forewing divided diagonally into a dark basal and a lighter distal part by an oblique pinkish grey fascia, which comprises the orbicular stigma, a tornal blotch and the space between them; basal streak black, with a pale ochreous subcostal blotch above; ante- and postmedian lines pinkish white, edged black; a broad pinkish white subterminal fascia from apex, sometimes merging with tornal blotch; subterminal line whitish, of two concave curves separated by two teeth which reach the termen; reniform stigma, costal half pinkish ochreous, dorsal half dark greyish; a pinkish brown blotch between reniform stigma and subterminal fascia. Hindwing pale fuscous; veins darker; discal spot obscure; terminal fascia only slightly darker than ground colour. Underside of fore- and hindwing variably tinged pink near margins.

Specimens from southern localities tend to be larger and darker than those from the north but the latter are more strongly suffused with pink. In ab. *dives* Haworth, which has been recorded from all parts of its range in Britain, the area from the reniform stigma to the subterminal fascia is almost white.

Similar species. L. w-latinum (Hufnagel) which has a conspicuous black bar from claviform stigma; pale orbicular stigma not connected to tornal blotch; no pinkish suffusion on upper- or underside. *Apamea remissa* (Hübner) f. *remissa* Hübner (Amphipyrinae) which has glabrous eyes; pale orbicular stigma not connected to tornal blotch; no pinkish suffusion on upper- or underside.

Life history
Ovum. Hemispherical, heavily ribbed and reticulate on the upper part, less so below the equator, the upper part thus appearing darker; pale grey at first, becoming even paler before hatching. Laid on the foodplant in large, untidy masses, hatching in about eight days.
Larva. Full-fed *c.*43mm long. Head greenish or greenish brown. Two principal colour forms. (**a**) Body coppery red; dorsal line slightly paler; subdorsal pattern of oblique brownish arcs; on each segment a black dot between dorsal and subdorsal lines; spiracular line darker brown; subspiracular band broad, paler than ground colour, tinged yellowish at extremities; spiracles white, ringed black; ventrally yellowish, legs and prolegs pinkish. (**b**) Body brownish green, subdorsal arcs greenish brown; ventrally bright green. In the early instars there is considerable colour variation, from yellowish green to red-brown, but in the last instar the majority are of the two forms described above.

Feeds by day or night on a wide variety of plants, including dock (*Rumex* spp.), rough-leaved willow (*Salix* spp.), birch (*Betula* spp.), oak (*Quercus* spp.), hazel (*Corylus avellana*), dyer's greenweed (*Genista tinctoria*), broom (*Sarothamnus scoparius*), bog myrtle (*Myrica gale*), golden rod (*Solidago virgaurea*) and bracken (*Pteridium aquilinum*). May be beaten from birch or sallow, often from small bushes or lower branches, in early September.

Pupa. Glossy, dark brown; thoracic segments rather stout, densely but finely sculptured; abdominal segments evenly tapering with punctation on anterior two-thirds; segment 8 concave ventrally to a blackish cremaster with two flattened lateral lobes and two short, stout, divergent apical spines with T-shaped tips. In a fragile cocoon, below ground.

Imago. Univoltine. Flies in June and July and will come to sugar and light. Frequently found at rest by day on fence-posts, tree-trunks and among foliage, from which it sometimes falls into the beating tray.

Distribution (Map 154)
This species has a discontinuous distribution similar to that of *Polia hepatica* (Clerck) and also inhabits lightly wooded heath and moorland. Locally common on acid soils in southern England and the Midlands; apparently more local on the Yorkshire moors; common in Scotland to Moray, including the Inner Hebrides; local but widely distributed in Ireland. Eurasiatic; from Portugal to central Scandinavia but generally local.

LACANOBIA W-LATINUM (Hufnagel)
The Light Brocade
Phalaena w-latinum Hufnagel, 1766, *Berlin.Mag.* **3**: 292.
Phalaena (Noctua) genistae Borkhausen, 1792, *Naturgesch. eur.Schmett.* **4**: 355.
Type locality: Germany; Berlin.

Description of imago (Pl.11, fig.36)
Wingspan 37–42mm. Antenna shortly ciliate, in male densely, in female sparsely. Forewing pale silvery brown with a darker brown median fascia which does not reach the dorsum, basal streak black, extending almost to claviform stigma; above it a greyish white subcostal blotch; ante- and postmedian lines black; subterminal fascia broad, from apex to tornus, pale silvery brown; subterminal line paler than ground colour, edged blackish, with two teeth reaching cilia; remainder of termen dark grey-brown; orbicular stigma pale fuscous, outlined black; reniform stigma even paler, sometimes tinged ochreous; claviform stigma grey-brown, merging with a heavy blackish streak which limits the median fascia, the dorsum being wholly of the ground colour. Hindwing pale fuscous; veins and obscure discal spot rather darker; terminal shade pale brownish fuscous; veins and fine terminal line darker brown.

Variation is slight, affecting the shape of the stigmata and the extent of the streak from the claviform stigma.

Similar species. L. contigua ([Denis & Schiffermüller]) q.v. *Apamea remissa* (Hübner) f. *remissa* Hübner (Amphipyrinae) which has glabrous eyes; subterminal fascia and dorsum much less pale in relation to darker parts of wing; colour of pale areas brownish fuscous rather than silvery brown.

Life history
Ovum. Hemispherical, ribbed and lightly reticulate, the upper half darker than the lower; whitish at first, darkening to greyish purple. Laid on the foodplant in large, regular batches, hatching in about ten days.
Larva. Full-fed *c.*50mm long, rather stout. Head dark brown. Body variable, subdued shades of brown and grey, shaded green or purplish; a row of rather darker dorsal diamonds; dorsal and subdorsal lines usually broken, paler than ground colour, darker-edged; spiracular line yellowish, edged black; spiracles white, outlined black; pale greenish drab ventrally.

Nocturnal and unusually intolerant of light. Feeds on a wide variety of plants, including broom (*Sarothamnus scoparius*), dyer's greenweed (*Genista tinctoria*), redshank (*Polygonum persicaria*) and knotgrass (*Polygonum aviculare*). Full-grown in August.

Pupa. Rather stout, rich dark brown; thoracic segments and

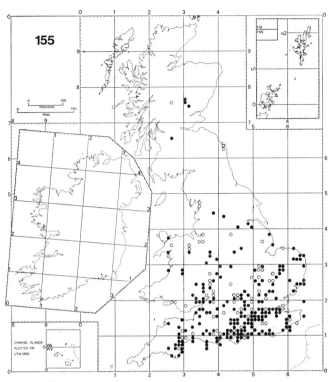

Lacanobia w-latinum

anterior part of each abdominal segment finely sculptured; abdominal segment 8 concave ventrally; cremaster wide and flattened with lateral projections and two stout, divergent apical spines. Below the soil surface, in a cocoon.

Imago. Univoltine. Occurs from the middle of May to the end of June. Frequently found at rest by day on tree-trunks or wooden fences and comes to flowers from dusk and later to sugar and light.

Distribution (Map 155)

Common and well distributed mainly on calcareous soils from the south coast to the London area, thence becoming progressively more local and less common to the Midlands; apparently scarce in Wales. Farther north there are scattered records from Yorkshire, north Lancashire and several from Scotland as far north as Perthshire. Baynes (1964) rejects the old Irish records. Eurasiatic; through western Europe to the Danish islands and south Sweden, rare in Finland.

LACANOBIA THALASSINA (Hufnagel)
The Pale-shouldered Brocade

Noctua thalassina Hufnagel, 1766, *Berlin.Mag.* **3**: 298.

Type locality: Germany; Berlin.

Description of imago (Pl.11, figs 37,38)

Wingspan 38–44mm. Antenna in both sexes shortly ciliate. Forewing purplish red-brown, sprinkled ochreous and black; basal streak black, almost to claviform stigma, with a pale ochreous subcostal blotch above; ante- and postmedian lines yellowish white, clearer dorsally; subterminal line complete with two teeth extending to termen; upper stigmata tinged ochreous, outlined black, the orbicular paler and more distinct than the reniform the distal margin of which is not clearly defined; claviform stigma blackish. Hindwing brownish fuscous, slightly paler basally; discal spot lunular, darker; veins slightly darker; cilia whitish.

The ground colour varies from greyish brown (ab. *humeralis* Haworth, fig.38) to purplish brown, but this variation does not appear to be of geographical significance.

Similar species. Apamea epomidion (Haworth) (Amphipyrinae) in which the eyes are glabrous; orbicular stigma oval, darker than reniform; subterminal line less conspicuous, with a convex curve where it is bidentate in *Lacanobia thalassina*.

Life history

Ovum. Hemispherical, lightly ribbed and reticulate; pale and glossy at first, gradually darkening overall. Laid on the underside of leaves in large, untidy masses, hatching in just over a week.

Larva. Full-fed *c.*38mm long. Head light brown. Body red-brown to grey-brown; dorsal line slightly paler; subdorsal line of oblique blackish bars suggesting incomplete chevrons; spiracular stripe yellowish brown, sometimes bordered blackish; spiracles white; pinacula conspicuously black; slightly paler ventrally.

Nocturnal and has been recorded from a variety of foodplants, including oak (*Quercus* spp.), willow (*Salix* spp.), hawthorn (*Crataegus* spp.), wild pear (*Pyrus communis*), crab apple (*Malus sylvestris*), barberry (*Berberis vulgaris*), groundsel (*Senecio vulgaris*), knotgrass (*Polygonum aviculare*), broom (*Sarothamnus scoparius*) and honeysuckle (*Lonicera periclimenum*). Full-fed in August.

Pupa. Rich dark brown and glossy; thoracic segments stout, finely sculptured; abdominal segments tapering sharply, each with an anterior band of punctation; segment 8 concave ventrally; cremaster blackish, flattened laterally, with two long apical spines. In a fragile subterranean cocoon.

Imago. Univoltine except in the south, where there is a small second brood. The main emergence is from late May

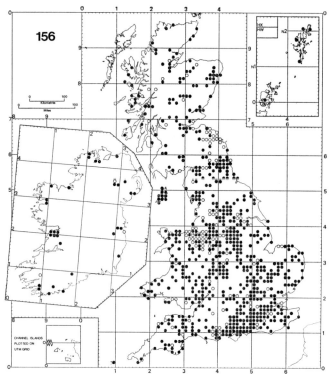
Lacanobia thalassina

to the middle of June (slightly later in the north). Often found at rest by day on tree-trunks or fences, but active only after dark, when it feeds at flowers and comes to sugar and light.

Distribution (Map 156)
The moth is fairly common over most of its range and usually inhabits wooded or lightly wooded country in the south, but occurs widely on moorland in the north. It occurs over most of mainland Britain, the Isle of Man, Inner Hebrides, Orkney and Shetland; widely distributed in Ireland. Eurasiatic; throughout western Europe, even a few records within the Arctic Circle.

LACANOBIA SUASA ([Denis & Schiffermüller])
The Dog's Tooth

Noctua suasa [Denis & Schiffermüller], 1775, *Schmett. Wien.*: 83.

Phalaena (Noctua) dissimilis Knoch, 1781, *Beitr. Insektengesch.* 1: 57.

Type locality: [Austria]; Vienna district.

Description of imago (Pl. 11, figs 39, 40)
Wingspan 32–42mm. Antenna ciliate, in female more shortly and sparsely than in male. The species varies greatly but is basically dimorphic, some specimens resembling the typical form (fig. 39), others being more or less unicolorous brown and approaching ab. *dissimilis* Knoch (fig. 40).

The 'typical' form. Forewing grey-brown to brown, somewhat paler subterminally, where there may be a rosy suffusion; basal streak black, short; above it a variably paler subcostal blotch; ante- and postmedian lines obscurely darker; subterminal line yellowish white with two teeth to the cilia, above and below which the termen is darker than the ground colour; upper stigmata pale, varying in relative prominence – in some specimens both are equally distinct, in others the reniform almost merges with the ground colour; claviform stigma black and rather thick. Hindwing pale fuscous; discal spot variable in size, rather darker; occasionally a narrow fuscous median fascia; terminal shade variable in width and darkness.

The '*dissimilis*' form. Forewing varying from dull red-brown to blackish brown; ante- and postmedian lines almost obsolete; subterminal line usually visible; basal streak usually present, but pale subcostal blotch reduced or absent; in paler specimens only, the outline of the upper stigmata may be visible. Hindwing much darker than in the 'typical' form, with broader terminal fascia.

There seems to be no geographical separation of the two main forms.

Similar species. L. *oleracea* (Linnaeus) q.v.

Life history
Ovum. Hemispherical, lightly ribbed and reticulate, greyish violet. Laid in batches, hatching in about ten days.

Larva. Full-fed *c*.45mm long. Head light brown. Body yellow-green, green or brownish grey; dorsal area filled by darker shading of oblique bars, dorsal line obscured; four black subdorsal dots on each segment, in line on thoracic segments, in pairs on abdominal segments; spiracular line yellow, bordered above with a row of black arcs; spiracles golden, ringed black; yellowish buff ventrally. The larva of *L. oleracea* is often very similar, but has a complete, less erratic, spiracular line.

Nocturnal in habits and feeds up rather quickly on dock (*Rumex* spp.), plantain (*Plantago* spp.) and knotgrass

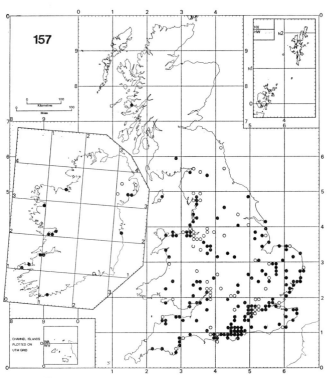

Lacanobia suasa

(*Polygonum aviculare*), being full-grown in about seven weeks.

Pupa. Thoracic segments stout, red-brown; wing-cases minutely sculptured; abdominal segments with anterior punctation, tapering rapidly; cremaster black, with two basal thorns and two very short, divergent, T-tipped apical spines. In a subterranean cocoon.

Imago. Univoltine in the north, bivoltine in the southern counties of England. The main emergence is in late May and through June, the second brood flying in September, sometimes in large numbers. The moth is less frequently seen by day than its relatives, but comes in good numbers to sugar and light.

Distribution (Map 157)

This species usually inhabits coastal areas and estuaries, the valleys of larger rivers and moorland, but is also found rather locally on the chalk, generally below 125m (400ft) altitude. Locally common from the south coast of England to southern Scotland and recently recorded from Mull; Isle of Man; widely distributed in Ireland. Eurasiatic; throughout western Europe to the Arctic.

LACANOBIA OLERACEA (Linnaeus)
The Bright-line Brown-eye

Phalaena (*Noctua*) *oleracea* Linnaeus, 1758, *Syst.Nat.* (Edn 10) **1**: 517.
Peucephila essoni Hampson, 1909, *Trans.ent.Soc.Lond.* **1909**: 46, pl.16, fig.1.
Type locality: [Sweden].

Description of imago (Pl.11, figs 41,42)

Wingspan 36–44mm. Antenna ciliate, in male more densely than in female. Forewing usually reddish brown, sometimes grey-brown or deep purplish brown; ante- and postmedian lines visible only as slightly darker shades; subterminal line complete, white, fairly straight except for two small teeth; orbicular stigma small and circular, usually ringed white; reniform stigma orange-brown; claviform stigma dark brown and obscure; in all but the darkest forms the main veins show as slightly darker longitudinal streaks, often sprinkled with white scaling. Hindwing pale fuscous, in female slightly darker; discal spot usually lunular; terminal fascia brownish fuscous.

Variation is considerable, affecting mainly the ground colour and, in consequence, the prominence of the upper stigmata, which in the darkest specimens are almost obscured; in extreme cases the subterminal line may be only just discernible. Dark forms occur throughout the range, but apparently are rare in the north.

Similar species. *L. suasa* ([Denis & Schiffermüller]) ab. *dissimilis* Knoch. The darkest forms only of these insects can be confused; even in these it is usually just possible to see the subterminal line and, in *L. suasa*, the short, black basal streak.

Life history

Ovum. Hemispherical, very lightly ribbed and reticulate; bright apple-green at first, becoming yellowish green. Laid on the foodplant in large untidy batches, hatching in about eight days.

Larva. Full-fed *c.*45mm long, rather small in relation to the size of the moth. Head light brown. Body pale green, yellow-brown or brown, in all forms densely and minutely dotted white; dorsal and subdorsal lines broken, grey-brown or grey-green; spiracular line complete and broad, pale yellow bordered grey above; spiracles white, ringed black; flesh-coloured ventrally.

Polyphagous, feeding mainly on herbaceous plants, but with an apparent natural preference for orache (*Atriplex* spp.) and goosefoot (*Chenopodium* spp.). The species has colonized gardens and market gardens where it is sometimes a pest of cultivated tomatoes, feeding internally on the fruit. Growth is rapid and in the south most larvae have pupated by late August.

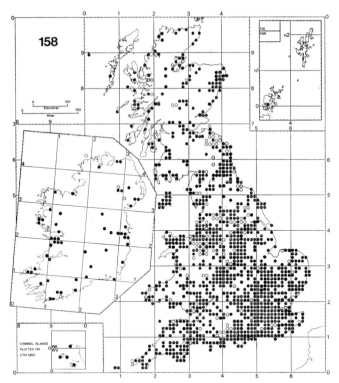

Lacanobia oleracea

Pupa. Thoracic segments stout, rich dark brown, heavily sculptured; abdominal segments sharply tapering, the posterior half of each paler than the anterior, segment 8 abruptly concave ventrally; cremaster blackish, triangular with two short, T-headed apical spines. Below ground, in a fragile cocoon.

Imago. Univoltine over most of Britain, with a small second brood in southern England. The main emergence is from late May, but can be as late as mid-July in northern Scotland. It is not often seen by day, but feeds at flowers such as red valerian, wood-sage and catmint soon after dusk and comes to sugar and light. Although it is more often found in gardens than in open country, the most favoured natural habitat seems to be the margins of salt-marshes.

Distribution (Map 158)
Great Britain to Orkney and Ireland; very common in most areas, less so in the extreme north. Eurasiatic; from south Europe to central Scandinavia and Finland.

LACANOBIA BLENNA (Hübner)
The Stranger

Noctua blenna Hübner, [1824], *Samml.eur.Schmett.* **4**: pl.152, fig.706.
Hadena peregrina Treitschke, 1825, *Schmett.eur.* **5**: 330.
Type locality: Europe.

Description of imago (Pl.11, fig.45)
Wingspan 36–40mm. Forewing yellowish white, suffused with pale yellow; stigmata faintly outlined darker, with a pale streak projecting obliquely outwards from the orbicular to join a pale postmedian band; subterminal fascia with two black identations. Hindwing whitish, with only veins and termen darker.

Similar species. *L. blenna* could be overlooked, especially at night, among *Hadena perplexa perplexa* ([Denis & Schiffermüller]) *q.v.*, which has a rather similar wing-pattern but is smaller and has the markings more variegated; and among pale specimens of *Luperina testacea* ([Denis & Schiffermüller]) (Amphipyrinae) which has a different wing-pattern and whiter hindwing.

Life history
Early stages not known in Britain.

Larva. Full-fed *c.*40mm long. Head brown. Body yellow-brown with two white points on the dorsum of each segment; white lateral lines bordered by reddish below and with dark streaks above; legs and ventral surface brownish grey. Described (1908, **1**: 174) and figured (1910, **4**: Nachtrag pl.3, fig.9) by Spuler.

Feeds near the shore on prickly saltwort (*Salsola kali*), *Mesembryanthemum*, orache (*Atriplex* spp.) and, especially, sea-beet (*Beta vulgaris maritima*), becoming full-grown in June.

Imago. Probably an occasional immigrant, possibly with temporary establishment ('Old Moth Hunter', 1952), but no recent record. There are only four authenticated British specimens, of which three were caught at Freshwater, Isle of Wight: the first in 1857 by F. Bond at sugar; the second near the same spot on 23 August 1859, also at sugar; a third reported in 1876 (Barrett, 1897, **4**: 179). A fourth, now in BMNH, was recorded on the downs near Lewes, Sussex, in 1868.

Distribution
Mediterranean-Asiatic, halophile; in western Europe it has been recorded in Spain and is found in France on the Mediterranean coast and also on the Atlantic coast in salt-marshes from the Gironde at least as far north as Carnac in Brittany.

PAPESTRA Sukhareva

Papestra Sukhareva, 1973, *Ent.Obozr.* **52**: 409.

Contains one species which occurs in Great Britain and Ireland.

The genus differs from *Lacanobia* Billberg in the structure of the sacculus of the male genitalia.

PAPESTRA BIREN (Goeze)
The Glaucous Shears

Phalaena (Noctua) biren Goeze, 1781, *Ent.Beyträge* **3** (3): 206.
Noctua glauca Hübner, [1809], *Samml.eur.Schmett.* **4**: pl.87, fig.410.
Hadena bombycina sensu auctt.

Type locality: not stated.

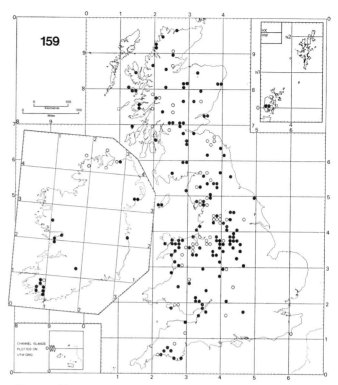

Papestra biren

Description of imago (Pl.11, figs 43,44)

Wingspan 32–40mm. Antenna ciliate, in female more shortly and sparsely than in male. Forewing rather narrow; shades of grey from silver to slate or purplish, distal part from postmedian line paler than remainder; subbasal, ante- and postmedian lines variably paler than ground colour, bordered black; subterminal line clear, yellowish white bordered proximally by a blackish line with a blackish inwardly dentate line in the dorsal half; upper stigmata whitish or greyish white, outlined black at least in part, reniform usually the more distinct; claviform stigma usually conspicuous, rather thick, yellowish white to pale ochreous, outlined black, usually with a small black triangle at the distal extremity; termen crossed by a number of fine pale streaks to the cilia. Hindwing pale fuscous with darker veins; lunular discal spot and terminal shade dark fuscous.

Specimens from the north and west tend to be darker than those from the southern part of the range, but this is not invariable, and very pale as well as very blue-grey forms occur in Ireland. An albino specimen has been recorded from Shropshire.

Similar species. (Northern Scotland and Scottish islands only) *Hadena confusa* (Hufnagel) ab. *ochrea* Tutt: worn specimens of this form are often very similar to *P. biren*, but can usually be separated by their very round, whitish or pale ochreous orbicular stigma, with a quadrate ochreous mark below.

Life history

Ovum. Hemispherical, ribbed and reticulate, micropyle on a small papilla; whitish near micropyle, greyish white below the equator. Laid in large batches on the foodplant, hatching in about 14 days.

Larva. Full-fed *c.*45mm long. Head pale brown. Body rich dark brown to grey-brown; dorsal line usually obscured by a series of darker diagonal subdorsal streaks; spiracular line variable, sometimes slightly, more often much, paler than ground colour; spiracles black, outlined white; pale brown to greenish brown ventrally. Green in the early instars, becoming brown as it approaches maturity.

The larva has a wide range of foodplants, including meadowsweet (*Filipendula ulmaria*), bog-myrtle (*Myrica gale*), colt's-foot (*Tussilago farfara*), knotgrass (*Polygonum aviculare*), smooth sow-thistle (*Sonchus oleraceus*), various moorland willows, especially *Salix aurita* and *S. repens*, and the common upland Ericaceae (*Vaccinium, Calluna, Erica, Arctostaphylos*). Has been reared in captivity on garden lettuce. Feeds rather slowly, usually well into August.

Pupa. Rather stout, reddish brown; thoracic segments finely sculptured; abdominal segments 5–7 with anterior dorsal ridges bearing short strong spines; posterior half of each darker and finely grained; cremaster short, rugose, flattened laterally and squared, with two small subapical spines and two larger apical spines. On the soil surface, in a long tubular cocoon of silk and plant litter, along which it is able to move. Humidity can be a problem in captivity as much of this stage in nature is spent in damp or very damp conditions and care is necessary to avoid desiccation or excess stagnant humidity.

Imago. Univoltine. Flies from mid-May and often rests on rocks, tree-trunks and fence-posts, usually low down. Fairly active by day and feeds, also at night, at the flowers of sallow, bilberry, bearberry and moss-campion. Although not greatly attracted to sugar, probably because of the abundance of natural food on moorland when it is on the wing, it comes to light in numbers.

Distribution (Map 159)
Western and northern; local in Cornwall, Devonshire and Somerset, more generally common on acid moorland from Monmouth northwards; Scotland to the Inner Hebrides and Orkney; well distributed in Ireland, most common in the south-west. Eurasiatic; in western Europe from Spain to Belgium, mainly montane, very local in the Netherlands, widespread in Denmark and throughout Scandinavia and Finland.

CERAMICA Guenée
Ceramica Guenée, 1852, *in* Boisduval & Guenée, *Hist.nat. Insectes* (Lépid.) 5: 343.

Similar to *Lacanobia* Billberg but the single British species lacks the strongly dentate subterminal line on the forewing present in that genus. In the male genitalia, the valva has two large costal projections, quite unlike any closely related species.

CERAMICA PISI (Linnaeus)
The Broom Moth
Phalaena (Noctua) pisi Linnaeus, 1758, *Syst.Nat.* (Edn 10) 1: 517.
Type locality: [Sweden].

Description of imago (Pl.12, figs 2–4)
Wingspan 33–42mm. Antenna in male shortly fasciculate, in female very sparsely and shortly ciliate. Forewing greyish brown to purplish brown; antemedian line clear, grey-tinged, edged darker grey; postmedian line usually incomplete and obscure, grey-tinged bordered darker grey proximally; subterminal line usually complete, pale yellow, widening into a deeper yellow pretornal blotch, a constant character; upper stigmata usually grey-tinged. Hindwing pale fuscous with greyish fuscous veins, lunular discal spot and terminal fascia; sometimes also a narrow median fascia.

There is considerable variation. In ab. *splendens* Staudinger (fig.3) the pattern is obscured and the forewing smooth and unicolorous, except for the subterminal line; this form is well distributed. Other less common aberrations show variation in the prominence of the cross-lines and stigmata. Geographical variation is shown by ab. *scotica* Tutt (fig.4), a largely northern form which is smaller, more brightly coloured and with greater contrast in the elements of the pattern.

Life history
Ovum. Flattened-spherical, ribbed and more lightly reticulate; pale yellowish at first, darkening slowly overall. Laid on the foodplant in untidy masses, hatching in about ten days.

Larva. Full-fed *c.*47mm long, of velvety texture, elongate and slightly tapering at extremities. Head light brown. Body variable, usually from pale green to blackish green, occasionally red-brown to purple-brown; dorsal and subspiracular lines broad, pale yellow, with black edging visible only in paler colour-forms; spiracles yellowish white, outlined black; pale yellowish white ventrally.

Polyphagous, feeding on deciduous trees and shrubs, herbaceous plants and ferns, especially bracken (*Pteridium*

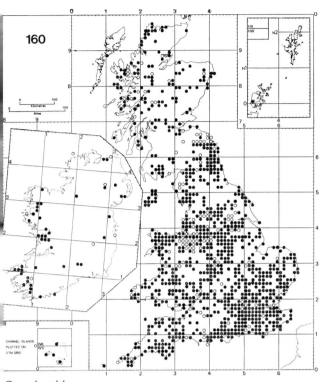

Ceramica pisi

aquilinum) in the Scottish Highlands; it is also recorded as being injurious to young larch (*Larix decidua*) in forest nurseries. Growth is rather slow and it is not full-fed until late August or September.

Pupa. Rather elongate, dark red-brown; anterior segments densely but finely punctate; abdominal segments evenly tapering, with a band of anterior punctation, the remainder lightly grained; cremaster elongate, heavily ribbed and punctate basally with two flat-tipped, slightly divergent apical spines. In a fragile subterranean cocoon.

Imago. Univoltine. Flies from mid-June, slightly later in the north. Hides by day and feeds at flowers and honey-dew by night, coming also to sugar and light.

Distribution (Map 160)

This species is more common in open country than in dense woodland or marsh; in many Scottish moorland areas it is abundant. Widely distributed over the whole of Great Britain and Ireland except Shetland. Eurasiatic; through Europe to the Arctic, and in Iceland.

HECATERA Guenée

Hecatera Guenée, 1852, *in* Boisduval & Guenée, *Hist.nat. Insectes* (Lépid.) **6**: 27.
Aethria Hübner, [1821], *Verz.bekannt.Schmett.*: 218.

A small genus with only two British species, one of which may now be extinct.

Superficially similar to *Hadena* Schrank, but thorax with only a posterior crest and female abdomen lacking pointed ovipositor.

Imago. Univoltine, but partly bivoltine in the south.

Ovum. Laid singly on stems or leaves of the foodplant.

Larva. Brightly and disruptively coloured. On flowers of Compositae by day and night.

Pupa. Elongate and slender, without ventral haustellum-sheath. Below ground, in a cocoon.

Economic importance. *Hecatera dysodea* ([Denis & Schiffermüller]) was formerly a pest on cultivated lettuce (*Lactuca* spp.).

HECATERA BICOLORATA (Hufnagel)
The Broad-barred White

Phalaena bicolorata Hufnagel, 1766, *Berlin.Mag.* **3**: 410.
Noctua serena [Denis & Schiffermüller], 1775, *Schmett. Wien.*: 84.

Type locality: Germany; Berlin.

Description of imago (Pl.12, figs 12,13)

Wingspan 28–35mm. Antenna shortly ciliate, in male densely, in female sparsely. Forewing (f. *leuconota* Eversmann, the form most usual in Britain) white, with a greyish black median fascia enclosing the stigmata; upper stigmata greyish fuscous, edged white, their relative prominence varying and depending on the clarity of the edging; claviform stigma blackish, usually occluded by the median fascia; subterminal line greyish black, broken, sometimes vestigial, its most constant component being a single (sometimes double) black, inward-pointing triangle, level with the dorsal lobe of the reniform stigma; termen narrowly grey. Hindwing of male whitish fuscous with darker veins, discal spot and terminal fascia; female fuscous with slightly darker terminal fascia.

In the typical form the forewing is bluish grey with a dark olive median fascia. British specimens vary considerably; those from Dungeness are abnormally white; in some specimens the ground colour is greyish white, while in others only the basal or subterminal area is darker than normal; or, yet again, the median fascia may be greyish brown.

Life history

Ovum. Hemispherical, heavily ribbed and reticulate, the micropylar area surrounded by a depression and appearing as a central mound; very pale yellow at first, quickly becoming pinkish and finally coppery red. Laid singly on the foodplant, apparently at random, no attempt being made (as in most *Hadena* species) to select buds or flowers, hatching in about six days.

Larva. Full-fed *c.*35mm long. Head similar to colour of body. Body green, brownish green, yellow-brown or brown; dorsal pattern complicated, consisting of darker diamonds which are linked at their lateral angles by still darker angulated subdorsal lines; spiracular stripe dull yellow to yellow-brown; spiracles pinkish brown, thickly outlined black.

Immediately after hatching the larva climbs to the flower heads of its foodplant and feeds entirely on the buds and flowers, both by day and night. It often rests on the flowers fully exposed in the afternoon sunshine. The recorded foodplants are all Compositae – oxtongue (*Picris* spp.), hawkweed (*Hieracium* spp.), hawk's-beard (*Crepis* spp.) and lettuce (*Lactuca* spp.). Sow-thistle (*Sonchus* spp.) is also recorded, but in captivity and with a choice available, other plant genera are preferred. Full-fed from mid-August.

Pupa. Elongate and slender, without haustellum-sheath; thoracic segments glossy, light reddish brown, with weak punctation; abdominal segments yellow-brown, each with central band of minute punctation; cremaster conical and grained, with many basal thorns and two fairly straight, T-headed apical spines. Just below the surface of the soil, in a fragile cocoon.

Imago. Univoltine with a very small second brood in southern England. Period of emergence is long, beginning at the end of May and continuing through June. Often found at rest by day and feeds at flowers from sunset; not often taken at sugar, but comes to light in abundance. The species is susceptible to rapid fluctuations in numbers, possibly because its foodplants are quick to colonize wasteland but are soon smothered in subsequent seasons.

Distribution (Map 161)

Rather common in south-east England and East Anglia, less so in south-west England and in Wales; becoming less common in the Midlands and more local in northern England. Locally common in Scotland north to Inverness and Aberdeenshire; Baynes (1964) describes it as scarce but widely distributed in Ireland. The Isles of Scilly; the Channel Islands. Eurasiatic; through western Europe to Denmark, south Norway, central Sweden and Finland.

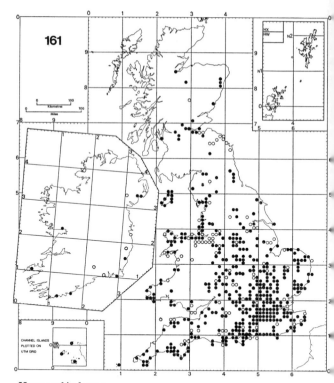

Hecatera bicolorata

HECATERA DYSODEA ([Denis & Schiffermüller])
The Small Ranunculus

Noctua dysodea [Denis & Schiffermüller], 1775, *Schmett. Wien.*: 72.

Phalaena (Noctua) chrysozona Borkhausen, 1792, *Natur-gesch.eur.Schmett.* **4**: 264.

Type locality: [Austria]; Vienna district.

Description of imago (Pl.12, fig.14)
Wingspan 32–36mm. Forewing grey-white, wholly suffused between subbasal and postmedian fasciae with pinkish brown mixed with orange; fasciae black, edged outwardly white. Hindwing pale grey, darker outwardly.
Variation slight.

Similar species. *H. bicolorata* (Hufnagel) has similar wing-pattern, but lacks orange speckling. *Polymixis flavicincta* ([Denis & Schiffermüller]) (Cuculliinae) has yellow-orange speckling but a different wing-pattern and is usually much larger.

Life history
Ovum. Hemispherical, flattened beneath, with strong but rather irregular ribs; white.

Larva. Full-fed *c.*30mm long. Body pale red-brown or dull greenish above; ventral surface and prolegs white; three fine blackish lines above black spiracles (figured by Buckler, 1895: pl.88).

Feeds in July and August on flowers and seeds of wild and garden lettuce (*Lactuca serriola*, *L. virosa* and *L. sativa*) and in some places was said to be a pest of lettuce grown in fields. Smooth hawk's-beard (*Crepis capillaris*) and sow-thistle (*Sonchus* spp.) are also mentioned.

Pupa. Slender, red-brown; cremaster conical with two diverging points (Forster & Wohlfahrt, 1964, **4**: 77–78). Overwinters just below the surface of the ground.

Imago. Certainly resident throughout the nineteenth century, but now probably extinct in Britain. Occurred in June and July. Most often seen at flowers, especially red valerian, at dusk, and occasionally at light and at rest on posts, in gardens, gravel pits and waste ground.

Distribution (Map 162)
Until about 1895 moths and larvae were locally common in East Anglia, around London, and in Kent and Surrey; they were scarcer westwards, but were reported from Dorset, Somerset, Glamorgan, Herefordshire and Northamptonshire. These limits correspond broadly with the distribution, both then and now, of the wild lettuces (Perring & Walters, 1962). Decline of the species was rapid. It had disappeared from most of its range by 1912, but there are later records, mostly of single specimens of the moth, in Essex (1918), Somerset (1935), Hertfordshire (1936 and

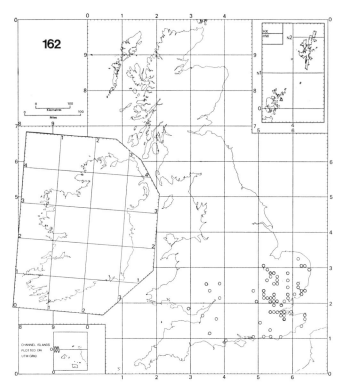

Hecatera dysodea

1937), and of one seen but not captured in Buckinghamshire (1941). A report of a moth bred from a larva found on prickly lettuce (*Lactuca serriola*) at Baldock, Hertfordshire in 1962 proved to be a misidentification (Reed, 1963; Skinner, 1964). *Hecatera dysodea* has been reported in Jersey, since 1960. Eurasiatic; in western Europe widespread in Portugal, Spain, France and Belgium, in the Netherlands common in the south but rare in the north; not known in Denmark or Fennoscandia. No convincing explanation of its extinction in Britain has been suggested, though decline in field cultivation of lettuce may have affected it locally; but it was near the northern limit of its range, and it is said to have become notably scarcer in north central Europe since about 1960.

HADENA Schrank

Hadena Schrank, 1802, *Fauna boica* **2** (2): 158.
Miselia Ochsenheimer, 1816, *Schmett.Eur.* **4**: 72.
Zeteolyga Billberg, 1820, *Enum.Ins.Mus.Blbg*: 87.
Harmodia Hübner, [1820], *Verz.bekannt.Schmett.*: 207.
Epia Hübner, [1821], *Ibid.*: 214.
Dianthoecia Boisduval, 1834, *Revue Ent.* (Silbermann) **2**: 246.
Anepia Hampson, 1918, *Novit.zool.* **25**: 116.

A compact genus with numerous Holarctic species whose members, except *H. luteago* ([Denis & Schiffermüller]), show close affinities in all stages. There are nine species in Great Britain and six in Ireland, all resident.

Imago. Small to medium size (wingspan 27–42mm). Thorax with anterior and posterior crests. Forewing short and rather triangular. Abdomen with several small crests, in female stout with conspicuously tapering ovipositor.

Ovum. Frequently dimpled overall. Laid singly on foodplant, usually on opening buds of flowers, hatching in four to seven days.

Larva. Head small. Body stout, tapering abruptly from abdominal segment 7; usually a shade of brown with darker dorsal chevrons; on Caryophyllaceae, all except *H. luteago*, which is an internal feeder, on the seeds.

Pupa. Rather slender, pale reddish brown, with haustellum-sheath longer than wing-cases; cremaster well developed, showing specific characters. Subterranean, in a cocoon.

Life history. Univoltine except in the south, where several species are bivoltine. The moth flies from dusk, feeding on the wing and alighting briefly to oviposit. The larvae feed inside seed capsules until too large, after which many hide low down by day, climbing after dusk to feed.

Breeding in captivity. The female needs space and a flowering spray of foodplant. Garden carnation is a useful substitute foodplant for all species and is easier to check for earwigs than sweet william (*Dianthus barbatus*). When airtight metal containers are used an interior cloth lining is useful to absorb excess moisture; full-grown larvae should be removed for pupation to prevent disturbance or sometimes cannibalism. Pupae should not be kept too damp during winter.

HADENA RIVULARIS (Fabricius)
The Campion

Noctua rivularis Fabricius, 1775, *Syst.Ent.*: 613.
Noctua cucubali [Denis & Schiffermüller], 1775, *Schmett Wien.*: 84.

Type locality: Germany; Saxony.

Description of imago (Pl.12, fig.15)
Wingspan 30–36mm. Antenna bipectinate, in male shortly in female very shortly. Forewing rich brown, variably suffused purplish; antemedian line brown, edged black postmedian line purplish, bordered proximally with black triangles and edged black distally; subterminal line pale golden, very erratic in course except for the final two to three millimetres, which are straight and reach the dorsum at a right angle or incline slightly toward the base (figure 17) upper stigmata purplish at centre, surrounded brown and outlined pale golden, touching or nearly touching dorsally claviform stigma blackish brown; underside, stigma surrounded by a pale suffusion. Hindwing pale fuscous with brownish fuscous lunular discal spot; sometimes a post median fascia and broad terminal fascia.

Most variation is in the alignment of the upper stigmata but the purple shading is rather evanescent and often lacking in worn specimens.

Similar species. *H. bicruris* (Hufnagel) which has upper stigmata usually clearly apart dorsally, subterminal line joining dorsum at an angle of about 45°, clearly inclined to and almost parallel with termen; underside without pale suffusion surrounding stigma.

Life history

Ovum. Rather flattened-hemispherical, dimpled overall pale yellowish white at first, beginning to darken within a few hours. Laid singly on and inside the opening buds and flowers of the foodplant, hatching in about six days.

Larva. Full-fed *c.*35mm long, thoracic and posterior abdominal segments rapidly tapering, central part of body stout. Head small, light brown. Body brownish green, the thoracic segments orange-tinged; dorsal line pale, an orange streak on each segment branching forward from it subdorsal line more obscurely pale with a similar orange

Figure 17 *Hadena rivularis* (Fabricius), forewing

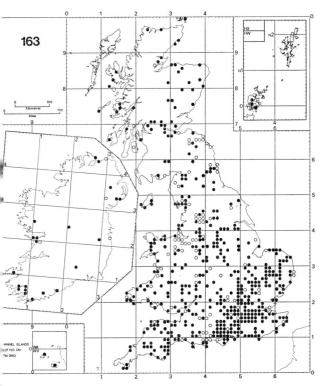

Hadena rivularis

brown streak branching off on each segment; subdorsal area sprinkled with whitish dots; spiracles oval, pale brown, outlined black. At first it is pale green with yellowish intersegmental rings and can be separated easily from all its relatives.

The young larva bores at once into the developing seed capsules and feeds internally until too large to remain hidden. Although it has been found on most species of *Silene* and *Lychnis*, ragged robin (*Lychnis flos-cuculi*) inland and sea-campion (*Silene maritima*) on the coast seem to be the main foodplants. Full-fed in August.

Pupa. Thoracic segments fairly stout, red-brown, with ventral haustellum-sheath; dorsal area heavily sculptured, wing-cases more delicately so; abdominal segments sharply tapering and finely punctate; cremaster blackish, rather conical, with several sharp basal projections and two stout, divergent apical spines, which originate close to each other, in-curving at the tips and each tapering to a fine point. In a fragile cocoon, just below the surface of the soil.

Imago. Univoltine, but partially bivoltine in the southern counties of England. The main emergence is from late May and through June, the second brood, where it occurs, in late August and September. Flies from early dusk, feeding on the wing at flowers such as campion, red valerian, woodsage or viper's bugloss; comes to light but not to sugar.

The female alights briefly to oviposit and in captivity seems to select opening buds in preference to older flowers.

Distribution (Map 163)

Widely distributed and locally common, apparently favouring damp habitats over most of Great Britain and Ireland, except Shetland. Eurasiatic; through western Europe to the Arctic, but scarce beyond 63°N.

HADENA PERPLEXA ([Denis & Schiffermüller])
The Tawny Shears

Noctua perplexa [Denis & Schiffermüller], 1775, *Schmett. Wien.*: 313.

Phalaena lepida Esper, 1790, *Schmett.* 4, 1 Abs.: 500, pl.152, fig.2.

Phalaena (Noctua) carpophaga Borkhausen, 1792, *Naturgesch.eur.Schmett.* 4: 422.

Type locality: [Austria]; Vienna district.

Description of imago (Pl.12, figs 5–11)

Wingspan 27–36mm. Antenna bipectinate, in male shortly, in female very shortly. Forewing with pointed apex, most variable in colour and in markings; whitish to muddy brown, some extreme examples of ab. *pallida* Tutt (fig.11) being almost immaculate except for the black outlines of the stigmata which can almost always be discerned. The specimens figured have been selected to show the range of normal variation and it is difficult to specify a completely constant character, the most nearly so being the dark, black-outlined claviform stigma, preceded by a pale subbasal blotch. Termen usually of ground colour or darker; subterminal line usually whitish, bordered proximally by two or three basally directed black wedges and often, at the dorsal extremity, a pale quadrate blotch. Hindwing whitish to ochreous fuscous, veins usually darker; discal spot variable in size and shape, terminal fascia variable in degree of contrast.

The variation shows a gradation from the palest forms on shingle in south-east England (Wightman, 1940b) to ochreous forms over much of the range, darkening to dull brown in south-west Wales and southern Scotland; in these areas the range almost merges with that of subsp. *capsophila* (Duponchel) which some specimens resemble closely.

Subsp. *capsophila* (Duponchel). The Pod Lover.

Dianthoecia capsophila Duponchel, [1842], *in* Godart & Duponchel, *Hist.nat.Lépid.Fr.* Suppl. 4: 100.

Size range similar. Forewing from dark fuscous to blackish, lacking brown tones; lines and upper stigmata finely outlined whitish; pale blotch before claviform stigma absent.

Similar species. *H. bicruris* (Hufnagel) which is larger; ground colour brown to dark brown; apex of forewing more obtuse; dorsal extremity of subterminal line parallel to termen.

Life history

Ovum. Hemispherical, ribbed and lightly reticulate, micropyle surrounded by a small depression; whitish at first, beginning to darken within hours. Laid singly in opening buds and flowers of foodplant, hatching in about seven days.

Larva. Full-fed *c*.38mm long. Head small, pale brown. Body bloated and glossy, large in relation to the size of the moth; pale yellowish grey or ochreous brown, with much paler dorsal, subdorsal and spiracular lines; spiracles golden, ringed black; ventrally putty-coloured. The larva of subsp. *capsophila* is reputed to be purplish brown (Barrett, 1897). This larva may be separated from all other British *Hadena* species by the longitudinally striped pattern.

The young larva bores into a developing seed capsule and feeds in concealment until about one-third grown. Only white-flowered campions are the natural foodplants – *Silene maritima*, *S. inflata*, *S. alba* and *S. nutans*, although Myers (1968) records that rock sea-spurrey (*Spergularia rupicola*) and cultivated *Dianthus* spp. are accepted in captivity. The larger larvae tend to hide by day and feed mainly by night, but in strong colonies a few can usually be found feeding in full view by day. Full-fed from late July to mid-August.

Pupa. Of typical *Hadena* shape, with haustellum-sheath; golden brown; dorsal surface of head rugose, with a small, rounded anterior protrusion; thoracic segments finely sculptured; abdominal segments coarsely grained, each with several fine dorsal hairs; cremaster blackish and roughened, with two stout, strongly divergent and apically curved spines. In a fragile cocoon, just below the surface of the ground.

Imago. Bivoltine in southern England, univoltine farther north; subsp. *capsophila* is univoltine. The main emergence takes place from mid-May to the end of June. Comes to flowers from early dusk to feed on the wing and lay eggs, for which the female alights only momentarily; although not attracted to sugar, it comes freely to light.

Distribution (Map 164)

Common on the coast and on calcareous soils inland to southern Scotland, approaching or merging slightly with subsp. *capsophila* in Cornwall and from Pembrokeshire northward; the Channel Islands. Subsp. *capsophila* has been recorded in Cornwall and northward on the coasts of Wales, Lancashire, south-west Scotland, Isle of Man and Ireland. Eurasiatic; occurring in various forms throughout western Europe to central Fennoscandia.

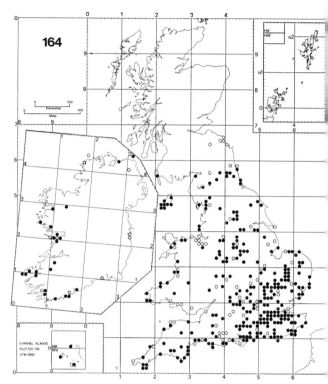

Hadena perplexa

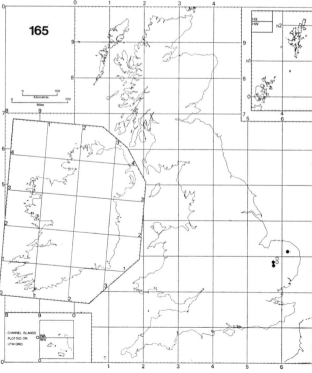

Hadena irregularis

HADENA IRREGULARIS (Hufnagel)
The Viper's Bugloss

Phalaena irregularis Hufnagel, 1766, *Berlin.Mag.* **3**: 394.
Phalaena (Noctua) echii Borkhausen, 1792, *Naturgesch.eur. Schmett.* **4**: 166.

Type locality: Germany; Berlin.

Description of imago (Pl.12, fig.16)

Wingspan 32–36mm. Antenna shortly bipectinate, in female more finely than in male. Forewing whitish to whitish ochreous, suffused to a varying degree reddish buff; subbasal and antemedian lines reddish brown; postmedian and subterminal lines reddish brown, the latter less clearly defined, both edged white distally; upper stigmata outlined red-brown, the orbicular conspicuously pale, the reniform, particularly the dorsal lobe, suffused reddish brown; a pale postreniform blotch; claviform stigma reduced to a slightly darker-outlined, reddish brown shade, followed by a small whitish dot; termen pale, sometimes whitish ochreous; fringe chequered ground colour and reddish buff. Hindwing pale fuscous with faintly darker discal spot; sometimes a complete postmedian line and usually a broad, slightly darker terminal fascia.

Variation is usually slight, although isolated extreme aberrations are known.

Life history

Ovum. Laid singly in flowers of the foodplant.

Larva. Full-fed *c.*35mm long, somewhat elongate and clearly less stout than most of the genus. Head light brown. Body yellowish brown with a row of grey-brown dorsal chevrons; dorsal, subdorsal and spiracular lines whitish or greyish, thinly edged grey-brown, rather obscure; subspiracular stripe ochreous; spiracles black; pale and rather glossy ventrally.

The only natural foodplant in Britain is the Spanish catchfly (*Silene otites*), the young larva feeding inside the seed capsules for several weeks; in later stages the larva probably feeds mostly at night, but may be found climbing and descending the foodplant at all hours. Much has been written about the difficulty of transferring it to substitute foodplants which is confusingly contradictory; larvae have been reared several times by Goater who states unequivocally that they will transfer readily at any stage of development to garden carnation flowers and will eat the whole flower except the calyx.

Pupa. Normal *Hadena* shape, with haustellum-sheath; thoracic segments lightly grained; abdominal segments punctate, the anterior part of each more heavily than the posterior; cremaster rugose, black, strongly tapering, with two fine, very short, divergent apical spines. In a subterranean cocoon.

Imago. Univoltine, with a very long period of emergence, from late May to July. Like other members of the genus it does not come to sugar, but flies from early dusk to feed and oviposit and comes to light.

Conservation. The species is listed as endangered and with good cause as it is extremely local and in its limited range suitable habitats are rapidly being destroyed; it is restricted to a single, very local foodplant and suffers greatly from parasites, a figure of 90 per cent infestation having been recorded. It is most important that any collecting should be kept to a minimum and that if more larvae than are needed reach maturity, the resulting moths should be returned to their native Breckland.

Distribution (Map 165)

England, Breckland district. South (1961: 187) gives a record for north Lincolnshire; the moth, now in the Lincoln Museum, was bred from a larva said to have been found on viper's bugloss (*Echium vulgare*); Spanish catchfly is not recorded from Lincolnshire. Eurasiatic; in France mainly western and southern, always local; in Holland, one or two localities on the dunes; three records only from south Sweden; absent from Denmark, north-west Germany and Belgium; in the Alps only in the southern valleys.

HADENA LUTEAGO BARRETTII (Doubleday)
Barrett's Marbled Coronet

Noctua luteago [Denis & Schiffermüller], 1775, *Schmett. Wien.*: 86.
Dianthoecia barrettii Doubleday, 1864, *Zoologist* **22**: 8915.

Type locality: Ireland; Dublin.

Description of imago (Pl.12, figs 17,18)

Wingspan 35–42mm. Antenna bipectinate, in male shortly, in female very shortly. Forewing rather ample, dull brown to dull slate-grey; obscure, more ochreous brown or paler greyish white suffusion in subbasal and median areas between upper stigmata and distal part of dorsum; upper stigmata somewhat paler than ground colour; subterminal line, the most consistently clear marking, whitish or pale ochreous. Hindwing fuscous; cilia whitish.

This obscurely coloured species shows considerable local variation, probably because of the isolation of the individual populations. Specimens from the east and south-east coasts of Ireland are usually dull brown; those from Co. Cork (ab. *turbata* Donovan, fig.18) slate-grey with more contrasting markings and those from Cornwall (ab. *ficklini* Tutt, fig.17) grey-brown. Specimens from Guernsey (ab. *lowei* Tutt) are ochreous, approaching the continental form *argillacea* Hübner.

Life history

Ovum. Flattened-hemispherical, lightly ribbed and reticulate, a depression surrounding the micropyle; yellowish and glossy. Laid singly on the foodplant, hatching in about seven days.

Larva. Full-fed *c.*30mm long. Head reddish pale brown. Prothoracic and anal plates yellowish brown. Body white, with a slight pinkish ochreous tinge, skin glossy and transparent; gut visible as a grey-brown dorsal line; pinacula black; spiracles black and conspicuous; legs pale red-brown.

The recorded foodplants are campions (*Silene maritima* and *S. vulgaris*), rock sea-spurrey (*Spergularia rupicola*) and sand-spurrey (*S. rubra*). The larva bores first into the leaves and leaf-axils, but gradually penetrates the stems and bores downwards to the roots, where it completes its growth by mid-September. Tenanted plants break off at the root-crown if pulled gently (Bowes, 1939). According to Allen (1949), the roots of garden lettuce may be used as a substitute foodplant in captivity.

Pupa. Normal *Hadena* shape, glossy reddish brown; wing- and appendage-cases finely sculptured; remainder finely punctate, except for a smooth band on the posterior part of each abdominal segment; cremaster elongate, with two short, parallel, apical spines. Among the roots of the foodplants.

Imago. Univoltine. Flies from early June to late July. Rarely seen by day, but flies very rapidly at dusk and feeds at flowers; does not visit sugar, but comes to light.

Distribution (Map 166)

Almost entirely coastal, although Kettlewell (1973) records larvae two miles inland in Cornwall. Extremely local in south-west England (Cornwall and Devon), Wales (Pembrokeshire, Caernarvonshire and Anglesey); a single record from the Isle of Wight; Ireland (Dublin area, Co. Waterford and Co. Cork); according to Chalmers-Hunt (1970) there is no record from the Isle of Man during the present century; the Channel Islands. The nominate subspecies is Eurasiatic; in western Europe in Portugal, Spain, south and central France, Switzerland, south Germany, Austria and Hungary.

HADENA COMPTA ([Denis & Schiffermüller])
The Varied Coronet

Noctua compta [Denis & Schiffermüller], 1775, *Schmett. Wien.*: 70.

Type locality: [Austria]; Vienna district.

Description of imago (Pl.12, fig.19)

Wingspan 28–34mm. Antenna in male shortly fasciculate, in female sparsely and shortly bipectinate. Forewing narrower than those of most of the genus; blackish brown, divided by a broad, white median fascia from costa through

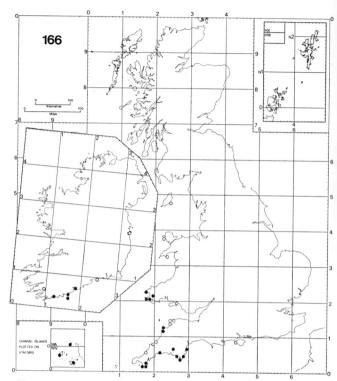

Hadena luteago

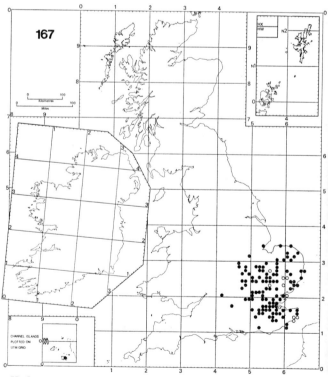

Hadena compta

the orbicular stigma to centre of dorsum; subbasal line black, outlined white; beyond this usually a short, orange-brown streak; reniform stigma white, centred brownish, adjacent to but not part of the median fascia; subterminal area dotted with fulvous scales; subterminal line incomplete, whitish, erratic in course; fringe chequered ground colour and white. Hindwing brownish fuscous, paler basally and dorsally; terminal line fine, slightly ochreous; cilia fuscous basally, white distally.

Variation is slight, affecting mainly the amount of brown tint in the ground colour and the alignment of the stigmata. *Similar species*. *H. confusa* (Hufnagel) which is larger, with more ample forewing; median fascia incomplete, diagonal, from orbicular stigma to tornus; whitish blotches at each extremity of subterminal line.

Life history

Ovum. Slightly elongate basally, lightly covered with hexagonal sculpturing; whitish initially, darkening to pale pinkish brown. Laid singly inside the flowers of sweet william (*Dianthus barbatus*), very loosely attached, hatching in about four days.

Larva. Full-fed *c*.33mm long. Head pale brown. Body usually drab yellowish brown, sometimes darker; dorsal stripe wide and greyish, sometimes with vestigial dorsal chevrons; laterally and ventrally pale yellowish brown. Obscurely marked and usually easily distinguished from the other two brown *Hadena* species likely to occur on the foodplant, *H. bicruris* (Hufnagel) and *H. confusa*, which both have clear dorsal chevrons.

Feeds almost exclusively on the seeds of sweet william in Britain, although there are records of it being found on bladder-campion (*Silene vulgaris*) in south-east Kent and Breckland, Norfolk. Feeds concealed until well grown, after which a number may often be found hiding by day beneath the foodplant. At all times larvae are hard to find inside the dense seed heads, and stale pieces of foodplant should be sorted several times before being thrown away; it is also necessary to take care not to introduce earwigs with fresh food. It may almost be taken for granted that a few larvae of *H. bicruris* will be introduced with the food; these can be recognized when young by the greyish ground colour, when larger by the well-defined dorsal pattern.

Pupa. Typical *Hadena* shape, bright red-brown and glossy; dorsal surface of thorax deeply and coarsely punctate; wing-cases lightly sculptured; dorsal surface of abdominal segments with anterior two-thirds deeply and coarsely punctate, ventral surface more lightly punctate to one-third; cremaster black, projecting, with two strong, divergent apical spines. In a cocoon, usually only 25–50mm (1–2in) below the surface of the soil.

Imago. Univoltine; a very small second brood may occur in hot, dry summers. Occurs in June and comes to flowers at dusk and later to light, being restricted mainly to gardens because of its foodplant.

Distribution (Map 167)

The early records of this species have been much discussed; opinions differ in detail, but there is general agreement that at least some of the early specimens were of doubtful provenance while others may have been immigrants. In view of later history, the Folkestone record of 1877 (Cockayne, 1947) is of interest. There has been no doubt of the species' status since 1948, when it was taken in some numbers in Dover and larvae were later found in gardens throughout the town. Since then it has spread quickly northward and is now (1978) well established through East Anglia and the counties immediately adjacent, but not far west in the southern part of the range. South-east England, including the London area, north to Lincolnshire and Leicestershire, westward to Sussex, Surrey, Berkshire and Gloucestershire; Jersey since 1960. Eurasiatic; from Portugal to Denmark, south Sweden and Finland.

HADENA CONFUSA (Hufnagel)
The Marbled Coronet

Phalaena confusa Hufnagel, 1766, *Berlin.Mag.* **3**: 414.
Phalaena nana Rottemburg, 1766, *Naturforscher, Halle* **9**: 132.
Noctua conspersa [Denis & Schiffermüller], 1775, *Schmett. Wien.*: 71.

Type locality: Germany; Berlin.

Description of imago (Pl.12, figs 20–23)

Wingspan 33–39mm. Antenna in male shortly fasciculate, in female very shortly bipectinate. Forewing blackish, mixed white basally and on much of the dorsum; a large white blotch formed by the two upper stigmata and a more or less quadrate mark below just fails to unite with a lesser white pretornal blotch, forming a broken diagonal fascia; a white apical blotch; subterminal line white (erratic); fringe chequered blackish and white. Hindwing fuscous, slightly paler basally; sometimes with a darker lunular discal spot.

The species varies little in its continental range, but greatly in the western and northern parts of Britain. The variation is based on melanic and ochreous suffusion which, although geographical in incidence, does not always show a regular gradation. The specimens from southern, central and eastern England are more or less typical; a progressive ochreous suffusion of the white areas and olivaceous suffusion of the black appears in south-west England, Wales and Scotland, although many specimens from Ireland and some from the Inner Hebrides are unusually black and white (fig.23); Baynes (1964), however, mentions slightly ochreous specimens from Co. Donegal. Orkney produces forms varying from those with light yellowish suffusion of the white markings only (fig.21) to others in which the whole forewing is heavily suffused with ochreous, and which approximate closely to the less extreme Shetland forms (fig.22). In Shetland, as well as an almost invariable ochreous influence, there is also a considerable degree of melanism.

Similar species. *H. compta* ([Denis & Schiffermüller]) *q.v.*

Life history

Ovum. Rather flattened-hemispherical, dimpled rather than ribbed, pale greenish white at first, darkening rapidly. Laid singly on buds and flowers of the foodplant, hatching in about seven days.

Larva. Full-fed *c.*36mm long, rather stout, tapering quickly to extremities. Head small, light brown. Body from dull yellow-brown to rich red-brown, usually of the darker shade in the north; dorsal line and chevrons darker, but becoming more obscure toward full growth, usually most distinct on the first three abdominal segments.

Feeds on the unripe seeds of sea-campion (*Silene maritima*), bladder-campion (*S. vulgaris*) and white campion (*S. alba*); red campion (*S. dioica*) and ragged robin (*Lychnis flos-cuculi*) are also recorded, but Haggett (1968) doubts whether red campion is ever eaten in the wild. Both this species and *H. bicruris* (Hufnagel) vary greatly in colour, but the latter species is not often found on sea-campion or bladder-campion, preferring red campion which *H. confusa* eschews. As the species is most abundant in areas of high rainfall, it is worth noting that most larvae leave the capsules in high wind or heavy rain. Full-fed by mid-August.

Pupa. Typical *Hadena* shape, with haustellum-sheath, reddish brown; wing-cases finely sculptured, remainder of thoracic segments more coarsely punctate; cremaster blackish, with two black apical spines, widely separate at base, each tapering to a pointed tip. In a fragile cocoon, just below the surface of the soil. Unlike other members of the genus it has no tendency to autumn emergence—rather the contrary—as it frequently overwinters more than once.

Imago. Univoltine. Occurs in late May and June, later in the north. Flies to flowers from dusk onwards and comes to light, but is not attracted to sugar.

Distribution (Map 168)

Widely but sparsely distributed on calcareous soils in England, common near the coast; fairly widely distributed in Scotland and abundant in the extreme north and the Western and Northern Isles; the Isles of Scilly; the Channel Islands. Local and mainly coastal in Ireland. Eurasiatic; through western Europe to Denmark and Finland, reaching central Scandinavia.

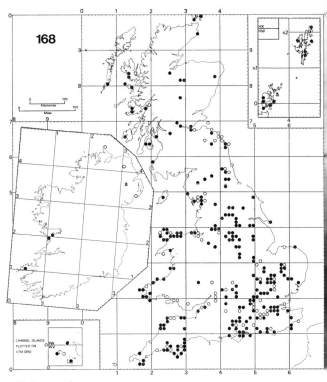

Hadena confusa

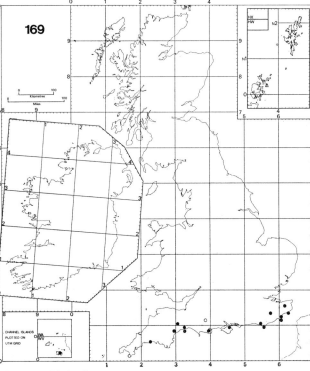

Hadena albimacula

HADENA ALBIMACULA (Borkhausen)
The White Spot

Phalaena (Noctua) albimacula Borkhausen, 1792, *Naturgesch.eur.Schmett.* **4**: 149.

Type locality: Europe.

Description of imago (Pl.12, fig.24)

Wingspan 30–38mm. Antenna in male fasciculate, in female shortly bipectinate. Forewing rich olivaceous brown with slightly darker median fascia; basal subcostal patch white, broken by black subbasal line; antemedian, postmedian and subterminal lines black, narrowly edged with white; orbicular and reniform stigmata white with brown centre, a small white dot between; a pure white, roughly quadrate mark between the upper stigmata and dorsum replacing the claviform stigma. Hindwing pale fuscous; discal spot and wide terminal fascia brownish fuscous; sometimes also an irregular median line.

Variation is usually in the shape of the individual components which make up the white median pattern; old or worn specimens tend to fade to rather yellowish brown.

Life history

Ovum. Laid on flowers of the foodplant.

Larva. Full-fed *c*.36mm long. Head light brown, with two darker brown vertical lines. Body ochreous brown, darker brown dorsally; dorsal line even darker, usually visible on thoracic segments, but replaced on abdominal segments by a series of roughly triangular dark brown dots which are flanked by numerous similarly coloured subdorsal and lateral spots; spiracles golden brown, ringed black. The spotted pattern distinguishes it from others of the genus.

Except when very small it is largely nocturnal. The only natural foodplant is Nottingham catchfly (*Silene nutans*) the small seed capsules of which offer little concealment, the larva hiding in the shingle by day and only climbing to feed soon after dusk; sweet william (*Dianthus barbatus*) and garden carnation (*Dianthus caryophyllus*) are satisfactory foodplants in captivity. This larva feeds more slowly than that of *H. perplexa* ([Denis & Schiffermüller]), the species most commonly found sharing its foodplant, and before it is fully fed in August the bulk of the available food has often been devoured by the more common species.

Pupa. Typical *Hadena* shape with haustellum-sheath, reddish brown; wing-cases smooth, with minimal sculpturing; dorsal surface of abdominal segments punctate; abdominal segments sculptured progressively more coarsely from the first; cremaster short, tapering to a ridged annulus; apical spines stout, divergent with T-shaped tip.

Imago. Univoltine. Flies in June and July. Often found at rest by day on fence-posts and the walls of wooden sheds but flies at dusk to feed at the flowers of campion and catchfly, wood-sage and viper's bugloss, and later comes to light.

Conservation. This species although listed as rare, seems to be in much less jeopardy than *H. irregularis* (Hufnagel); it has several well-separated stations of which the principal, Dungeness, with large areas of shingle protected from commercial exploitation by the power stations' perimeter fence, seems at present safe. But since all the known colonies are on the south coast, a small climatic change may exterminate the species.

Distribution (Map 169)

Very local on some shingle beaches from Kent to Cornwall; the Channel Islands. Eurasiatic; in western Europe from the Iberian peninsula north-eastwards to southern Fennoscandia, becoming rarer and more coastal in the northern part of its range.

HADENA BICRURIS (Hufnagel)
The Lychnis

Phalaena bicruris Hufnagel, 1766, *Berlin.Mag.* **3**: 302.
Noctua capsincola [Denis & Schiffermüller], 1775, *Schmett. Wien.*: 84.

Type locality: Germany; Berlin.

Description of imago (Pl.12, fig.25)

Wingspan 30–40mm. Antenna bipectinate, in female more shortly. Forewing greyish brown to dark brown; subbasal line incomplete, yellowish white outlined with blackish brown; a blackish subbasal dorsal blotch; ante- and postmedian lines paler than ground colour, darker-bordered; subterminal line complete and erratic, yellowish white, the final 2–3mm before the dorsum so inclined from the perpendicular as to be more or less parallel to the termen (figure 18); upper stigmata outlined whitish or yellowish white, then blackish, usually clearly separated dorsally; claviform stigma blackish, followed by a somewhat triangular, yellowish fuscous blotch; underside, stigma without surrounding suffusion. Hindwing pale fuscous; discal spot and wide terminal fascia brownish fuscous.

There is some minor variation in the wing pattern and specimens from northern localities are darker and more

Figure 18 *Hadena bicruris* (Hufnagel), forewing

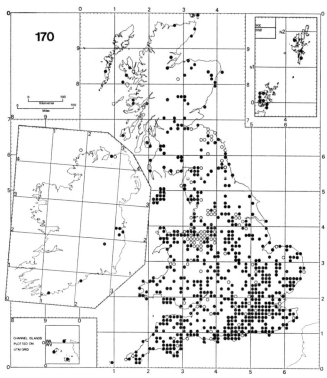

Hadena bicruris

richly coloured than those from the south. The very spectacular aberration figured in Newman (1869) has been shown (Cockayne, 1950a) to be a specimen of *H. rivularis* (Fabricius).

Similar species. *H. rivularis* q.v.

Life history

Ovum. Almost spherical, with flattened base, dimpled overall; greenish white with slightly darker ring around micropyle, darkening within a few hours. Laid singly on a flower or bud of the foodplant, hatching in five to six days.

Larva. Full-fed c.40mm long. Head small, light brown. Body stout, tapering to extremities, from dull putty-colour to brownish ochreous; dorsal line fine, paler; dorsal chevrons darker than ground colour although variable in degree; spiracular line wide, paler; spiracles pale flesh-colour, outlined with black; ventrally pale ochreous with darker freckling.

Feeds on the seeds of most species of *Silene*, *Lychnis* and *Dianthus* although with apparent preference for red campion (*Silene dioica*) and sweet william (*Dianthus barbatus*). When too large to remain concealed inside a seed-capsule it will often feed in full view by day, the head inside and the bloated abdominal segments protruding. There must be considerable competition for food in strong colonies, since by mid-July it can be difficult to find un-eaten capsules to feed captive larvae; at such times the leaves are nibbled, but without enthusiasm. Full-fed from early July; the second brood, where it occurs, from mid-August.

Pupa. Typical *Hadena* shape with haustellum-sheath, bright reddish brown; thoracic segments glossy, lightly sculptured; abdominal segments with broad anterior band of punctation; cremaster blackish and conical, with several fairly large basal thorns and two short, stout, pointed apical spines. In a fragile subterranean cocoon.

Imago. Bivoltine in the south, univoltine from the Midlands northward. Occurs from early June to September, feeding at flowers from early dusk and coming to light; not attracted to sugar. There seem to be no ecological preferences and the species is found in woodland, marsh, sand-dunes, heathland and gardens.

Distribution (Map 170)

Common throughout Great Britain and Ireland. Eurasiatic; through western Europe to central Scandinavia and Finland.

HADENA CAESIA MANANII (Gregson)
The Grey

Noctua caesia [Denis & Schiffermüller], 1775, *Schmett. Wien.*: 82.

Dianthoecia caesia mananii Gregson, 1866, *Entomologist* **3**: 103.

Type locality: Isle of Man.

Description of imago (Pl.12, figs 26,27)

Wingspan 36–42mm. Antenna bipectinate, in male very shortly, in female extremely shortly. Forewing with obtusely pointed apex; bluish grey to bluish black, of rough-looking texture; all markings obscure; ante- and postmedian lines faint or obsolete; subterminal line sometimes present as a row of separate grey or greyish ochreous elongate dots; upper stigmata obscure or absent, often, however, a pale suffusion between their presumed positions. Hindwing greyish fuscous, slightly paler basally.

Local variation is considerable. Chalmers-Hunt (1970) comments on the constancy of the appearance of all Manx specimens examined by him and many Irish specimens resemble those from the Isle of Man. Specimens from the Burren, Co. Clare are more strongly bluish and a blackish blue form occurs in the extreme south-west of Co. Kerry (fig.27). The Hebridean and Scottish mainland forms are as yet little known, but are said to be more variegated. The nominate subspecies in Europe is much brighter, with paler ground colour, enlivened with ochreous and yellowish markings in the basal and median areas.

Life history

Ovum. Unknown.

Larva. Full-fed *c.*47mm long. Head light brown to reddish brown. Body brownish grey to grey, tinged ochreous dorsally; heavily freckled dorsally and laterally with darker dots forming diamond-shaped clusters which may extend below the spiracles; spiracles pale brown, ringed black; ventrally paler grey.

Feeds on the seeds of sea- and bladder-campion (*Silene maritima* and *S. vulgaris*). Growth is very rapid and the earliest larvae pupate well before the moths cease to fly. It has been noted that larvae from Ireland do not readily accept their foodplant when it is obtained from other districts; Myers (1968) speculates on the reasons for this.

Pupa. Normal *Hadena* shape, the haustellum-sheath particularly conspicuous, bright red-brown; the wing-cases and appendages finely sculptured; remainder of thoracic segments densely punctate; abdominal segments smooth, with an anterior band of fine punctation; cremaster short and wide, with two small, fine apical spines. In a fragile cocoon, below the surface of the soil.

Imago. Has a very long, uninterrupted period of emergence, from late May to August, and there is some doubt whether the species is univoltine or partly bivoltine; in all probability there is at least a small second brood in the southern part of its range. Flies rapidly at dusk and is extremely difficult to intercept, but comes freely to light.

Distribution (Map 171)

Although the European distribution of this species is essentially montane, the ecology of the British subspecies is extremely specialized, as it occurs very locally indeed over a comparatively large area, always (*teste* Baynes, 1970) within 50m of high water-mark. Ireland, locally common on coasts, Cos Waterford, Kerry, Clare and Donegal; Baynes (1970) queries Greer's inland record from Co. Tyrone. Isle of Man, very local in the south, mainly on the east coast. Scotland, Inner Hebrides: Islay, Rhum, Canna, Mull. The only records from the British mainland are from Ardnamurchan, Argyll and four specimens, still extant, labelled Cumberland, 1899. Abroad probably Eurasiatic; the typical form, several subspecies and many widely differing forms occur in the mountainous parts of Spain, France, Switzerland and the Balkans; in south Norway and south-east Sweden it is mainly coastal.

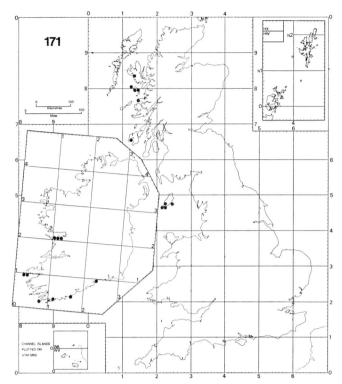

Hadena caesia

ERIOPYGODES Hampson

Eriopygodes Hampson, 1905, *Cat.Lepid.Phalaenae Br.Mus.* **5**: 353.

This genus contains several Palaearctic species, one of which has recently been discovered in Britain.

Frons smooth; palpus with second joint covered with long hair. Abdomen without crests.

ERIOPYGODES IMBECILLA (Fabricius)
The Silurian

Noctua imbecilla Fabricius, 1794, *Ent.syst.* **3**: 113.
Type locality: Germany; Kiel.

Description of imago (Pl.12, fig.1,♀; Vol.**10**: Pl.6, fig.1,♂)

Wingspan 24–27mm, female the smaller, with narrower wings and long, heavy abdomen. Antenna in male strongly dentate, in female ciliate, scaled and thickened medially. Forewing reddish brown, in female chocolate-coloured, with basal, median and outer areas somewhat darker; two cross-lines widely separated at the costa but much closer together at the dorsum; orbicular and claviform stigmata not visible, reniform stigma angular and whitish distally. Hindwing dark greyish brown.

Life history
Although the early stages have not yet been found in the wild in Britain, moths were first reared here by B. Goater in 1977 (see below).

Ovum. Hemispherical and distinctly ribbed below central rosette; yellowish white, with upper part speckled carmine, dark red and white (Forster & Wohlfahrt, 1964, **4**: 87).

Larva. Head small, yellowish brown heavily lined with rust-brown, with two rust-brown streaks. Prothoracic plate shiny, light yellowish or brownish grey. Body yellowish to brownish grey, lighter dorsally, with a row of connected oval marks along the fine whitish mediodorsal line, subdorsal line whitish; subdorsum divided by a fine pale line into darker upper and somewhat lighter lower halves; small ochreous yellow spiracles on the edge of the subdorsal area broadly ringed with black; a broad spiracular band sprinkled light brown, edged above and below by narrow white lines; pinacula greyish brown, often inconspicuous (Nordström *et al.*, 1941; figured by Spuler, 1910).

In Germany it overwinters, feeding on many herbaceous plants and soft grasses until May. It can be found by night, especially on bedstraw (*Galium* spp.) (Bergmann, 1954). In Finland scabious (*Knautia* spp.) and dead-nettle (*Lamium* spp.) are recorded as foodplants (Mikkola & Jalas, 1977).

Pupa. Stout, red-brown; cremaster spatulate, with four converging bristles. Subterranean.

Rearing. In captivity, eggs laid in mid-July in small groups on dead grass-stems and debris hatched at room temperature in nine days. The larvae fed up rapidly, preferring withered leaves of dandelion (*Taraxacum officinale*), but also taking chickweed (*Stellaria media*), *Viola* spp. and other plants. They were very sluggish, resting fully stretched by day and feeding only at night. The pupae were lively, wriggling vigorously when disturbed. Moths emerged 16–20 October from which pairings and many eggs were obtained. Moths from these began to emerge in the following February (Goater, 1978).

Imago. Only recently discovered as a resident British species (Horton & Heath, 1973; Horton, 1976). It was, however, first mentioned as British by Curtis (1829), but without detail beyond indications that he had neither found the species himself nor possessed a British specimen. Wood (1839) included it among 'doubtful British species' with the comment 'supposed to be British by Mr. Curtis'. The first authenticated British capture, a male, was made at mercury-vapour light at *c.*500m (1,500ft) on a mountain in Monmouthshire before 23.30 hours on 30 July 1972, a year which was, for most species, a very late season. Searches in the following years were unsuccessful until the last few days of June 1976, when Horton obtained many of both sexes at light and also flying in hot sunshine in a gully near the original locality. In mid-July 1977 many more were seen, all the females and some males during the afternoon between 12.30 and 17.30 hours, when their flight was straight and direct, buzzing about a metre (two or three feet) above the ground; males were only found plentifully at light between 02.00 hours and dawn. None were seen at sugar or at flowers (Goater).

Conservation. As this species is at present known from only one locality it is desirable that collecting of all stages should be kept to a minimum.

Distribution
The species is clearly very local, but terrain with a similar flora to that in its Monmouth locality is widespread in south Wales. Eurasiatic; in Europe the species is widespread at moderate altitudes in the Pyrenees, the French Massif Central, Bohemia and in the Alps; in Norway it occurs as far as 62°N in the mountains, and at low altitudes in south and central Sweden and commonly in south Finland.

CERAPTERYX Curtis
Cerapteryx Curtis, 1833, *Br.Ent.* **10**: 451.

There are two Palaearctic species, one of which is widely distributed in Great Britain and Ireland and occasionally a pest on hill pasture. The genus is close to *Tholera* Hübner, but with the haustellum fully developed; frons flat; antenna in male strongly bipectinate; palpus short, second segment hairy, third segment short. Abdomen with lateral hair-tufts.

CERAPTERYX GRAMINIS (Linnaeus)
The Antler Moth
Phalaena (Bombyx) graminis Linnaeus, 1758, *Syst.Nat.* (Edn 10) **1**: 506.
Type locality: [Sweden; Stockholm].

Description of imago (Pl.12, figs 30–35)
Wingspan, male 27–32mm, female 35–39mm. Female much larger than male, usually paler in colour. Antenna bipectinate, in male strongly, in female very shortly and sparsely. Forewing very variable in colour and pattern; pale olive-brown to dark reddish brown; pattern consisting essentially of longitudinal streaks, the paler of them particularly variable in length and position; costa of ground colour; a fine, pale ochreous subcostal streak usually reaching at least to orbicular stigma; a gradually widening strip of ground colour enclosing upper stigmata; a median-basal streak, pale ochreous, at least to reniform stigma where it forks, sometimes several times, and may reach the termen; a narrow strip of the ground colour, containing a short antemedian pale ochreous streak and the black

claviform stigma; vein 1b (2A+3A) clearly whitish; a black subdorsal streak of varying length; dorsum of ground colour, sometimes paler; median fascia darker than ground colour; subterminal line of short, blackish longitudinal dashes or dots; upper stigmata coloured as paler streaks. Hindwing greyish fuscous, variably paler basally; sometimes a small discal spot.

The species varies in almost every detail. The darkest males are brownish mahogany, the palest females pale greyish ochreous. The most conspicuous feature is the median-basal streak which may be of a brighter colour than the other pale features; at its simplest it is a narrow, sometimes broken streak which widens slightly at the reniform stigma, but at its most extreme development it forms a large blotch, filling the basal and median areas except the costa and dorsum, with branches extending nearly to the cilia. The most extreme forms occur in Ireland and the Scottish Highlands, where the species is particularly abundant.

Life history

Ovum. Almost spherical, with small, flattened base, micropyle recessed, delicately sculptured; creamy white at first, becoming pinkish in a few days and undergoing several colour-changes during the winter to purplish and finally leaden grey before hatching, usually, in late March. The eggs are broadcast during low flight. The ovum is extremely tough-shelled, bounces if dropped on a hard surface and may be delicately touched with a needle point without being punctured. According to Hedges (1949), the larva is fully formed within a week.

Larva. Full-fed *c.*35mm long, rather stout. Head brown. Body bronzy grey-brown and glossy, with conspicuous paler apical margins to each segment; dorsal, subdorsal and spiracular lines wide, supraspiracular line fine, all very pale yellowish brown; spiracles black; ventrally brown with paler freckling. Barrett (1897) gives the abdominal bands as a character for separation from *Tholera decimalis* (Poda), but that species has similar, although rather narrower, bands.

Feeds nocturnally on grasses, preferring those of harder texture (*Nardus*, *Molinia* and *Festuca*) as well as *Scirpus* and *Juncus* spp. From time to time there are population explosions, when the larvae feed by day and night and advance in 'armies' to devastate large areas of grassland, particularly on upland acid soils.

Pupa. Rather short and stout, reddish brown; wing-cases and appendages coarsely wrinkled; dorsal surface of thoracic segments finely sculptured; abdominal segments with anterior band of fine punctation; cremaster elongate, black and longitudinally striate, with two close-set apical spines which are short and blunt-tipped. In a well-formed oval earthen chamber among grass roots.

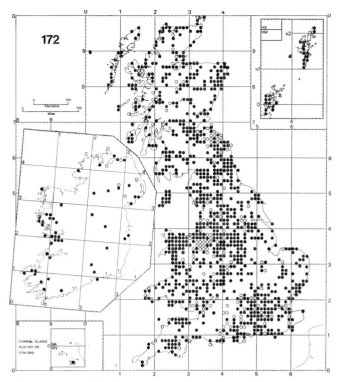

Cerapteryx graminis

Imago. Univoltine. The moth is on the wing from mid-July to early September in the north, being most abundant from early August. It has a short mid-morning flight and in calm weather will feed throughout the day at flowers of ragwort, thistle and hogweed and other Umbelliferae. It is a very early visitor to sugar, often arriving before an m.v. lamp has reached full brilliance: it can become rather a nuisance on the sheet by disturbing other insects and sometimes ovipositing copiously.

Distribution (Map 172)

Throughout Great Britain and Ireland, most abundantly on acid soils, but able to thrive in any open country, including downland. Eurasiatic; throughout Europe to the Arctic, and in Iceland.

THOLERA Hübner
Tholera Hübner, [1821], *Verz.bekannt.Schmett.*: 215.
Neuronia Hübner, [1821], *ibid.*: 215.
Charaeas Stephens, 1829, *Ill.Br.Ent.* (Haust.) **2**: 108.
Epineuronia Rebel, 1901, *in* Staudinger & Rebel, *Cat.Lepid. pal.Faunengeb.* (1): 155.

There are two Palaearctic species, both of which occur in the British Isles. Larger species than *Cerapteryx* Curtis.

Imago. Frons flat; antenna in male bipectinate; labial palpus short, densely hairy below; haustellum short. Thorax with posterior and anterior crests. Abdomen with long, dense, lateral cilia.

Ovum. Almost spherical. Laid singly at random.

Larva. Very similar to *Cerapteryx*, bright green with pale lines at first, dull olive-brown and very stout when full-fed.

Pupa. Rather short and stout; among grass roots in a roomy cocoon.

Life history. Univoltine. Overwinters as ovum.

THOLERA CESPITIS ([Denis & Schiffermüller])
The Hedge Rustic

Noctua cespitis [Denis & Schiffermüller], 1775, *Schmett. Wien.*: 82.

Type locality: [Austria]; Vienna district.

Description of imago (Pl.12, figs 28,29)
Wingspan 34–40mm. Antenna in male bipectinate, in female very shortly so. Forewing blackish brown, darker subcostally; ante- and postmedian lines black, edged with pale fuscous, the former rather inconspicuous; subterminal line yellowish white, usually complete but sometimes reduced to a row of dots; upper stigmata finely outlined whitish, reniform variable in shape; claviform stigma visible, if at all, only as a darker shade. Hindwing of male whitish with greyish fuscous terminal line; remainder of wing-pattern variable, sometimes a narrow greyish fuscous subterminal fascia, but there may be three more complete or vestigial fasciae in median and postmedian areas; hindwing of female fuscous, paler basally.

Life history
Ovum. Almost spherical, rather coarsely ribbed; creamy white at first, undergoing colour changes through the winter to pinkish, purplish and finally leaden grey. Laid singly, at random, hatching in late March or April.

Larva. Full-fed *c.*35mm long, similar in ground colour and dorsal markings to that of *Cerapteryx graminis* (Linnaeus) and *Tholera decimalis* (Poda), but with three faint, pale lines between subdorsal and subspiracular lines, where the other species have only one; ventrally faintly green-tinged.

Nocturnal in habit and feeds on grasses, preferring the 'harder' species such as mat-grass (*Nardus stricta*), tufted and wavy hair-grasses (*Deschampsia cespitosa* and *D. flexuosa*). In the early instars it is green with whitish longitudinal stripes and feeds high on the grass-blades, whence it may be swept by night in May and early June, but after attaining the dull colouring of the later instars, it feeds at ground level until full-fed in July.

Pupa. Short and rather stout, dark reddish brown, glossy; wing-cases heavily sculptured; abdominal segments with an anterior band of punctation; cremaster concave ventrally, laterally flattened and coarsely grained with two short, pointed, slightly divergent apical spines. In a rather large, firm-sided cell among the grass roots.

Imago. Univoltine. Flies from mid-August and is largely nocturnal although, according to Curtis (1934), an afternoon flight has been noted. The male comes to light freely, the female rather rarely; both sexes feed at flowers, such as ragwort, and will come to sugared flower-heads or leaves.

Distribution (Map 173)
Locally common in England, Wales, Isle of Man and Ireland; in Scotland to Aberdeenshire; recorded also from the Inner and Outer Hebrides; slightly more northerly in range than *T. decimalis*, although, within their common range, *T. cespitis* is the more local species; the Isles of Scilly, the Channel Islands. Eurasiatic; in western Europe from Portugal to Denmark, central Scandinavia and Finland.

THOLERA DECIMALIS (Poda)
The Feathered Gothic

Phalaena (*Geometra*) *decimalis* Poda, 1761, *Ins.Mus.Graec.*: 92.

Bombyx popularis Fabricius, 1775, *Syst.Ent.*: 577.

Type locality: [Greece].

Description of imago (Pl.12, figs 36,37)
Wingspan 38–48mm. Antenna bipectinate, in male strongly, in female very shortly. Forewing brown, densely irrorate with yellowish white scales, rather glossy; ante- and postmedian lines double, black; subterminal line complete, creamy white with an internal row of blackish wedges; veins whitish, having the appearance of rays; outlining of upper stigmata yellowish white; claviform stigma of ground colour, narrowly outlined black. Hindwing of male whitish; discal spot small and obscure; subterminal fascia of variable width, fuscous; tornus whitish; hindwing of female pale

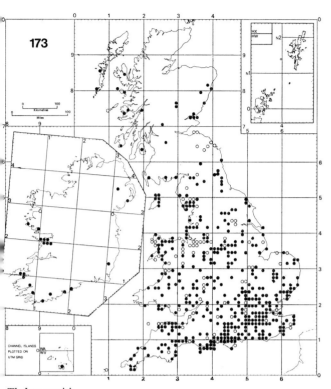

Tholera cespitis

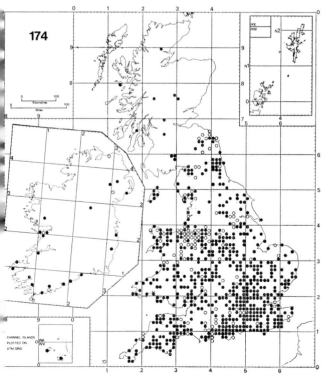

Tholera decimalis

fuscous, terminal fascia brownish, wider than in male. Female usually appreciably larger than male, with much larger abdomen.

Similar species. *Naenia typica* (Linnaeus) (Noctuinae) and *Heliophobus reticulata* (Goeze) q.v.

Life history

Ovum. Almost spherical, with small, flattened base, heavily but obtusely ribbed and finely reticulate, glossy and with tough, resilient cuticle; yellowish white, darkening quickly to pinkish brown, thereafter gradually to blue-grey and finally dark grey; becoming transparent, with the larva visible inside during March. Laid singly at random, hatching March to April.

Larva. Full-fed c.45mm long, very similar to those of *Cerapteryx graminis* (Linnaeus) and of *Tholera cespitis* ([Denis & Schiffermüller]) q.v.; the characters given by Buckler (1891) for separation from *Cerapteryx graminis* 'dorsally more richly coloured and ventrally darker in *Tholera decimalis*' are probably only of practical use when both species are available for comparison.

The young larva is green with whitish dorsal and lateral lines until about half-grown. It is nocturnal in habit and feeds on the blades of 'hard' grasses at first; later, when the dull mature colouring is assumed, it remains at root level, biting through the stems. Full-fed in July.

Pupa. Rather short and stout, reddish brown; first thoracic segment coarsely wrinkled, metathorax and abdominal segments 1–3 sculptured, abdominal segments 4–7 each with anterior band of punctation; cremaster large, elongate, rather flattened laterally and longitudinally grained, with two parallel apical spines.

Imago. Univoltine. Flies from mid-August. The male comes to light, sometimes in large numbers, the female much less frequently; neither sex seems to be attracted to flowers or to sugar.

Distribution (Map 174)

Although well distributed, this species is more exacting than *Cerapteryx graminis* in choice of habitat and is most attached to rough pasture, parkland and the fringes of woodland, preferring light, sandy soils. Generally common over much of England and Wales; Scotland, very local; Ireland; the Isles of Scilly and the Channel Islands. Eurasiatic; range similar to that of *Tholera cespitis*, but more general and much commoner.

PANOLIS Hübner

Panolis Hübner, [1821], *Verz.bekannt.Schmett.*: 214.
Ilarus Boisduval, 1828, *Eur.Lepid.Index method.*: 76.

Represented by a single species which occurs throughout most of Great Britain and Ireland.

Frons flat; antenna in male fasciculate; palpus very short, with long hairs; head and thorax densely hairy; legs hairy; abdomen with long lateral fringes.

PANOLIS FLAMMEA ([Denis & Schiffermüller])
The Pine Beauty

Noctua flammea [Denis & Schiffermüller], 1775, *Schmett. Wien.*: 87.
Phalaena (Noctua) griseovariegata Goeze, 1781, *Ent.Beyträge* **3** (3): 250.
Phalaena (Noctua) piniperda Panzer, 1786, *Naturgesch. Forstphaläne* (2): 51, pl.1, figs 1–12.

Type locality: [Austria]; Vienna district.

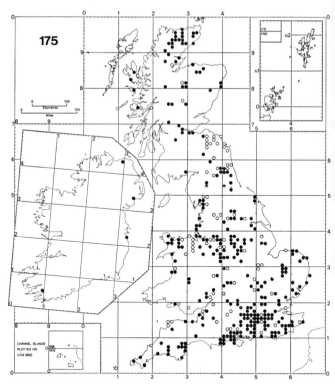

Panolis flammea

Description of imago (Pl.12, figs 41–44)

Wingspan 32–40mm. Antenna in male fasciculate, in female finely ciliate. Forewing narrow, termen oblique and evenly curved to tornus, light reddish brick-coloured, often suffused ochreous; ante- and postmedian lines brownish red, edged whitish, the former proximally, the latter distally, about four times as far apart at costa as at dorsum; subterminal line obscure; termen ochreous, broken by whitish streaks; upper stigmata whitish, variably centred greenish ochreous, their dorsal extremities usually joined by a fine white streak, both very variable in shape and alignment; the reniform stigma almost touching the costa and often elongate towards the apex; claviform stigma of ground colour, more or less obscurely outlined darker. Hindwing pinkish brown; cilia white.

Ab. *griseovariegata* Goeze (figs 43,44). Forewing variably suffused grey and greenish grey. Hindwing grey-brown with only the tips of the cilia white. The two forms and intermediates co-exist throughout the range.

Life history

Ovum. Flattened-hemispherical, heavily ribbed and more lightly reticulate, micropyle raised from a surrounding depression; pale yellow-green at first, darkening through yellow-brown to greyish brown before hatching. Laid singly or in rows beneath or at the base of the foliage of the foodplant, hatching in about eight days.

Larva. Full-fed c.40mm long, elongate and rather slender. Head and legs conspicuously red-brown. Body dark green, occasionally dark brown; longitudinal stripes all white, dorsal, wide, subdorsal, narrower and sometimes double; spiracular, wide, bordered orange-yellow below; spiracles pinkish, ringed black; ventrally light green with three fine, whitish longitudinal stripes.

Feeds on pine needles (*Pinus sylvestris* and *P. pinaster* are both recorded). It is of a rather nondescript pale green at first, but very soon develops the typical striped appearance and is then very difficult to see among the pine needles; should it fall, it remains rigid and the resemblance is maintained. Full-fed about the middle of July.

Pupa. Rather stout; on abdominal segment 4 a reniform dorsal depression with blackish raised margins; dull reddish brown; thoracic segments lightly sculptured ventrally, lightly punctate dorsally; abdominal segments lightly and sparsely punctate; cremaster blackish, concave ventrally, with four subapical bristles and two fairly straight, well-separated apical spines. In a very flimsy cocoon among the surface litter on the ground, or on the trunk in a crack in the bark.

Imago. Univoltine. Flies from late March to early May, the date of the first appearance being much influenced by weather. Rests by day on the trunks or branches of pine trees and feeds at sallow bloom or, more rarely, sugar by night; comes abundantly to light.

Distribution (Map 175)

The range of this species has been greatly increased by large-scale planting of pine and, as a population can survive among the relatively low concentrations of pines planted in parks and gardens, it is now established over a large area. Occasionally a serious pest in plantations, as in north Sutherland in 1978. Common where there are pines in England, Wales and Scotland to Sutherland and Skye in the Inner Hebrides; the Isles of Scilly; less common in Ireland, although records cover much of the country. Eurasiatic; in western Europe from Spain to central Scandinavia and Finland, wherever there are pine forests.

XANTHOPASTIS Hübner

Xanthopastis Hübner, [1821], *Verz.bekannt.Schmett.*: 211.

A New-World genus with one tropical and one subtropical species of which one example has been recorded in Britain.

XANTHOPASTIS TIMAIS (Cramer)

Phalaena timais Cramer, 1780, *Uitl.Kapellen* 3: 148, pl.275, fig.B.

Type locality: probably Central America (see Nye, 1975: 504).

A specimen was exhibited by A. S. Corbet at a meeting of the South London Entomological & Natural History Society on 22 May 1946, found among insects taken in the London area and brought to him for naming. The specimen has not since been traced. Presumably it was accidentally imported, as the species is native in the Americas, from south United States to Argentina, where the larva is believed to feed on Amaryllidaceae.

BRITHYS Hübner

Brithys Hübner, [1821], *Verz.bekannt.Schmett.*: 226.

A very small genus occurring mainly in the Oriental and Ethiopian regions but also represented in the Mediterranean region. Larvae of one species were once found in England.

BRITHYS CRINI PANCRATII (Cyrillo)
The Kew Arches

Bombyx crini Fabricius, 1775, *Syst.Ent.*: 587.
Noctua pancratii Cyrillo, 1787, *Ent.Neap.*: 8.
Type locality: Italy; Naples.

Description of imago (Pl.13, fig.1)

Wingspan 40mm. Thorax black. Forewing dark brown from base to subterminal fascia, beyond which is a narrow pinkish white band containing a series of brighter pink spots; terminal fascia dentate; reniform stigma sharply outlined with base extending inwards, pale but pinkish-centred; cilia brown. Hindwing clear white, with dark suffusion near tornus slight or absent in male, more extensive in female. Abdomen white, with some brown suffusion on crest and the last two segments brown.

Subsp. *pancratii* differs from the nominate subspecies by the generally darker colour of the forewing, with much smaller development of the pinkish band and inframarginal spots; on the hindwing by the much slighter suffusion of black; on the abdomen by absence of brown suffusion on the middle segments. The genitalia are, however, identical (Bretherton & Hayes, 1976). The figure in South (1961) appears to represent *B. crini crini*.

Life history

Ovum. Undescribed.

Larva. Chequered black and white, ground colour of the last abdominal somite white, looking like a miniature chess-board. Larvae of subsp. *pancratii* have the prothoracic plate dark brown and spots on the anal segment coalescent; in subsp. *crini* the plate is straw-coloured and the four anal spots are separate. In October 1933, 24 half-grown larvae were found by J. A. C. Greenwood on a bulbous edging plant, *Zephyranthes candida*, in Kew Gardens, Surrey. They were close together, as if from a single female. Twenty pupated, and from those kept at 15°C (60°F) in a moist glasshouse, moths emerged in March 1934; from others kept at normal temperatures moths emerged in mid-June. Fertile eggs were obtained from the latter. The young larvae bored into the fleshy leaves of *Z. candida* and later ate down the stems into the bulbs; some were fed successfully on narcissus and snowdrop, but bluebell (*Endymion non-scriptus*) was refused. They were full-grown at the end of

July, and produced many moths after a pupal stage lasting a fortnight. *Zephyranthes candida* is native in South America, where *B. crini* does not occur. The plant had been long established at Kew, and none had been recently imported there; but larvae have not been seen except in 1933, and no moth is known to have been caught in Britain. The larvae found at Kew may have been the progeny of an immigrant or introduced female, or perhaps were imported in their early stages in other bulbs (Greenwood, 1934).

Pupa. Dark sienna brown, smooth and glossy; cremaster with two short points. Subterranean.

Distribution

B. crini pancratii is a shore resident, locally common, in Portugal, Spain, and on the Mediterranean coasts of France and Italy, where the larvae feed on sand-lily (*Pancratium maritimum*); southwards its range extends throughout Africa, where it is inland as well as coastal. The range of *B. crini crini* is from India through the Far East to Japan and Australia. So far as known, there is a wide geographical gap between these subspecies.

EGIRA Duponchel

Egira Duponchel, 1845, *Cat.méth.Lépid.Eur.* (2): 162.
Xylomyges Guenée, 1852, *in* Boisduval & Guenée, *Hist.nat. Insectes* (Lépid.) **5**: 147.

There are several Palaearctic species, of which one is found very locally in England and south Wales.

Frons flat; labial palpus porrect, short and densely hairy. Venation of hindwing with 3 (Cu_1) and 4 (M_3) shortly stalked (figure 19). First abdominal segment with dorsal crest.

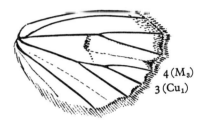

Figure 19 *Egira conspicillaris* (Linnaeus), hindwing

EGIRA CONSPICILLARIS (Linnaeus)
The Silver Cloud

Phalaena (*Noctua*) *conspicillaris* Linnaeus, 1758, *Syst.Nat.* (Edn 10) **1**: 515.
Type locality: Europe.

Description of imago (Pl.12, figs 38–40)

Wingspan 36–42mm. Basal seven-eighths of antenna in male very shortly fasciculate, in female simple. Typical form (figs 38,39) sexually dimorphic. Forewing narrow, termen oblique, apex pointed. *Male.* Ochreous grey to bluish grey, dorsum paler except basally; ante- and postmedian lines represented by darker, angulated shades; subterminal line a row of darker elongate dots; stigmata paler than ground colour, outlined darker, rather obscure; veins conspicuously blackish, giving a striate appearance. *Female.* Darker than male; a blackish grey basal streak which widens to cover the median fascia and narrows beyond the postmedian line, enclosing the orbicular stigma, but not the reniform; subcostal area narrowly blue-grey to the apex. Hindwing white; veins blackish; discal spot grey, lunular; terminal line fine, black.

References to 'intermediate forms' or ab. *intermedia* Tutt are to the female of the typical form, which was only recognized in 1937 (Cockayne, 1937); the earlier literature should be read with this in mind.

Ab. *melaleuca* Vieweg (fig.40) is much more common in England than the typical form. The sexes are similar; and the colour of the hindwing is not a reliable character for their separation. Forewing blackish brown, costa slightly paler; a broad, whitish subdorsal stripe from the centre of the wing to the tornus, where it converges or is contiguous with a whitish subterminal fascia from the apex; basal part of dorsum pale fuscous.

Life history

Ovum. Hemispherical, lightly ribbed and reticulate; greenish white at first, changing through pale blue-grey, pinkish and purplish before hatching. Laid in large batches, hatching in about ten days. Batches have been found on the flower stems of dock (*Rumex* spp.) and were very conspicuous when first laid.

Larva. Full-fed *c*.43mm long. Head pale brown. Body greenish brown to ochreous brown with many minute greyish freckles; dorsal and subdorsal lines fine and pale, indistinctly black-edged; spiracular line reddish ochreous, edged darker above; spiracles pinkish, edged black.

The natural foodplant is unknown but captive larvae have been reared on a large number of unrelated plants, including common bird's-foot trefoil (*Lotus corniculatus*) (flowers), lesser trefoil (*Trifolium dubium*), greater plantain (*Plantago major*), knotgrass (*Polygonum aviculare*), blackthorn (*Prunus spinosa*) and the suckers of English elm (*Ulmus procera*). Full-fed in July.

Pupa. Short and stout; dark purplish brown; thoracic segments glossy and finely sculptured; abdominal segments coarsely punctate with clear intersegmental divisions; cremaster blunt, with several minute basal bristles and two slightly longer inner spines. In a cocoon, below ground.

Imago. Univoltine. Flies in April and early May, the exact period varying from year to year, but usually for only about two weeks in any season. Comes to light and feeds at sugar and flowers, especially those of blackthorn and plum. During its short season it may be found at rest in the day time by careful searching of tree-trunks and fences, where it closely resembles a flake of wood.

Distribution (Map 176)

Centred on the lower Severn valley. Locally common in Somerset, Gloucestershire, Worcestershire and Herefordshire, with outlying records from Devon, Monmouth and Warwickshire. The origin of individuals which occur from time to time in the eastern home counties is uncertain; they may be of a near-extinct indigenous population, or possibly immigrants from northern France or Belgium, but in view of the paucity of other immigrant species in April, proof by analogy would be difficult. Eurasiatic; in western Europe from Spain to the extreme south of the Netherlands, widespread, but rare in central and eastern Germany.

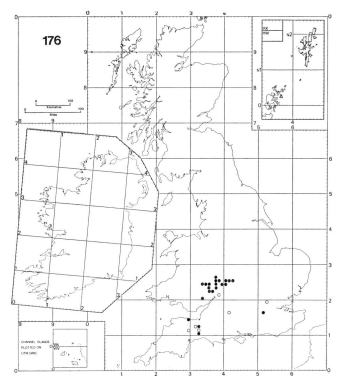

Egira conspicillaris

ORTHOSIA Ochsenheimer

Orthosia Ochsenheimer, 1816, *Schmett.Eur.* **4**: 79.
Orthoa Billberg, 1820, *Enum.Ins.Mus.Blbg*: 85.
Monima Hübner, [1821], *Verz.bekannt.Schmett.*: 229.
Cuphanoa Hübner, [1821], *ibid.*: 230.
Semiophora Stephens, 1829, *Ill.Br.Ent.* (Haust.) **2**: 138.
Taeniocampa Guenée, 1839, *Annls Soc.ent.Fr.* **8**: 477.

There are numerous Palaearctic species, of which nine are widely distributed in the British Isles; all appear from early spring onwards.

Imago. Wingspan 30–45mm. Frons flat; haustellum well developed; antenna variable, scape usually with hair-tuft; labial palpus long, descending. Thorax with crests. Abdomen without crests, short, rather flattened. Most species are very variable in colour and pattern.

Ovum. Usually laid in untidy masses on the foodplant or adjacent vegetation.

Larva. Elongate. Usually feeding on deciduous trees or shrubs.

Pupa. Rather variable; some species (*e.g. O. stabilis* ([Denis & Schiffermüller])) have a stout thorax, tapering rapidly, others such as *O. munda* ([Denis & Schiffermüller]) are more evenly proportioned overall; cremaster well developed with definite, although sometimes obscure, specific characters. Subterranean, in a cocoon.

Life history. Univoltine. The moths fly from early spring and feed at sallow and plum blossom or exudations of sap. As the imago is formed inside the pupa from early autumn, mild weather causes unusually early emergence from time to time.

Economic importance. All the common species of this genus contribute to early summer defoliation of forest trees; this, however, is rarely more than a temporary nuisance. *O. incerta* (Hufnagel) may damage fruit buds in apple orchards and seedlings in forest nurseries.

ORTHOSIA CRUDA ([Denis & Schiffermüller])
The Small Quaker

Noctua cruda [Denis & Schiffermüller], 1775, *Schmett. Wien.*: 77.
Bombyx pulverulenta Esper, 1786, *Schmett.* **3**: 386.
Type locality: [Austria]; Vienna district.

Description of imago (Pl.12, figs 48,49)

Wingspan 28–32mm. Antenna bipectinate, in male pectinations more than three times diameter of shaft, in female half diameter of shaft. Forewing pale greyish ochreous, variably overlaid reddish brown; basal line indicated by one or more black dots; ante- and postmedian lines indicated by black dots, seldom enough to show the full course of either; subterminal line of pale dots; upper stigmata variable, orbicular usually indistinctly darker, sometimes paler outlined; reniform stigma usually obscurely darker than ground colour, the dorsal lobe sometimes finely outlined with yellowish brown. Hindwing from pale fuscous to brownish fuscous; veins and sometimes discal spot slightly darker. Abdomen of female tapers to pointed ovipositor.

There seems to be no purely local variation, although melanic specimens (ab. *haggarti* Tutt) have been recorded rarely from south-west Scotland and Surrey.

Life history

Ovum. Flattened-hemispherical, lightly ribbed and reticulate; whitish at first, darkening quickly and becoming slate-grey immediately before hatching. Laid freely on gauze in a pillbox in captivity, although the shape of the ovipositor suggests that the eggs are normally inserted into crevices; hatching in about seven to ten days.

Larva. Full-fed *c.*30mm long, stout. Head and prothoracic plate black. Body from pale green to olive-green (occasionally brown), each segment darker anteriorly; dorsal and subdorsal lines complete, yellowish, with two black dots between each segment; spiracular line wide, usually fairly bright pale yellow; spiracles white, outlined black; ventral surface dull yellowish green.

The larva feeds mainly on oak (*Quercus* spp.) or rough-leaved willow (*Salix* spp.) but hazel (*Corylus avellana*), hawthorn (*Crataegus* spp.) and rose (*Rosa* spp.) are also recorded. In common with many other species which defoliate trees in May and June, the larva is reported to be an occasional cannibal. Full-fed in about the middle of June.

Pupa. Short and stout, dark brown; thoracic segments finely sculptured; anterior part of abdominal segments densely but finely punctate; cremaster short, with two strongly divergent apical spines arising close to each other. Below ground, in an earthen cocoon, the moth being fully formed and ready for emergence by late autumn.

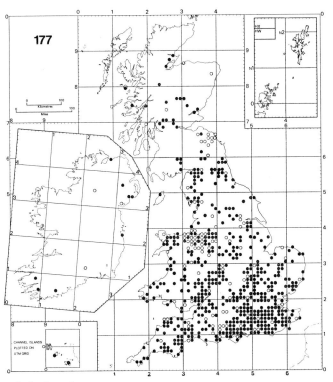

Orthosia cruda

Imago. Univoltine. Appears, exceptionally, in mild weather from December, but the main emergence usually takes place from the middle of March. In some years this is before the sallows are fully in bloom and, if so, it will then come to sugar; usually, however, the sallows are the greater attraction. Also comes to light.

Distribution (Map 177)

It is widely distributed in and near deciduous woodland throughout England and Wales; Scotland to Easter Ross, Argyll and the Inner Hebrides (Rhum); the Channel Islands. Recorded locally in Ireland. Eurasiatic; in Europe most common in the temperate areas, from Portugal, as far north as southern Sweden and just reaching Norway.

ORTHOSIA MINIOSA ([Denis & Schiffermüller])
The Blossom Underwing

Noctua miniosa [Denis & Schiffermüller], 1775, *Schmett. Wien.*: 88.

Type locality: [Austria]; Vienna district.

Description of imago (Pl.12, figs 50,51)

Wingspan 32–38mm. Antenna bipectinate, in male pectinations more than three times diameter of shaft, in female one and a half times diameter of shaft. Forewing greyish, fawn basally; median fascia reddish fawn, distal part slightly more reddish-tinged than basal area; ante- and postmedian lines greyish, paler-edged; subterminal line of dots and wedges unicolorous with median fascia; upper stigmata greyish, paler-outlined, reniform usually the more conspicuous; orbicular stigma coloured as median fascia or reduced to a small, dark spot. Hindwing whitish, delicately flushed roseate; discal spot grey-brown and clear; postmedian line greyish brown, clear near dorsum, attenuating towards termen; terminal line fine, grey-brown; occasionally a slight fuscous shade between postmedian and terminal lines.

The species varies less than most other members of the genus and mainly in the colour of the median fascia and the contrast between it and the adjacent areas; the degree of elongation of the components of the subterminal line also varies.

Similar species. *O. stabilis* ([Denis & Schiffermüller]) has subterminal line complete and pale; upper stigmata clearly outlined whitish; median fascia, when present, narrow and dark; hindwing fuscous.

Life history

Ovum. Flattened–hemispherical, ribbed and more lightly reticulate, the immediate vicinity of micropyle coarsely grained. Laid on oak twigs (*Quercus* spp.) in small batches of about 15, usually near to a leaf-bud, hatching in about 14 days.

Larva. Full-fed *c.*37mm long. Head pale brown, heavily spotted blackish brown. Body blue-grey, rarely pinkish brown; dorsal line primrose-yellow; subdorsal line, when present, similar, edged with black; dorsal line flanked by large black spots; rows of smaller black dots on each side of subdorsal line; spiracular line wide, yellow, spotted black; ventral surface pale pinkish brown.

The larvae live gregariously in a web when young, feeding on the young foliage of oak (*Quercus* spp.); when well grown they separate, some continuing to feed on oak leaves and the galls of *Neuroterus quercusbaccarum* (Linnaeus) and *Biorhiza pallida* (Olivier), while others descend to the ground and become polyphagous on low plants; growth is completed quickly by the middle of June. Larvae are said to turn cannibal if not provided with sufficiently succulent food.

Pupa. Rather stout, reddish brown; thoracic segments fairly heavily sculptured; abdominal segments rapidly tapering, with clear intersegmental divisions; cremaster sharply concave ventrally, with two stout, divergent apical spines. In a subterranean cocoon.

Imago. Univoltine. The moth flies from late March and comes to sallow blossom, sugar infrequently and light. The ecology of this moth is more specialized than that of most species of the genus; it prefers damp places and larvae are more often found on scrub oak in clearings and hedges than on mature, high-canopy oaks.

Distribution (Map 178)
Mainly southern and western in England, but locally common to Cumbria; local in Wales and rare in Ireland. Eurasiatic; in western Europe, Portugal and Spain, western, central and northern France, Belgium, the Netherlands, just reaching south Sweden and Norway.

ORTHOSIA OPIMA (Hübner)
The Northern Drab

Noctua opima Hübner, [1809], *Samml.eur.Schmett.* **4**: pl.90, fig.424.
Orthosia advena sensu auctt.
Type locality: Europe.

Description of imago (Pl.12, figs 52,53)
Wingspan 34–40mm. Antenna in male fasciculate, in female very shortly bipectinate. Forewing with pointed apex, emphasized by course of subterminal line; grey, variably tinged brown; ante- and postmedian lines indistinct; median fascia brown; subterminal line straight, whitish, darker-edged proximally and very distinct; upper stigmata may be partly or completely within the median fascia, visible only as paler outlines; terminal dots blackish. Hindwing greyish fuscous; terminal line fine, dark brown; cilia paler.

Ab. *brunnea* Tutt (fig.53) is as common as the typical form. Forewing almost unicolorous dark brown; median fascia only slightly darker; ante- and postmedian lines and outlining of upper stigmata variably paler; subterminal line as in the typical form. Hindwing brownish fuscous, cilia less pale than in the typical form. Although the typical form varies in the darkness of ground colour and in the width and degree of contrast of the median fascia, ab. *brunnea* may approach it in specimens with a paler basal area. Ab. *nigra* Lempke is even darker brown than ab. *brunnea*.

Similar species. *O. incerta* (Hufnagel). Some of the darker, more unicolorous forms of this species slightly resemble *O. opima* ab. *brunnea*. They differ as follows: male antenna bipectinate; apex of forewing not pointed; subterminal line

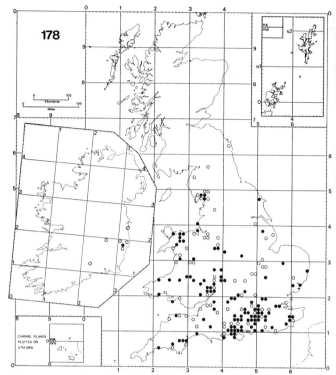

Orthosia miniosa

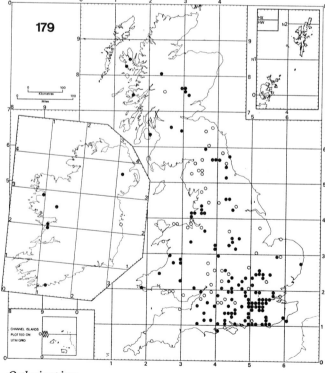

Orthosia opima

less conspicuous and less straight, particularly at extremities; cilia of hindwing pale fuscous.

Life history

Ovum. Rather conical, ribbed and finely reticulate; creamy-white at first and very conspicuous, gradually becoming purplish brown and finally leaden grey before hatching. Laid in large, untidy masses, often on dead herbage or withered grasses, hatching in about seven days.

Larva. Full-fed *c.*45mm long. Head light chestnut-brown. Body bright purplish brown; dorsal and subdorsal lines fine, paler; spiracular stripe broad, pale yellow; spiracles white, heavily outlined with black; ventral surface apple-green.

Feeds on a variety of plants, including various willows (*Salix* spp.), birch (*Betula* spp.), dyer's greenweed (*Genista tinctoria*), ragwort (*Senecio jacobaea*), hound's tongue (*Cynoglossum officinale*) and marram (*Ammophila arenaria*); a number of alternative foodplants, including garden plum (*Prunus* spp.), have been accepted in captivity. For much of its life the larva is olive-brown, freckled darker dorsally and yellowish green ventrally, the striking purple-brown, yellow and green colouring appearing only in the last instar; it is full-fed in late June or July. It has the reputation of being rather unpredictable and difficult to rear in captivity, sometimes feeding up rapidly without casualties, at other times, under apparently similar conditions, dying, usually when about half to three-quarters grown.

Pupa. Rather stout; rich dark brown; thoracic segments finely punctate; abdominal segments gradually tapering; segment 8 abruptly hollowed before a compact cremaster with two short, stout apical spines, which diverge at first, but in-curve at the tips. Below ground, in an earthen cocoon.

Imago. Univoltine. Flies from mid-April and feeds at such sallow bloom as remains and blackthorn blossom, as well as coming to sugared herbage and to light.

Distribution (Map 179)

Although local, its known habitats include sandhills, chalk downland and acid heathland; in Scotland in marshes. England, locally common to the Scottish border; Scotland, rare and local, mainly in the southern counties, but also in Mull and Skye; Wales, local; Ireland, rare, but well distributed. Eurasiatic; in western Europe, France, north-eastern and upland only, rare; local in Belgium and the Netherlands, but common in Denmark, south Sweden and inland, just reaching south Norway.

ORTHOSIA POPULETI (Fabricius)
The Lead-coloured Drab
Bombyx populeti Fabricius, 1781, *Spec.Ins.* **2**: 201.
Type locality: Norway.

Description of imago (Pl.12, figs 45–47)

Wingspan 34–40mm. Antenna bipectinate, in male strongly, in female very shortly. Forewing rather narrow, apex blunt, termen slightly rounded; grey-brown to purplish brown; cross-lines often indistinct, but any or all may be complete and slightly darker than ground colour; upper stigmata variable, sometimes obsolete, both, or only the reniform, outlined with pale yellowish grey; orbicular stigma only occasionally more conspicuous than reniform; their juxta-position varies and they may touch; often two or more black, sometimes elongate, spots before the subterminal line. Hindwing pale fuscous; discal spot obscure; cilia whitish.

Specimens from Scotland (fig.47) tend to be darker, more richly coloured and with more clearly contrasting markings.

Similar species. *O. incerta* (Hufnagel) (small unicolorous forms). Forewing more ample, termen not rounded; characteristic subterminal line straight except for indentations at each extremity; pectinations of male antenna less than twice diameter of shaft; underside, fore- and hindwings with distinct stigmata and median lines. *O. stabilis* ([Denis & Schiffermüller]). Subterminal line straight and conspicuously pale.

Life history

Ovum. Flattened-hemispherical, ribbed and more delicately reticulate; glossy greyish white at first, gradually darkening before hatching. Laid in small batches on the twigs of the foodplant, hatching in about eight days.

Larva. Full-fed *c.*40mm long, rather slender. Head brown or greenish, with a black spot on each side. Body very pale green, even lighter dorsally; dorsal line white and clear; subdorsal line also white, fine and rather indistinct; spiracles concolorous with ground colour, closely outlined black; supra- and subspiracular lines fine, pale and undulating; ventral surface concolorous with ground colour.

At first the larva feeds inside the catkins of aspen (*Populus tremula*) and occasionally other species of poplar (*Populus* spp.); later it spends the day between two spun leaves, usually high up, feeding nocturnally. Other fairly pale green larvae of somewhat similar appearance which inhabit aspen are *Ipimorpha subtusa* ([Denis & Schiffermüller]) (Amphipyrinae), which is brighter green and has the head, prothoracic plate and legs black, with many black dots on the body, living in a single folded leaf; and *Tethea or* ([Denis & Schiffermüller]) (Thyatiridae), which is stouter,

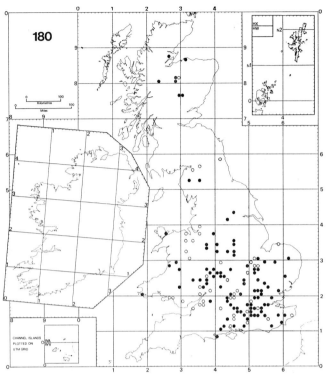

Orthosia populeti

with large, brown head and is still small when *Orthosia populeti* is full-fed in June.

Pupa. Thorax moderately stout, abdominal segments gradually tapering; reddish brown; cremaster rugose, sharply concave ventrally to a smooth papilla with two fine, tapering, slightly out-curving bristles. In a subterranean cocoon.

Imago. Univoltine. Occurs from early April and comes to sallow bloom and light, but rarely to sugar.

Distribution (Map 180)

Occurs mainly among sizeable stands of aspen, but numerous single records from individual light-traps suggest that it wanders widely. Local in southern England; more generally distributed in the north Midlands, although scarce in Yorkshire; in Scotland locally common to Inverness-shire and Easter Ross. Baynes (1964) rejects the Irish records quoted by Barrett and South. Eurasiatic; in western Europe, in France mainly central and northern, local in Belgium and Holland, common in Denmark, reaching the southern parts of Norway, Sweden and Finland to about 60°N.

ORTHOSIA GRACILIS ([Denis & Schiffermüller])
The Powdered Quaker

Noctua gracilis [Denis & Schiffermüller], 1775, *Schmett Wien.*: 76.

Type locality: [Austria]; Vienna district.

Description of imago (Pl.13, figs 2–6)

Wingspan 35–42mm. Antenna bipectinate, in male pectinations equal to diameter of shaft, in female more sparsely and less than half diameter of shaft. Forewing very pale greyish white, tinged ochreous, pale brown or pinkish often with many random black scales; there are also numerous very much darker local forms (see below) ante- and postmedian lines, when present, indicated by blackish dots; subterminal line complete, fairly straight usually conspicuously pale; upper stigmata vary and may be paler than adjacent area or only indicated by paler outlines reniform stigma usually darker than orbicular, its dorsal lobe sometimes darker than ground colour; claviform stigma absent or faintly outlined paler. Hindwing pale greyish white; discal spot fuscous; sometimes a median line of fuscous dots, which may be elongate; terminal shade fuscous, very variable in width.

There is extreme local variation; large parts of the populations in Scotland and Ireland, as well as isolated colonies in Wales and England (New Forest, Somerset and formerly Kent), producing specimens which are deep pink (ab. *rufescens* Tutt, fig.3), reddish brown to brown (ab. *brunnea* Tutt, fig.5) and even purplish-tinted. These dark forms usually occur on bogs, often side by side with colonies of normal colouring with which they do no mix as they are seven to ten days earlier in all stages. The larvae feed on bog-myrtle (*Myrica gale*) although this doe not seem to have been so in the now extinct Kentish colony. A most interesting characteristic of these dark *O. gracilis* is that those from each area show subtle differences which allow separation from superficially similar forms from elsewhere. Cockayne *in* Turner (1926–50, **2**: 222 224) analyses and gives a tabulation of these differences Whitish specimens (ab. *alba* Cockayne) have been recorded from the Fens and from Perthshire and a melanic specimen from Yorkshire.

Life history

Ovum. Hemispherical, coarsely ribbed and lightly reticulate to the shoulder, area of micropyle covered with irregular network; yellowish white at first, darkening to dull reddish and finally grey before hatching. Laid in large, untidy masses on twigs or withered vegetation, hatching in about ten days.

Larva. Full-fed *c.*42mm long, rather stout. Head an posterior part of the prothoracic plate brown, anterior part concolorous with ground colour. Body greenish brown or

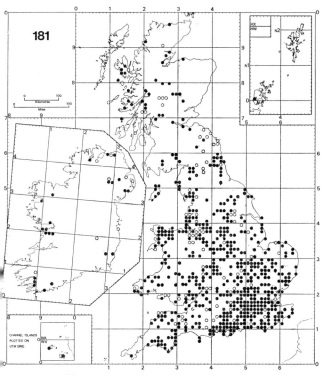

Orthosia gracilis

dark green; dorsal and subdorsal lines finer, paler, a whitish-ringed black pinaculum between them on each segment; spiracular line broad, of paler green or brown than ground colour, rather indistinct, heavily bordered with black above; spiracles white, ringed with black; ventral surface pale green.

Nocturnal in habit, the larva spends the day in the spun terminal shoots of its foodplants, among the more frequently recorded of which are willows (*Salix* spp.), bog-myrtle (*Myrica gale*), meadowsweet (*Filipendula ulmaria*), yellow loosestrife (*Lysimachia vulgaris*) and purple loosestrife (*Lythrum salicaria*). When too large to hide thus, it descends by day, but may be found still feeding well after dawn when it is most conspicuous (Cockayne, 1951). Hammond (1952) recommends *Lysimachia* as the best foodplant in captivity and suggests that in the absence of this, smooth-leaved willow should be offered. The species suffers heavily from parasitization, although this is probably local in incidence; there are contradictory accounts of its tendency to cannibalism, which it may well acquire only by necessity.

Pupa. Short and stout; dark red-brown; wing-cases coarsely sculptured, the veins showing clearly; abdominal segments each with an anterior band of fine punctation; cremaster small and blackish, with two divergent, well separated apical spines. In a cocoon below the surface of the ground.

Imago. Univoltine. Flies from late April, usually too late for sallow bloom except in Scotland, but comes to blackthorn or plum blossom as well as to sugar and light. The species is widely distributed, but the principal foodplants are all associated with marshland, and it is here that it is most common.

Distribution (Map 181)

Widely distributed and locally common in England, Wales and Ireland; in Scotland, where its larvae feed mainly on bog-myrtle, it is found locally to Sutherland and in the Inner Hebrides; Orkney (Hoy); the Channel Islands. Eurasiatic; in western Europe from Portugal to south Scandinavia and Finland.

ORTHOSIA STABILIS ([Denis & Schiffermüller])
The Common Quaker
Noctua stabilis [Denis & Schiffermüller], 1775, *Schmett. Wien.*: 76.

Type locality: [Austria]: Vienna district.

Description of imago (Pl.13, figs 7–12)

Wingspan 34–40mm. Antenna bipectinate, in male pectinations three times diameter of shaft, in female three-quarters diameter of shaft. Forewing variable in colour, from pale greyish ochreous through various shades of brown to sepia or to brick-red; albino and melanic specimens are known, but are very rare; antemedian line, when present, darker than ground colour, erratic and broken; postmedian line, when present, of darker dots; subterminal line complete, almost straight, yellowish white, bordered darker than ground colour proximally; upper stigmata large, usually clearly outlined pale yellowish white, as is the distal part of the claviform. Hindwing fuscous, paler in the forms with ochreous and reddish forewings than in those with brown forewings.

In ab. *fasciata* Lenz (fig.10) a narrow, ill-defined, darker median shade extends from the reniform stigma to the dorsum; in ab. *marginata* Cockayne the terminal area is paler than the rest of the forewing, the veins showing clearly whitish. Both forms occur fairly commonly throughout the range of colouring. The species varies similarly through its range and there is no purely local variation.

Similar species. O. *miniosa* ([Denis & Schiffermüller]) and O. *populeti* (Fabricius) *q.v.*

Life history

Ovum. Hemispherical, lightly ribbed and reticulate, a faintly dimpled area surrounding the micropyle; soft pale green at first, darkening rapidly and becoming leaden grey before hatching. Laid in rather untidy masses on the twigs of the foodplant, hatching in about ten days.

Larva. Full-fed *c.*38mm long, rather stout. Head large, pale green to blue-green. Body bright apple-green, thickly sprinkled with yellow dots; dorsal, subdorsal and spiracular lines yellow, the subdorsal fine; a transverse yellow anal bar; spiracles white, outlined black; intersegmental divisions yellow when the larva is full-fed.

Feeds on most deciduous trees, especially oak (*Quercus* spp.), rough-leaved willow (*Salix* spp.) and elm (*Ulmus* spp.), at first inside developing leaf-buds, later between spun terminal shoots and finally completely exposed. Full-fed by late June.

Pupa. Rather stout, thickest at the rear extremity of the wing-cases; dark reddish brown; thoracic segments finely sculptured; abdominal segments tapering sharply, minutely punctate, darker intersegmentally; spiracles large and prominent; cremaster blackish, concave ventrally, with two short, slightly divergent apical spines. Below the surface of the ground, in a cocoon, the moth being fully formed by late autumn.

Imago. Univoltine. Occasionally emerges during mild weather in mid-winter, but the main emergence is from mid-March. This species and *O. gothica* (Linnaeus) are the commonest of the visitors to sallow bloom, but the period of emergence lasts into May, and the later-emerging moths will come to blackthorn blossom and sugar.

Distribution (Map 182)
Although most abundant in and near woodland, the species seems to have no ecological limitations. Common throughout England, Wales, and southern Scotland; rather less so in the Inner Hebrides and the northern Scottish mainland, to Caithness; Ireland; the Isles of Scilly; the Channel Islands. Eurasiatic; through Europe to south Norway and Sweden; very rare in Finland.

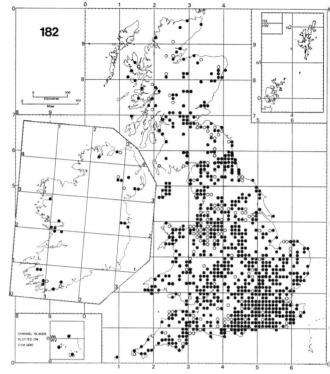

Orthosia stabilis

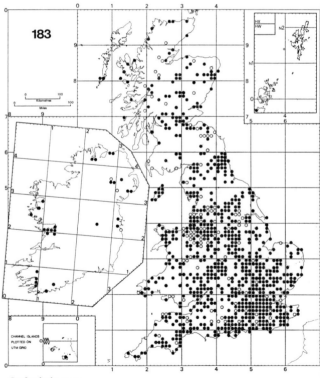

Orthosia incerta

ORTHOSIA INCERTA (Hufnagel)
The Clouded Drab

Phalaena incerta Hufnagel, 1766, *Berlin.Mag.* **3**: 298.
Noctua instabilis [Denis & Schiffermüller], 1775, *Schmett. Wien.*: 76.

Type locality: Germany; Berlin.

Description of imago (Pl.13, figs 14–20)

Wingspan 32–42mm. Antenna bipectinate, in male pectinations less than twice diameter of shaft, in female less than quarter diameter of shaft. Forewing very variable, from almost unicolorous blackish brown to bright rufous, pale grey or greyish ochreous, with varying degrees of lighter relief or dark brown to red-brown irroration: in the darker forms the upper stigmata are often outlined paler and the subterminal line is frequently very conspicuous; in paler forms the dorsal lobe of the reniform stigma is darker than the costal, the veins may show as dark longitudinal streaks and there is often a darker, angulate median fascia: the most constant character is the pale, often yellowish white, subterminal line, more or less straight except for abrupt concavities at each extremity; even this, however, is not always visible. Hindwing pale fuscous; lunular discal spot and veins fuscous; terminal fascia sometimes present, variable in width, fuscous.

The nomenclature of the forms of this species has been much bedevilled by synonyms and the short list below, in which the forewings are described, is based on the major categories into which the long series in BMNH are separated:

ab. *atra* Tutt, dull, unicolorous blackish brown;
ab. *melaleuca* Lenz, blackish brown, subterminal line and outlining of upper stigmata pale;
ab. *olivacea* Warren, pale grey, olive-tinged;
ab. *nebulosus* Haworth, darker grey, with rufous markings;
ab. *virgata-brunnea* Tutt, brownish grey with darker median fascia;
ab. *angustus* Haworth, unicolorous reddish black;
ab. *rufa* Tutt, unicolorous, extreme reddish.

Despite the polymorphism of this species, it is almost always recognizable on sight.

Variation is fairly constant throughout the range, with darker forms predominating in the south and east, paler forms in the north and west.

Similar species. *O. munda* ([Denis & Schiffermüller]) ab. *immaculata* Staudinger has pectinations of antenna about twice as long in both sexes; costa of forewing more strongly arched postmedially.

Life history

Ovum. Hemispherical with slightly puckered surface, lightly ribbed and reticulate; pale yellowish white at first, gradually darkening. Laid on the foodplant in irregular masses, hatching in about ten days.

Larva. Full-fed *c.*45mm long, elongate and cylindrical. Head yellowish brown. Body pale bluish green, thickly and minutely dotted white, thus having a frosted appearance; wide dorsal and fine subdorsal lines white; spiracular stripe broad, white, bordered with black above; ventral surface darker, less bluish green than dorsal and lateral surfaces.

Polyphagous on deciduous trees and shrubs, with a preference for oak (*Quercus* spp.) and rough-leaved willow (*Salix* spp.). Feeds at first in developing leaf-buds; later hides by day in spun terminal shoots and feeds by night; when approaching full growth feeds fully exposed, but well camouflaged from above or below the foodplant. Full-fed in June.

Pupa. More elongate than most members of the genus with evenly tapering abdominal segments; dark red-brown and glossy; thoracic segments finely sculptured; abdominal segments with narrow anterior band of heavy punctation; cremaster hollowed ventrally, small, with two very short, non-tapering, slightly divergent, apical spines. Below ground, in a very fragile cocoon, the moth being formed by early autumn.

Imago. Univoltine. Exceptionally, individuals may emerge between October and February, but the main emergence is in late March. It is more attracted to natural food – sallow bloom or birch sap – than to sugar, but is attracted to light.

Distribution (Map 183)

Although mainly a woodland species, it is very widely distributed. Curtis (1934) comments on its inexplicable local scarcity in certain apparently suitable areas. Common over most of mainland Britain to north Caithness; Isle of Man; Inner and Outer Hebrides; the Channel Islands; Ireland. Eurasiatic; through western Europe to south Norway, central Sweden and Finland.

ORTHOSIA MUNDA ([Denis & Schiffermüller])
The Twin-spotted Quaker
Noctua munda [Denis & Schiffermüller], 1775, *Schmett. Wien.*: 76.
Type locality: [Austria]; Vienna district.

Description of imago (Pl.13, figs 29–31)
Wingspan 38–44mm. Antenna bipectinate, in male pectinations more than three times diameter of shaft, in female one and a quarter times diameter of shaft. Forewing pale greyish ochreous to reddish ochreous; usually a black subbasal dot near costa, rarely a complete subbasal line; ante- and postmedian lines obsolete, indicated by dots or complete but inconspicuous; subterminal line usually reduced to two black median dots, but these may be obsolete (ab. *immaculata* Staudinger (fig.30)), be joined by others nearer to costa and dorsum, or, rarely, form part of a complete row; as their number increases, the central pair tend to elongate; upper stigmata variable, usually outlined paler, reniform clearer than orbicular. Hindwing fuscous, paler basally, with darker discal spot; fine brownish terminal line and pale cilia.
  There seems to be no purely local variation.
Similar species. O. *incerta* (Hufnagel) *q.v.*

Life history
Ovum. Hemispherical, ribbed and reticulate; creamy white at first, gradually darkening to leaden grey. Laid on the foodplant in untidy masses, hatching in about ten days.
Larva. Full-fed *c.*50mm long. Head light brown. Prothoracic plate pale grey-brown with a black transverse bar which is broken by paler dorsal and subdorsal lines. Body pale brown, with darker freckling, especially on the anterior part of each abdominal segment; dorsal line yellowish, lined brown; subdorsal line undulating, pale but inconspicuous; spiracular stripe broad, yellowish brown, bordered above by a thick, undulating black line; spiracles pale brown, outlined black.
  Feeds on a wide variety of plants, including rough-leaved willow (*Salix* spp.), elm (*Ulmus* spp.), oak (*Quercus* spp.), blackthorn (*Prunus spinosa*), aspen (*Populus tremula*), hop (*Humulus lupulus*) and honeysuckle (*Lonicera periclymenum*), preferring young, tender foliage; it has the reputation of being a cannibal. Full-fed in mid-June.
Pupa. Thoracic segments not very stout, abdominal segments evenly tapering; reddish brown and glossy; eyes and spiracles conspicuous; thoracic segments finely sculptured ventrally, punctate dorsally; abdominal segments punctate, more heavily on the anterior part of each; cremaster small, evenly rounded, with two short, very divergent spines which originate close together. In a subterranean cocoon.

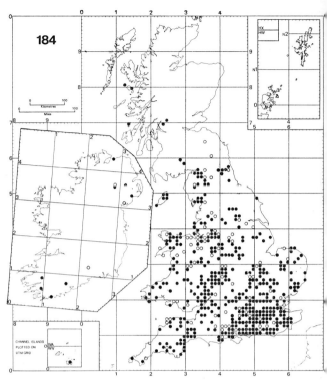

Orthosia munda

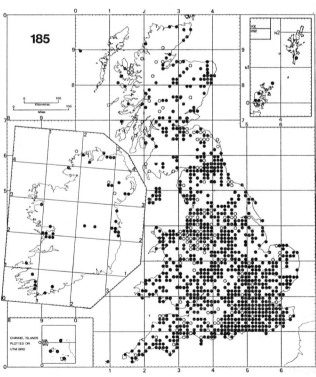

Orthosia gothica

Imago. Univoltine. Like some other members of the genus, individuals may emerge in mild weather during winter, but the main emergence coincides with the flowering of the sallows. As well as feeding at sallow bloom, this species comes to sugar fairly freely.

Distribution (Map 184)

Common and well distributed, especially in deciduous woodland, in the southern and midland counties of England and in Wales; becoming progressively less so farther north, but found in the Isle of Man, southern Scotland to the Forth-Clyde line and recorded from Rhum and Canna in the Inner Hebrides; the Channel Islands. Uncommon in Ireland. Eurasiatic; in Europe, northern France, Belgium, the Netherlands, Denmark, southern Sweden, and the U.S.S.R.

ORTHOSIA GOTHICA (Linnaeus)
The Hebrew Character
Phalaena (Noctua) gothica Linnaeus, 1758, *Syst.Nat.* (Edn 10) **1**: 516.
Type locality: Europe.

Description of imago (Pl.13, figs 21–28)

Wingspan 30–40mm. Antenna bipectinate, in male pectinations three times diameter of shaft, in female equal to diameter of shaft. Forewing purplish brown to reddish brown, usually tinged grey; subbasal line incomplete, black; antemedian, postmedian and subterminal lines pale greyish, the last often broadly bordered paler distally; upper stigmata paler than ground colour, the dorsal part of orbicular and the costal lobe of reniform often outlined with pale yellow; costal half of median area largely filled by a complex of black markings from which the vernacular name is derived; at its most developed it is a rectangular block, excavated costally by the orbicular stigma, clearly separate from the heavy, black claviform stigma which reaches the postmedian line; the shape and relative proportions of the median design are extremely variable. Hindwing pale fuscous to fuscous, discal spot and veins obscurely darker; cilia whitish.

In ab. *gothicina* Herrich-Schäffer (figs 25,26) the forewing is reddish brown without grey tint and the median pattern is paler than the ground colour. Hindwing whitish, sometimes with clear pale fuscous ante- and postmedian lines as well as discal spot and veins. The name is in general use for a number of forms of varying ground colour in which the median pattern is pale rather than dark. Tutt (1892) named this reversal of the normal pattern ab. *obsoleta* adding the individual ground colour variation as a suffix. Such forms are most common on moorland in the north and west including Ireland, but are found locally farther south and even on calcareous soils in Hampshire (Goater, 1974). Local variation is restricted to the presence or absence of *gothicina/obsoleta* forms, but the moth shows great and constant variation in markings throughout its range. Melanic and albino forms occur only rarely.

Life history

Ovum. Hemispherical, lightly ribbed and reticulate, a slightly dimpled band surrounding the micropyle; usually pale greenish white at first, occasionally ochreous, darkening rapidly. Laid in untidy masses on almost any vegetation, withered or living, hatching in about ten days.

Larva. Full-fed *c.*50mm long, fairly stout, slightly tapering towards the head, intersegmental divisions very clear. Head pale green. Body yellowish green to bright green; dorsal and subdorsal lines white, the former distinct; spiracular line wide, from chalky white to only slightly paler than ground colour, bordered with black above; spiracles white, outlined black.

Polyphagous, preferring deciduous trees and shrubs, but well able to live on herbaceous plants such as dock (*Rumex* spp.), clover (*Trifolium* spp.) and common nettle (*Urtica dioica*); on the northern moors meadowsweet (*Filipendula ulmaria*), especially the flower-buds, is a favourite foodplant. Full-grown by mid-June (over a month later in Ireland and northern Scotland).

Pupa. Thorax stout, abdominal segments tapering rapidly; dark reddish brown; wing-cases finely sculptured; anterior part of each abdominal segment finely punctate; cremaster black and conical, with two divergent, pointed spines. Below ground, in a fragile cocoon.

Imago. Univoltine. Individuals may emerge in mild weather during winter, but the main emergence is in late March and April, later in Scotland and Ireland.

Distribution (Map 185)

There seems to be no ecological preference and the species is found in most types of habitat throughout its range. Great Britain and its off-shore islands (including St. Kilda) to Shetland; the Channel Islands. Common throughout Ireland, where it is said to fly much later, even to mid-June. The exact frequency of ab. *gothicina sensu lato* is hard to assess, owing to the tendency of lepidopterists to note its presence, but ignore the normal specimens occurring with it; Vine Hall (1962) shows that 6 per cent of a large sample of specimens from the Spey valley were of this form. Eurasiatic; throughout western Europe to the Arctic.

MYTHIMNA Ochsenheimer

Mythimna Ochsenheimer, 1816, *Schmett.Eur.* **4**: 78.
Leucania Ochsenheimer, 1816, *ibid.*, **4**: 81.
Philostola Billberg, 1820, *Enum.Ins.Mus.Blbg*: 87.
Hyphilare Hübner, [1821], *Verz.bekannt.Schmett.*: 239.

A large genus, with numerous Holarctic species, represented by 13 species resident in Great Britain and eight also in Ireland, with four others which are immigrant and may, perhaps, become temporarily established.

Imago. Medium size. Frons flat; haustellum well developed; labial palpus thickly scaled, the terminal segment short and thick. Thoracic crests sometimes present. Legs almost hairless, thinly scaled. Forewing in the majority of species with the cross-lines and stigmata reduced or absent. Several species with ventral tufts at the base of the first abdominal segment. Female genitalia with extensile anal papillae for insertion into leaf-axils of Gramineae.

Ovum. In most species almost spherical, with only light sculpturing. Laid in neat rows in the axils or between blades of Gramineae.

Larva. Cylindrical, usually pale brown with longitudinal lines of various subdued colours; feeding on Gramineae, most species hibernating while small.

Pupa. Slender to extremely slender; cremaster rather weakly developed and of little value in identification. Usually subterranean, in a cocoon.

Life history. Variable. Univoltine, bivoltine or partly bivoltine. Winter passed as larva, usually small, but sometimes in a cocoon, pupating only in spring.

Economic importance. The immigrant species *M. unipuncta* (Haworth) and *M. loreyi* (Duponchel) cause serious damage to cereal crops in the tropics.

MYTHIMNA TURCA (Linnaeus)
The Double Line

Phalaena (Noctua) turca Linnaeus, 1761, *Fauna Suecica* (Edn 2): 322.
Type locality: Sweden.

Description of imago (Pl.13, fig.13)

Wingspan 44–52mm. Antenna in male very finely ciliate, in female very fine, simple. Male with large tuft of red-brown hairs on hindtibia. Forewing thickly scaled, ochreous brown to red-brown, lightly irrorate with black scales; subbasal line incomplete, black, or obsolete; ante- and postmedian lines grey, widely separated and variable in course, the former usually slightly angulate, the latter fairly straight; reniform stigma a short, fine, vertical white line, edged completely or only distally with grey-brown; other stigmata absent; terminal line fine and faint, grey-brown. Hindwing dark grey-brown, suffused ochreous, reddish or purple; cilia crimson to purple.

Although the basic pattern of the forewing is simple, there is much variation in the ground colour and in the course of the ante- and postmedian lines. Tutt (1891: 33) notes that the clarity of the lines is in inverse proportion to the darkness of the ground colour.

Life history

Ovum. Rather elongate; white and glossy, very smooth in texture. Laid in rows side by side in folded blades of the foodplant.

Larva. Full-fed *c.*48mm long, cylindrical and stout. Head rather large, pale brown; posterior abdominal segments tapering abruptly. Body rich ochreous, darker dorsally; intersegmental divisions paler; dorsal line fine and pale; subdorsal line very pale and inconspicuous; above it a reddish brown shading forms X-shaped designs on the abdominal segments; spiracles pinkish, outlined black; although there is no clear spiracular line, the colour becomes pale ochreous at that level.

The larva is nocturnal, feeding on wood-rush (*Luzula* spp.), cock's-foot (*Dactylis glomerata*) or wood meadow-grass (*Poa nemoralis*). It can be induced by constant warmth to complete feeding and pupate in the early autumn in captivity but in the wild it overwinters when small and resumes feeding in about the middle of April. Full-fed in May.

Pupa. Robust; glossy red-brown; thorax stout, with well-defined eyes, wing-cases and appendages; abdominal segments evenly tapering from 6, finely punctate, spiracles conspicuous; cremaster evenly rounded, without dorsolateral bristles but with two medium-sized, parallel, central bristles, which taper to an apical hook.

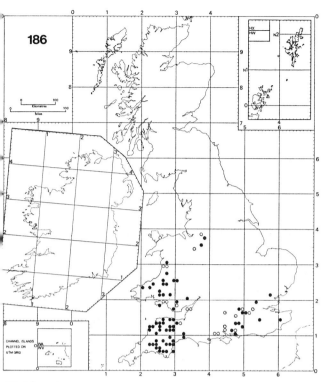

Mythimna turca

Imago. Univoltine. The moth flies from mid-June to mid-July. It is rarely seen by day, but comes to sugar and light.

Distribution (Map 186)

Almost entirely a woodland species which seems to have vanished from many of its old haunts in the eastern part of southern England and is now found most commonly in suitable areas of Cornwall, Devonshire and Somerset, south-west and central Wales and Cheshire. It occurs less commonly in the home counties and the central Chilterns. Eurasiatic; widely spread through western Europe from Spain to the Danish islands and the extreme south of Sweden, but always local.

MYTHIMNA CONIGERA ([Denis & Schiffermüller])
The Brown-line Bright-eye

Noctua conigera [Denis & Schiffermüller], 1775, *Schmett. Wien.*: 84.

Type locality: [Austria]; Vienna district.

Description of imago (Pl.13, figs 34,35)

Wingspan 32–38mm. Antenna setose-ciliate, in female more shortly than in male. Forewing with pointed apex; from pale golden brown to dark brownish orange (ab. *suffusa* Tutt (fig.35)); termen, costa and immediate vicinity of reniform stigma darker than remainder of wing; antemedian line sharply angulate; postmedian line straight, close to and more or less parallel to termen, both reddish brown; veins conspicuous between postmedian and fine, dark terminal line; orbicular stigma obscure, pale fulvous in typical form, suffused orange in ab. *suffusa*; reniform stigma elongate S-shaped, costal half coloured as orbicular, dorsal half white, variable in shape. Hindwing pale fuscous to fuscous, postmedian line slightly darker; terminal line fine, brownish; cilia concolorous with wing-base. Anal tuft of male golden brown.

Life history

Ovum. Hemispherical, whitish and glossy, with faintly sculptured reticulation, which is visible only in good light from an oblique angle. Laid side by side in single or double rows in folded blades of grass, hatching in about 14 days.

Larva. Full-fed c.40mm long. Head light brown with two blackish vertical lines. Body ochreous; dorsal line complete, whitish, edged grey; subdorsal line blackish, rather thick, broken at the posterior end of each segment, thinly lined yellowish below; supraspiracular line wide, pale ochreous, thinly edged with black; subspiracular line dull yellowish; ventral surface slightly darker. Very similar to the larva of *M. ferrago* (Fabricius); both vary, but in general *M. conigera* is smaller, somewhat paler in ground colour, with clearer vertical lines on the head.

Feeds on grasses, especially cock's-foot (*Dactylis glomerata*) and common couch (*Agropyron repens*) but will accept annual meadow-grass (*Poa annua*) in captivity. Feeds slowly until autumn and overwinters while still small, but feeds up quickly in the late spring.

Pupa. Rather slender; reddish brown and glossy; thorax finely sculptured; abdominal segments evenly tapering, more heavily sculptured on anterior half; cremaster swollen basally, concave terminally, with several fine dorsolateral bristles and two short, straight, blackish apical spines. Just below the surface of the soil, in a cocoon of silk and earth.

Imago. Univoltine. Flies from mid-June, later in the north. The moth is an early dusk visitor to such flowers as privet, wood-sage, campion, ragwort and red valerian, and also comes to sugar and light.

Distribution (Map 187)
Although without strong ecological preferences, the species is probably most common at the edges of woods, in clearings and in wooded lanes. Generally common south of a line from the Mersey to the Humber; locally common farther north to Caithness; recorded also from the Isle of Man, Inner Hebrides and Orkney. Widely distributed in southern Ireland; the Channel Islands. Eurasiatic; widely distributed and usually common in western Europe from Spain to central Scandinavia.

MYTHIMNA FERRAGO (Fabricius)
The Clay
Noctua ferrago Fabricius, 1787, *Mant.Ins.* **2**: 160.
Phalaena (Noctua) lythargyria Esper, 1788, *Schmett.* **4**: 341.
Type locality: Germany; Kiel.

Description of imago (Pl.13, figs 36,37)
Wingspan 36–44mm. Antenna in male setose-ciliate, in female very shortly setose-ciliate. Forewing rather narrow; ochreous grey (ab. *grisea* Haworth) to rusty red-brown, variably sprinkled black; ante- and postmedian lines indicated by black dots, the latter more completely; sometimes a short line of dots from the reniform stigma to the dorsum; a row of black terminal dots, followed by a fine, ochreous terminal line; orbicular stigma obscure; reniform stigma elongate, costal half pale ochreous, indistinct, dorsal half a whitish dot or short vertical line; often a slightly darker apical shade or darker suffusion along vein 6 (M_1) from base to reniform stigma. Hindwing from fuscous to dark grey; cilia ochreous, sometimes pink-tinged, often with a ferruginous central line. Abdomen of male with black ventral tuft on first segment.

Similar species. *M. albipuncta* ([Denis & Schiffermüller]) is smaller with the forewing less elongate; dorsal half of reniform stigma white, not whitish, and more completely circular than in *M. ferrago*; ante- and postmedian lines complete, obscurely paler than ground colour; hindwing whitish to pale fuscous.

Life history
Ovum. Spherical, covered with very fine, reticulate sculpturing; whitish and glossy. Laid side by side in single or double rows in a folded grass blade (often withered outer growth) or inserted into the leaf-axils; time of hatching variable, but may be as little as five days.
Larva. Full-fed *c.*48mm long. Very similar to that of *M. conigera* ([Denis & Schiffermüller]) but larger when full-fed, rather darker in colour and with more heavily

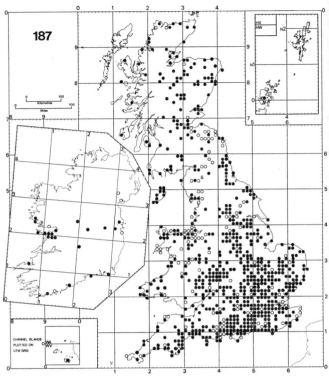

Mythimna conigera

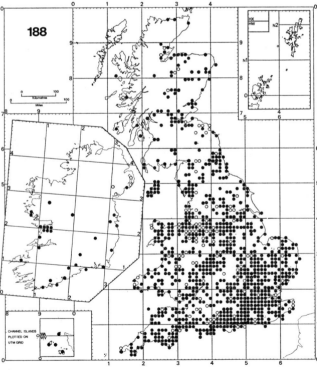

Mythimna ferrago

marked subdorsal line; as both species vary, it is difficult to separate them in the field.

The larva feeds slowly until hibernation and although grasses are the principal food, it also feeds on common chickweed (*Stellaria media*), dandelion (*Taraxacum officinale*) and plantain (*Plantago* spp.). After hibernation growth is rapid and it is full-fed by the middle of May.

Pupa. Slender; reddish brown and glossy; thorax finely sculptured, with prominent eyes, wing-cases and appendages; abdominal segments slightly yellowish-tinged; spiracles clear; an anterior band of fine punctation on each segment; segment 8 finely scobinate and concave ventrally to a rounded black cremaster with four fine subapical bristles and two short, parallel, apically hooked spines. In a cocoon of silk and earth, below the surface of the soil.

Imago. Univoltine, although a few individuals may emerge in early autumn. Flies from late June to early August and comes to flowers, honey-dew, sugar and light.

Distribution (Map 188)

It thrives in most habitats, but is less common near the coast than inland, and most abundant in and near woodland. Common over most of England, Wales and Ireland; Scotland to Argyll and the Inner Hebrides in the west, but to Caithness in the east; so far not recorded from the Outer Hebrides or northern islands; present in the Channel Islands. Eurasiatic; in western Europe generally common from Portugal to southern Scandinavia.

MYTHIMNA ALBIPUNCTA ([Denis & Schiffermüller])
The White-point

Noctua albipuncta [Denis & Schiffermüller], 1775, *Schmett. Wien*.: 84.

Type locality: [Austria]; Vienna district.

Description of imago (Pl.13, figs 39,40)

Wingspan 34–38mm. Antenna setose-ciliate, more shortly in female than male. Forewing varies from pale ochreous to red-brown; markings obscure except for a conspicuous rounded white dot in the reniform stigma; paler ante- and postmedian lines and a short oblique apical streak obscurely indicated. Hindwing grey-white, outwardly suffused darker. Abdomen of male with ventral basal tuft of dark hairs.

Ab. *grisea* Tutt, thought to be an old, pre-industrial melanic, is said to form one half of some broods (Kettlewell, 1973: 341).

Similar species. *M. ferrago* (Fabricius) *q.v.* is usually larger, has the reniform spot less rounded and the subterminal fascia clearly dotted. *M. unipuncta* (Haworth) is also larger and has the forewing more pointed with a dark oblique apical streak.

Life history

Ovum. Round, glossy; pale yellow at first, brown later. Laid in seed-heads of cock's-foot (*Dactylis glomerata*) and other grasses.

Larva. Full-fed *c*.38mm long. Body colour varies from pinkish brown to putty, with abdominal rings rather shorter than they are broad; a pair of dusky lines bordering the narrow white dorsal line; narrow subdorsal lines, with black dashes etched rather than streaked. In early instars the subdorsal lines are usually absent. Described and figured by Haggett (1963: 195–197, pl.11, figs 1–3).

The natural life-cycle in the wild is uncertain; but a presumably hibernated larva found on 4 April 1937 produced a moth on 11 July (Chalmers-Hunt, 1964, **2**: 195). In captivity moths can be reared from the egg in two to three months, according to temperature.

Pupa. Reddish brown. Subterranean, in a fragile cocoon.

Imago. Immigrant, varying greatly in numbers; probably also breeding from early arrivals and possibly establishing itself intermittently for successive years on warm shingle beaches and cliffs. The species was relatively common in 1938, 1948–51 (about 200 reported in 1950), and 1962, since when it has been scarce; most numerous from late August to October, with a few in most years in June and July. Moths are most often found at sugar or light, occasionally on grass-heads late at night.

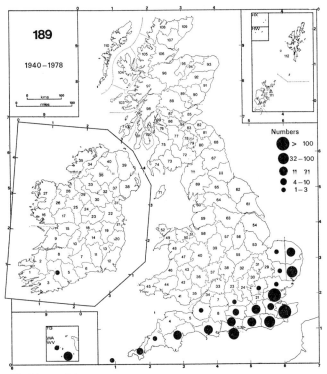
Mythimna albipuncta

MYTHIMNA VITELLINA (Hübner)
The Delicate
Noctua vitellina Hübner, [1808], *Samml.eur.Schmett.* 4: pl.81, fig.379.
Type locality: [Europe].

Description of imago (Pl.13, figs 32,33)

Wingspan 36–43mm. Antenna shortly setose-ciliate, very shortly ciliate in female. Forewing ground colour varies from pale straw (f. *pallida* Warren, fig. 33) to rich orange-red (f. *saturatior* Dannehl), with three wavy darker fasciae; orbicular and reniform stigmata also slightly darker, the latter sometimes pinkish above and with a small black dot at its base. Hindwing pale greyish white, suffused darker on veins, especially in female The full range of variation is figured by Haggett (1964: 78–92, pl.5).

Life history

Ovum. Hemispherical, with a wide flat base; glossy, pale yellow at first, darkening later (figured by Stokoe & Stovin, 1958, 1: pl.67). Laid in close, flat rows inside the sheaths of cock's-foot (*Dactylis glomerata*) and other grasses; in captivity indoors hatching after ten days.

Larva. Full-fed *c*.37mm long. Head large, honey-coloured with two curved dark stripes. Body in early instars either dark green with weak lines, or putty-coloured; in later instars colour very variable, from grey-green through sandy to rosy purplish; dorsal lines double, interrupted by dark segmental rings; subdorsal lines slender, edged with white; ochreous grey ventrally (Haggett, 1963: 191–193, pl.10, figs 3–7).

The larva is hardly known wild in England, but it has often been bred from the egg in captivity. Feeds nocturnally on many grasses, retiring at once into the surface debris if exposed to light; a cannibal in its first instar. Larvae which hatched 28–31 October and were reared indoors at 15–20°C (60–70°F) developed at varying rates, pupated after five to six weeks, and produced moths between 20 December and 10 January; some which were kept at 10–15°C (50–60°F) in their last instar and pupal stage emerged as much darker moths 4–7 February, but about one third died.

Pupa. *C*.18mm long, pale reddish brown, with segmental rings and wings darker; head blunt and squarish; cremaster rounded, with two points. In a slight cocoon among grass roots, on or just below the soil surface.

Imago. A frequent immigrant, probably breeding from early arrivals, and possibly surviving mild winters as a larva in sheltered places near the south-west coast. First recorded at Brighton in 1856; since 1945 it has been most common in 1953, 1961, 1962 (250 reported), 1967, 1976 to 1978, but not recorded at all in 1957, or in 1963 following a hard winter. The moth frequents grassy places on cliffs and

Distribution (Map 189)

Recorded in all coastal counties from Cornwall to Norfolk, and occasionally inland to Berkshire, Surrey and Hertfordshire, but mainly easterly, especially in the Isle of Wight, east Kent, Essex and Suffolk. The species was first reported in Ireland in 1973; possibly established in Jersey. Abroad it is Mediterranean-Asiatic; in western Europe, widespread both coastally and inland in Portugal, Spain, France and Belgium, resident in the south Netherlands, but only a very rare immigrant in Denmark and south Sweden.

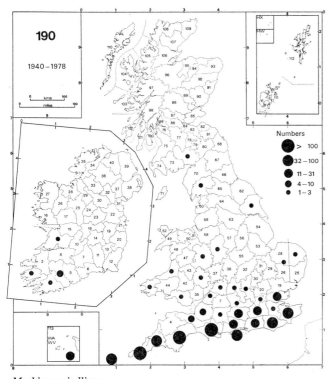
Mythimna vitellina

shingle beaches; it comes to light, but more readily to sugar and to flowers of wood-sage. Most specimens have been reported between late August and early October, but about 5 per cent of the total from late May to July, and a few in November. Cockayne (1936b: 25–28) and Haggett (1964) have shown experimentally that pale moths result from rapid development in warmth and darker forms from exposure to lower temperatures, around 15°C (60°F) or less, especially in the pupal stage; genetic factors may also be involved. At still lower temperatures young larvae will not begin to feed, and such partly grown larvae as survive produce moths from which only infertile eggs are obtained. It is therefore believed that f. *pallida* is a primary immigrant, while some at least of the darker moths found on the coast in autumn may be its locally-bred descendants. The frequent recurrence of moths in successive years in some coastal localities suggests that there may be occasional winter survival; but the results of low temperature rearing in captivity point strongly against this. Moreover, many of the moths recorded in May, June and July can be associated in date and place with arrivals of other immigrant species, especially *Spodoptera exigua* (Hübner) (Amphipyrinae), *Rhodometra sacraria* (Linnaeus) and *Orthonama obstipata* (Fabricius) (Geometridae), winter survival of which in Britain is still more unlikely.

Distribution (Map 190)

In contrast to *Mythimna albipuncta* ([Denis & Schiffermüller]), mainly south-western; three-quarters of the records since 1945 are from Cornwall, Devon, Dorset and Somerset. It occurs, however, sometimes commonly, eastwards near the coast as far as Kent (especially at Dungeness) and Essex, and occasionally inland as far as Herefordshire, Shropshire and Warwickshire. There are single records from Cumbria and, in 1978, from south Lincolnshire and Yorkshire. Irish records are few and confined to the south-west and Co. Clare. Mediterranean-Asiatic and subtropical; it occurs in the Canary Islands, Morocco, Portugal and Spain; in France it is widespread from the south up the Atlantic coast to Britanny, but is rare farther north, as in Belgium; in the Netherlands it was first seen in 1924, but not often since; it is not known in Denmark or Fennoscandia.

MYTHIMNA PUDORINA ([Denis & Schiffermüller])
The Striped Wainscot

Noctua pudorina [Denis & Schiffermüller], 1775, *Schmett. Wien.*: 85.

Noctua impudens Hübner, [1803], *Samml.eur.Schmett.* 4: pl.47, fig.329.

Type locality: Germany; Saxony.

Description of imago (Pl.13, fig.38)

Wingspan 36–43mm. Antenna in both sexes shortly and densely setose-ciliate, in female the cilia very short indeed. Forewing ample, apex rounded, termen rather convex; of satiny texture; whitish ochreous, suffused roseate near costa and dorsum, slightly darker near termen; many random black scales, more in female than in male; no line visible; sometimes a small median subcostal dot represents the reniform stigma; veins paler, showing more clearly in distal part of wing. Hindwing greyish, tinged with pink except near dorsum, which is pale fuscous; cilia pinkish ochreous.

Variation lies mainly in the amount and distribution of the black scales which sometimes form short lines or streaks. Pale specimens often prove merely to be worn; moths become damaged quickly and are difficult to take in really good condition. No other 'wainscot' closely resembles *M. pudorina*; it can always be distinguished by the wing-shape and texture.

Life history

Ovum. Hatches in about ten days.

Larva. Full-fed c.45mm long, stout. Head small, reddish brown. Body yellowish brown; dorsal line clear, whitish, obscurely edged greyish; subdorsal line black and thick, unbroken or almost so, bordered white below; lateral stripes fine, only slightly paler than ground colour; spiracles small, black; ventral surface greyish brown.

The larva overwinters when very small. Wightman (1943) lists a number of natural foodplants, including common reed (*Phragmites australis*), hairy wood-rush (*Luzula pilosa*), purple moor-grass (*Molinia caerulea*), flea-sedge (*Carex pulicaris*) and cock's-foot (*Dactylis glomerata*); he mentions also that the larva does not climb to feed but lies prone on the surface of the ground, raising only the anterior segments. Full-fed in early May.

Pupa. Slender; red-brown; cremaster with four fine subapical bristles and two long, closely based apical spines. On the surface of the soil, among vegetable litter.

Imago. Univoltine. Flies from late June. Rarely seen by day, but feeds after dark at flowering grasses such as floating sweet-grass (*Glyceria fluitans*) and will come to sugared reeds and grasses, as well as being attracted abundantly to light.

Distribution (Map 191)

The species inhabits marshes, fens and the wetter parts of acid moorland, but wanders for short distances. Southern England, East Anglia and locally to Yorkshire; Wales, mainly in the south; Ireland, in the south-west. Eurasiatic; in western Europe from Spain to Denmark, the extreme south of Sweden and the Baltic islands; local in damp woodland.

MYTHIMNA STRAMINEA (Treitschke)
The Southern Wainscot

Leucania straminea Treitschke, 1825, *Schmett.Eur.* 5 (2): 297.

Type locality: Germany; Darmstadt.

Description of imago (Pl.13, figs 50,51)

Wingspan 32–40mm. Antenna setose-ciliate, in male more densely. Forewing with apex pointed and termen straight; whitish ochreous, with varying amounts of black scaling, mainly in basal and median areas; veins clear, whitish with fine reddish brown interneural streaks giving a finely striate appearance; postmedian line of black dots, often reduced to only two; a row of minute black terminal dots; reniform stigma a faint, pale subcostal suffusion; a reddish brown basal streak may reach the termen. Hindwing off-white, suffused greyish near dorsum; discal spot and postmedian line of ill-defined greyish fuscous dots; veins slightly darker than ground colour. Underside with a clear discal spot on each wing. Abdomen of male with conspicuous anal tuft.

There is relatively little variation, but specimens suffused reddish (ab. *rufolinea* Tutt) or black (ab. *nigrostriata* Tutt) have been found in Kent.

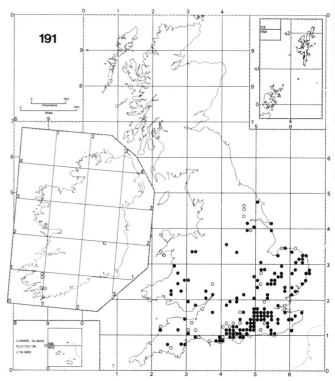

Mythimna pudorina

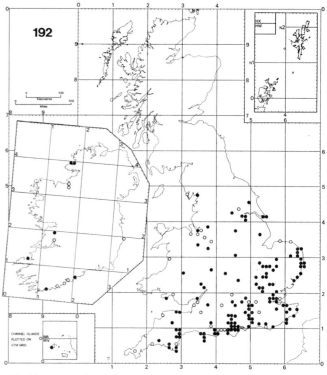

Mythimna straminea

Similar species. M. pallens (Linnaeus) lacks the straight termen, pointed apex and row of postmedian dots on the forewing, the postmedian line of dots on the hindwing and the underside discal spots. *M. obsoleta* (Hübner) has a more narrow forewing and an even more pointed apex; subterminal line of dots usually complete, parallel to termen; hindwing veins are more conspicuously dark; postmedian line absent; the underside is immaculate. *M. litoralis* (Curtis), *M. putrescens* (Hübner) and *M. loreyi* (Duponchel) all have pure white hindwings, as does the unrelated but not dissimilar *Simyra albovenosa* (Goeze) (Acronictinae). *Mythimna impura* (Hübner) underside has, at most, ill-defined discal spots on either wing.

Life history
Ovum. Globular, with strongly flattened base, upper surface reticulate; reddish gold. Laid in rows on leaves of *Phragmites* and *Phalaris* spp.
Larva. Full-fed *c.*45mm long, of smooth texture, more slender than most of the genus. Head yellowish brown. Body orange-ochreous; dorsal line dark grey; subdorsal line pale grey; lateral and spiracular lines obscured by darker freckling; spiracles white, outlined black; anal claspers extended clearly behind.

The larva is nocturnal in habit, hiding by day in hollow reed stems. It overwinters when small, resumes feeding in April and is full-fed toward the end of May.
Pupa. Cylindrical, long and slender (Tweedie, 1969); reddish brown; wing-cases finely sculptured; dorsal surface of thoracic segments and anterior part of each abdominal segment finely punctate; cremaster reduced and obtusely rounded, with two short, well-separated black spines surrounded by a number of hooked bristles. In a cocoon, usually between folded leaves, but occasionally below the surface of the soil.
Imago. Univoltine. Flies in July. The moth hides low down amongst reeds by day and feeds at flowering grasses and sugared reeds by night, coming also to light.

Distribution (Map 192)
The ecology of this moth is obscure and the reasons for its presence in certain reed beds (often quite small) and absence from apparently similar localities nearby are not understood. Locally common among *Phragmites* south of a line from the Bristol Channel to the Wash; less common farther north, but occurs locally in mid-Wales, Cheshire, Yorkshire and Cumbria. Most Irish records are from the south; the Channel Islands. Eurasiatic; in western Europe from western and southern France to Denmark and south Sweden, patchily distributed and in some places rare.

MYTHIMNA IMPURA (Hübner)
The Smoky Wainscot
Noctua impura Hübner, [1808], *Samml.eur.Schmett.* **4**: pl.85, fig.396.
Type locality: Europe.

Description of imago (Pl.13, figs 41,42)
Wingspan 31–38mm. Antenna setose-ciliate, in male more strongly. Forewing of female usually longer, with more pointed apex; straw-coloured, paler subcostally; veins whitish, interneural shading fine, pale red-brown; reniform stigma a black median dot; postmedian line two clear black dots, often with additional lesser black specks; a grey-brown basal shade reaches the median area; variable black subdorsal scaling. Hindwing brownish fuscous to greyish fuscous; veins darker; cilia whitish. Underside of forewing with greyish to blackish subcostal shading; a small blackish discal spot sometimes present on each wing. Abdomen of male with anal tuft.

Ab. *punctina* Haworth has darker red-brown interneural shading, the veins more contrasting than in the typical form. This trend is intensified in very rare brownish or blackish forms. Ab. *scotica* Cockayne (fig.42) is smaller, with paler forewing and blackish grey hindwing.
Similar species. M. pallens (Linnaeus), male, *M. loreyi* (Duponchel), *M. litoralis* (Curtis), *M. putrescens* (Hübner) and *Simyra albovenosa* (Goeze) (Acronictinae) all have white hindwings. *Mythimna straminea* (Treitschke) has a dark discal spot on the underside of each forewing. *M. obsoleta* (Hübner) has a complete row of subterminal dots parallel to termen on the forewing. *M. pallens*, female, has the hindwing usually less dark and the underside without median and subcostal shading.

Life history
Ovum. Globular, covered with very fine reticulate sculpturing; whitish and glossy, darkening gradually to slate-grey. Laid side by side in rows in the leaf-axils, folded blades or withered leaves of grasses, hatching in about a week.
Larva. Full-fed *c.*40mm long, cylindrical. Head light brown. Body pale ochreous; dorsal and subdorsal lines white, the former edged grey-brown, the latter bordered dark grey above; a clear black pinaculum immediately above subdorsal line in the middle of each segment; a smaller one, slightly anterior, half-way between dorsal and subdorsal lines; subspiracular line broad, grey; spiracles pinkish, outlined black; ventral surface paler than ground colour.

The larva feeds by night on various grasses, including cock's-foot (*Dactylis glomerata*), annual meadow-grass (*Poa annua*), tufted hair-grass (*Deschampsia cespitosa*), common reed (*Phragmites australis*), hairy wood-rush (*Luzula pilosa*) and field wood-rush (*L. campestris*), hibernating while small. It resumes feeding in April and is full-grown by the end of May.

Pupa. Short and rather stout; reddish brown; abdominal segments hollowed ventrally to a compact cremaster with four fine subbasal bristles and two parallel, hooked apical spines. In a fragile cocoon on or just below the surface of the soil.

Imago. Mainly univoltine, although there may be a very small second generation in southern England. The moth flies from about midsummer, a month later in the north. It feeds at flowering grasses, ragwort, wood-sage and knapweed, sometimes by day, and comes in numbers to sugar and light.

Distribution (Map 193)

Although uncommon on acid moorland or in dense woodland, this species seems to be able to thrive in most other kinds of habitat and in the extreme northern part of its range is particularly abundant on sand-dunes. Widely distributed in England, Wales, Ireland and southern Scotland; more local farther north, but locally common to Shetland; the Isles of Scilly; the Channel Islands. Ab. *scotica* is not treated here as a subspecies, since it occurs with varying frequency throughout the range, although most commonly in the west and north. Holarctic; in western Europe generally common to southern Scandinavia.

MYTHIMNA PALLENS (Linnaeus)
The Common Wainscot

Phalaena (Noctua) pallens Linnaeus, 1758, *Syst.Nat.* (Edn 10) **1**: 510.

Type locality: Europe.

Description of imago (Pl.13, figs 44–46)

Wingspan 32–40mm. Antenna setose-ciliate, in female the cilia extremely short. Forewing rather narrow, with moderately pointed apex; pale straw-coloured to reddish ochreous (ab. *ectypa* Hübner, fig.45) with whitish veins and, in the darker forms, heavy interneural shading; the whole wing finely striate; reniform stigma a black dot; often a pale streak from reniform stigma to termen; postmedian line sometimes indicated by two very small black dots; cilia concolorous with ground colour. Hindwing in male white, veins slightly darker; in female varying from white to pale fuscous, the colouring being due to darker pigmentation along the veins, which may unite to form a terminal shade of varying width. Although Barrett (1899) states that ab. *ectypa* is most common in Ireland, it seems to occur over most of Britain; there is, however, a tendency for specimens from marshy areas to be mainly of the paler forms.

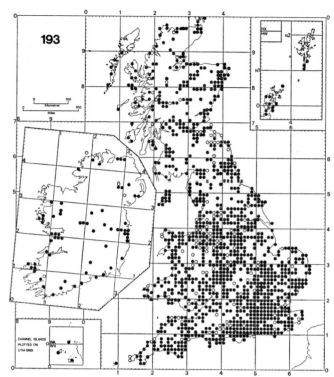

Mythimna impura

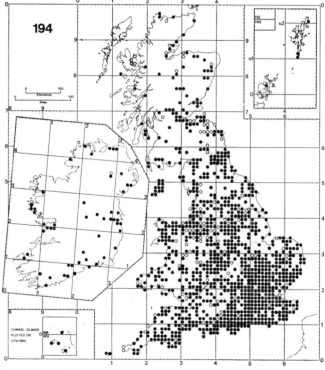

Mythimna pallens

Similar species. M. impura (Hübner) always has darker hindwing in male, usually also in female; postmedian dots are more distinct; underside of forewing has greyish subcostal and median shading. *M. straminea* (Treitschke) has forewing with pointed apex and straight termen; postmedian dots are more distinct; hindwing has postmedian line of dots and a discal spot on the underside of each wing. *M. obsoleta* (Hübner) has forewing with pointed apex; postmedian line of dots parallel to termen; terminal line of small dots; hindwing veins clearly defined. *M. litoralis* (Curtis) has forewing with clear white basal streak to termen. *M. loreyi* (Duponchel) has forewing with black basal streak; postmedian line of dots angulate subcostally. *M. putrescens* (Hübner) has forewing with black basal streak accompanied by a purplish brown shade at least to postmedian line.

Life history

Ovum. Hemispherical, very finely sculptured with a reticulate pattern; glossy and whitish at first, darkening quickly. Laid side by side in single or double rows in the leaf-axils or folded blades of grasses, hatching in about a week.

Larva. Full-fed *c.*42mm long, slender and cylindrical. Head small, yellowish brown. Body pale ochreous, the longitudinal lines contrasting less than in most of the genus; dorsal line whitish, edged grey, finer than in *M. impura*; subdorsal line whitish and rather obscure, edged grey above, grey-brown below; supraspiracular line broad, greyish; subspiracular line finer, ochreous and merging with greyish ochreous ventral surface; spiracles whitish, outlined black.

The larvae feed by night on grasses, the most frequently recorded being cock's-foot (*Dactylis glomerata*), common couch (*Agropyron repens*), tufted hair-grass (*Deschampsia cespitosa*) and annual meadow-grass (*Poa annua*). At first they feed together in small companies of five to ten in spun terminal blades, but later they hide among the roots by day. The first brood of the bivoltine populations grows rapidly and is full-fed by August; the others overwinter when small and feed up in April and May.

Pupa. Moderately stout; yellowish brown to reddish brown; thoracic segments smooth; abdominal segments rather sharply tapering, each with an anterior band of very fine punctation; cremaster slightly concave ventrally with two small, slightly in-curving apical spines. Below the surface of the ground in an earthen cocoon.

Imago. Bivoltine in the south, flying in early July and September, the relative abundance of the two broods varying from year to year; univoltine from the Midlands northward, flying from mid-July. The moth hides by day but is active from late dusk, feeding at flowers, flowering grasses and sugar; it also comes freely to light.

Distribution (Map 194)

Widely distributed and common, especially in marshy areas, up to a level of about 400m (1,200ft). Great Britain and Ireland, very common to lowland Scotland, more local farther north. Holarctic; in western Europe, generally common to central Fennoscandia.

MYTHIMNA FAVICOLOR (Barrett)
Mathew's Wainscot

Leucania favicolor Barrett, 1896, *Entomologist's mon.Mag.* **32**: 100.

Type localities: England; Suffolk and Essex.

Description of imago (Pl.13, figs 47,48)

Wingspan 34–42mm. Antenna shortly setose-ciliate, in female the cilia extremely short. Forewing in female more ample and with more pointed apex; texture smooth and satiny, without striation; pale buff, through shades of fulvous to rich red-brown; reniform stigma a black dot; postmedian line usually of two or more black dots, but occasionally almost complete. Hindwing of smooth texture, whitish; veins pale grey, the adjacent area sometimes suffused greyish; cilia white.

The status of this moth has been much debated and is still by no means unanimously agreed, although Cockayne (1952) considered it to be a distinct species. It differs from *M. pallens* (Linnaeus) in superficial appearance (greater wingspan, more ample, smoother wings, greater development of the postmedian line), ecology (restricted to saltmarshes) and, to a lesser extent, life history and behaviour. On the other hand, the genitalia are similar, the flight periods overlap to some extent, mixed pairing has been observed in the wild (Sperring, 1949) and each form has been reared from eggs laid by a female apparently of the other form. The relationship must be extremely close.

Similar species. M. pallens has a finely striate forewing. *M. litoralis* (Curtis) has a white basal streak to termen on forewing.

Life history

Ovum. Undescribed.

Larva. Full-fed *c.*40mm long, cylindrical and fairly stout. Body pinkish brown, finely sprinkled with grey dorsally; dorsal and subdorsal lines fine, white, edged with grey-brown; lateral stripe pinkish, bordered white below; supraspiracular stripe broad, slate-grey; spiracles pale, outlined black; subspiracular stripe grey-brown; ventral surface greyish. Descriptions of this larva vary in detail as may be expected in this variable genus; larvae of *M. favicolor* are in general more strongly pigmented than those of *M. pallens* which might be found feeding with them (Haggett, 1960b).

Feeds on common saltmarsh-grass (*Puccinellia maritima*) and probably on other grasses, as the female will lay on cock's-foot (*Dactylis glomerata*) or annual meadow-grass (*Poa annua*) and larvae can be reared successfully on either. After overwintering when small, it feeds up in spring.

Pupa. Very similar to that of *M. pallens* q.v. That of *M. favicolor* is slightly larger and the apical spines of the cremaster are relatively longer.

Imago. Univoltine, with slight evidence of an occasional very small second brood (Tutt, 1904); such a second brood occurred in 1976 which was a year of exceptional drought. Occurs in June and July, the earliest moths emerging slightly before *M. pallens*. Feeds at flowering grasses and rushes, as well as coming to sugared foliage when, according to several observers, it is more alert and quick to fly than *M. pallens*. Also comes to light and, according to Sperring (*op. cit.*), pairs may be found *in copula* on grass stems from 23.00 hours.

Distribution (Map 195)

Very locally common in salt-marshes from southern Suffolk to East Kent and from West Sussex to western Hampshire. There are isolated records from the Isle of Wight and inland from the New Forest, Hampshire, Surrey and East Anglia, but the insect has never been found inland regularly. It is unknown outside England.

MYTHIMNA LITORALIS (Curtis)
The Shore Wainscot

Leucania litoralis Curtis, 1827, *Br.Ent.* **4** : 157.

Type locality: England; Christchurch, Dorset.

Description of imago (Pl.13, fig.43)

Wingspan 36–42mm. Antenna shortly setose-ciliate, in female cilia very short. Forewing long and narrow with fairly pointed apex; pale golden brown, paler subcostally; basal streak reaching almost to termen, white, edged greyish in basal and median areas, usually with lesser postmedian branches towards dorsum; several white streaks in subterminal area. Hindwing pure white, veins slightly darker. Abdomen of male with dark reddish brown ventrobasal tuft, partly concealed by paler thoracic hairs.

Variation is slight, but the ground colour may tend to fulvous and occasionally there is a fuscous subdorsal shade; the white subterminal streaks vary in number and thickness.

Similar species. *M. favicolor* (Barrett) has a more slender forewing with less pointed apex; no white basal streak.

Life history

Ovum. Relatively large, globular, with fine reticulate sculpturing; pale yellow at first, darkening through yellowish

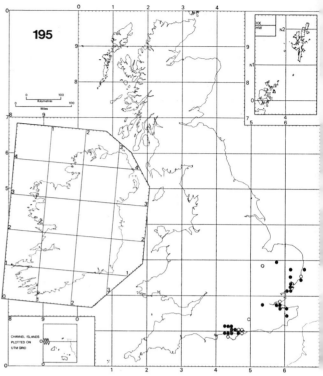

Mythimna favicolor

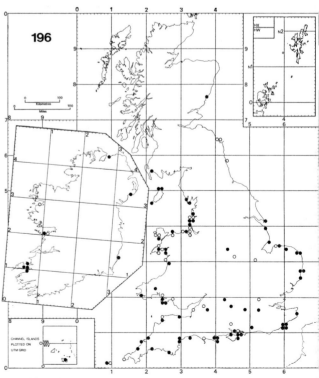

Mythimna litoralis

orange to purplish red. Laid side by side in rows in a folded blade of the foodplant, hatching in about ten days.

Larva. Full-fed *c.*45mm long. Head relatively large, light brown. Body variable, greyish, tinged brown, green or pink; dorsal and subdorsal lines white, edged darker grey; two lateral pairs of narrow alternately paler and darker lines; spiracular line wide, pale; spiracles white, ringed black, ventral surface ochreous green to green. It is not possible to state precisely the colour of the 'dark' and 'pale' lines, as variation is great, but the ecology and foodplant should make recognition easy.

The larva feeds by night on marram (*Ammophila arenaria*) hiding in the sand by day, and overwintering while small. It completes feeding quickly in spring and will accept annual meadow-grass (*Poa annua*), mat-grass (*Nardus stricta*) or tufted hair-grass (*Deschampsia cespitosa*) in captivity; for successful breeding in captivity it must be provided with a considerable depth of dune sand for concealment by day and for pupation.

Pupa. Slender and elongate; pale yellow-brown; wing-cases dull and finely sculptured; abdominal segments glossy, with two fine apical spines surrounded by several slightly smaller bristles. In sand, among the roots of marram.

Imago. Univoltine. Occurs in a single prolonged emergence, from late June to the end of August. It hides by day among marram or under dune overhangs, feeds by night at flowering grasses or sugared marram and comes to light.

Distribution (Map 196)

Locally common on coastal sandhills of England, Wales and Ireland; Scotland, to Angus in the east, Galloway in the west; the Channel Islands; the Isles of Scilly. In Europe, coastal only, from Portugal to Jutland and the Danish islands.

MYTHIMNA L-ALBUM (Linnaeus)
The L-album Wainscot

Phalaena (Noctua) l-album Linnaeus, 1767, *Syst.Nat.* (Edn 12) **1**: 850.

Type locality: Portugal.

Description of imago (Pl.13, fig.49)

Wingspan 34–40mm. Antenna shortly setose-ciliate, in female cilia very short. Forewing greyish ochreous, suffused brown in median and subterminal areas; basal streak short, black; dorsobasal streak black, fine and short; conspicuous white median streak angled distally towards costa; apical streak wide, slightly paler than ground colour; termen dark brown, terminal row of black dots, followed by fine pale ochreous and brownish lines; veins whitish in subapical and terminal areas. Hindwing pale fuscous, with darker veins and terminal line. Abdomen of male with conspicuous ventrobasal tuft.

There is some variation in the ground colour (Wightman, 1940a) which may approach grey-brown or red-brown; the median streak is variable in shape and length, sometimes abnormally developed, or even reduced to a white dot (ab. *o-album* Milman).

Life history

Ovum. Globular and glossy, with light reticulate sculpturing; whitish at first, darkening to slate-grey. Laid side by side in single or double rows in leaf-axils and folded blades of grasses, hatching in about 10 to 14 days.

Larva. Full-fed *c.*33mm long, of rather fragile build and very glossy appearance owing to exceptionally thin skin. Head pale brown. Body greyish ochreous; dorsal line whitish; subdorsal line blackish brown, almost complete and conspicuous, usually enabling separation from larvae of *M. pallens* (Linnaeus) and *M. impura* (Hübner); lateral line grey-brown; subspiracular line pinkish ochreous. For detailed descriptions, see Cockayne (1938), Haggett (1960b).

The larvae of the first brood feed rapidly; they pupate in about five weeks, the moths emerging three weeks later; those of the second brood go immediately into hibernation, although in captivity and kept at about 20°C (70°F) they will feed up quickly and produce adults by mid-December. The overwintered larvae feed nocturnally until late May. The natural foodplant is uncertain; Cockayne (*loc.cit.*) suggests tall fescue (*Festuca arundinacea*) but in captivity larvae accept most common grasses.

Pupa. Elongate and glossy; pale reddish brown; abdominal segments sharply tapering; cremaster a central papilla with four small subapical bristles and two divergent apical spines. Below ground in a cocoon of silk and earth.

Imago. Bivoltine, the first brood flying in July, the second from mid-September to mid-October, this latter often

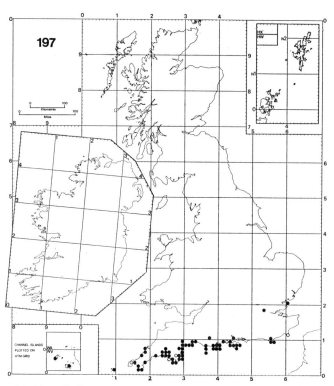
Mythimna l-album

considerably the more abundant. The summer brood feeds at flowers, the autumn at ivy-bloom and blackberries; both come to sugar and light. After midnight, pairs may be found *in copula* on grass-stems. Females will lay freely in captivity if given a reasonable amount of room and a few stalks of grass.

Distribution (Map 197)
First recorded in Devonshire in 1901 and a rare migrant until the mid-1930s, when it became well established on the coasts of Cornwall and Devonshire; now locally common, but only exceptionally more than a mile or so inland, in Dorset, the Isle of Wight and Hampshire; much less common farther east, although it has reached Sussex and Kent and has been recorded in Essex. The single specimen recorded from central London seems unlikely to have reached there unaided. Eurasiatic; in Europe it is widespread and not exclusively coastal, in Portugal, Spain, France and Belgium spreading northwards since 1930; established in the Netherlands since about 1947 and an occasional immigrant to Denmark since 1950.

MYTHIMNA UNIPUNCTA (Haworth)
The White-speck, or American Wainscot
Noctua unipuncta Haworth, 1809, *Lepid.Br.*: 174.
Leucania extranea Guenée, 1852, *in* Boisduval & Guenée, *Hist.nat.Insectes* (Lépid.) **5**: 77.
Type locality: England.

Description of imago (Pl.13, figs 52,53)
Wingspan 41–48mm. Forewing reddish or yellowish brown, speckled black; a fine white line extends from the base to a clear white speck in the reniform stigma, and there is a dark, oblique, apical dash. In the female, the black scaling is reduced and the apical dash is sometimes almost absent. Hindwing grey, darker outwardly and on veins.

Variation in colour is considerable; the reniform and orbicular stigmata are sometimes almost invisible against the pale ground colour, but often bright orange and connected by reddish suffusion. In ab. *nigrosuffusa* Richardson, thorax and forewing are almost black.

Similar species. *M. ferrago* (Fabricius) *q.v.* lacks the median line and apical dash. *M. loreyi* (Duponchel) has the hindwing white.

Life history
Ovum. Hemispherical, with flat base extending as a rim; pale and glossy. Laid either in small groups on blades of grass or in rows packed beneath the sheath (Stokoe & Stovin, 1958, **1**: pl.67, fig.3).

Larva. Full-fed *c*.40mm long (sixth instar). Head brown with darker stripes. Prothoracic plate with three conspicuous white stripes. Body in early instars usually whitish or grey-green, later varying greatly from putty to dark purplish brown above, but always pale ventrally; white dorsal stripe and white or pink subdorsal lines usually prominent (figured by Haggett, 1963: 193–195, pl.11, figs 4–10).

The larva has only once been reported wild in Britain. In captivity it feeds nocturnally on most grasses and is easily reared from the egg to moth in warmth in about 50 days.

Pupa. Brown, with curved cremaster. In a flimsy cocoon placed vertically low down among grass stems.

Imago. An erratic immigrant, possibly now also resident in sheltered spots in the Isles of Scilly and on the south-west coasts of England and Ireland. Barrett (1899) knew of only a dozen captures, and until 1956 it was reported infrequently and only in single figures, except for a large number in 1928, mostly in south-west Ireland. In 1957 about 80 were seen in the Isles of Scilly and about 20 elsewhere. Since then it has been found every year, with an abundance of over 400 recorded in 1966 and again in 1978. Most moths have been taken in light-traps, but they feed, warily, at sugar and ivy-blossom; most common from late

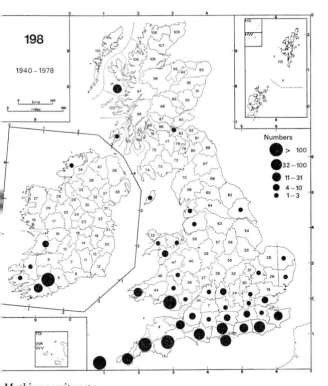

Mythimna unipuncta

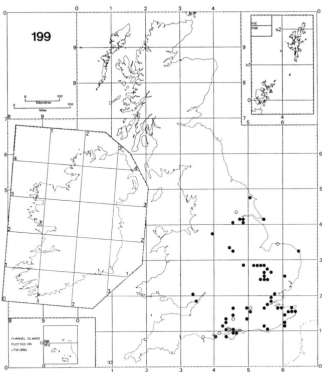

Mythimna obsoleta

August to early October, but recorded in all months except February.

Distribution (Map 198)

Mainly south-western in Britain and Ireland, but in recent years extending casually up the west coast to the Isle of Canna, Inner Hebrides, more commonly along the south coast to Kent and Essex, and inland to Warwickshire, Northamptonshire and Cambridgeshire. In 1978 one was taken near Edinburgh. The species is cosmopolitan, of Neotropical origin; a pest of crops in many countries. It is probably resident from the Canary Islands to southern France, and has recently spread northwards and eastwards; it was first reported in Normandy in 1970, and only in recent years in Belgium, the Netherlands, on Bornholm in the Baltic in 1969, and once even in Iceland.

MYTHIMNA OBSOLETA (Hübner)
The Obscure Wainscot

Noctua obsoleta Hübner, [1803], *Samml.eur.Schmett.* **4**: pl.48, fig.233.

Type locality: Europe.

Description of imago (Pl.13, fig.54)

Wingspan 36–40mm. Antenna shortly setose-ciliate, in female the cilia very short. Forewing narrow, with rather pointed apex; in female termen more oblique than in male; brownish ochreous to greyish ochreous, variably and finely irrorate with blackish; veins white, closely and finely bordered with blackish brown, giving a very finely striate appearance; reniform stigma represented by a white median dot on vein 6 (M_1), which is outstandingly clear to this point; postmedian line of blackish dots oblique, slightly convex, more or less parallel to termen. Hindwing whitish; veins and variable terminal suffusion pale grey-brown. Hübner's illustration represents a form not found in Britain, though occasional reddish-tinted specimens have been found which approach it in colour.

Similar species. *M. straminea* (Treitschke), *M. impura* (Hübner) and *M. pallens* (Linnaeus) *q.v.*

Life history

Ovum. According to Stokoe & Stovin (1958), globular and glossy whitish, very finely striate. Laid in rows on leaves of the foodplant, hatching in about ten days.

Larva. Full-fed c.40mm long, cylindrical and fairly plump. Head pale brown. Body ochreous grey; dorsal line fine, white; subdorsal line similar, edged with brownish black; subspiracular line greyish white, with a narrow, pale grey line above; spiracles greyish, thickly ringed black; ventral surface pale putty-colour. Anal claspers extended when at rest.

The larva feeds by night on the common reed (*Phragmites australis*) and hides by day in hollow stems. Full-fed by autumn, it overwinters in a reed stem where it pupates in April or early May. There is, however, a record (Todd, 1964) of successful rearing from the egg on knotgrass (*Polygonum aviculare*), the larvae having refused *Phragmites*.

Pupa. Slender, streaked with red-brown; cremaster with two lateral bristles and two apical spines.

Imago. Univoltine. Occurs from late May to early July; according to Tutt (1902) the moth emerges between 19.00 and 20.00 hours, and after the wings have expanded, sits head downward and well camouflaged on a reed stem until dusk. Feeds at flowering grasses and comes to sugar and light.

Distribution (Map 199)

This species is extremely local. It inhabits marshes and fens, forming colonies which may be restricted to a few acres; individuals are taken in light-traps from time to time well away from their normal haunts. England; very local in most counties south of the Thames and Severn, the Isles of Scilly, the Fens and East Anglia. There is a record from Cheshire and the species has also been recorded from the Midlands, Lincolnshire and south Yorkshire; Wales, Monmouth. Eurasiatic; in Europe from Spain to Denmark and south Sweden, and near Oslo in Norway.

MYTHIMNA COMMA (Linnaeus)
The Shoulder-striped Wainscot

Phalaena (Noctua) comma Linnaeus, 1761, *Fauna Suecica* (Edn 2): 316.

Type locality: Sweden.

Description of imago (Pl.13, figs 55,56)

Wingspan 35–42mm. Antenna shortly setose-ciliate, in female cilia very short. Male with pale golden ventral scent-brushes on metathorax. Forewing brownish ochreous, paler subcostally, finely irrorate black; veins whitish, thickly lined pale brown to red-brown, giving a coarsely striate appearance; basal streak black, usually to centre of wing; subdorsal streak fine and short, black; reniform stigma a white median dot on vein 6 (M_1); numerous black streaks of varying length and thickness in subterminal area; terminal line of black dots. Hindwing pale fuscous to brownish fuscous, variably darker near margins; veins usually slightly darker than ground colour.

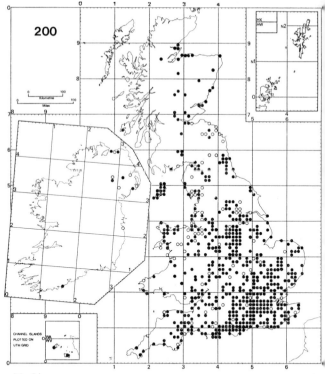

Mythimna comma

Life history

Ovum. Globular, glossy, with very fine reticulate sculpturing; pale greenish white at first, darkening slowly. Laid in rather untidy rows in leaf-axils or folded blades of grasses, hatching in about six days. The egg is often dented without detriment by contact with neighbouring eggs.

Larva. Full-fed c.43mm long, cylindrical, rather stout. Head light reddish brown with two darker vertical lines. Body greenish ochreous to reddish ochreous; dorsal and subdorsal lines white, the former fine, heavily edged with grey-brown, the latter with a row of dark dashes above, which Buckler (1891) gives as a character for identification, but others of the genus have similar markings; supraspiracular stripe wide, pale brown; subspiracular line narrow, pale ochreous; spiracles pale, heavily outlined with black, this last seeming a more reliable character; ventral surface pale ochreous.

The larva feeds by night; cock's-foot (*Dactylis glomerata*) is probably its main foodplant, but it has been recorded feeding on a number of Gramineae as well as dock (*Rumex* spp.). Full-grown in September and goes well below ground to overwinter in a cocoon in which it eventually pupates the following April; because of this the species is very difficult to rear in captivity.

Pupa. Short and stout; yellow-brown; cremaster with two central and two dorsolateral, slightly hooked spines, the inner only slightly larger than the outer.

Imago. Univoltine, although occasional late autumn specimens occur. The moth flies from the end of May to the latter part of June. It hides by day, but feeds at flowers from late dusk and comes to sugar and light.

Distribution (Map 200)

Although very common in fens and marshes, this species seems to have no actual ecological limitations and is found in most types of habitat, including woodland. Common over most of Britain to southern Scotland; Ireland. More local in eastern Scotland as far north as Ross-shire; the Channel Islands. Holarctic; in Europe, generally common, reaching the Arctic.

MYTHIMNA PUTRESCENS (Hübner)
The Devonshire Wainscot

Noctua putrescens Hübner, [1824], *Samml.eur.Schmett.* **4**: pl.156, figs 730,731.

Type locality: Europe.

Description of imago (Pl.13, fig.57)

Wingspan 32–36mm. Antenna shortly setose-ciliate, in female the cilia very short. Forewing in female slightly more ample than in male; pale straw-coloured, with fine blackish irroration costally; veins whitish, finely outlined with purplish brown; basal streak fine, black, of variable length, sometimes almost to postmedian line; reniform stigma a clear white dot; dorsum and subterminal area variably covered with fine purplish brown suffusion; postmedian line of small black dots; often a pale suffusion from reniform stigma to apex. Hindwing rather iridescent white; veins inconspicuously pale brown; terminal line of small blackish dots.

Life history

Ovum. Globular, with extremely fine reticulate sculpturing; pale and glossy. Laid in rows in leaf-axils or folded blades of the foodplant, hatching in about ten days.

Larva. Full-fed *c.*42mm long. Head mottled greyish, a vertical black and a white line on each half. Body varying from greyish green to pale fawn; dorsal line whitish (silvery white, according to Curzon (1886)), clearly edged with black; subdorsal line complete, whitish, shaded darker above and bordered with black below; on each segment a black pinaculum between dorsal and subdorsal lines; lateral stripe broad, from pinkish buff to pale grey; supraspiracular stripe whitish, edged with black; spiracles whitish, ringed with black; ventrally paler.

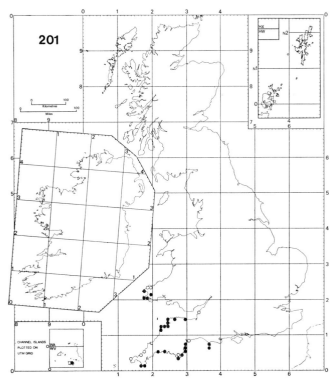

Mythimna putrescens

The larva feeds nocturnally on grasses (Curzon (*loc.cit.*) mentions a preference for 'fine hair-grass') through the autumn and winter, becoming full-fed from late January to late February, when it constructs a subterranean cocoon in which it remains for several months before pupating towards the middle of May.

Pupa. Undescribed.

Imago. Univoltine. Flies from late July. The moth comes at dusk to flowers such as red valerian and wood-sage; later it visits sugar and light.

Distribution (Map 201)

The species is extremely local and confined to the coast in the mildest areas of England and Wales; it is rarely found more than a few hundred yards inland. England, local on the coasts of Devon and Cornwall, with occasional records from Somerset and Dorset; Wales, Pembrokeshire; the Channel Islands. Mediterranean; occurring in southern Europe and north Africa and on the French coast as far north as Finistère.

MYTHIMNA COMMOIDES (Guenée)

Leucania commoides Guenée, 1852, *in* Boisduval & Guenée, *Hist.nat.Insectes* (Lépid.) **5**: 86.

Type locality: U.S.A.; New York.

The only British record is of four specimens 'taken in a spot bordering on Romney Marsh, during the first week of August 1873', and later identified by Doubleday (Parry, 1873: 522–523). This North American species somewhat resembles *M. comma* (Linnaeus). Chalmers-Hunt (1964, **2**: 198) regarded the record as 'very doubtfully genuine'.

MYTHIMNA LOREYI (Duponchel)
The Cosmopolitan

Noctua loreyi Duponchel, 1827, *in* Godart & Duponchel, *Hist.nat.Lépid.Fr.* **7** (1): 81, pl.105, fig.7.

Type locality: France; Provence.

Description of imago (Pl.13, fig.58)
Wingspan 34–44mm. Antenna shortly setose-ciliate, in female very shortly. Forewing varies from pale yellowish to dark brown; stigmata indistinct, but with a clear white point in the reniform stigma and a dark median line from the base; a postmedian series of black dots. Hindwing pure white. Male has basal tufts of dark scales below the abdomen. It has been shown by experimental rearing that darkening of the ground colour is related to timing and duration of exposure of the pupae to low temperatures, and that in extreme cases near obsolescence of the median streak may also be caused thereby. The range of colour forms so produced is figured by Myers (1977).
Similar species. *M. ferrago* (Fabricius), *M. albipuncta* ([Denis & Schiffermüller]) and *M. unipuncta* (Haworth) q.v. all have dark hindwings, as do other *Mythimna* species for which *M. loreyi* might be mistaken when at rest.

Life history
Ovum. Flattened and often misshapen. Laid in captivity in sheaths of bents (*Agrostis* spp.), hatching in eight days.
Larva. Full-fed *c*.43mm long, cylindrical, with some flattening dorsoventrally round the thorax. It closely resembles the larva of *M. unipuncta*, but the head is dark brown rather than yellowish; the space enclosed by the double grey dorsal line is of the ground colour and does not stand out as a white line as it commonly does in *M. unipuncta*. The usual *Mythimna* chevrons on each segment, above the dorsal line, are black and very distinct (Myers *in litt.* and 1975). Figured by Spuler (1910: Nachtrag pl.5, fig.11).

Apart from a larva said to have been found in a garden in the Isle of Man before 1872 (Hodgkinson, 1872), the early stages are unknown in the wild in the British Isles, and in captivity the species was reared from the egg, probably for the first time, in 1975.

Imago. A scarce immigrant of which before 1975 there were fewer than thirty records: Sussex (two in 1862, the first British records); Cornwall and the Isles of Scilly (12), South Devon (6), Hampshire (4), Kent (1), Co. Cork (2), in all months from June to November, but most in September. Fewer were taken at light than at sugar (at which it sits quietly with closed wings unlike the very wary *M. unipuncta*). In 1975 three were taken at the Lizard on 24 and 28 August, together with other immigrant species (Skinner, 1975: 276), and half a dozen more a few days later elsewhere in Cornwall and the Isles of Scilly; also one on 3 September in South Devon. In Ireland, six were trapped singly between 4 and 18 September and eight more between 4 and 8 October at Fountainstown, Co. Cork (Myers *in litt.* and 1975). Moths were seen at the Lizard in 1976 and one in Dorset in 1978. Single captures at the same light-trap at Bodinnick, Cornwall, in 1964, 1966, 1968 and 1969, and of two nearby at Polperro in 1968, together with the moth's recurrence at the Lizard in 1976, may indicate local establishment; also the long time-spread of the Irish records in 1975 suggests that some moths may have bred nearby, perhaps from an immigrant earlier in the season.

Distribution (Map 202)
The distribution in the British Isles is at present essentially south-western, resembling that of *M. vitellina* (Hübner) and *M. unipuncta* but much less extensive. The species is cosmopolitan, of subtropical origin. It occurs in the Canary Islands, Azores, Madeira, Morocco, Portugal, Spain and in south France from the Garonne to the Mediterranean; in north Italy it is believed to be immigrant only.

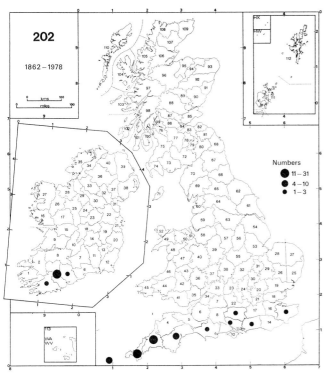
Mythimna loreyi

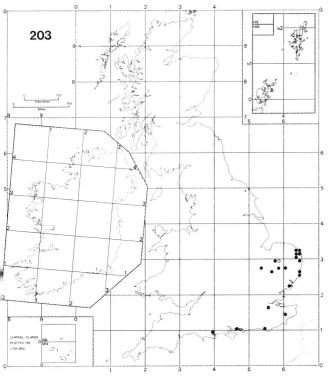
Senta flammea

SENTA Stephens
Senta Stephens, 1834, *Ill.Br.Ent.* (Haust.) **4**: 297.
Meliana Curtis, 1836, *Br.Ent.* **13**: 587.

A small genus containing two closely allied species, of which one is found locally in England.

Frons projecting, rounded with a sclerotized plate below; terminal segment of labial palpus very short. Thorax slender, with scales adpressed, without crests. Tibiae and tarsi with scales adpressed. Abdomen slender, with scales adpressed, without crests.

SENTA FLAMMEA (Curtis)
The Flame Wainscot
Melia flammea Curtis, 1828, *Br.Ent.* **5**: 201.
Type locality: England; Lewisham, Kent.

Description of imago (Pl.13, fig.59)
Wingspan 32–40mm. Antenna shortly ciliate, in male densely, in female sparsely. Thorax narrow. Forewing long and narrow, with very oblique termen and very pointed apex; pale straw-coloured, often shaded with grey-brown or reddish brown and finely but distinctly striate with pale brown to purplish brown; basal streak dark brown, its costal margin narrowly edged whitish, extending almost to postmedian line of dark brown dots, which is more or less parallel to termen; subdorsal part of wing often irrorate darker than rest of wing. Hindwing whitish to pale fuscous; veins slightly darker; sometimes a row of very small, brownish terminal dots; cilia white. Abdomen long, slender and delicate with conspicuous, pale, yellow-centred anal tuft.

Variation affects mainly the ground colour of the forewing, the darkness of the striations, the length and degree of marginal shading of the basal streak and the clarity and shape of the dots which make up the postmedian line.

Life history
Ovum. Globular, slightly grained and relatively small; whitish and glossy. Laid in untidy double rows on leaves of the foodplant, hatching in about two weeks.

Larva. Full-fed *c.*33mm long, elongate and cylindrical, the anal claspers protruding when at rest. Head dark and glossy. Body greyish ochreous, with numerous paler ochreous lines of varying width; dorsal line narrow and clearly outlined darker, the rest merging with the ground colour; supraspiracular line broad and dark; spiracular stripe pale; spiracles whitish, finely outlined black; ventrally pale ochreous.

Feeds by night on leaves of common reed (*Phragmites australis*) and hides by day in hollow stems, sometimes several to a stem. Full-fed in October or November.

Pupa. Cylindrical and very elongate, showing little taper before abdominal segment 6; red-brown at first, darkening to blackish brown; abdominal segments with a dorsal band of anterior punctation; cremaster with two minute spines and two very small, fine curled bristles. In a hollow stem which is thoroughly lined with silk and sealed above with a fine silk membrane. It is believed that marshland birds eat many pupae during the winter.

Imago. Univoltine. The moth flies from mid-May. It appears soon after dusk for a short flight, after which it settles on the reeds, but flies more widely and freely later in the night, when it comes to light. It seems not to feed or visit sugar, although the haustellum is fully developed.

Distribution (Map 203)
This species has a rather specialized ecology and frequents only the drier parts of its already limited haunts. Very local in Norfolk, Suffolk, Cambridgeshire, Hampshire and Dorset. Light-trap records from Sussex suggest the possibility of an as yet undiscovered population in the marshes to the north of the South Downs or of occasional immigration from France. Eurasiatic; in Europe it is found in France, south-west and north, including Normandy; Belgium, the Netherlands, Denmark and the extreme south of Sweden, local and often scarce.

GRAPHANIA Hampson
Graphania Hampson, 1905, *Cat.Lepid.Phalaenae Br.Mus.* 5: 468.

A genus with numerous species confined to New Zealand, one of which has once been taken in England.

GRAPHANIA DIVES (Philpott)
Melanchra dives Philpott, 1930, *Rec.Auckland Inst.Mus.* 1: 1.
Type localities: New Zealand; Flagstaff Hill and Waitati, Dunedin.

The only British specimen is said to have been taken alive by day on Spurn Head, Yorkshire by J. N. Thornton of Leeds, as far as can be remembered on a warm afternoon in July 1950 (Smith, 1954: 20). The species is native to New Zealand. This example, which was not identified for some time, may have been accidentally introduced, perhaps from a passing ship. Figured in South (1961: pl.72, fig.11).

References

Adkin, R., 1895. (Report of South London Entomological & Natural History Society meeting, 28 November 1895). *Entomologist's Rec. J. Var.* **7**: 192.

Allen, P. B. M., 1949. *Larval foodplants*, 126 pp. London.

Barrett, C. G., 1892–1900. *The Lepidoptera of the British Islands*, **1–6**. London.

Baynes, E. S. A., 1964. *A revised catalogue of Irish macrolepidoptera*, 110 pp. Hampton.

——, 1970. *Supplement to a revised catalogue of Irish macrolepidoptera*, 28 pp. Hampton.

Beck, N., 1960. Die Larvalsystematik der Eulen (Noctuidae). *Abh. Larvalsyst. Insekten* **4**: 406 pp.

Bergmann, A., 1954. *Die Gross-Schmetterlinge Mitteldeutschlands*, **4**: xx, 580 pp. Jena.

Boursin, C., 1963a. Une espèce de *Noctua* L. (*Triphaena* O.) européenne et française, méconnue depuis 173 ans. *Noctua interposita* Huebner, 1789, nec 1790 (Lep. Noctuidae). *Bull. mens. Soc. linn. Lyon* **32**: 72–79.

——, 1963b. Eine seit 173 Jahren verkannte europäische *Noctua* L. (*Triphaena* O.) Art: *Noctua interposita* Hübner, 1789, nec 1790 (Lep. Noctuidae). *Z. wien. ent. Ges.* **48**: 193–205.

——, 1964. Les Noctuidae Trifinae de France et de Belgique. *Bull. mens. Soc. linn. Lyon* **33**: 204–240.

Bowes, A. J. L., 1939. Collecting notes, 1938. *Entomologist's Rec. J. Var.* **51**: 107–110.

Bretherton, R. F., 1957. The history and status of *Spaelotis ravida* (Schiffermüller) the Stout Dart (Lep., Caradrinidae) in Britain (Lep., Noctuidae). *Entomologist's Gaz.* **8**: 3–19, 195–198.

——, 1969. On rearing *Agrotis ipsilon* Hufnagel (The Dark Sword-grass). *Ibid.* **20**: 83–85.

——, 1972. Eastern immigrants and resident natives: a survey of some British Lepidoptera. *Proc. Trans. Br. ent. nat. Hist. Soc.* **5**: 95–119.

——, 1974. Day-flying Lepidoptera attracted to light. *Entomologist's Rec. J. Var.* **86**: 93–94.

—— & Hayes, A. H., 1976. *Brithys pancratii* (Cyrillo), 1787 a subspecies of *Brithys crini* (Fabricius 1775) (Lep., Noctuidae). *Entomologist's Gaz.* **27**: 226–228.

Buckler, W., 1865. Description of larva of *Agrotis ravida*. *Entomologist's mon. Mag.* **2**: 115–116.

——, 1891–1899. *The larvae of the British butterflies and moths*, **4–8**. London.

Chalmers-Hunt, J. M., 1962–1970, 1976. *The butterflies & moths of Kent (Sphingidae to Plusiidae)*, **2**, Suppl. West Wickham. (Published as supplements to *Entomologist's Rec. J. Var.*)

——, 1970. The butterflies and moths of the Isle of Man. *Trans. Soc. Br. Ent.* **19**: 1–171.

Cockayne, E. A., 1936a. Notes on the life-history of *Agrotis ipsilon*, Rott. (*suffusa*, Hb.). *Entomologist's Rec. J. Var.* **48**: 2–4.

——, 1936b. Notes on breeding *Leucania vitellina*, Hb. and *Leucania albipuncta*, F. *Ibid.* **48**: 25–28.

——, 1937. The genetics and status of *Xylomania* (*Xylomiges*) *conspicillaris*, L. and ab. *intermedia*, Tutt, and ab. *melaleuca*, View. *Ibid.* **49**: 81–82.

——, 1938. Notes on the life history of *Leucania l-album*, L. *Ibid.* **50**: 13–18.

——, 1947. Two unrecorded rarities: *Hadena* (*Dianthoecia*) *compta*, F., and *Leucania loreyi*, Dup. *Ibid.* **59**: 58.

——, 1950a. *Hadena cucubali*, Schiff., ab. *bondii*, Turner. *Ibid.* **62**: 35.

——, 1950b. *Diarsia florida* Schmidt (Lep. Agrotinae): a species new to Britain. *Entomologist* **83**: 173–174.

——, 1951. A larval habit of *Orthosia gracilis* Sch. and *Xylena vetusta* Hub. *Entomologist's Rec. J. Var.* **63**: 142–143.

——, 1952. The chromosome number of *Leucania favicolor* Barrett and *Leucania pallens* Linnaeus. *Ibid.* **64**: 220–221.

——, in Turner. See Turner H. J., 1926–1950.

Curtis, J., 1829[–1831]. *A guide to the arrangement of British insects*, 256 pp. London.

Curtis, W. P., 1934. A list of the Lepidoptera of Dorset. Introduction and Part 1. *Trans. Soc. Br. Ent.* **1**: 185–286.

Curzon, E. R., 1886. Note on the larvae of *L. putrescens*, *H. hispida* and *S. anomala*. *Young Nat.* **7**: 54–55.

David, W. A. L. & Gardiner, B. O. C., 1966. Rearing *Mamestra brassicae* (L.) on semi-synthetic diets. *Bull. ent. Res.* **57**: 137–142.

Dickson, R., 1976. *A lepidopterist's handbook*, 138 pp., 34 figs. The Amateur Entomologist's Society, Hanworth.

Doubleday, H., 1850. Occurrence of *Opigena fennica* in England. *Zoologist* **8**: 2971.

Durden, L. A., 1974. *Ochropleura fennica* (Tauscher) (Lep., Noctuidae): a second British record. *Entomologist's Gaz.* **25**: 51.

Emmet, A. M., 1978. Exhibits meeting 8 September, 1977. *Proc. Trans. Br. ent. nat. Hist. Soc.* **11**: 43.

Forster, W. & Wohlfahrt, T. A., 1963–1971. *Die Schmetterlinge Mitteleuropas*, **4**: 329 pp., 32 pls. Stuttgart.

French, R. A. & Hurst, G. W., 1969. Moth immigrations in the British Isles in July 1968. *Entomologist's Gaz.* **20**: 37–44.

Gardiner, B. O. C., 1968. On *Coenophila subrosea* (Stephens), (Lep., Noctuidae). *Ibid.* **19**: 251–255.

Goater, B., 1969. Variation in *Agrotis exclamationis* Linnaeus. *Proc. Trans. Br. ent. nat. Hist. Soc.* **2**: 55–67.

———, 1974. *The butterflies and moths of Hampshire and the Isle of Wight*, xiv, 439 pp. Faringdon.

———, 1978. On rearing *Eriopygodes imbecilla* (F.) (Lepidoptera: Noctuidae). *Entomologist's Gaz.* **29**: 107–108.

Goodson, A. L., 1951. Expedition to Askham Bog in search of *Diarsia florida* (Schmidt) (Marsh Square-spot). *Ibid.* **2**: 71–74.

Greenwood, J. A. C., 1934. Larvae of a species of *Brithys* (Lep. Noctuidae) in Kew Gardens. *Entomologist* **67**: 15.

Haggett, G. M., 1954. The pupa of *Hada nana* Hufn. (*dentina* Esp.). *Entomologist's Gaz.* **5**: 223–224, 1 fig.

———, 1957. Larvae of the British Lepidoptera not figured by Buckler. *Proc. Trans. S. Lond. ent. nat. Hist. Soc.* **1955**: 152–163, 3 col. pls.

———, 1960a. The early stages of *Spaelotis ravida* Hübner (Lep., Noctuidae). *Entomologist's Gaz.* **11**: 161–163.

———, 1960b. Larvae of the British Lepidoptera not figured by Buckler. Part IV. *Proc. Trans. S. Lond. ent. nat. Hist. Soc.* **1959**: 207–214, 2 col. pls.

———, 1963. Larvae of the British Lepidoptera not figured by Buckler. Part VII. *Ibid.* **1962**: 191–198, 2 col. pls.

———, 1964. Researches into colour variation of the moth *Leucania vitellina* Hübner. *Ibid.* **1963**: 78–92, 1 col. pl.

———, 1968. Larvae of the British Lepidoptera not figured by Buckler. Part VIII. *Proc. Trans. Br. ent. nat. Hist. Soc.* **1**: 57–109, 10 col. pls.

Hammond, H. E., 1952. A list of previously unrecorded foodplants of lepidopterous larvae, with additional notes on preferences, etc. *Entomologist's Gaz.* **3**: 59–68.

Hawkins, C. N., 1954. Exhibits – meeting 13 August, 1952. Exhibits – meeting 12 November, 1952. *Proc. Trans. S. Lond. ent. nat. Hist. Soc.* **1952–53**: 12, 55.

Hedges, A. V., 1949. Technique of breeding Lepidoptera. *Ibid.* **1947–48**: 75.

———, 1954. A remarkable case of convergence in the pupa and cocoon of two agrotid moths. *Entomologist's Rec. J. Var.* **66**: 129–131, 2 pls.

Hellins, J., 1868. Note on *Agrotis suffusa*. *Entomologist's mon. Mag.* **4**: 255.

Hodgkinson, J. B., 1872. Rare Lepidoptera taken in the Isle of Man. *Ibid.* **9**: 44.

Hoffmeyer, S., 1962. *De Danske Ugler* (Edn 2), 387 pp., 33 col. pls. Aarhus.

Horton, G. A. N., 1976. The discovery of *Eriopygodes imbecilla* (Fabricius) (Lep.: Noctuidae) as a resident British species. *Entomologist's Rec. J. Var.* **88**: 246–248.

——— & Heath, J. 1973. *Eriopygodes imbecilla* (Fabricius) (Lep., Noctuidae), a species new to Britain. *Entomologist's Gaz.* **24**: 219–222.

Huggins, H. C., 1957. Rearing *Mamestra albicolon* Hb. and *Heliothis dipsacea* Linn. *Entomologist's Rec. J. Var.* **69**: 174–175.

———, 1964. Dingle 1963. *Ibid.* **76**: 18–20.

———, 1967. The Dingle peninsula, 1967. *Ibid.* **79**: 280–284.

Kettlewell, H. B. D., 1961. Selection experiments on melanism in *Amathes glareosa* Esp. (Lepidoptera). *Heredity, Lond.* **16**: 415–434.

———, 1973. *The evolution of melanism*, xv, 423 pp., 39 pls. (3 col.). Oxford.

——— & Cadbury, C. J., 1963. Investigations on the origins of non-industrial melanism. *Entomologist's Rec. J. Var.* **75**: 149–160.

——— & Gibson, C., 1973. *Amathes* (*Paradiarsia*) *glareosa* (Esper) f. *edda* Staudinger on the mainland of Scotland. *Ibid.* **85**: 240.

Kloet, G. S. & Hincks, W. D., 1972. A check list of British insects: Lepidoptera. (Edn 2). *Handbk Ident. Br. Insects* **11** (2): viii, 153 pp.

Lees, F. H., 1954. An account of rearing *Heliophobus anceps* Schiff. *Entomologist's Rec. J. Var.* **66**: 4–5.

Lempke, B. J., 1962. Catalogus de Nederlandse Macrolepidoptera (Negende supplement). *Tijdschr. Ent.* **105**: 142–232.

Lhomme, L., 1923–1935. *Catalogue des Lépidoptères de France et de Belgique*, **1**: 140–344. Le Carriol.

Littlewood, F., 1941. On rearing Lepidoptera. *Entomologist* **74**: 88–94, 101–106, 124–130, 161–165, 177–183, 205–209, 230–235, 255–260, 270–274, 1 fig.

Long, R., 1965. Notes on recent additions to the Macrolepidoptera of Great Britain and the Channel Islands. *Entomologist's Gaz.* **16**: 17–19.

Marchant, M. E., 1978. *Ochropleura fennica* (Tauscher) (Lep., Noctuidae): Eversmann's Rustic in Nottinghamshire. *Entomologist's Rec. J. Var.* **90**: 248–249.

Marsden, C. & Young, M. R., 1978. *Ochropleura fennica* (Tauscher) (Eversmann's Rustic) in Aberdeenshire. *Ibid.* **90**: 84.

Meyrick, E., 1928. *A revised handbook of the British Lepidoptera*, 914 pp. London.

Mikkola, K. & Jalas, I., 1977. *Suomen Perhoset. Yökköset*, **1**: 255 pp., 18 pls. Helsinki.

Millward, G. D. 1907. Another note on *Naenia typica*. *Entomologist's Rec. J. Var.* **19**: 23.

Myers, A. A., 1968. The diet of the *Hadena rivularis* Fab. (Lep., Noctuidae) group of species. *Entomologist* **101**: 147–149.

——, 1975. Temporary residence of *Mythimna loreyi* (Duponchel) in S.W. Ireland with a note on the occurrence of other migratory Lepidoptera. *Entomologist's Rec. J. Var.* **87**: 302.

——, 1977. The effect of temperature during pupal actiphase on imaginal colouration in *Mythimna loreyi* (Duponchel) (Lep., Noctuidae). *Entomologist's Gaz.* **28**: 75–79, 1 col. pl.

Newman, E., 1869. *The natural history of British moths*, viii, 486 pp. London.

Nordström, F., Wahlgren, E. & Tullgren, A., 1935–1941. *Svenska Fjärilar*, (1): lv, 86 pp., 66 figs; (2): 353 pp., 369 figs, 1 map, 50 col. pls. Stockholm.

Nye, I. W. B., 1975. *The generic names of moths of the world*, **1**: 568 pp. London.

Old Moth Hunter, An, 1952. *Hadena blenna* Hübner in England. *Entomologist's Rec. J. Var.* **64**: 174–177.

Parry, G., 1873. Supposed occurrence of *Leucania commoides* in Kent. *Entomologist* **6**: 522–523.

Perring, F. H. & Walters, S. M., eds, 1962. *Atlas of the British flora* (Edn 1), xxiv, 432 pp. London.

Pierce, C. W., 1971. Needham Market Lepidoptera in 1970. *Suffolk nat. Hist.* **15** (4): 372–375.

Pierce F. N., 1909. *The genitalia of the group Noctuidae of the Lepidoptera of the British Islands*, xii, 88 pp., 32 pls. Liverpool.

——, 1942. *The genitalia of the group Noctuidae of the Lepidoptera of the British Islands. An account of the morphology of the female reproductive organs*, 64 pp., 15 pls. Oundle.

Redway, D. B., 1973. A further record of *Amathes alpicola* (Zetterstedt) (Lep., Noctuidae) in Ireland. *Entomologist's Gaz.* **24**: 296.

—— & Heath, J., 1973. *Amathes alpicola* (Zetterstedt) (Lep., Noctuidae) in Ireland and on the Pennines. *Ibid.* **24**: 6.

Reed, M. D., 1963. The small ranunculus moth (*Hadena dysodea* Schiff.) in Hertfordshire. *Bull. amat. Ent. Soc.* **22**: 67–68.

Revell, R. J., 1965. The rosy marsh moth, *Coenophila subrosea* Stephens in Wales. *Entomologist's Gaz.* **16**: 162.

Russell, A. G. B. & de Worms, C. G. M., 1944. A new locality for *Amathes alpicola* Zett. (Lep. Agrotidae). *Entomologist* **77**: 1–4.

Scorer, A. G., 1913. *The entomologist's log-book, and dictionary of the life histories and food plants of the British Macrolepidoptera*, vii, 374 pp. London.

Shorey, H. H. & Hale, R. L., 1965. Mass-rearing of the larvae of nine noctuid species on a simple artificial medium. *J. econ. Ent.* **58**: 522–524.

Singh, M. P. & Kevan, D. K. McE., 1956. Notes on three common British species of agrotid moth. 1. Longevity and oviposition. *Entomologist's Rec. J. Var.* **68**: 233–235.

Singh, P., 1977. *Artificial diets for insects, mites and spiders*, vii, 594 pp. Auckland.

Skinner, B. F., 1964. The small ranunculus moth in Hertfordshire: a correction. *Bull. amat. Ent. Soc.* **23**: 34–35.

——, 1975. *Mythimna loreyi* Dup. (The Cosmopolitan), *Colias croceus* Geoff. and other immigrants in Cornwall in 1975. *Entomologist's Rec. J. Var.* **87**: 276.

Smith, S. G., 1947. Records of butterflies and moths in the counties of Cheshire, Flintshire, Denbighshire, Caernarvonshire, Anglesey and Merionethshire. *Rep. Lancs. Chesh. ent. Soc.* **1947**: 32.

——, 1948. The butterflies and moths of Cheshire, Flintshire, Denbighshire, Caernarvonshire, Anglesey and Merionethshire. *Proc. Chester Soc. nat. Sci. Lit. Art* **2**: 228.

——, 1954. Capture of a New Zealand agrotid in England. *Entomologist's Rec. J. Var.* **66**: 20.

South, R., 1961. *The moths of the British Isles* (Edn 4), **1**: 427 pp., 148 pls. London.

Sperring, A. H., 1949. A search for *H. suasa*, Hayling Island. *Entomologist's Rec. J. Var.* **61**: 94.

Spuler, A., 1908–1910. *Die Schmetterlinge Europas*, 1–4. Stuttgart.

Stephens, J. F., 1829. *A systematic catalogue of British insects*, xxxiv, 388, 416 pp. London.

Stokoe, W. J. & Stovin, G. H. T., 1958. *The caterpillars of British moths* (Edn 2), **1**: 408 pp., 90 pls. London.

Styles, J. H., 1960. *Syndemis musculana* Hübner (Lep., Tortricidae) in conifer plantations and forest nurseries in the British Isles. *Entomologist's Gaz.* **11**: 144–148, 2 figs.

Symes, H., 1957. Notes on the treatment of hibernating larvae. *Entomologist's Rec. J. Var.* **69**: 208–211.

Todd, R. G., 1964. Food plant of *Leucania obsoleta* Hübn. *Ibid.* **76**: 268.

Turner, H. J., 1926–1950. *The British Noctuae and their varieties* (*J. W. Tutt*). *Supplementary notes*, **1–4**. London.

Tutt, J. W., 1891–1892. *The British Noctuae and their varieties*, **1–4**. London.

——, 1902. *Practical hints for the field lepidopterist*, **2**. London.

——, 1904. Variation of *Leucania favicolor*. *Entomologist's Rec. J. Var.* **16**: 252–254.

Tweedie, M. W. F., 1969. Exhibit – 97th A.G.M., 23rd January, 1969. *Proc. Trans. Br. ent. nat. Hist. Soc.* **2**: 43.

Urbahn, E., 1970. Das alte *Diarsia rubi-florida*-Problem neu untersucht (Lep. Noct.). *Z. wien. ent. Ges.* **54**: 8–22.

Vallins, F. T., 1951. (No title). *Proc. Trans. S. Lond. ent. nat. Hist. Soc.* **1950–51**: 12, 46.

Vine Hall, J. H., 1962. Local variation in certain Lepidoptera occurring in the Spey valley. *Entomologist's Gaz.* **13**: 68–71.

Wightman, A. J., 1940a. Random notes on British Noctuae. *Entomologist's Rec. J. Var.* **52**: 117–120.

——, 1940b. *Harmodia* (*Dianthoecia*) *lepida*, Esp. (*carpophaga*, Bork.). *Ibid.* **52**: 126–128.

——, 1943. Foodplant of *Leucania impudens*. *Ibid.* **55**: 100–101.

——, 1969. *Rhyacia lucipeta* (Denis & Schiffermüller) (Lep., Noctuidae), a new migrant to the British Isles. *Entomologist's Gaz.* **20**: 50.

Williams, C. B., 1958. *Insect migration*, xiii, 235 pp., 49 figs, 24 (8 col.) pls. London.

Withers, B. G., 1974. *Xestia alpicola* (Zetterstedt) (Lep., Noctuidae) in Westmorland. *Entomologist's Gaz.* **25**: 87–88.

Wood, W., 1839. *Index entomologicus*, 266 pp., 54 pls. London.

Worms, de, C. G. M., 1968. The recent discovery in Wales of the rosy marsh moth *Coenophila subrosea* (Stephens) (Lep., Noctuidae). *Entomologist's Gaz.* **19**: 83–89, 1 col. pl.

Young, M. R., 1976. *Xestia alpicola* (Zetterstedt) (Lep., Noctuidae) in Northumberland. *Ibid.* **27**: 274.

Addenda and Notes

Addenda and Notes

THE PLATES

Plate 1: Sphingidae

Figs 1–10, × 1

1 *Mimas tiliae* (Linnaeus) ♂ Lime Hawk-moth. *Page 26*
2 *Mimas tiliae* (Linnaeus) ab. *brunnea* Tutt ♂ Lime Hawk-moth. *Page 26*
3 *Mimas tiliae* (Linnaeus) ab. *centripuncta* Clark ♂ Lime Hawk-moth. *Page 26*
4 *Daphnis nerii* (Linnaeus) ♂ Oleander Hawk-moth. *Page 32*
5 *Hyles gallii* (Rottemburg) ♂ Bedstraw Hawk-moth. *Page 34*
6 *Laothoe populi* (Linnaeus) ♂ Poplar Hawk-moth. *Page 28*
7 *Hyles euphorbiae* (Linnaeus) ♂ Spurge Hawk-moth. *Page 33*
8 *Hyles lineata livornica* (Esper) ♂ Striped Hawk-moth. *Page 35*
9 *Smerinthus ocellata* (Linnaeus) ♂ Eyed Hawk-moth. *Page 27*
10 *Hippotion celerio* (Linnaeus) ♂ Silver-striped Hawk-moth. *Page 38*

Plate 1: Sphingidae

Figs 1–10, ×1

Plate 2: Sphingidae

Figs 1–9, × 1

1 *Agrius convolvuli* (Linnaeus) ♂ Convolvulus Hawk-moth. *Page 21*
2 *Hemaris fuciformis* (Linnaeus) ♂ Broad-bordered Bee Hawk-moth. *Page 30*
3 *Acherontia atropos* (Linnaeus) ♂ Death's-head Hawk-moth. *Page 23*
4 *Hemaris tityus* (Linnaeus) ♂ Narrow-bordered Bee Hawk-moth. *Page 29*
5 *Macroglossum stellatarum* (Linnaeus) ♂ Humming-bird Hawk-moth. *Page 31*
6 *Sphinx ligustri* Linnaeus ♂ Privet Hawk-moth. *Page 24*
7 *Deilephila porcellus* (Linnaeus) ♂ Small Elephant Hawk-moth. *Page 37*
8 *Hyloicus pinastri* (Linnaeus) ♂ Pine Hawk-moth. *Page 25*
9 *Deilephila elpenor* (Linnaeus) ♀ Elephant Hawk-moth. *Page 36*

Plate 2: Sphingidae

Figs 1–9, × 1

Plate 3: Notodontidae

Figs 1–33, × 1

1 *Phalera bucephala* (Linnaeus) ♂ Buff-tip. *Page 41*
2 *Furcula furcula* (Clerck) ♂ Sallow Kitten. *Page 43*
3 *Furcula furcula* (Clerck) ♂ Sallow Kitten (Surrey). *Page 43*
4 *Furcula bifida* (Brahm) ♂ Poplar Kitten. *Page 44*
5 *Furcula bicuspis* (Borkhausen) ♂ Alder Kitten. *Page 43*
6 *Cerura vinula* (Linnaeus) ♂ Puss Moth. *Page 42*
7 *Cerura vinula* (Linnaeus) ♀ Puss Moth. *Page 42*
8 *Stauropus fagi* (Linnaeus) ♂ Lobster Moth. *Page 45*
9 *Stauropus fagi* (Linnaeus) ♂ Lobster Moth. *Page 45*
10 *Notodonta dromedarius* (Linnaeus) ♂ Iron Prominent. *Page 46*
11 *Notodonta dromedarius* (Linnaeus) ♂ Iron Prominent (Scotland). *Page 46*
12 *Tritophia tritophus* ([Denis & Schiffermüller]) ♂ Three Humped Prominent. *Page 47*
13 *Eligmodonta ziczac* (Linnaeus) ♂ Pebble Prominent. *Page 48*
14 *Eligmodonta ziczac* (Linnaeus) ♂ Pebble Prominent (Scotland). *Page 48*
15 *Eligmodonta ziczac* (Linnaeus) ♂ Pebble Prominent (Ireland). *Page 48*
16 *Peridea anceps* (Goeze) ♂ Great Prominent. *Page 50*
17 *Peridea anceps* (Goeze) ♂ Great Prominent. *Page 50*
18 *Pheosia gnoma* (Fabricius) ♂ Lesser Swallow Prominent. *Page 51*
19 *Pheosia gnoma* (Fabricius) ♂ Lesser Swallow Prominent (Scotland). *Page 51*
20 *Pheosia tremula* (Clerck) ♂ Swallow Prominent. *Page 52*
21 *Pheosia tremula* (Clerck) ♂ Swallow Prominent (Scotland). *Page 52*
22 *Ptilodon capucina* (Linnaeus) ♂ Coxcomb Prominent. *Page 53*
23 *Ptilodon capucina* (Linnaeus) ♂ Coxcomb Prominent. *Page 53*
24 *Ptilodontella cucullina* ([Denis & Schiffermüller]) ♂ Maple Prominent. *Page 54*
25 *Odontosia carmelita* (Esper) ♂ Scarce Prominent. *Page 55*
26 *Pterostoma palpina* (Clerck) ♂ Pale Prominent. *Page 56*
27 *Leucodonta bicoloria* ([Denis & Schiffermüller]) ♂ White Prominent. *Page 57*
28 *Ptilophora plumigera* ([Denis & Schiffermüller]) ♂ Plumed Prominent. *Page 58*
29 *Ptilophora plumigera* ([Denis & Schiffermüller]) ♂ Plumed Prominent. *Page 58*
30 *Drymonia dodonaea* ([Denis & Schiffermüller]) ♂ Marbled Brown. *Page 59*
31 *Drymonia dodonaea* ([Denis & Schiffermüller]) ♂ Marbled Brown. *Page 59*
32 *Drymonia dodonaea* ([Denis & Schiffermüller]) ♂ Marbled Brown. *Page 59*
33 *Drymonia dodonaea* ([Denis & Schiffermüller]) ♂ Marbled Brown. *Page 59*

Plate 3: Notodontidae

Figs 1–33, ×1

Plate 4: Notodontidae, Lymantriidae

Figs 1–35, × 1

1. *Drymonia ruficornis* (Hufnagel) ♂ Lunar Marbled Brown. *Page 60*
2. *Drymonia ruficornis* (Hufnagel) ♂ Lunar Marbled Brown. *Page 60*
3. *Drymonia ruficornis* (Hufnagel) ♀ Lunar Marbled Brown. *Page 60*
4. *Drymonia ruficornis* (Hufnagel) ♀ Lunar Marbled Brown. *Page 60*
5. *Gluphisia crenata vertunea* Bray ♂ Dusky Marbled Brown. *Page 61*
6. *Clostera pigra* (Hufnagel) ♂ Small Chocolate-tip. *Page 62*
7. *Clostera pigra* (Hufnagel) ♀ Small Chocolate-tip. *Page 62*
8. *Clostera anachoreta* ([Denis & Schiffermüller]) ♂ Scarce Chocolate-tip. *Page 63*
9. *Diloba caeruleocephala* (Linnaeus) ♂ Figure of Eight Moth. *Page 64*
10. *Diloba caeruleocephala* (Linnaeus) ♀ Figure of Eight Moth. *Page 64*
11. *Clostera curtula* (Linnaeus) ♂ Chocolate-tip. *Page 63*
12. *Laelia coenosa* (Hübner) ♂ Reed Tussock. *Page 68*
13. *Laelia coenosa* (Hübner) ♀ Reed Tussock. *Page 68*
14. *Orgyia recens* (Hübner) ♂ Scarce Vapourer. *Page 69*
15. *Orgyia recens* (Hübner) ♀ Scarce Vapourer. *Page 69*
16. *Orgyia antiqua* (Linnaeus) ♂ Vapourer. *Page 70*
17. *Orgyia antiqua* (Linnaeus) ♀ Vapourer. *Page 70*
18. *Dasychira pudibunda* (Linnaeus) ♂ Pale Tussock. *Page 72*
19. *Dasychira pudibunda* (Linnaeus) ♂ Pale Tussock. *Page 72*
20. *Dasychira pudibunda* (Linnaeus) ♀ Pale Tussock. *Page 72*
21. *Dasychira fascelina* (Linnaeus) ♂ Dark Tussock. *Page 71*
22. *Dasychira fascelina* (Linnaeus) ♀ Dark Tussock. *Page 71*
23. *Euproctis chrysorrhoea* (Linnaeus) ♂ Brown-tail. *Page 73*
24. *Euproctis chrysorrhoea* (Linnaeus) ♀ Brown-tail. *Page 73*
25. *Euproctis similis* (Fuessly) ♂ Yellow-tail. *Page 74*
26. *Euproctis similis* (Fuessly) ♀ Yellow-tail. *Page 74*
27. *Leucoma salicis* (Linnaeus) ♂ White Satin Moth. *Page 75*
28. *Leucoma salicis* (Linnaeus) ♀ White Satin Moth. *Page 75*
29. *Arctornis l-nigrum* (Müller) ♂ Black V Moth. *Page 76*
30. *Lymantria dispar* (Linnaeus) ♂ Gipsy Moth. *Page 77*
31. *Lymantria dispar* (Linnaeus) ♀ Gipsy Moth. *Page 77*
32. *Lymantria monacha* (Linnaeus) ♂ Black Arches. *Page 76*
33. *Lymantria monacha* (Linnaeus) ♂ Black Arches. *Page 76*
34. *Lymantria monacha* (Linnaeus) ♀ Black Arches. *Page 76*
35. *Lymantria monacha* (Linnaeus) f. *eremita* Ochsenheimer ♀ Black Arches. *Page 76*

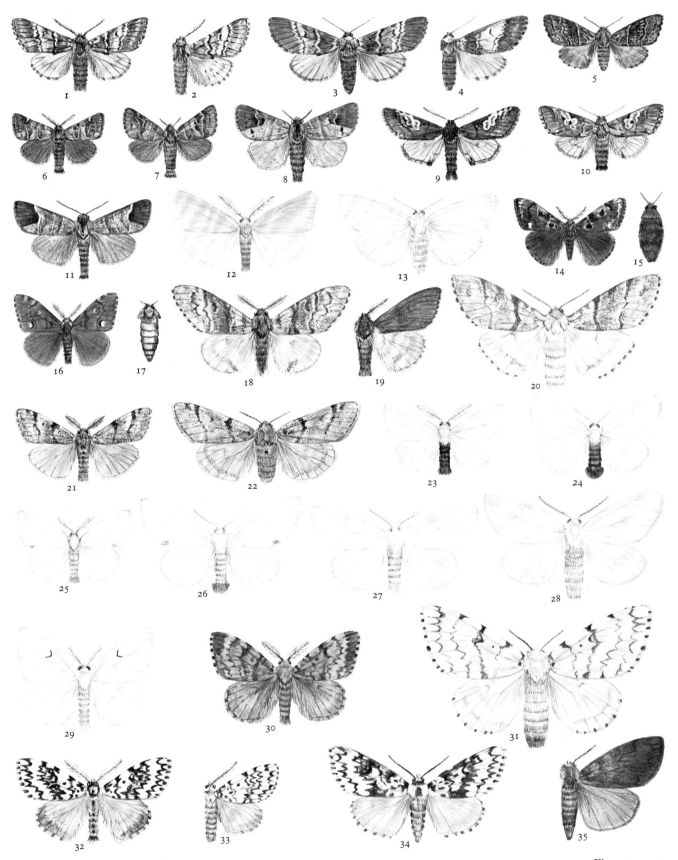

Plate 4: Notodontidae, Lymantriidae   Figs 1–35, ×1

Plate 5: Arctiidae

Figs 1–41, × 1

1 *Thumatha senex* (Hübner) ♂ Round-winged Muslin. *Page 81*
2 *Setina irrorella* (Linnaeus) ♂ Dew Moth. *Page 82*
3 *Setina irrorella* (Linnaeus) f. *signata* Borkhausen ♂ Dew Moth. *Page 82*
4 *Setina irrorella* (Linnaeus) ♀ Dew Moth. *Page 82*
5 *Miltochrista miniata* (Forster) ♂ Rosy Footman. *Page 83*
6 *Nudaria mundana* (Linnaeus) ♂ Muslin Footman. *Page 84*
7 *Atolmis rubricollis* (Linnaeus) ♂ Red-necked Footman. *Page 85*
8 *Cybosia mesomella* (Linnaeus) ♂ Four-dotted Footman. *Page 86*
9 *Cybosia mesomella* (Linnaeus) ♂ Four-dotted Footman (Kent). *Page 86*
10 *Pelosia muscerda* (Hufnagel) ♂ Dotted Footman. *Page 87*
11 *Pelosia obtusa* (Herrich-Schäffer) ♂ Small Dotted Footman. *Page 87*
12 *Eilema sororcula* (Hufnagel) ♂ Orange Footman. *Page 88*
13 *Eilema griseola* (Hübner) ♂ Dingy Footman. *Page 89*
14 *Eilema griseola* f. *stramineola* Doubleday ♂ Straw-coloured Footman. *Page 89*
15 *Eilema caniola* (Hübner) ♂ Hoary Footman. *Page 90*
16 *Eilema pygmaeola* (Doubleday) ♂ Pigmy Footman. *Page 91*
17 *Eilema pygmaeola pallifrons* (Zeller) ♂ Pigmy Footman. *Page 91*
18 *Eilema complana* (Linnaeus) ♂ Scarce Footman. *Page 91*
19 *Eilema deplana* (Esper) ♂ Buff Footman. *Page 93*
20 *Eilema deplana* (Esper) ♀ Buff Footman. *Page 93*
21 *Lithosia quadra* (Linnaeus) ♂ Four-spotted Footman. *Page 95*
22 *Lithosia quadra* (Linnaeus) ♀ Four-spotted Footman. *Page 95*
23 *Eilema sericea* (Gregson) ♂ Northern Footman. *Page 92*
24 *Eilema lurideola* ([Zincken]) ♂ Common Footman. *Page 94*
25 *Spiris striata* (Linnaeus) ♂ Feathered Footman. *Page 96*
26 *Spiris striata* (Linnaeus) ♀ Feathered Footman. *Page 96*
27 *Coscinia cribraria* (Linnaeus) ♂ Speckled Footman. *Page 96*
28 *Utetheisa pulchella* (Linnaeus) ♂ Crimson Speckled. *Page 97*
29 *Parasemia plantaginis* (Linnaeus) ♂ Wood Tiger. *Page 98*
30 *Parasemia plantaginis* (Linnaeus) f. *hospita* [Denis & Schiffermüller] ♂ Wood Tiger. *Page 98*
31 *Parasemia plantaginis* (Linnaeus) ab. *matronalis* Freyer ♂ Wood Tiger. *Page 98*
32 *Parasemia plantaginis* (Linnaeus) ♀ Wood Tiger. *Page 98*
33 *Parasemia plantaginis* (Linnaeus) ab. *rufa* Tutt ♀ Wood Tiger. *Page 98*
34 *Arctia caja* (Linnaeus) ♂ Garden Tiger. *Page 99*
35 *Arctia caja* (Linnaeus) ab. *fumosa* Hörhammer ♂ Garden Tiger. *Page 99*
36 *Arctia caja* (Linnaeus) ab. *brunnescens* Stättermayer ♂ Garden Tiger. *Page 99*
37 *Arctia caja* (Linnaeus) ab. *petriburgensis* Cockayne ♂ Garden Tiger. *Page 99*
38 *Arctia caja* (Linnaeus) ab. *nigrescens* Lambillion ♂ Garden Tiger. *Page 99*
39 *Arctia caja* (Linnaeus) ab. *decrescens-lutescens* Cockayne ♂ Garden Tiger. *Page 99*
40 *Arctia caja* (Linnaeus) ab. ♀ Garden Tiger. *Page 99*
41 *Arctia caja* (Linnaeus) ab. ♀ Garden Tiger. *Page 99*

Plate 5: Arctiidae

Figs 1–41, ×1

Plate 6: Arctiidae, Nolidae

Figs 1–34, × 1; 35–47, × 1.5

1 *Arctia villica* (Linnaeus) ♂ Cream-spot Tiger. *Page 100*
2 *Arctia villica* (Linnaeus) ab. *ursula* Schultz ♂ Cream-spot Tiger. *Page 100*
3 *Arctia villica* (Linnaeus) ab. *wardi* Mathew ♂ Cream-spot Tiger. *Page 100*
4 *Arctia villica* (Linnaeus) ♀ Cream-spot Tiger. *Page 100*
5 *Diacrisia sannio* (Linnaeus) ♂ Clouded Buff. *Page 101*
6 *Diacrisia sannio* (Linnaeus) ab. *maerens* Strand ♂ Clouded Buff. *Page 101*
7 *Diacrisia sannio* (Linnaeus) ab. *immarginata* Niepelt ♂ Clouded Buff. *Page 101*
8 *Diacrisia sannio* (Linnaeus) ♀ Clouded Buff. *Page 101*
9 *Spilosoma urticae* (Esper) ♂ Water Ermine. *Page 104*
10 *Spilosoma lubricipeda* (Linnaeus) ♂ White Ermine. *Page 102*
11 *Spilosoma lubricipeda* (Linnaeus) ♂ White Ermine. *Page 102*
12 *Spilosoma lubricipeda* (Linnaeus) ab. *brunnea* Oberthür ♂ White Ermine. *Page 102*
13 *Spilosoma lubricipeda* (Linnaeus) ♀ White Ermine (Ireland). *Page 102*
14 *Spilosoma lubricipeda* (Linnaeus) ab. *godarti* Oberthür ♀ White Ermine. *Page 102*
15 *Spilosoma lutea* (Hufnagel) ♂ Buff Ermine. *Page 103*
16 *Spilosoma lutea* (Hufnagel) ♂ Buff Ermine. *Page 103*
17 *Spilosoma lutea* (Hufnagel) ab. *zatima* Stoll ♂ Buff Ermine. *Page 103*
18 *Spilosoma lutea* (Hufnagel) ab. *zatima* Stoll ♂ Buff Ermine. *Page 103*
19 *Spilosoma lutea* (Hufnagel) ♀ Buff Ermine. *Page 103*
20 *Spilosoma lutea* (Hufnagel) ab. *intermedia* Standfuss ♀ Buff Ermine. *Page 103*
21 *Diaphora mendica* (Clerck) ♂ Muslin Moth. *Page 105*
22 *Diaphora mendica* (Clerck) f. *rustica* Hübner ♂ Muslin Moth (Ireland). *Page 105*
23 *Diaphora mendica* (Clerck) f. *rustica* Hübner ♂ Muslin Moth (Ireland). *Page 105*
24 *Diaphora mendica* (Clerck) ♀ Muslin Moth. *Page 105*
25 *Phragmatobia fuliginosa fuliginosa* (Linnaeus) ♂ Ruby Tiger. *Page 106*
26 *Phragmatobia fuliginosa borealis* (Staudinger) ♂ Ruby Tiger. *Page 106*
27 *Euplagia quadripunctaria* (Poda) ♂ Jersey Tiger. *Page 107*
28 *Euplagia quadripunctaria* (Poda) ab. *lutescens* Staudinger ♂ Jersey Tiger. *Page 107*
29 *Tyria jacobaeae* (Linnaeus) ♂ Cinnabar. *Page 109*
30 *Tyria jacobaeae* (Linnaeus) ♀ Cinnabar. *Page 109*
31 *Callimorpha dominula* (Linnaeus) ♂ Scarlet Tiger. *Page 108*
32 *Callimorpha dominula* (Linnaeus) f. *bimacula* Cockayne ♂ Scarlet Tiger. *Page 108*
33 *Callimorpha dominula* (Linnaeus) f. *ocellata* Kettlewell ♂ Scarlet Tiger. *Page 108*
34 *Callimorpha dominula* (Linnaeus) f. *crocea* Schultz ♀ Scarlet Tiger. *Page 108*
35 *Meganola strigula* ([Denis & Schiffermüller]) ♂ Small Black Arches. *Page 113*
36 *Meganola strigula* ([Denis & Schiffermüller]) ♀ Small Black Arches. *Page 113*
37 *Nola cucullatella* (Linnaeus) ♂ Short-cloaked Moth. *Page 117*
38 *Nola cucullatella* (Linnaeus) ♂ Short-cloaked Moth. *Page 117*
39 *Nola aerugula* (Hübner) ♂ Scarce Black Arches. *Page 119*
40 *Nola aerugula* (Hübner) ♂ Scarce Black Arches. *Page 119*
41 *Nola aerugula* (Hübner) ♂ Scarce Black Arches. *Page 119*
42 *Meganola albula* ([Denis & Schiffermüller]) ♂ Kent Black Arches. *Page 115*
43 *Meganola albula* ([Denis & Schiffermüller]) ♂ Kent Black Arches. *Page 115*
44 *Meganola albula* ([Denis & Schiffermüller]) ♀ Kent Black Arches. *Page 115*
45 *Nola confusalis* (Herrich-Schäffer) ♂ Least Black Arches. *Page 118*
46 *Nola confusalis* (Herrich-Schäffer) ♀ Least Black Arches. *Page 118*
47 *Nola confusalis* (Herrich-Schäffer) ♀ Least Black Arches. *Page 118*

Plate 6: *Arctiidae, Nolidae*

Figs 1–34, × 1; 35–47, × 1.5

Plate 7: Noctuidae Noctuinae

Figs 1–54, × 1

1. *Euxoa obelisca grisea* (Tutt) ♂ Square-spot Dart. *Page 130*
2. *Euxoa obelisca grisea* (Tutt) ♀ Square-spot Dart. *Page 130*
3. *Euxoa tritici* (Linnaeus) ♂ White-line Dart (Orkney). *Page 132*
4. *Euxoa tritici* (Linnaeus) ♀ White-line Dart (Orkney). *Page 132*
5. *Euxoa tritici* (Linnaeus) ♀ White-line Dart (Suffolk). *Page 132*
6. *Euxoa tritici* (Linnaeus) ♀ White-line Dart. *Page 132*
7. *Euxoa tritici* (Linnaeus) ab. *rhabdota* Edelsten ♂ White-line Dart (S. Devon). *Page 132*
8. *Euxoa nigricans* (Linnaeus) ♂ Garden Dart (Scotland). *Page 134*
9. *Euxoa nigricans* (Linnaeus) ♂ Garden Dart (S. England). *Page 134*
10. *Euxoa nigricans* (Linnaeus) ♀ Garden Dart (S. England). *Page 134*
11. *Euxoa cursoria* (Hufnagel) ♂ Coast Dart (Suffolk). *Page 136*
12. *Euxoa cursoria* (Hufnagel) ♂ Coast Dart (Suffolk). *Page 136*
13. *Euxoa cursoria* (Hufnagel) ♀ Coast Dart (Suffolk). *Page 136*
14. *Euxoa cursoria* (Hufnagel) ab. *distincta* Tutt ♀ Coast Dart. *Page 136*
15. *Euxoa cursoria* (Hufnagel) ♂ Coast Dart (Morayshire). *Page 136*
16. *Euxoa cursoria* (Hufnagel) ♀ Coast Dart (Morayshire). *Page 136*
17. *Euxoa cursoria* (Hufnagel) ab. *sagitta* Hübner ♂ Coast Dart (Shetland). *Page 136*
18. *Euxoa cursoria* (Hufnagel) ♂ Coast Dart (Shetland). *Page 136*
19. *Agrotis cinerea* ([Denis & Schiffermüller]) ♂ Light Feathered Rustic. *Page 137*
20. *Agrotis cinerea* ([Denis & Schiffermüller]) ♂ Light Feathered Rustic. *Page 137*
21. *Agrotis cinerea* ([Denis & Schiffermüller]) ♂ Light Feathered Rustic. *Page 137*
22. *Agrotis cinerea* ([Denis & Schiffermüller]) ♀ Light Feathered Rustic. *Page 137*
23. *Agrotis cinerea* ([Denis & Schiffermüller]) ♀ Light Feathered Rustic. *Page 137*
24. *Agrotis cinerea* ([Denis & Schiffermüller]) ♀ Light Feathered Rustic. *Page 137*
25. *Agrotis vestigialis* (Hufnagel) ♂ Archer's Dart (Sussex). *Page 138*
26. *Agrotis vestigialis* (Hufnagel) ♀ Archer's Dart (Sussex). *Page 138*
27. *Agrotis vestigialis* (Hufnagel) ♂ Archer's Dart (Studland, Dorset). *Page 138*
28. *Agrotis vestigialis* (Hufnagel) ♀ Archer's Dart (Studland, Dorset). *Page 138*
29. *Agrotis vestigialis* (Hufnagel) ♂ Archer's Dart (Breckland). *Page 138*
30. *Agrotis vestigialis* (Hufnagel) ♀ Archer's Dart (Breckland). *Page 138*
31. *Agrotis segetum* ([Denis & Schiffermüller]) ♂ Turnip Moth. *Page 139*
32. *Agrotis segetum* ([Denis & Schiffermüller]) ♂ Turnip Moth. *Page 139*
33. *Agrotis segetum* ([Denis & Schiffermüller]) ♀ Turnip Moth. *Page 139*
34. *Agrotis segetum* ([Denis & Schiffermüller]) ab. *subatratus* Haworth ♀ Turnip Moth. *Page 139*
35. *Agrotis segetum* ([Denis & Schiffermüller]) ♂ Turnip Moth. *Page 139*
36. *Agrotis segetum* ([Denis & Schiffermüller]) ♀ Turnip Moth. *Page 139*
37. *Agrotis clavis* (Hufnagel) ♂ Heart and Club. *Page 141*
38. *Agrotis clavis* (Hufnagel) ♂ Heart and Club. *Page 141*
39. *Agrotis clavis* (Hufnagel) ♂ Heart and Club. *Page 141*
40. *Agrotis clavis* (Hufnagel) ♀ Heart and Club. *Page 141*
41. *Agrotis clavis* (Hufnagel) ♀ Heart and Club. *Page 141*
42. *Agrotis clavis* (Hufnagel) ♂ Heart and Club. *Page 141*
43. *Agrotis clavis* (Hufnagel) ab. *venosa* Tutt ♀ Heart and Club. *Page 141*
44. *Agrotis exclamationis* (Linnaeus) ♂ Heart and Dart. *Page 141*
45. *Agrotis exclamationis* (Linnaeus) ♀ Heart and Dart. *Page 141*
46. *Agrotis exclamationis* (Linnaeus) ♀ Heart and Dart. *Page 141*
47. *Agrotis exclamationis* (Linnaeus) ♂ Heart and Dart. *Page 141*
48. *Agrotis exclamationis* (Linnaeus) ab. *catenata* Wize ♀ Heart and Dart. *Page 141*
49. *Agrotis exclamationis* (Linnaeus) ♀ Heart and Dart. *Page 141*
50. *Agrotis exclamationis* (Linnaeus) ♀ Heart and Dart. *Page 141*
51. *Agrotis exclamationis* (Linnaeus) ab. *juncta* Tutt. ♂ Heart and Dart. *Page 141*
52. *Agrotis exclamationis* (Linnaeus) ab. *lineolatus* Tutt ♀ Heart and Dart. *Page 141*
53. *Agrotis trux lunigera* Stephens ♂ Crescent Dart. *Page 142*
54. *Agrotis trux lunigera* Stephens ♀ Crescent Dart. *Page 142*

Plate 7: Noctuidae Noctuinae

Figs 1–54, ×1

Plate 8: Noctuidae Noctuinae

Figs 1–40, × 1

1 *Agrotis puta* (Hübner) ♂ Shuttle-shaped Dart. *Page 144*
2 *Agrotis puta* (Hübner) ♀ Shuttle-shaped Dart. *Page 144*
3 *Agrotis puta insula* Richardson ♂ Shuttle-shaped Dart (Isles of Scilly). *Page 144*
4 *Agrotis puta insula* Richardson ♀ Shuttle-shaped Dart (Isles of Scilly). *Page 144*
5 *Ochropleura plecta* (Linnaeus) ♀ Flame Shoulder. *Page 152*
6 *Ochropleura plecta* (Linnaeus) ab. *rubricosta* Fuchs ♀ Flame Shoulder. *Page 152*
7 *Agrotis ripae* (Hübner) ab. *obotritica* Schmidt ♂ Sand Dart (Hampshire). *Page 145*
8 *Agrotis ripae* (Hübner) ♂ Sand Dart (Hampshire). *Page 145*
9 *Agrotis ripae* (Hübner) ♀ Sand Dart (Hampshire). *Page 145*
10 *Agrotis ripae* (Hübner) ab. *desillii* Pierret ♀ Sand Dart (Sussex). *Page 145*
11 *Agrotis ripae* (Hübner) ab. *weissenbornii* Freyer ♀ Sand Dart. *Page 145*
12 *Agrotis ripae* (Hübner) ab. *brunnea* Tutt ♀ Sand Dart ('Erdington'). *Page 145*
13 *Agrotis ripae* (Hübner) ♀ Sand Dart (Rosslare, Co. Wexford). *Page 145*
14 *Agrotis ipsilon* (Hufnagel) ♂ Dark Sword Grass. *Page 143*
15 *Agrotis ipsilon* (Hufnagel) ♀ Dark Sword Grass. *Page 143*
16 *Agrotis crassa* (Hübner) ♂ Great Dart. *Page 146*
17 *Actinotia polyodon* (Clerck) ♀ Purple Cloud. *Page 148*
18 *Axylia putris* (Linnaeus) ♂ Flame. *Page 149*
19 *Ochropleura praecox* (Linnaeus) ♂ Portland Moth. *Page 150*
20 *Ochropleura fennica* (Tauscher) ♀ Eversmann's Rustic. *Page 151*
21 *Ochropleura flammatra* ([Denis & Schiffermüller]) ♀ Black Collar. *Page 151*
22 *Rhyacia lucipeta* ([Denis & Schiffermüller]) ♀ Southern Rustic. *Page 157*
23 *Noctua orbona* (Hufnagel) ♂ Lunar Yellow Underwing. *Page 159*
24 *Noctua interjecta caliginosa* (Schawerda) ♂ Least Yellow Underwing. *Page 163*
25 *Noctua pronuba* (Linnaeus) ab. *ochreabrunnea* Tutt ♂ Large Yellow Underwing. *Page 157*
26 *Noctua pronuba* (Linnaeus) ab. *innuba* Tutt ♂ Large Yellow Underwing. *Page 157*
27 *Noctua pronuba* (Linnaeus) ab. *ochrea* Tutt ♀ Large Yellow Underwing. *Page 157*
28 *Noctua pronuba* (Linnaeus) ab. *caerulescens* Tutt ♀ Large Yellow Underwing. *Page 157*
29 *Noctua pronuba* (Linnaeus) ab. *rufa* Tutt ♀ Large Yellow Underwing. *Page 157*
30 *Noctua pronuba* (Linnaeus) ab. *postnigra* Turner ♀ Large Yellow Underwing. *Page 157*
31 *Noctua fimbriata* (Schreber) ♂ Broad-bordered Yellow Underwing. *Page 161*
32 *Noctua fimbriata* (Schreber) ♀ Broad-bordered Yellow Underwing. *Page 161*
33 *Noctua janthina* ([Denis & Schiffermüller]) ♂ Lesser Broad-bordered Yellow Underwing. *Page 162*
34 *Noctua janthina* ([Denis & Schiffermüller]) ab. *rufa* Tutt ♂ Lesser Broad-bordered Yellow Underwing. *Page 162*
35 *Noctua comes* (Hübner) ♀ Lesser Yellow Underwing. *Page 159*
36 *Noctua comes* (Hübner) ab. *sagittifer* Cockayne ♂ Lesser Yellow Underwing. *Page 159*
37 *Noctua comes* (Hübner) ♂ Lesser Yellow Underwing. *Page 159*
38 *Noctua comes* (Hübner) ab. *rufa* Tutt ♂ Lesser Yellow Underwing. *Page 159*
39 *Noctua comes* (Hübner) ab. *nigrescens* Tutt ♀ Lesser Yellow Underwing. *Page 159*
40 *Noctua comes* (Hübner) ♂ Lesser Yellow Underwing. *Page 159*

Plate 8: Noctuidae Noctuinae

Figs 1–40, ×1

Plate 9: Noctuidae Noctuinae

Figs 1–40, × 1

1 *Standfussiana lucernea* (Linnaeus) ♂ Northern Rustic. *Page 154*
2 *Standfussiana lucernea* (Linnaeus) ab. *renigera* Stephens ♀ Northern Rustic (Shetland). *Page 154*
3 *Rhyacia simulans* (Hufnagel) ♂ Dotted Rustic (S. England) *Page 156*
4 *Rhyacia simulans* (Hufnagel) ♂ Dotted Rustic. *Page 156*
5 *Rhyacia simulans* (Hufnagel) f. *suffusa* Tutt ♀ Dotted Rustic (Orkney). *Page 156*
6 *Graphiphora augur* (Fabricius) ♂ Double Dart (S. England). *Page 165*
7 *Graphiphora augur* (Fabricius) ♀ Double Dart (Scotland). *Page 165*
8 *Eugraphe subrosea* (Stephens) ♂ Rosy Marsh Moth (Wales). *Page 166*
9 *Eugraphe subrosea* (Stephens) ♀ Rosy Marsh Moth (Wales). *Page 166*
10 *Spaelotis ravida* ([Denis & Schiffermüller]) ♂ Stout Dart. *Page 164*
11 *Spaelotis ravida* ([Denis & Schiffermüller]) ♀ Stout Dart. *Page 164*
12 *Paradiarsia glareosa* (Esper) ♂ Autumnal Rustic (S. England). *Page 169*
13 *Paradiarsia glareosa* (Esper) ♀ Autumnal Rustic (S. England). *Page 169*
14 *Paradiarsia glareosa* (Esper) ♂ Autumnal Rustic (Inverness-shire). *Page 169*
15 *Paradiarsia glareosa* (Esper) ab. *edda* Staudinger ♂ Autumnal Rustic (Shetland). *Page 169*
16 *Lycophotia porphyrea* ([Denis & Schiffermüller]) ♂ True Lover's Knot (S. England). *Page 170*
17 *Lycophotia porphyrea* ([Denis & Schiffermüller]) ♂ True Lover's Knot (Shetland). *Page 170*
18 *Paradiarsia sobrina* (Duponchel) ♂ Cousin German. *Page 167*
19 *Diarsia mendica* (Fabricius) ♂ Ingrailed Clay (S. England). *Page 172*
20 *Diarsia mendica* (Fabricius) ♂ Ingrailed Clay (S. England). *Page 172*
21 *Diarsia mendica* (Fabricius) ♂ Ingrailed Clay (S. England). *Page 172*
22 *Diarsia mendica* (Fabricius) ♀ Ingrailed Clay (S. England). *Page 172*
23 *Diarsia mendica* (Fabricius) ♂ Ingrailed Clay (Perthshire). *Page 172*
24 *Diarsia mendica* (Fabricius) ♀ Ingrailed Clay (Perthshire). *Page 172*
25 *Diarsia mendica thulei* (Staudinger) ♂ Ingrailed Clay (Shetland). *Page 172*
26 *Diarsia mendica thulei* (Staudinger) ♂ Ingrailed Clay (Shetland). *Page 172*
27 *Diarsia mendica thulei* (Staudinger) ♀ Ingrailed Clay (Shetland). *Page 172*
28 *Diarsia mendica thulei* (Staudinger) ♀ Ingrailed Clay (Shetland). *Page 172*
29 *Diarsia brunnea* ([Denis & Schiffermüller]) ♂ Purple Clay. *Page 174*
30 *Eugnorisma depuncta* (Linnaeus) ♂ Plain Clay. *Page 153*
31 *Diarsia dahlii* (Hübner) ♂ Barred Chestnut. *Page 173*
32 *Diarsia dahlii* (Hübner) ♀ Barred Chestnut. *Page 173*
33 *Diarsia rubi* (Vieweg) ♂ Small Square Spot (spring brood). *Page 176*
34 *Diarsia rubi* (Vieweg) ♂ Small Square Spot (summer brood). *Page 176*
35 *Diarsia rubi* (Vieweg) ♀ Small Square Spot (Shetland). *Page 176*
36 *Peridroma saucia* (Hübner) ♂ Pearly Underwing. *Page 171*
37 *Peridroma saucia* (Hübner) f. *margaritosa* Haworth ♀ Pearly Underwing. *Page 171*
38 *Peridroma saucia* (Hübner) f. *nigricosta* Tutt ♂ Pearly Underwing. *Page 171*
39 *Diarsia florida* (Schmidt) ♂ Fen Square Spot. *Page 176*
40 *Diarsia florida* (Schmidt) ab. *flava* Walker ♀ Fen Square Spot. *Page 176*

Plate 9: Noctuidae Noctuinae

Figs 1–40, × 1

Plate 10: Noctuidae Noctuinae

Figs 1–40, × 1

1 *Xestia alpicola alpina* (Humphreys & Westwood) ♂ Northern Dart (Perthshire). *Page 178*

2 *Xestia alpicola alpina* (Humphreys & Westwood) ♀ Northern Dart (Perthshire). *Page 178*

3 *Xestia alpicola alpina* (Humphreys & Westwood) ♂ Northern Dart (Aviemore, Inverness-shire). *Page 178*

4 *Xestia alpicola alpina* (Humphreys & Westwood) ♀ Northern Dart (Aviemore, Inverness-shire). *Page 178*

5 *Xestia c-nigrum* (Linnaeus) ♀ Setaceous Hebrew Character. *Page 179*

6 *Xestia ditrapezium* ([Denis & Schiffermüller]) ♂ Triple-spotted Clay. *Page 180*

7 *Xestia triangulum* (Hufnagel) ♀ Double Square-spot. *Page 181*

8 *Xestia ashworthii* (Doubleday) ♂ Ashworth's Rustic. *Page 182*

9 *Xestia ashworthii* (Doubleday) ♀ Ashworth's Rustic. *Page 182*

10 *Xestia baja* ([Denis & Schiffermüller]) ♀ Dotted Clay. *Page 183*

11 *Xestia rhomboidea* (Esper) ♀ Square-spotted Clay. *Page 184*

12 *Xestia castanea* (Esper) ♂ Neglected Rustic (Perthshire). *Page 185*

13 *Xestia castanea* (Esper) ♀ Neglected Rustic (Shropshire). *Page 185*

14 *Xestia castanea* (Esper) ♀ Neglected Rustic (Aviemore, Inverness-shire). *Page 185*

15 *Xestia castanea* (Esper) ab. *neglecta* Hübner ♀ Neglected Rustic. *Page 185*

16 *Xestia xanthographa* ([Denis & Schiffermüller]) ♂ Square-spot Rustic. *Page 187*

17 *Xestia xanthographa* ([Denis & Schiffermüller]) ♂ Square-spot Rustic. *Page 187*

18 *Xestia xanthographa* ([Denis & Schiffermüller]) ♀ Square-spot Rustic. *Page 187*

19 *Xestia xanthographa* ([Denis & Schiffermüller]) ♂ Square-spot Rustic (Shetland). *Page 187*

20 *Xestia xanthographa* ([Denis & Schiffermüller]) ♂ Square-spot Rustic (Perthshire). *Page 187*

21 *Xestia sexstrigata* (Haworth) ♂ Six-striped Rustic. *Page 187*

22 *Xestia agathina* (Duponchel) ♂ Heath Rustic. *Page 188*

23 *Xestia agathina* (Duponchel) ♂ Heath Rustic. *Page 188*

24 *Xestia agathina* (Duponchel) ab. *rosea* Tutt ♂ Heath Rustic. *Page 188*

25 *Xestia agathina* (Duponchel) ab. *scopariae* Millière ♂ Heath Rustic (S.-W. Yorkshire). *Page 188*

26 *Xestia agathina hebridicola* (Staudinger) ♂ Heath Rustic. *Page 188*

27 *Naenia typica* (Linnaeus) ♀ Gothic. *Page 189*

28 *Eurois occulta* (Linnaeus) ♂ Great Brocade. *Page 191*

29 *Eurois occulta* (Linnaeus) ♀ Great Brocade (Shetland). *Page 191*

30 *Eurois occulta* (Linnaeus) f. *passetii* Thierry-Mieg ♀ Great Brocade (Perthshire). *Page 191*

31 *Anaplectoides prasina* ([Denis & Schiffermüller]) ♀ Green Arches. *Page 192*

32 *Anaplectoides prasina* ([Denis & Schiffermüller]) ♀ Green Arches. *Page 192*

33 *Anaplectoides prasina* ([Denis & Schiffermüller]) ab. *demuthi* Richardson ♂ Green Arches. *Page 192*

34 *Mesogona acetosellae* ([Denis & Schiffermüller]) ♀ Pale Stigma. *Page 195*

35 *Cerastis rubricosa* ([Denis & Schiffermüller]) ♂ Red Chestnut (S. England). *Page 193*

36 *Cerastis rubricosa* ([Denis & Schiffermüller]) ab. *mucida* Esper ♂ Red Chestnut (Scotland). *Page 193*

37 *Cerastis rubricosa* ([Denis & Schiffermüller]) ab. *pallida* Tutt ♂ Red Chestnut. *Page 193*

38 *Cerastis leucographa* ([Denis & Schiffermüller]) ♂ White-marked. *Page 193*

39 *Cerastis leucographa* ([Denis & Schiffermüller]) ♀ White-marked. *Page 193*

40 *Cerastis leucographa* ([Denis & Schiffermüller]) ab. *suffusa* Tutt ♀ White-marked. *Page 193*

Plate 10: *Noctuidae Noctuinae* Figs 1–40, ×1

Plate 11: Noctuidae Hadeninae

Figs 1–45, × 1

1 *Anarta myrtilli* (Linnaeus) f. *rufescens* Tutt ♀ Beautiful Yellow Underwing. *Page 200*
2 *Anarta myrtilli* (Linnaeus) ♀ Beautiful Yellow Underwing. *Page 200*
3 *Anarta cordigera* (Thunberg) ♂ Small Dark Yellow Underwing. *Page 202*
4 *Anarta cordigera* (Thunberg) ♂ Small Dark Yellow Underwing. *Page 202*
5 *Anarta cordigera* (Thunberg) ab. *suffusa* Tutt ♀ Small Dark Yellow Underwing. *Page 202*
6 *Anarta melanopa* (Thunberg) ♂ Broad-bordered White Underwing. *Page 203*
7 *Anarta melanopa* (Thunberg) ab. *rupestralis* Hübner ♂ Broad-bordered White Underwing. *Page 203*
8 *Anarta melanopa* (Thunberg) ab. *wistromi* Lampa ♂ Broad-bordered White Underwing. *Page 203*
9 *Discestra trifolii* (Hufnagel) ♂ Nutmeg. *Page 204*
10 *Discestra trifolii* (Hufnagel) ♂ Nutmeg. *Page 204*
11 *Discestra trifolii* (Hufnagel) ab. *saucia* Esper ♂ Nutmeg. *Page 204*
12 *Discestra trifolii* (Hufnagel) ♀ Nutmeg. *Page 204*
13 *Hada nana* (Hufnagel) ♂ Shears. *Page 206*
14 *Hada nana* (Hufnagel) ♂ Shears. *Page 206*
15 *Hada nana* (Hufnagel) ab. *ochrea* Tutt ♀ Shears. *Page 206*
16 *Hada nana* (Hufnagel) ab. *leucostigma* Haworth ♂ Shears. *Page 206*
17 *Polia bombycina* (Hufnagel) ♂ Pale Shining Brown. *Page 207*
18 *Polia bombycina* (Hufnagel) ab. *unicolor* Tutt ♀ Pale Shining Brown. *Page 207*
19 *Polia hepatica* (Clerck) ♀ Silvery Arches. *Page 208*
20 *Polia hepatica* (Clerck) ♂ Silvery Arches (Scotland). *Page 208*
21 *Sideridis albicolon* (Hübner) ♂ White Colon. *Page 212*
22 *Sideridis albicolon* (Hübner) ♀ White Colon. *Page 212*
23 *Polia nebulosa* (Hufnagel) ♂ Grey Arches. *Page 209*
24 *Polia nebulosa* (Hufnagel) ab. *pallida* Tutt ♂ Grey Arches. *Page 209*
25 *Polia nebulosa* (Hufnagel) ab. *thompsoni* Arkle ♂ Grey Arches. *Page 209*
26 *Polia nebulosa* (Hufnagel) ab. *plumbosa* Mansbridge ♀ Grey Arches. *Page 209*
27 *Lacanobia contigua* ([Denis & Schiffermüller]) ♀ Beautiful Brocade. *Page 217*
28 *Pachetra sagittigera britannica* Turner ♂ Feathered Ear. *Page 210*
29 *Pachetra sagittigera britannica* Turner ♀ Feathered Ear. *Page 210*
30 *Mamestra brassicae* (Linnaeus) ♀ Cabbage Moth. *Page 214*
31 *Mamestra brassicae* (Linnaeus) ♂ Cabbage Moth. *Page 214*
32 *Heliophobus reticulata marginosa* (Haworth) ♂ Bordered Gothic. *Page 213*
33 *Heliophobus reticulata marginosa* (Haworth) ♀ Bordered Gothic. *Page 213*
34 *Heliophobus reticulata hibernica* Cockayne ♀ Bordered Gothic. *Page 213*
35 *Melanchra persicariae* (Linnaeus) ♂ Dot. *Page 216*
36 *Lacanobia w-latinum* (Hufnagel) ♂ Light Brocade. *Page 218*
37 *Lacanobia thalassina* (Hufnagel) ♂ Pale-shouldered Brocade. *Page 219*
38 *Lacanobia thalassina* (Hufnagel) ab. *humeralis* Haworth ♂ Pale-shouldered Brocade. *Page 219*
39 *Lacanobia suasa* ([Denis & Schiffermüller]) ♀ Dog's Tooth. *Page 220*
40 *Lacanobia suasa* ([Denis & Schiffermüller]) ab. *dissimilis* Knoch ♀ Dog's Tooth. *Page 220*
41 *Lacanobia oleracea* (Linnaeus) ♂ Bright-line Brown-eye. *Page 221*
42 *Lacanobia oleracea* (Linnaeus) ♀ Bright-line Brown-eye. *Page 221*
43 *Papestra biren* (Goeze) ♂ Glaucous Shears. *Page 223*
44 *Papestra biren* (Goeze) ♀ Glaucous Shears (Ireland). *Page 223*
45 *Lacanobia blenna* (Hübner) ♂ Stranger. *Page 222*

Plate 11: Noctuidae Hadeninae

Figs 1–45, ×1

Plate 12: Noctuidae Hadeninae

Figs 1–53, × 1

1 *Eriopygodes imbecilla* (Fabricius) ♀ Silurian. *Page 237*
2 *Ceramica pisi* (Linnaeus) ♂ Broom Moth. *Page 224*
3 *Ceramica pisi* (Linnaeus) ab. *splendens* Staudinger ♂ Broom Moth. *Page 224*
4 *Ceramica pisi* (Linnaeus) ab. *scotica* Tutt ♀ Broom Moth. *Page 224*
5 *Hadena perplexa capsophila* (Duponchel) ♂ Pod Lover (Co. Antrim). *Page 229*
6 *Hadena perplexa capsophila* (Duponchel) ♂ Pod Lover (Co. Kerry). *Page 229*
7 *Hadena perplexa* ([Denis & Schiffermüller]) ♂ Tawny Shears. *Page 229*
8 *Hadena perplexa* ([Denis & Schiffermüller]) ♂ Tawny Shears. *Page 229*
9 *Hadena perplexa* ([Denis & Schiffermüller]) ♀ Tawny Shears. *Page 229*
10 *Hadena perplexa* ([Denis & Schiffermüller]) ♀ Tawny Shears. *Page 229*
11 *Hadena perplexa* ([Denis & Schiffermüller]) ab. *pallida* Tutt ♀ Tawny Shears. *Page 229*
12 *Hecatera bicolorata* (Hufnagel) ♂ Broad-barred White. *Page 225*
13 *Hecatera bicolorata* (Hufnagel) ♀ Broad-barred White. *Page 225*
14 *Hecatera dysodea* ([Denis & Schiffermüller]) ♂ Small Ranunculus. *Page 227*
15 *Hadena rivularis* (Fabricius) ♂ Campion. *Page 228*
16 *Hadena irregularis* (Hufnagel) ♂ Viper's Bugloss. *Page 231*
17 *Hadena luteago barrettii* (Doubleday) ab. *ficklini* Tutt ♂ Barrett's Marbled Coronet (Cornwall). *Page 231*
18 *Hadena luteago barrettii* (Doubleday) ab. *turbata* Donovan ♀ Barrett's Marbled Coronet (Co. Cork). *Page 231*
19 *Hadena compta* ([Denis & Schiffermüller]) ♂ Varied Coronet. *Page 232*
20 *Hadena confusa* (Hufnagel) ♂ Marbled Coronet. *Page 233*
21 *Hadena confusa* (Hufnagel) ♂ Marbled Coronet (Orkney). *Page 233*
22 *Hadena confusa* (Hufnagel) ♂ Marbled Coronet (Shetland). *Page 233*
23 *Hadena confusa* (Hufnagel) ♂ Marbled Coronet (Hebrides). *Page 233*
24 *Hadena albimacula* (Borkhausen) ♂ White Spot. *Page 235*
25 *Hadena bicruris* (Hufnagel) ♀ Lychnis. *Page 235*
26 *Hadena caesia mananii* (Gregson) ♂ Grey (Isle of Man). *Page 236*
27 *Hadena caesia mananii* (Gregson) ♂ Grey (Co. Kerry). *Page 236*
28 *Tholera cespitis* ([Denis & Schiffermüller]) ♂ Hedge Rustic. *Page 240*
29 *Tholera cespitis* ([Denis & Schiffermüller]) ♀ Hedge Rustic. *Page 240*
30 *Cerapteryx graminis* (Linnaeus) ♂ Antler Moth. *Page 238*
31 *Cerapteryx graminis* (Linnaeus) ♂ Antler Moth. *Page 238*
32 *Cerapteryx graminis* (Linnaeus) ♂ Antler Moth. *Page 238*
33 *Cerapteryx graminis* (Linnaeus) ♂ Antler Moth. *Page 238*
34 *Cerapteryx graminis* (Linnaeus) ♀ Antler Moth. *Page 238*
35 *Cerapteryx graminis* (Linnaeus) ♀ Antler Moth. *Page 238*
36 *Tholera decimalis* (Poda) ♂ Feathered Gothic. *Page 240*
37 *Tholera decimalis* (Poda) ♀ Feathered Gothic. *Page 240*
38 *Egira conspicillaris* (Linnaeus) ♂ Silver Cloud. *Page 244*
39 *Egira conspicillaris* (Linnaeus) ♀ Silver Cloud. *Page 244*
40 *Egira conspicillaris* (Linnaeus) ab. *melaleuca* Vieweg ♂ Silver Cloud. *Page 244*
41 *Panolis flammea* ([Denis & Schiffermüller]) ♀ Pine Beauty. *Page 242*
42 *Panolis flammea* ([Denis & Schiffermüller]) ♀ Pine Beauty. *Page 242*
43 *Panolis flammea* ([Denis & Schiffermüller]) ab. *griseovariegata* Goeze ♀ Pine Beauty. *Page 242*
44 *Panolis flammea* ([Denis & Schiffermüller]) ab. *griseovariegata* Goeze ♀ Pine Beauty. *Page 242*
45 *Orthosia populeti* (Fabricius) ♂ Lead-coloured Drab. *Page 249*
46 *Orthosia populeti* (Fabricius) ♂ Lead-coloured Drab. *Page 249*
47 *Orthosia populeti* (Fabricius) ♂ Lead-coloured Drab (Scotland). *Page 249*
48 *Orthosia cruda* ([Denis & Schiffermüller]) ♂ Small Quaker. *Page 246*
49 *Orthosia cruda* ([Denis & Schiffermüller]) ♀ Small Quaker. *Page 246*
50 *Orthosia miniosa* ([Denis & Schiffermüller]) ♂ Blossom Underwing. *Page 247*
51 *Orthosia miniosa* ([Denis & Schiffermüller]) ♂ Blossom Underwing. *Page 247*
52 *Orthosia opima* (Hübner) ♀ Northern Drab. *Page 248*
53 *Orthosia opima* (Hübner) ab. *brunnea* Tutt ♀ Northern Drab. *Page 248*

Plate 12: Noctuidae Hadeninae

Figs 1–53, ×1

Plate 13: Noctuidae Hadeninae

Figs 1–59, × 1

1 *Brithys crini pancratii* (Cyrillo) ♂ Kew Arches. *Page 243*
2 *Orthosia gracilis* ([Denis & Schiffermüller]) ♂ Powdered Quaker. *Page 250*
3 *Orthosia gracilis* ([Denis & Schiffermüller]) ab. *rufescens* Tutt ♂ Powdered Quaker. *Page 250*
4 *Orthosia gracilis* ([Denis & Schiffermüller]) ♀ Powdered Quaker. *Page 250*
5 *Orthosia gracilis* ([Denis & Schiffermüller]) ab. *brunnea* Tutt ♀ Powdered Quaker. *Page 250*
6 *Orthosia gracilis* ([Denis & Schiffermüller]) ♀ Powdered Quaker. *Page 250*
7 *Orthosia stabilis* ([Denis & Schiffermüller]) ♂ Common Quaker. *Page 251*
8 *Orthosia stabilis* ([Denis & Schiffermüller]) ♂ Common Quaker. *Page 251*
9 *Orthosia stabilis* ([Denis & Schiffermüller]) ♀ Common Quaker. *Page 251*
10 *Orthosia stabilis* ([Denis & Schiffermüller]) ab. *fasciata* Lenz ♀ Common Quaker. *Page 251*
11 *Orthosia stabilis* ([Denis & Schiffermüller]) ♀ Common Quaker. *Page 251*
12 *Orthosia stabilis* ([Denis & Schiffermüller]) ♀ Common Quaker. *Page 251*
13 *Mythimna turca* (Linnaeus) ♂ Double Line. *Page 256*
14 *Orthosia incerta* (Hufnagel) ♂ Clouded Drab. *Page 253*
15 *Orthosia incerta* (Hufnagel) ♂ Clouded Drab. *Page 253*
16 *Orthosia incerta* (Hufnagel) ♂ Clouded Drab. *Page 253*
17 *Orthosia incerta* (Hufnagel) ♂ Clouded Drab. *Page 253*
18 *Orthosia incerta* (Hufnagel) ♂ Clouded Drab. *Page 253*
19 *Orthosia incerta* (Hufnagel) ♀ Clouded Drab. *Page 253*
20 *Orthosia incerta* (Hufnagel) ♀ Clouded Drab. *Page 253*
21 *Orthosia gothica* (Linnaeus) ♂ Hebrew Character. *Page 255*
22 *Orthosia gothica* (Linnaeus) ♂ Hebrew Character. *Page 255*
23 *Orthosia gothica* (Linnaeus) ♂ Hebrew Character. *Page 255*
24 *Orthosia gothica* (Linnaeus) ♀ Hebrew Character. *Page 255*
25 *Orthosia gothica* (Linnaeus) ab. *gothicina* Herrich-Schäffer ♂ Hebrew Character. *Page 255*
26 *Orthosia gothica* (Linnaeus) ab. *gothicina* Herrich-Schäffer ♀ Hebrew Character. *Page 255*
27 *Orthosia gothica* (Linnaeus) ♀ Hebrew Character. *Page 255*
28 *Orthosia gothica* (Linnaeus) ♀ Hebrew Character. *Page 255*
29 *Orthosia munda* ([Denis & Schiffermüller]) ♀ Twin-spotted Quaker. *Page 254*
30 *Orthosia munda* ([Denis & Schiffermüller]) ab. *immaculata* Staudinger ♂ Twin-spotted Quaker. *Page 254*
31 *Orthosia munda* ([Denis & Schiffermüller]) ♂ Twin-spotted Quaker. *Page 254*
32 *Mythimna vitellina* (Hübner) ♂ Delicate. *Page 260*
33 *Mythimna vitellina* (Hübner) f. *pallida* Warren ♀ Delicate. *Page 260*
34 *Mythimna conigera* ([Denis & Schiffermüller]) ♂ Brown-line Bright-eye. *Page 257*
35 *Mythimna conigera* ([Denis & Schiffermüller]) ab. *suffusa* Tutt ♂ Brown-line Bright-eye. *Page 257*
36 *Mythimna ferrago* (Fabricius) ♂ Clay. *Page 258*
37 *Mythimna ferrago* (Fabricius) ♀ Clay. *Page 258*
38 *Mythimna pudorina* ([Denis & Schiffermüller]) ♂ Striped Wainscot. *Page 261*
39 *Mythimna albipuncta* ([Denis & Schiffermüller]) ♂ White-point. *Page 259*
40 *Mythimna albipuncta* ([Denis & Schiffermüller]) ♀ White-point. *Page 259*
41 *Mythimna impura* (Hübner) ♂ Smoky Wainscot. *Page 263*
42 *Mythimna impura* (Hübner) ab. *scotica* Cockayne ♂ Smoky Wainscot. *Page 263*
43 *Mythimna litoralis* (Curtis) ♀ Shore Wainscot. *Page 266*
44 *Mythimna pallens* (Linnaeus) ♂ Common Wainscot. *Page 264*
45 *Mythimna pallens* (Linnaeus) ab. *ectypa* Hübner ♀ Common Wainscot. *Page 264*
46 *Mythimna pallens* (Linnaeus) ♀ Common Wainscot. *Page 264*
47 *Mythimna favicolor* (Barrett) ♂ Mathew's Wainscot. *Page 265*
48 *Mythimna favicolor* (Barrett) ♂ Mathew's Wainscot. *Page 265*
49 *Mythimna l-album* (Linnaeus) ♂ L-album Wainscot. *Page 267*
50 *Mythimna straminea* (Treitschke) ♂ Southern Wainscot. *Page 262*
51 *Mythimna straminea* (Treitschke) ♂ Southern Wainscot. *Page 262*
52 *Mythimna unipuncta* (Haworth) ♂ White-speck (American Wainscot). *Page 268*
53 *Mythimna unipuncta* (Haworth) ♀ White-speck (American Wainscot). *Page 268*
54 *Mythimna obsoleta* (Hübner) ♂ Obscure Wainscot. *Page 269*
55 *Mythimna comma* (Linnaeus) ♂ Shoulder-striped Wainscot. *Page 270*
56 *Mythimna comma* (Linnaeus) ♀ Shoulder-striped Wainscot. *Page 270*
57 *Mythimna putrescens* (Hübner) ♂ Devonshire Wainscot. *Page 271*
58 *Mythimna loreyi* (Duponchel) ♂ Cosmopolitan. *Page 272*
59 *Senta flammea* (Curtis) ♂ Flame Wainscot. *Page 273*

Plate 13: Noctuidae Hadeninae

Figs 1–59, ×1

General Index

Principal entries are given in **bold type**. Plate references are shown as (B:2; 10:22–26). The index also includes references to figures in the text, as 113 (text fig.8), and keys.

See separate index to host plants.

abducta, Spaelotis ravida ab. 164
acetosellae, Mesogona 130, **195** (10:34)
Acherontia 23
Acontiinae 14, 120 (B:13), 122, 126
Acronictinae 14, 120 (B:7,8), 122, 123, 124, 125, 263
Actebia 150
Actinotia **148**
advena, Noctua 207
advena, Orthosia 248
aerugula, Nola 112, **119** (6:39–41)
aethiops, Anarta cordigera ab. 202
Aethria 225
agathina, Xestia 128, 170, **188** (B:2; 10:22–26)
Agrius 21
Agrochola 122
Agronoma 137
Agrotis 122, 125, **137**
alba, Orthosia gracilis ab. 250
albescens, Agrotis ipsilon ab. 143
albicolon, Sideridis 199, 204, **212** (11:21,22), 215
albimacula, Hadena 197, **235** (12:24)
albipuncta, Mythimna 197, 258, **259** (13:39,40), 261, 272
albovenosa, Simyra 263
albula, Meganola 112, 113 (text fig.8), **115** (6:42–44)
alder kitten, the 43
alpicola, Xestia 129, 178, 179
alpina, Xestia alpicola **178** (10:1–4), 179
Amathes 178
American wainscot 268
Ammogrotis 166
Amphipoea 122, 124
Amphipyrinae 14, 15, 120 (B:9,10), 126, 212, 217, 218, 219, 222, 249, 261
anachoreta, Clostera 40, 62, **63** (4:8)
Anaplectoides 191, **192**
Anarta 122, **200**
anceps, Peridea 40, 47, **50** (3:16,17)
ancilla, Dysauxes 111
Anepia 228
anglica, Euprepia cribraria 96
angustus, Orthosia incerta ab. 253
annexa, Agrotis 147
antiqua, Orgyia 67, 69, **70** (4:16,17)
antler moth, the 238
Apamea 124
Aplecta 207
apricaria, Pluvialis 169
aquilina, Euxoa 133
Archanara 123
archer's dart, the 138
Arctia **99**

Arctiidae 9, 11 (text fig.2), 12, 13, 15, 16, **78** (A:9–12), 79 (text figs 6,7) and key
Arctiidae, eversible structures 11 (text fig.2), 12
Arctiinae 78, 79, **96** (A: 11, 12)
Arctornis **76**
arenaria, Coscinia cribraria **96**, 97
argillacea, Hadena luteago barrettii f. 231
armigera, Helicoverpa 125 (B:12)
'army worms' 125
ashworthii, Xestia 129, **182** (10:8,9)
Ashworth's rustic 182
Atolmis **85**
atra, Orthosia incerta ab. 253
atropos, Acherontia 9, 12, 21, **23** (2:3)
augur, Graphiphora 128, 130, 164, **165** (9:6,7)
aureola, Bombyx 88
auricoma, Acronicta 14
auriflua, Bombyx 74
autumnal rustic, the 169
Axylia **149**

baja, Xestia 127, **183** (10:10)
bankiana, Deltote 14
Barathra 214, 217
barred chestnut, the 173
barrettii, Hadena luteago **231** (12:17,18)
Barrett's marbled coronet 231
beautiful brocade, the 217
beautiful yellow underwing, the 200
bedstraw hawk-moth, the 34
bella, Phalaena (Noctua) 176
bella, Utetheisa **98**
bicolorata, Hecatera 199, **225** (12:12,13), 227
bicoloria, Leucodonta 40, **57** (3:27)
bicruris, Hadena 124, 199, 228, 229, 233, 234, **235** (text fig.18; 12:25)
bicuspis, Furcula 40, **43** (3:5), 44
bifida, Furcula 40, 43, **44** (3:4)
bilitura, Euxoa 147
biloba, Autographa 121 (text fig.11a)
bimacula, Callimorpha dominula f. 108 (6:32)
biren, Papestra 199, **223** (11:43,44)
biren, Phalaena 7
bivittata, Coscinia cribraria **96** (5:27)
black arches, the 76
black collar, the 151
black V moth, the 76
blenna, Lacanobia **222** (11:45)
blossom underwing, the 247
bombycina, Hadena 223
bombycina, Polia 199, **207** (11:17,18)
bombyliformis, Sphinx 29, 30
bordered gothic, the 213
borealis, Phragmatobia fuliginosa **106** (6:26)
brassicae, Mamestra 124, 196, 199, 204, 212, **214** (text fig.16; 11:30,31)

bright-line brown-eye, the 221
britannica, Arctia villica **100** (6:1–4), 101
britannica, Pachetra sagittigera **210** (11:28,29)
Brithys **243**
broad-barred white, the 225
broad-bordered bee hawk-moth, the 30
broad-bordered white underwing, the 203
broad-bordered yellow underwing, the 161
broom moth, the 224
Brotis 130
brown-line bright-eye, the 257
brown-tail, the 73
brunnea, Agrotis ripae ab. 145 (8:12)
brunnea, Diarsia 127, 166, 173, **174** (9:29)
brunnea, Mimas tiliae ab. 26 (1:2)
brunnea, Noctua fimbriata ab. 161
brunnea, Noctua pronuba ab. 157
brunnea, Orthosia gracilis ab. 250 (13:5)
brunnea, Orthosia opima ab. 248 (12:53)
brunnea, Spilosoma lubricipeda ab. 102 (6:12)
brunnescens, Arctia caja ab. 99 (5:36)
bucephala, Phalera 40, **41** (3:1)
buff ermine, the 103
buff footman, the 93
buff-tip, the 41

cabbage moth, the 214
caca, Ceramidia **111**
caeruleocephala, Diloba 40, **64** (4:9,10)
caerulescens, Noctua pronuba ab. 157 (8:28)
caesia, Hadena 198, 236
caja, Arctia 13, 80, **99** (5:34–41)
calcatrippae, Noctua 119
caliginosa, Noctua interjecta 163 (8:24)
Calligenia 83
Callimorpha **108**
Callopis 96
camelina, Phalaena (Bombyx) 53
campion, the 228
candelarum, Xestia ashworthii 182, 183
caniola, Eilema 79, **90** (5:15), 91
canus, Larus 169
capsincola, Noctua 235
capsophila, Hadena perplexa 200, **229** (12:5,6), 230
capucina, Ptilodon 40, **53** (3:22,23), 54
carmelita, Odontosia 40, **55** (3:25)
carnica, Xestia alpicola 179
carolina, Sphinx 24
carpophaga, Phalaena (Noctua) 229
castanea, Xestia 129, **185** (10:12–15)
catenata, Agrotis exclamationis ab. 141 (7:48)
Catephia 126
Catocala 14, 123

Catocalinae 120 (C:7,8), 122, 125
Celama 116
Celerio 33
celerio, Hippotion 21, **38** (1:10)
centonalis, Pyralis 119
centripuncta, Mimas tiliae ab. 26 (1:3)
Ceramica **224**
Ceramidia **111**
Cerapteryx **238**, 240
Cerastia 193
Cerastis 122, 126, **193**
Cerura **42**, 43
cespitis, Tholera 199, **240** (12:28,29), 241
Chaonia 59
chaonia, Bombyx 60
Charaeas 240
Charelia 200
chenopodii, Noctua 204
chlamitulalis, Nola 120
Chloantha 148
Chloephorinae 14, 120 (C:1,2), 122, 126
chocolate-tip, the 63
Choerocampa 36
chrysitis, Diachrysia 14 (C:6)
chrysorrhoea, Bombyx 74
chrysorrhoea, Euproctis 67, **73** (4:23,24), 74
chrysozona, Phalaena (Noctua) 227
cinerea, Agrotis 129, **137** (7:19–24)
cingulata, Agrius 12, **21**
cinnabar, the 112
clara, Noctua interjecta ab. 163
clavis, Agrotis 128, **141** (7:37–43)
clay, the 258
clorana, Earias 14
Clostera **62**
clouded buff, the 101
clouded drab, the 253
c-nigrum, Xestia 127, 151, **179** (10:5)
coast dart, the 136
Coenophila 166
coenosa, Laelia 67, **68** (4:12,13)
columbina, Nola confusalis f. 118
comes, Noctua 127, **159** (8:35–40), 160
comma, Mythimna 15, 197, **270** (13:55,56), 272
commoides, Mythimna **272**
common footman, the 94
common quaker, the 251
common wainscot, the 264
complana, Eilema 79, 90, **91** (5:18), 92, 94
complana, Lithosia 92
complanula, Lithosia 92, 94
compta, Hadena 197, **232** (12:19), 234
concolor, Dasychira pudibunda ab. 67, 72
confluens, Spaelotis ravida ab. 164
confusa, Hadena 122, 197, **233** (12:20–23), 234
confusalis, Nola 112, 113, 114 (text fig.9), 116 (text fig.10), **118** (6:45–47)
conigera, Mythimna 197, **257** (13:34,35), 258
Conistra 122
connuba, Noctua comes ab. 160

conspersa, Noctua 233
conspicillaris, Egira 197, **244** (text fig.19; 12:38–40)
contigua, Lacanobia 198, 208, **217** (11:27), 218
convolvuli, Agrius 9, **21** (2:1), 24
convolvulus hawk-moth, the 21
cordigera, Anarta 197, 200, **202** (11:3–5), 203
corticea, Noctua 141
coryli, Colocasia 14 (C:4)
Coscinia **96**
cosmopolitan, the 272
costata, Agrotis exclamationis ab. 141
cousin German, the 167
coxcomb prominent, the 53
Crambus 91
crassa, Agrotis 128, **146** (8:16)
crassalis, Hypena 125 (C:12)
cream-spot tiger, the 100
crenata, Gluphisia 40, 60, 61
crescent dart, the 142
cribraria, Coscinia 12, 13, 80, 96 (5:27)
cribrum, Phalaena (Bombyx) 96
crimson speckled, the 97
crini, Brithys 198, 243, 244
crini, Brithys crini 243, 244
crocea, Callimorpha dominula f. 108 (6:34)
cruda, Orthosia 197, 198, **246** (12:48,49)
Cryphia 122
Ctenuchidae **111**
cucubali, Noctua 228
cuculla, Bombyx 54
cucullatella, Nola 112, 115, **117** (A:13; 6:37,38)
Cucullia 122
Cucullinae 14, 15, 120 (B:5,6), 124, 125, 227
cucullina, Ptilodontella 40, 53, **54** (3:24)
Cuphanoa 246
cursoria, Euxoa 122, 128, 133 (text fig.14), 134, 135 (text fig.15), **136** (7:11–18), 145, 146
curtula, Clostera 40, 62, **63** (4:11)
Cybosia **86**
Cycnia 105

dahlii, Diarsia 127, 167, **173** (9:31,32), 174, 175
Danaidae 15
Daphnis **32**
dark sword grass, the 143
dark tussock, the 71
Dasychira **71**
death's-head hawk-moth, the 23
decimalis, Tholera 189, 196, 198, 213, 239, **240** (12:36,37), 241
decrescens-lutescens, Arctia caja ab. 99 (5:39)
deflorata, Ecpantheria 107, **110**
Deilephila **36**
Deiopeia 97
delicate, the 260
demolinii, Lophostethus 12
demuthi, Anaplectoides prasina ab. 192 (10:33)
denticulatus, Bombyx 137
dentina, Noctua 206

deplana, Eilema 79, 80, 91, **93** (5:19,20)
depressa, Noctua 93
deprivata, Agrotis **147**
depuncta, Eugnorisma 128, 129, **153** (9:30)
derivalis, Paracolax 14
desillii, Agrotis ripae ab. 145 (8:10)
Devonshire wainscot, the 271
dew moth, the 82
Diacrisia **101**
Dianthoecia 228
Diaphora **105**
Diarsia **172**, 173, 178
Dicranura 42
dictaea, Phalaena (Bombyx) 52
dictaeoides, Bombyx 51
Dilina 27
Diloba **64**, 126
dingy footman, the 89
Discestra **204**
dispar, Lymantria 67, 76, **77** (4:30,31)
dissimilis, Lacanobia suasa ab. 220 (11:40), 221
dissimilis, Phalaena (Noctua) 220
distinctacaerulescens, Noctua pronuba ab. 157
distincta, Euxoa cursoria ab. 136 (7:14)
ditrapezium, Xestia 127, 179, **180** (10:6), 181, 184
dives, Graphania **274**
dives, Lacanobia contigua ab. 217
dodonaea, Drymonia 40, **59** (3:30–33), 60
dog's tooth, the 220
dominula, Callimorpha 13, 80, **108** (6:31–34)
dotted clay, the 183
dotted footman, the 87
dotted rustic, the 156
dot, the 216
double dart, the 165
double line, the 256
double square-spot, the 181
dromedarius, Notodonta 40, **46** (A:6; 3:10,11), 47
drupiferarum, Sphinx 12, **25**
druraei, Sphinx 21
Drymonia **59**
dusky marbled brown, the 61
Dysauxes **111**
dysodea, Hecatera 200, 225, **227** (12:14)

Earias 126
echii, Phalaena (Noctua) 231
Ecpantheria **110**
ectypa, Mythimna pallens ab. 264 (13:45)
edda, Paradiarsia glareosa ab. 169 (9:15)
Egira **244** (text fig.19)
Eilema 13, **88**
elephant hawk-moth, the 36
Eligmodonta **48**
elpenor, Deilephila 9, 10 (text fig.1), 12, 16, 21, **36** (A:3,4; 2:9)
Emmelia 126
Emydia 96
Endrosa 82
Epia 228

Epicallia 99
Epineuronia 240
Epipsilia 156
Episema 64
epomidion, Apamea 219
eremita, Lymantria monacha f. 67, **76** (4:35)
Eriopygodes 237
essoni, Peucephila 221
Euchromia 111
Eugnorisma **153**
Eulepia 96
euphorbiae, Hyles 12, 21, **33** (1:7), 34, 35
Euplagia **107**
Euproctis **73**
Eurois **191**
Euschesis 157
Euthemonia 101
Euxoa 122, **130**, 131 (text fig.13), 137
eversible structures **9**
eversmann's rustic 151
Exarnis 151
exclamationis, Agrotis 121, 127, **141** (7:44–52)
exigua, Spodoptera 124, 125, 261
extranea, Leucania 268
eyed hawk-moth, the 27
Eyprepia 99

fagana, Pseudoips 14 (C:2)
fagi, Stauropus 40, **45** (3:8,9)
fascelina, Dasychira 67, **71** (4:21,22)
fasciata, Orthosia stabilis ab. 251 (13:10)
favicolor, Mythimna 196, 197, **265** (13:47,48), 266
feathered ear, the 210
feathered footman, the 96
feathered gothic, the 240
Feltia **147**
fennica, Ochropleura 127, **151** (8:20)
fen square spot, the 176
ferrago, Mythimna 197, 257, **258** (13:36,37), 259, 268, 272
festiva, Noctua 172
ficklini, Hadena luteago barrettii ab. 231 (12:17)
figure of eight moth, the 64
fimbria, Phalaena (Noctua) 161
fimbriata, Noctua 126, **161** (8:31,32)
flame shoulder, the 152
flame, the 149
flame wainscot, the 273
flammatra, Ochropleura 130, **151** (8:21), 179, 181
flammea, Panolis 197, **242** (12:41–44)
flammea, Senta 196, **273** (13:59)
flava, Diarsia florida ab. 177 (9:40)
flavescens, Tyria jacobaeae ab. 109
flavicincta, Polymixis 227
florida, Diarsia 122, 127, **176** (9:39,40), 177
four-dotted footman, the 86
four-spotted footman, the 95
fraxini, Catocala 120 (C:8)

fuciformis, Hemaris 20, 29, **30** (2:2)
fuliginaria, Parascotia 122, 123
fuliginosa, Phragmatobia 12, 13, 79 (text fig.7), 80, **106** (6:25,26)
fulminea, Bombyx 210
fumosa, Arctia caja ab. 99 (5:35)
Furcula 7, 42, **43**, 49
furcula, Furcula 40, **43** (3:2,3), 44
furcula, Phalaena 43
furva, Apamea 15, 212

galii, Sphinx 34
gallii, Hyles 21, **34** (1:5), 35
gangis, Creatonotus 12, 16
garden dart, the 134
garden tiger, the 99
genistae, Phalaena (Noctua) 218
Geometridae 14, 15, 16, 122, 261
Georyx 137
gipsy moth, the 77
glareosa, Paradiarsia 122, 129, **169** (9:12–15)
glauca, Noctua 223
glaucous shears, the 223
Gluphisia **61**
gnoma, Pheosia 40, **51** (3:18,19), 52
Gnophria 85
godarti, Spilosoma lubricipeda ab. 102 (6:14)
gonostigma, Orgyia 69
gonostigma, Phalaena 70
gothica, Orthosia 197, 252, **255** (B:3; 13:21–28)
gothicina, Orthosia gothica ab. 197, 255 (13:25,26)
gothic, the 189
gracilis, Orthosia 198, 200, **250** (13:2–6)
graminis, Cerapteryx 125, 197, **238** (12:30–35), 240, 241
grammica, Phalaena (Bombyx) 96
Graphania **274**
Graphiphora 165
great brocade, the 191
great dart, the 146
great prominent, the 50
green arches, the 192
Gregson's dart 139
grey arches, the 209
grey, the 236
grisea, Agrotis ripae ab. 145
grisea, Euxoa obelisca **130** (7:1,2), 131 (text fig.13), 132, 133
grisea, Mythimna albipuncta ab. 259
grisea, Mythimna ferrago ab. 258
griseola, Eilema 11 (text fig.2), 13, 79, **89** (5:13,14)
griseovariegata, Panolis flammea ab. 242 (12:43,44)
griseovariegata, Phalaena (Noctua) 242
gueneei, Luperina nickerlii 145
gulls, common 169
gynandromorphs 28
Gypsitea 193

Hada 206
Hadena 123, 124, 225, 226, **228**, 230, 231, 232, 233, 234, 235, 236, 237
Hadeninae 14, 15, 120, 122, 125, **196** (B:3,4) and key

GENERAL INDEX

Haemorrhagia 29
haggarti, Orthosia cruda ab. 246
Halisidota **107**
Hammatophora 41
Hapalia 150
Harmodia 228
Harpyia 7, 43, **49**
heart and club, the 141
heart and dart, the 141
heath rustic, the 188
hebrew character, the 255
hebridicola, Xestia agathina **188** (10:26), 189
Hecatera **225**
hedge rustic, the 240
Heliophobus **213**
Heliothinae 120 (B:11,12), 122, 125, 126, 149
helveola, Lithosia 93
helvola, Bombyx 93
Hemaris 29
Hemeria 29
hemerobia, Bombyx 84
hepatica, Polia 199, **208** (11:19,20), 218
hera, Phalaena (Noctua) 107
herbida, Noctua 192
hermelina, Phalaena 44
Herse 21
heterozygote 209
hibernica, Heliophobus reticulata **213** (11:34), 214
hippophaes, Deilephila 35
hippophaes, Hyles **35**
Hippotion 38
hoary footman, the 90
hoegi, Noctua pronuba ab. 158
homozygote 209
'hop-dog' 72
Hoplitis 49
hospita, Parasemia plantaginis f. 98 (5:30)
humeralis, Lacanobia thalassina ab. 219 (11:38)
humming-bird hawk-moth, the 31
Hybocampa 7, 49
Hyles **33**
Hyloicus 25
Hypeninae 14, 120 (C:11,12), 122, 125
hyperborea, Hadena 178
Hyphilare 256
hypothous, Daphnis 33

Ichthyura 62
Ilarus 242
imbecilla, Eriopygodes 196, 197, 198, **237** (12:1 ♀; Vol. **10**, 6:1 ♂)
immaculata, Orthosia munda ab. 198, 253, 254 (13:30)
immarginata, Diacrisia sannio ab. 101 (6:7)
impudens, Noctua 261
impura, Mythimna 197, **263** (13:41,42), 265, 267, 269
incerta, Orthosia 198, 246, 248, 249, **253** (13:14–20), 254
ingrailed clay, the 172
innuba, Noctua pronuba ab. 157 (8:26)
inquilinus, Phalaena 38
instabilis, Noctua 253
insula, Agrotis puta **144** (8:3,4), 145

insulana, Earias 125
insularum, Parasemia plantaginis **98**, 99
interjecta, Noctua 127, 162, 163
intermedia, Egira conspicillaris ab. 244
intermedia, Spilosoma lutea ab. 103 (6:20)
interposita, Noctua 159
interpunctella, Plodia 16
interrogationis, Syngrapha 151
ipsilon, Agrotis 128, 139, 140, **143** (8:14,15), 147, 172
iron prominent, the 46
irregularis, Hadena 198, 199, **231** (12:16), 235
irrorata, Lithosia 82
irrorella, Setina 13, 80, **82** (5:2–4)
isabella, Pyrrharctia 107

jacobaeae, Tyria 13, 80, **109** (6:29,30)
jaculifera, Agrotis 147
janthina, Noctua 127, **162** (8:33,34), 163
Jersey tiger, the 107
jotunensis, Xestia ashworthii 183
juncta Agrotis exclamationis ab. 141 (7:51)

Kent black arches, the 115
Kew arches, the 243

Lacanobia 7, **217**, 223, 224
Lacinipolia **205**
lacteola, Eilema caniola f. 90
Laelia 68
l-album, Mythimna 197, **267** (13:49)
L-album wainscot, the 267
Lampra 157
Laothoe **28**
large dark prominent, the 47
large yellow underwing, the **157**
laudabilis, Lacinipolia **205**
lead-coloured drab, the 249
least black arches, the 118
least yellow underwing, the 163
Leiocampa 51
lepida, Phalaena 229
lesser broad-bordered yellow underwing, the 162
lesser swallow prominent, the 51
lesser yellow underwing, the 159
lethe, Euchromia 111
Lethia 24
Leucania 256
Leucodonta 57
leucogaster, Ochropleura 152
leucographa, Cerastis 129, **193** (10:38–40)
Leucoma **75**
leuconota, Hecatera bicolorata f. 225
leucophaea, Noctua 210
leucostigma, Hada nana ab. 206 (11:16)
light brocade, the 218
light feathered rustic, the 137
ligustri, Sphinx 12, 20 (text fig.3), 21, **24** (2:6), 25
lime hawk-moth, the 26

lineata, Hyles 21, 35
lineolatus, Agrotis exclamationis ab. 141 (7:52)
Liparis 73
Lithophane 122
Lithosia 95
Lithosiinae 13, 78, 79, **81** (A:9,10)
Lithosis 95
litoralis, Mythimna 197, 263, 265, **266** (13:43)
littoralis, Spodoptera 125
livornica, Hyles lineata 12, **35** (1:8)
l-nigrum, Arctornis 67, 75, **76** (4:29)
lobster moth, the 45
Lophopteryx 53
loreyi, Mythimna 15, 196, 256, 263, 265, 268, **272** (13:58)
lowei, Hadena luteago barrettii ab. 231
lubricipeda, Spilosoma 11 (text fig.2), 12, 13, 16, 80, **102** (6:10–14), 103, 104
lucernea, Standfussiana 129, **154** (9:1,2), 157, 164, 205
lucipeta, Rhyacia 130, 156, **157** (8:22)
lueneburgensis, Aporophyla 122
lunar marbled brown, the 60
lunar yellow underwing, the 159
lunigera, Agrotis trux **142** (7:53,54), 146
lunula, Calophasia 123
Luperina 124
lurideola, Eilema 13, 79 (text fig.6), 90, 91, 92, **94** (A:9,10; 5:24)
lutarella, Eilema 91
lutarella, Lithosia 91
luteago, Hadena 199, 228, 231
lutea, Spilosoma 12, 16, 80, 102, **103** (A:12; 6:15–20)
lutescens, Euplagia quadripunctaria ab. 107 (6:28)
lutulenta, Aporophyla 122
lychnis, the 235
Lycophotia **170**
Lymantria **76**
Lymantriidae 66 (text fig.5; A:7,8), key 67
Lytaea 178
lythargyria, Phalaena (Noctua) 258

Macroglossinae 20, **29** (A:3,4)
Macroglossum **31**
maerens, Diacrisia sannio ab. 101 (6:6)
Mamestra **214**, 216
mananii, Hadena caesia **236** (12:26,27)
Manduca 24
maple prominent, the 54
marbled brown, the 59
marbled coronet, the 233
margaritosa, Peridroma saucia f. 171 (9:37)
marginata, Orthosia stabilis ab. 251
marginosa, Heliophobus reticulata **213** (11:32,33), 214
maritimus, Chilodes 122
Mathew's wainscot 265
matronalis, Parasemia plantaginis ab. 98 (5:31)

mediterranean hawk-moth, the 34
Meganola **113** (text fig.8), 114 (text fig.9)
Megasema 178
melaleuca, Egira conspicillaris ab. 196, 244 (12:40)
melaleuca, Orthosia incerta ab. 253
Melanchra **216**
melanopa, Anarta 198, 202, **203** (11:6–8)
Meliana 273
mendica Diaphora 12, 80, 102, 104, **105** (6:21–24)
mendica, Diarsia 127, 166, 167, **172** (9:19–28), 173, 174, 176
menthastri, Bombyx 102
Mesogona **195**
mesomella, Cybosia 80, **86** (5:8,9)
Metaxyja 130
meticulosa, Phlogophora 10 (text fig.1), 14, 16 (B:9)
Metopsilus 36
milhauseri, Bombyx 43
milhauseri, Harpyia 49
Miltochrista **83**
Mimas **26**
miniata, Miltochrista 13, 80, **83** (5:5)
miniosa, Orthosia 198, 200, **247** (12:50,51), 251
Miselia 228
mista, Cerastis rubricosa ab. 193
mista, Paradiarsia sobrina ab. 167
moeschleri, Halisidota 107
molesta, Cydia 16
molybdeola, Lithosia 92
monacha, Lymantria 67, **76** (A:7; 4:32–35)
moneta, Polychrysia 14
Monima 246
monoglypha, Apamea 16
mucida, Cerastis rubricosa ab. 193 (10:36)
munda, Bombyx 84
mundana, Nudaria 13, 79, 81, **84** (5:6)
munda, Orthosia 197, 198, 246, **254** (13:29–31)
muralis, Cryphia 122
muscerda, Pelosia 13, 79, **87** (5:10)
musicola, Ceramidia 111
muslin footman, the 84
muslin moth, the 105
myrtilli, Anarta 198, **200** (11:1,2), 202
Mythimna 15, 122, 123, 196, **256**, 272

Naenia **189**
nana, Hada 199, **206** (11:13–16)
nana, Phalaena 233
narrow-bordered bee hawk-moth, the 29
nebulosa, Polia 191, 199, 208, **209** (11:23–26)
nebulosus, Orthosia incerta ab. 253
neglecta, Noctua 185
neglecta, Xestia castanea ab. 185 (10:15)
neglected, or grey rustic, the 185
negrana, Tyria jacobaeae ab. 109
Nelopa 84
Nemeophila 98

nerii, Daphnis 12, 21, **32** (1:4), 33
nerii, Sphinx 32
Neuronia 240
nicaea, Hyles **34**
nickerlii, Luperina 122
nigra, Orthosia opima ab. 248
nigrescens, Arctia caja ab. 99 (5:38)
nigrescens, Noctua comes ab. 159 (8:39)
nigrescens, Noctua fimbriata ab. 161
nigricans, Euxoa 128, 129, 133, **134** (7:8–10), 135 (text fig.15), 164, 187
nigricosta, Peridroma saucia f. 171 (9:38)
nigrostriata, Mythimna straminea ab. 262
nigrosuffusa, Mythimna unipuncta ab. 268
nitens, Noctua 207
ni, Trichoplusia 14, 16, 125
Noctua **157**
Noctuidae 9, 10 (text fig.1), 13, 14, 15, 16, 64, **120** (B; C), 121 (text figs 11,12), 122, 124, key 125, 196
Noctuidae, economic importance 125
Noctuidae, eversible structures 10 (text fig.1), 13
Noctuidae, life history and behaviour 122
Noctuidae, rearing 123
Noctuinae 13, 14, 15, 120, 123, 125, **126** (B:1,2) and key, 196, 205, 229, 213, 241
Nola 114 (text fig.9), **116** (text fig.10)
Nolidae **112** (A:13) and key, 116
northern dart, the 178
northern drab, the 248
northern footman, the 92
northern rustic, the 154
Notodonta **46**
Notodontidae 7, **39** (text fig.4; A:5,6), key 40, 123, 126
Nudaria **84**
nupta, Catocala 10 (text fig.1), (C:7)
nutmeg, the 204
Nygmia 73

oak processionary moth 65
o-album, Mythimna l-album ab. 267
obelisca, Euxoa 128, 130
obotritica, Agrotis ripae ab. 145 (8:7)
obscura, Apamea remissa ab. 212
obscura, Noctua 164
obscure wainscot, the 269
obsoleta, Mythimna 15, 196, 263, 265, **269** (13:54)
obsoleta, Orthosia gothica ab. 255
obstipata, Orthonama 261
obtusa, Pelosia 79, **87** (5:11)
occulta, Eurois 128, 151, **191** (10:28–30), 209
ocellata, Callimorpha dominula f. 108 (6:23)
ocellata, Smerinthus 12, 21, 26, **27** (A:2; 1:9), 28

ochracea, Diarsia florida ab. 177
ochreabrunnea, Noctua pronuba ab. 157 (8:25)
ochrea, Hada nana ab. 206 (11:15)
ochrea, Hadena confusa ab. 223
ochrea, Noctua pronuba ab. 157 (8:27)
Ochropleura **150**
ochrorenis, Melanchra persicariae ab. 216
oculea, Mesapamea secalis ab. 149
Odontosia **55**
Oeonistis 95
Ogygia 150
oleander hawk-moth, the 32
oleracea, Lacanobia 199, 220, **221** (11:41,42)
Oligia 122, 124
olivacea, Orthosia incerta ab. 253
oo, Dicycla 122
Ophiderinae 14, 120 (C:9,10), 122, 126
opima, Orthosia 198, 200, **248** (12:52,53)
orange footman, the 88
orbona, Noctua 123, 126, 158, **159** (8:23), 160
orbona, Triphaena 159
Orgyia 66, **69**
oriental fruit-moth 15
orkneyensis, Diarsia mendica **172**, 173
or, Tethea 249
Orthoa 246
Orthosia 122, 193, **246**

Pachetra **210**
pale prominent, the 56
pale shining brown, the 207
pale-shouldered brocade, the 219
pale stigma, the 195
pale tussock, the 72
pallens, Mythimna 196, 197, 263, **264** (13:44–46), 265, 266, 267, 269
pallida, Cerastis rubricosa ab. 193 (10:37)
pallida, Hadena perplexa ab. 229 (12:11)
pallida, Mythimna vitellina f. 260 (13:33), 261
pallida, Polia nebulosa ab. 209 (11:24)
pallifrons, Eilema pygmaeola **91** (5:17)
palpina, Pterostoma 40, **56** (A:5; 3:26)
Panaxia 108
pancratii, Brithys crini **243** (13:1), 244
Panolis **242**
Pantheinae 14, 120 (C:4), 122, 125, 196
Papestra 7, **223**
papyratia, Phalaena (Bombyx) 104
Paradiarsia **167**
Parasemia **98**
passetii, Eurois occulta f. 191 (10:30)
pavonia, Saturnia 28
pearly underwing, the 171
pebble prominent, the 48
Pelosia **87**

peregrina, Hadena 222
perfusca, Diarsia dahlii ab. 174
Peridea **50**
Peridroma **171**
Periphanes 125
perplexa, Hadena 199, 222, **229** (12:5–11), 235
persicariae, Melanchra 197, 198, **216** (11:35)
perspicillaris, Phalaena (Noctua) 148
petriburgensis, Arctia caja ab. 99 (5:37)
Peucephila 217
phaeorrhoeus, Bombyx 73
Phalera **41**
phegea, Syntomis **111**
Pheosia **51**
Philea 82
Philostola 256
phoebe, Phalaena (Bombyx) 47
Phragmatobia **106**
Piesta 88
pigmy footman, the 91
pigra, Clostera 40, **62** (4:6,7), 63
pilicornis, Cerastis rubricosa ab. 193
pinastri, Hyloicus 21, **25** (2:8)
pine beauty, the 242
pine hawk-moth, the 25
pine processionary moth 65
piniperda, Phalaena (Noctua) 242
pisi, Ceramica 198, 199, **224** (12:2–4)
pityocampa, Thaumetopoea 65
plain clay, the 153
plantaginis, Parasemia 13, 80, **98** (A:11; 5:29–33)
plecta, Ochropleura 10 (text fig.1), 15, 129, **152** (8:5,6)
plover, golden 169
plumbeolata, Lithosia 89
plumbosa, Polia nebulosa ab. 209 (11:26)
plumed prominent, the 58
plumigera, Ptilophora 40, **58** (3:28,29)
Plusia 14
Plusiinae 14, 120 (C:5,6), 122, 126
pod lover, the 229
Polia 206, **207**, 210
polyodon, Actinotia 129, **148** (8:17)
poplar hawk-moth, the 28
poplar kitten, the 44
popularis, Bombyx 240
populeti, Orthosia 197, 198, **249** (12:45–47), 250, 251
populi, Laothoe 12, 21, 27, **28** (A:1; 1:6)
porcellus, Deilephila 12, 21, 36, **37** (2:7)
porphyrea, Lycophotia 128, **170** (9:16,17), 188
porphyrea, Peridroma 171
Porthesia 73
Porthetria 76
Portland moth, the 150
postnigra, Noctua pronuba ab. 158 (8:30)
powdered quaker, the 250
praecox, Ochropleura 128, **150** (8:19)
prasina, Anaplectoides 127, **192** (10:31–33)

primulae, Phalaena (Noctua) 172
privet hawk-moth, the 24
proboscidalis, Hypena 121 (text fig.11), (C:11)
processionea, Thaumetopoeidae 65
pronuba, Noctua 127, **157** (B:1; 8:25–30), 159, 161
Psammophila 137
Psilura 76
Pterostoma **56**
Ptilodon **53**, 54
Ptilodontella **54**
Ptilodontis 56
Ptilophora **58**
pudibunda, Dasychira 66 (text fig.5), 67, **72** (4:18–20)
pudorina, Mythimna 15, 197, **261** (13:38)
pulchella, Utetheisa 11 (text fig.2), 13, 80, **97** (5:28), 98
pulchra, Noctua 97
pulverulenta, Bombyx 246
punctina, Mythimna impura ab. 263
purple clay, the 174
purple cloud, the 148
purpurascens, Drymonia dodonaea f. 59
puss moth, the 42
puta, Agrotis 128, **144** (8:1–4)
putrescens, Mythimna 15, 196, 263, 265, **271** (13:57)
putris, Axylia 129, **149** (8:18)
Pygaera 62
pygarga, Lithacodia 14
pygmaeola, Eilema 80, **91** (5:16,17), 93
Pyralidae 91
pyrophila, Noctua 156
Pyrrharctia **107**
Pyrrhia 126

quadra, Lithosia 11 (text fig.2), 13, 79, 80, **95** (5:21,22)
quadrifine moths 121 (text figs 11a, b)
quadripunctaria, Euplagia 11 (text fig.2), 13, 80, **107** (6:27,28)
quinquemaculatus, Manduca 12, 24

radius, Bombyx 144
ravida, Spaelotis 123, 129, 154, **164** (9:10,11), 165
recens, Orgyia 67, **69** (4:14,15), 70
reclusa, Bombyx 62
red chestnut, the 193
red-necked footman, the 85
reed tussock, the 68
remissa, Apamea 10 (text fig.1), 15
remissa, Apamea remissa f. 217, 218
renigera, Lacinipolia **205**
renigera, Ochropleura 205
renigera, Standfussiana lucernea ab. 154 (9:2)
reticulata, Heliophobus 189, 198, **213** (11:32–34), 241
rhabdota, Euxoa tritici ab. 133 (7:7)
rhomboidea, Xestia 129, **184** (10:11)
Rhyacia 156
riffelensis, Xestia alpicola 179

GENERAL INDEX

ripae, *Agrotis* 127, 128, 129, 136, 137, **145** (8:7–13)
rivularis, *Hadena* 200, **228** (text fig.17; B:4; 12:15), 236
roboris, *Noctua* 60
robsoni, *Polia nebulosa* ab. 209
Roeselia 116
rosea, *Bombyx* 83
rosea, *Paradiarsia glareosa* ab. 169
rosea, *Xestia agathina* ab. 188 (10:24)
rossica, *Callimorpha dominula* f. 108
rosy footman, the 83
rosy marsh moth, the 166
rotunda, *Nudaria* 81
round-winged muslin, the 81
rubi, *Diarsia* 122, 127, 173, **176** (9:33–35), 177
rubricollis, *Atolmis* 13, 79, **85** (5:7)
rubricosa, *Cerastis* 129, **193** (10:35–37), 194
rubricosta, *Ochropleura plecta* ab. 152 (8:6)
ruby tiger, the 106
rufa, *Cerastis leucographa* ab. 194
rufa, *Noctua comes* ab. 160 (8:38)
rufa, *Noctua fimbriata* ab. 161
rufa, *Noctua janthina* ab. 162 (8:34)
rufa, *Noctua pronuba* ab. 157 (8:29)
rufa, *Orthosia incerta* ab. 253
rufa, *Parasemia plantaginis* ab. 98 (5:33)
rufescens, *Anarta myrtilli* f. 200 (11:1)
rufescens, *Orthosia gracilis* ab. 250 (13:3)
ruficornis, *Drymonia* 40, 59, **60** (4:1–4), 61
rufolinea, *Mythimna straminea* ab. 262
rupestralis, *Anarta melanopa* ab. 203 (11:7)
russula, *Phalaena* (*Bombyx*) 101
rustica, *Diaphora mendica* f. 80, 105 (6:22,23)

sacraria, *Rhodometra* 261
sagitta, *Euxoa cursoria* ab. 136 (7:17)
sagittifer, *Noctua comes* ab. 160 (8:36)
sagittigera, *Pachetra* 199, 210, 211
salicinus, *Bombyx* 75
salicis, *Leucoma* 67, **75** (A:8; 4:27,28), 76
sallow kitten, the 43
sand dart, the 145
sannio, *Diacrisia* 12, 80, **101** (6:5–8)
saponariae, *Phalaena* (*Noctua*) 213
Sarrothripinae 120 (C:3), 126
saturatior, *Mythimna vitellina* f. 260
saucia, *Discestra trifolii* ab. 204 (11:11)
saucia, *Peridroma* 15, 121 (text fig.12), 124, 128, 129, 139, 140, 143, 146, **171** (9:36–38), 172
scarce black arches, the 119
scarce chocolate-tip, the 63

scarce footman, the 91
scarce prominent, the 55
scarce vapourer, the 69
scarlet tiger, the 108
Scoliopteryx 126
scopariae, *Xestia agathina* ab. 188 (10:25)
Scotia 137
scotica, *Ceramica pisi* ab. 224 (12:4)
scotica, *Mythimna impura* ab. 263 (13:42), 264
Scotophila 170
Segetia 178
segetum, *Agrotis* 128, **139** (7:31–36), 141, 142, 143, 146, 147
Semiophora 246
senex, *Thumatha* 13, 79, **81** (5:1), 84
Senta 273
serena, *Noctua* 225
sericea, *Eilema* 79, 88, 91, **92** (5:23), 94
setaceous hebrew character, the 179
Setina 82
sexstrigata, *Xestia* 130, **187** (10:21)
sexta, *Manduca* 9, 12, 24
shears, the 206
shore wainscot, the 266
short-cloaked moth, the 117
shoulder-striped wainscot, the 270
shuttle-shaped dart, the 144
Sideridis 212
signata, *Setina irrorella* f. 82 (5:3)
Silurian, the 237
silver cloud, the 244
silver-striped hawk-moth, the 38
silvery arches, the 208
similis, *Euproctis* 67, 73, **74** (4:25,26)
simulans, *Rhyacia* 123, 130, 154, **156** (9:3–5), 157, 165
Simyra 125
six-striped rustic, the 187
small black arches, the 113
small chocolate-tip, the 62
small dark yellow underwing, the 202
small dotted footman, the 87
small elephant hawk-moth, the 37
small quaker, the 246
small ranunculus, the 227
small square spot, the 176
Smerinthus 27
smoky wainscot, the 263
sobrina, *Paradiarsia* 129, **167** (9:18), 187
solani, *Noctua fimbriata* ab. 161
sororcula, *Eilema* 79, **88** (5:12)
southern rustic, the 157
southern wainscot, the 262
Spaelotis 164
sparshalli, *Trichiocercus* 65
speckled footman, the 96
Spectrum 24
Sphingidae 9, 10 (text fig.1), 12, 16, **20** (text fig.3; A:1–4) and key, 123
Sphingidae, eversible structures 9, 10 (text fig.1)

Sphinginae 20, **21** (A:1,2)
Sphinx 24
Spilosoma **102**
spinifera, *Agrotis* 127, **139**
Spiris 96
splendens, *Ceramica pisi* ab. 224 (12:3)
sponsa, *Catocala* 14
spurge hawk-moth, the 33
square-spot dart, the 130
square-spot rustic, the 187
square-spotted clay, the 184
stabilis, *Orthosia* 198, 200, 246, 247, 249, **251** (13:7–12)
Standfussiana 154
Stauropus 45
stellatarum, *Macroglossum* 20, **31** (2:5)
stigmatica, *Noctua* 184
Stilpnotia 75
stout dart, the 164
straminea, *Mythimna* 197, **262** (13:50,51), 263, 265, 269
stramineola, *Eilema griseola* f. 80, 88, 89 (5:14), 90
stramineola, *Lithosia* 89
stranger, the 222
straw-coloured footman, the 89
striata, *Spiris* 13, 80, **96** (5:25,26)
striped hawk-moth, the 35
striped wainscot, the 261
strigula, *Agrotis* 170
strigula, *Meganola* 112, **113** (6:35,36), 114 (text fig.9), 118
suasa, *Lacanobia* 198, **220** (11:39,40), 221
subatratus, *Agrotis segetum* ab. 139 (7:34)
subcaerulea, *Eugraphe subrosea* 167
subgothica, *Feltia* 147
subrosea, *Eugraphe* 129, **166** (9:8,9)
subsequa, *Noctua* 159
subterranea, *Feltia* **147**
subtusa, *Ipimorpha* 249
suffusa, *Anarta cordigera* ab. 202 (11:5)
suffusa, *Cerastis leucographa* ab. 194 (10:40)
suffusa, *Lycophotia porphyrea* ab. 170 (9:17)
suffusa, *Mythimna conigera* ab. 257 (13:35)
suffusa, *Noctua* 143
suffusa, *Noctua pronuba* ab. 158
suffusa, *Paradiarsia sobrina* ab. 167
suffusa, *Rhyacia simulans* f. 156 (9:5)
swallow prominent, the 52
sweet-potato hornworm, the 21
Syntomis **111**
Systropha 88

Taeniocampa 246
tawny shears, the 229
Telmia 130
tephra, *Dasychira* 71
tephrina, *Agrotis cinerea* 138
testacea, *Luperina* 222
thalassina, *Lacanobia* 198, **219** (11:37,38)
Thaumetopoeidae **65**
Tholera 238, **240**

thompsoni, *Polia nebulosa* ab. 209 (11:25)
three humped prominent, the 47
thulei, *Diarsia mendica* 169, **172** (9:25–28), 173
Thumatha **81**
Thyatiridae 249
tiliae, *Mimas* 12, 21, **26** (1:1–3), 28
timais, *Xanthopastis* **243**
tincta, *Phalaena* (*Noctua*) 208
tityus, *Hemaris* 20, **29** (2:4), 30
tobacco hornworm, the 24
tomato hornworm, the 24
Tortricidae 16
torva, *Notodonta* **47**
trabealis, *Emmelia* 14
trapezina, *Cosmia* 122
tremula, *Pheosia* 40, 49, 51, **52** (3:20,21)
trepida, *Bombyx* 50
triangulum, *Xestia* 127, 151, 179, 180, **181** (10:7)
trifine moths 121 (text fig.12)
trifolii, *Discestra* 199, **204** (11:9–12), 212, 215
trigemina, *Abrostola* 14
trimacula, *Bombyx* 59
Triphaena 157
triplasia, *Abrostola* 14
triple-spotted clay, the 180
tritici, *Euxoa* 121, 128, 131 (text fig.13), **132** (7:3–7), 133 (text fig.14), 134, 136, 137, 147
Tritophia **47**
tritophus, *Phalaena* (*Bombyx*) 47
tritophus, *Tritophia* 40, **47** (3:12)
trituberculana, *Celama* 119
true lover's knot, the 170
trux, *Agrotis* 127, 128, 142
tuberculana, *Celama* 119
turbata, *Hadena luteago barrettii* ab. 231 (12:18)
turca, *Mythimna* 197, **256** (13:13)
turfosalis, *Hypenodes* 120
turnip moth, the 139
twin-spotted quaker, the 254
typica, *Naenia* 121, 130, **189** (10:27), 213, 241
Tyria **109**

umbrosa, *Noctua* 187
uncula, *Eustrotia* 14, 121 (B:13)
unicolor, *Eilema deplana* f. 93
unicolor, *Melanchra persicariae* ab. 216
unicolor, *Polia bombycina* ab. 207 (11:18)
unipuncta, *Mythimna* 125, 197, 256, 259, **268** (13:52,53), 272
ursula, *Arctia villica* ab. 100 (6:2)
urticae, *Spilosoma* 12, 13, 80, 102, **104** (6:9)
Utetheisa **97**

valligera, *Noctua* 138
vapourer, the 70
varia, *Phalaena* (*Noctua*) 170
varied coronet, the 232
variegata, *Phalaena* 58
vau-nigra, *Leucoma* 76
venosa, *Agrotis clavis* ab. 141 (7:43)
venosa, *Diaphora mendica* ab. 105

vertunea, *Gluphisia crenata* **61** (4:5)
vestigialis, *Agrotis* 127, 128, 133, **138** (7:25–30), 139
villica, *Arctia* 13, 80, 100 (6:1–4)
vinula, *Cerura* 40, **42** (3:6,7)
viper's bugloss, the 231
virescens, *Heliothis* 16
virescens, *Noctua fimbriata* ab. 161
virgata-brunnea, *Orthosia incerta* ab. 253
viridis, *Ceramidia* 111
vitellina, *Mythimna* 198, 200, **260** (13:32,33), 272
v-nigrum, *Bombyx* 76

wardi, *Arctia villica* ab. 100 (6:3)
water ermine, the 104
weissenbornii, *Agrotis ripae* ab. 145 (8:11)
white colon, the 212
white ermine, the 102
white-line dart, the 132
white-marked, the 193
white-point, the 259
white prominent, the 57
white satin moth, the 75
white-speck, the 268
white spot, the 235
wild-cherry sphinx, the 25
wistromi, *Anarta melanopa* ab. 203 (11:8)
w-latinum, *Lacanobia* 198, 217, **218** (11:36)
wood tiger, the 98
'woolly bear' 99

Xanthia 122, 123
xanthographa, *Xestia* 129, 134, 167, 185, **187** (10:16–20)
Xanthopastis 243
Xestia **178**
Xylena 122
Xylomyges 244

yellow-tail, the 74
ypsilon, *Agrotis* 143

zatima, *Spilosoma lutea* ab. 103 (6:17,18)
Zeteolyga 228
ziczac, *Eligmodonta* 39 (text fig.4), 40, **48** (3:13–15)
Zoote 99

Index of Host Plants

Acer campestre 54, 58
Acer pseudoplatanus 54, 161, 209
Agropyron junceiforme 137
Agropyron repens 159, 257, 265
Agrostis 272
Aira praecox 137
alder 26, 43, 46, 87, 90
algae 85
alison, yellow 204
Allium cepa 204
Alnus glutinosa 26, 43, 46
Alyssum saxatile 204
Amaryllidaceae 243

Ammophila arenaria 249, 267
Andromeda polifolia 166
Anthyllis vulneraria 90
apple 27, 64, 94, 117, 246
apple, crab 64, 190, 219
apple, wild 27
Arbutus unedo 202, 203
Arctium 179
Arctostaphylos 223
Arctostaphylos uva-ursi 202, 203
Artemisia 150
Arum 162
ash 23, 24
aspen 44, 47, 48, 52, 56, 61, 63, 249, 250, 254
Atriplex 146, 162, 204, 212, 221, 222

barberry 219
bearberry 202, 203, 224
bedstraw 30, 31, 34, 35, 36, 37, 38, 139, 238
beech 45, 59, 85, 88, 89, 118, 124, 185
beet 140, 152
bell-heather 96, 98, 101, 169, 170, 178, 186, 201
bent 272
Berberis vulgaris 219
Beta vulgaris maritima 222
Betula 26, 43, 45, 46, 51, 53, 55, 57, 70, 74, 160, 161, 165, 168, 173, 175, 181, 183, 191, 208, 209, 218, 249
Betula pendula 59, 178
bilberry 168, 173, 175, 179, 192, 194, 203, 224
bindweed, field 22
Biorhiza pallida 247
birch 26, 43, 45, 46, 51, 53, 55, 57, 70, 74, 160, 161, 165, 168, 173, 175, 181, 183, 191, 208, 209, 218, 249, 253
birch, scrub 169
birch, silver 59, 178
blackberries 268
blackthorn 64, 73, 74, 109, 117, 118, 165, 181, 183, 190, 245, 249, 251, 252, 254
bladder-campion 214, 233, 234, 237
bluebell 169, 187, 243
bog-bilberry 166
bog-myrtle 77, 166, 183, 191, 208, 218, 223, 250, 251
bog-rosemary 166
borage 98
Borago officinalis 98
bracken 175, 218, 224
bramble 162, 173, 174, 175, 176, 181, 183, 187, 209
Brassica 125, 215
broom 71, 218, 219
buckthorn 87
buckwheat, cultivated 133
Buddleia 122, 136, 142, 145, 156, 158, 160, 162, 176, 180, 188
bugle 30
burdock 136, 160, 185, 188
bur-reed, branched 68

cabbage 140, 192
Cakile maritima 146, 212

Calluna 122, 223
Calluna vulgaris 71, 86, 160, 166, 168, 169, 170, 173, 176, 178, 182, 186, 189, 201
Calystegia soldanella 212
Campanula rotundifolia 155, 182
campion 122, 206, 208, 209, 214, 229, 232, 235, 258
campion, red 234, 236
campion, white 230, 234
campions, white-flowered 230
canary-grass, reed 159
Carex pulicaris 262
carnation, garden 228, 231, 235
carrot 140
carrot, sliced 142, 143, 146, 183, 185
carrot, wild 141
Caryophyllaceae 214, 228
catchfly 235
catchfly, Nottingham 235
catchfly, Spanish 231
catmint 222
Cerastium 133, 150
Chenopodiaceae 204
Chenopodium 141, 142, 149, 204, 212, 221
chickweed 105, 194, 238
chickweed, common 179, 185, 206, 212, 259
cinquefoil, creeping 159
Cladium mariscus 68
clover 134, 255
clover, hare's-foot 92
clover, white 90
cock's-foot 159, 169, 256, 257, 259, 260, 262, 263, 265, 266, 270
cocoa 101
colt's-foot 223
comfrey 108
Compositae 206, 208, 225, 226
Convolvulus 22
Convolvulus arvensis 22
Corylus avellana 41, 45, 46, 53, 162, 181, 218, 246
cotton 144
couch, common 159, 257, 265
cowberry 203
cowslip 159
crack-willow 166
Crataegus 64, 69, 73, 74, 117, 160, 162, 165, 173, 174, 181, 183, 208, 219, 246
Crataegus monogyna 71, 178
Crepis 169, 206, 226
Crepis capillaris 227
crowberry 178, 203
Cynoglossum officinale 149, 249

Dactylis glomerata 159, 169, 256, 257, 259, 260, 262, 263, 265, 266, 270
dandelion 96, 103, 105, 106, 107, 139, 143, 144, 156, 164, 176, 181, 184, 190, 206, 208, 212, 238, 259
Daucus carota 141
dead-nettle 107, 238
dead-nettle, white 179
Deschampsia cespitosa 96, 240, 263, 265, 267
Deschampsia flexuosa 240
dewberry 115

Dianthus 236
Dianthus barbatus 228, 233, 235, 236
Dianthus caryophyllus 235
Dianthus, cultivated 230
Dicranoweisia cirrata 81
Digitalis purpurea 160, 182
dock 35, 103, 105, 106, 108, 153, 156, 160, 164, 165, 173, 174, 179, 183, 185, 187, 188, 194, 208, 212, 218, 220, 245, 255, 271
dock, water 104
dogwood 94, 181

Echium vulgare 231
elder 216
elm 41, 76, 124, 252, 254
elm, English 26, 245
Empetrum 178
Empetrum nigrum 203
Endymion non-scriptus 169, 187, 243
Epilobium 34, 36, 37, 38, 151, 179
Erica 122, 223
Ericaceae 200, 223
Erica cinerea 96, 98, 101, 169, 170, 178, 186, 201
Erica tetralix 186
Euphorbia 33, 34

Fagopyrum esculentum 133
Fagus sylvatica 45, 59, 118
fen-sedge, great 68
fescue, tall 267
Festuca 239
Festuca arundinacea 267
Festuca ovina 138, 155, 169
figwort, water 187
Filipendula ulmaria 223, 251, 255
flea-sedge 262
forget-me-not 98
foxglove 160, 182
Fragaria 115, 151
Fragaria ananassa 187
Fraxinus excelsior 23, 24
Fuchsia 34, 36, 38
fungi 122, 123

Galium 30, 31, 34, 35, 36, 37, 38, 132, 133, 139, 152, 169, 182, 193, 238
Galium mollugo 149, 187
Genista tinctoria 218, 249
Glechoma hederacea 107
Glyceria fluitans 262
goldenrod 106, 182, 218
goosefoot 204, 212, 221
Gramineae 124, 188, 256, 271
grape-vine 32, 36, 38
grass 96, 110, 139, 155, 156, 158, 159, 163, 169, 187, 211, 212, 238, 239, 240, 241, 257, 259, 260, 262, 263, 265, 267, 268, 270, 271
grass, flowering 262, 263, 264, 265, 266, 267, 270
grass, marsh 176
grass, roots 147
grass, wild 134
greenweed, dyer's 218, 249
ground-ivy 107
groundsel 110, 152, 179, 193, 219

INDEX OF HOST PLANTS

hair-grass, early 137
hair-grass, tufted 240, 263, 265, 267
hair-grass, wavy 240
harebell 155, 182
hawk's-beard 206, 226
hawk's-beard, smooth 227
hawkweed 206, 226
hawthorn 64, 69, 71, 73, 74, 117, 160, 165, 173, 174, 178, 181, 183, 208, 219, 246
hazel 41, 45, 46, 53, 181, 218, 246
heather 71, 86, 92, 96, 97, 99, 102, 106, 122, 132, 134, 137, 139, 150, 155, 160, 166, 168, 169, 170, 173, 174, 176, 178, 182, 186, 188, 189, 201
heather, cross-leaved 186
Hebe 162
Hedera 122
hedge-bedstraw 149, 187
Helianthemum 132, 160
Helianthemum chamaecistus 182
Heracleum 134
herbs 134
herbs, cultivated 158
herbs, small 133
herbs, wild 158
Hieracium 182, 206, 226
Hippophae rhamnoides 35
hogweed 239
Homalothecium sericeum 81, 87
honeysuckle 30, 36, 37, 219, 254
Honkenya peploides 137
hop 72, 254
hound's-tongue 149, 249
Humulus lupulus 72, 254
Hypericum 148

Impatiens 36
Ipomoea 22
Ipomoea batatas 21
ivy 122, 180, 268

Juncus 122, 163, 239

knapweed 143, 204, 264
Knautia 238
Knautia arvensis 29
knotgrass 92, 143, 173, 203, 204, 206, 208, 212, 214, 218, 219, 220, 223, 245, 270

Labrador-tea 166
Lactuca 145, 149, 208, 225, 226
Lactuca sativa 139, 227
Lactuca serriola 227
Lactuca virosa 227
Lamium 107, 149, 153, 238
Lamium album 179
larch 216, 225
Larix decidua 216, 225
lavender 163
Ledum groenlandicum 166
lettuce 139, 226
lettuce, cultivated 152, 208, 225
lettuce, field 227
lettuce, garden 223, 227, 232
lettuce, prickly 227
lettuce, wild 227
lichen 81, 83, 85, 86, 87, 88, 89, 91, 92, 93, 94, 95, 118, 122, 178
lichen, black and yellow 82

lichen, orange-coloured 84
Ligustrum vulgare 24
lilac 24
lily, martagon 208
lime 26, 41, 76, 141, 145, 181
Linaria vulgaris 34
Lonicera 22
Lonicera periclimenum [sic] 219
  see *L. periclymenum*
Lonicera periclymenum 30, 219, 254
loosestrife, purple 37, 251
loosestrife, yellow 104, 251
Lotus 119, 150
Lotus corniculatus 90, 92, 245
lousewort 104
Luzula 256
Luzula campestris 264
Luzula pilosa 203, 262, 263
Luzula sylvatica 175
Lychnis 122, 229, 236
Lychnis flos-cuculi 229, 234
Lycium barbarum 23
Lycopersicum 24
lyme-grass 146, 212
Lysimachia 251
Lysimachia vulgaris 104, 251
Lythrum salicaria 37, 251

madder, wild 31
mallow, common 163
Malus 27, 94, 117
Malus sylvestris 64, 190, 219
Malva sylvestris 163
maple, field 54, 58
marram 91, 134, 137, 139, 146, 150, 212, 249, 267
mat-grass 240, 267
Matricaria 162
meadow-grass, annual 169, 211, 257, 263, 265, 266, 267
meadow-grass, wood 211, 256
meadowsweet 223, 251, 255
Medicago 119
Mentha 104
Mesembryanthemum 222
mint 104
Molinia 239
Molinia caerulea 262
moor-grass, purple 262
moss 81, 87, 88, 89, 92, 95, 108, 110, 203
moss-campion 224
mustard 140
Myosotis 98
Myrica gale 77, 166, 183, 191, 208, 218, 223, 250, 251

narcissus 243
Nardus 239
Nardus stricta 240, 267
Nerium oleander 32
nettle 108, 216
nettle, common 255
Neuroterus quercusbaccarum 247
Nicotiana 22

oak 26, 41, 45, 50, 59, 60, 69, 74, 76, 85, 88, 89, 94, 95, 109, 113, 115, 118, 124, 188, 194, 195, 218, 219, 246, 247, 252, 253, 254
oak, high-canopy 248
oak, scrub 195, 248
oleander 32

onion 147, 204
Ononis repens 212
orache 204, 212, 221, 222
oxtongue 226

Pancratium maritimum 244
Parthenocissus 38
pear 117
pear, wild 219
Pedicularis sylvatica 104
Peltigera canina 81, 83, 87, 89, 95
periwinkle, greater 32, 187
periwinkle, lesser 32
Petunia 22
Phalaris 263
Phalaris arundinacea 159
Phragmites 122, 263, 270
Phragmites australis 68, 262, 263, 270, 273
Picea abies 25
Picris 226
pine 65, 76, 242, 243
pine, Scots 25, 77
Pinus 76
Pinus pinaster 242
Pinus sylvestris 25, 242
Plantago 105, 107, 134, 142, 143, 152, 169, 174, 176, 179, 185, 188, 194, 220, 259
Plantago lanceolata 187
Plantago major 245
plantain 105, 107, 143, 174, 185, 188, 194, 220, 259
plantain, greater 245
plum 27, 64, 117, 245, 246, 251
plum, garden 249
Poa annua 169, 211, 257, 263, 265, 266, 267
Poa nemoralis 211, 256
Polygonum 141, 145, 149, 192, 212
Polygonum aviculare 92, 143, 173, 203, 204, 206, 208, 212, 214, 218, 219, 221, 223, 245, 270
Polygonum persicaria 218
poplar 27, 28, 42, 44, 45, 47, 48, 52, 53, 56, 61, 63, 75, 249
Populus 27, 28, 42, 44, 47, 48, 52, 53, 56, 61, 63, 75, 249
Populus tremula 56, 63, 249, 254
potato 23, 24
potato, sliced 185
Potentilla reptans 159
primrose 160, 163, 173, 181, 183, 188
Primula 153, 155, 162, 181
Primula veris 159
Primula vulgaris 160, 163, 173, 181, 183, 188
privet 24, 141, 190, 211, 257
Prunus 26, 27, 117, 249
Prunus spinosa 64, 73, 74, 109, 117, 118, 165, 181, 183, 190, 245, 254
Pteridium aquilinum 175, 218, 224
Puccinellia maritima 266
Pyrus communis 117, 219

Quercus 26, 41, 45, 50, 59, 60, 69, 74, 76, 94, 95, 109, 113, 118, 188, 194, 218, 219, 246, 247, 252, 253, 254

ragged robin 229, 234

ragwort 132, 134, 136, 137, 139, 150, 152, 154, 155, 162, 163, 174, 176, 180, 181, 184, 185, 187, 188, 204, 239, 240, 249, 258, 264
ragwort, common 110
Ranunculus 159
raspberry 115
redshank 218
reed 68, 122, 262, 263, 270, 274
reed, common 68, 262, 263, 270, 273
restharrow, common 212
Rhacomitrium lanuginosum 203
rhododendron 30, 37
ribwort plantain 187
rock-rose, common 182
Rosa 246
Rosaceae 115, 117
rose 246
rosebay 185
Rubia peregrina 31
Rubus 192
Rubus caesius 115
Rubus fruticosus 162, 173, 174, 175, 181, 183, 187, 209
Rubus idaeus 115
Rumex 35, 103, 105, 106, 108, 141, 142, 145, 149, 152, 153, 160, 162, 163, 164, 165, 169, 173, 174, 175, 176, 179, 181, 183, 185, 187, 188, 190, 192, 193, 194, 208, 212, 218, 220, 245, 255, 271
Rumex hydrolapathum 104
rush 122, 176, 181, 266

St. John's-wort 148
Salix 27, 28, 41, 42, 44, 48, 52, 53, 56, 62, 69, 71, 74, 75, 76, 86, 94, 122, 150, 160, 161, 162, 163, 165, 166, 169, 173, 174, 175, 181, 182, 183, 185, 188, 190, 191, 192, 193, 194, 208, 209, 216, 218, 219, 246, 249, 251, 252, 253, 254
Salix alba 201
Salix aurita 223
Salix caprea 109, 203
Salix cinerea 186
Salix fragilis 166
Salix repens 62, 77, 150, 179, 223
sallow 28, 41, 42, 44, 48, 69, 71, 74, 75, 76, 86, 87, 89, 90, 94, 122, 124, 160, 161, 165, 173, 174, 175, 181, 185, 188, 190, 191, 192, 193, 194, 208, 218, 224, 242, 246, 247, 248, 249, 250, 251, 252, 253, 255
sallow, grey 186
sallow-thorn 35
Salsola kali 146, 222
saltmarsh-grass, common 266
saltwort, prickly 146, 222
Sambucus 216
sand-couch 137
sand-lily 244
sand-spurrey 232
Sarothamnus scoparius 71, 218, 219
Saxifraga 155
scabious 238
scabious, devil's-bit 29
scabious, field 29, 92

Scirpus 239
Scrophularia auriculata 187
sea-beet 222
sea-bindweed 212
sea-buckthorn 35
sea-campion 229, 234, 237
sea-rocket 146, 212
sea-sandwort 137
sea-spurrey, rock 143, 230, 232
sedge 68
Sedum acre 155
seedlings, forest nursery 246
Senecio jacobaea 110, 249
Senecio vulgaris 110, 152, 179, 193, 219
sheep's-fescue 138, 155, 169
Silene 122, 229, 236
Silene alba 230, 234
Silene dioica 234, 236
Silene inflata [sic] 214, 230, for bladder campion see *S. vulgaris*
Silene maritima 229, 230, 232, 234, 237
Silene nutans 230, 235
Silene otites 231
Silene vulgaris 214, 230, 232, 233, 234, 237
snowberry 30
snowdrop 243
Solanaceae 23, 24
Solanum tuberosum 23, 24
Solidago virgaurea 106, 182, 218
Sonchus 164, 190, 208, 226, 227
Sonchus oleraceus 223
sow-thistle 164, 208, 226, 227

sow-thistle, smooth 223
Sparganium erectum 68
Spergula 133
Spergularia rubra 232
Spergularia rupicola 143, 230, 232
spruce, Norway 25, 77
spurge 33, 34
Stellaria 105, 133, 139, 150, 162, 176, 181, 194
Stellaria media 179, 185, 206, 212, 238, 259
stitchwort 139
stonecrop, biting 155
strawberry 115
strawberry, garden 187
strawberry-tree 202, 203
Succisa pratensis 29
swede 140
sweet-grass, floating 262
sweet-potato 21
sweet william 228, 233, 235, 236
Swida sanguinea 94, 181
sycamore 54, 161, 209
Symphoricarpos rivularis 30
Symphytum 108
Syringa vulgaris 24

tansy 136, 160, 162, 165, 184, 188
Taraxacum 145, 149, 151
Taraxacum officinale 96, 103, 105, 106, 107, 139, 143, 144, 164, 176, 181, 185, 190, 206, 208, 212, 238, 259
Taxus baccata 93

teaplant, Duke of Argyll's 23
thistle 92, 94, 143, 239
thyme 182
thyme, wild 138
Thymus drucei 138, 182
Tilia 26, 41, 76
toadflax, common 34
tomato 24, 125, 221
traveller's-joy 94
trefoil, common bird's-foot 90, 92, 245
trefoil, lesser 245
Trifolium 119, 134, 141, 255
Trifolium arvense 92
Trifolium dubium 245
Trifolium repens 90
turnip 140
turnip, sliced 142
Tussilago farfara 223

Ulmus 41, 76, 162, 252, 254
Ulmus procera 26, 245
Umbelliferae 134, 239
Urtica 108, 216
Urtica dioica 255

Vaccinium 202, 223
Vaccinium myrtillus 168, 173, 175, 179, 192, 194, 203
Vaccinium uliginosum 166
Vaccinium vitis-idaea 203
valerian 158, 180
valerian, red 30, 36, 141, 142, 155, 156, 214, 222, 227, 229, 258, 271

Verbascum 141, 179
vetch, kidney 90
Vinca 160
Vinca major 32, 187
Vinca minor 32
vines 147
Viola 137, 238
viper's bugloss 208, 214, 229, 231, 235
Virginia creeper 38
Vitis vinifera 32, 36, 38

Weisia 87
wheat 134, 144
willow 27, 44, 48, 52, 53, 56, 182, 219, 223, 249, 251
willow, creeping 62, 77, 150, 179
willow, goat 109, 203
willowherb 34, 36, 37, 38, 151
willow, rough-leaved 208, 209, 216, 218, 246, 252, 253, 254
willow, smooth-leaved 251
willow, white 201
wood-rush 256
wood-rush, field 264
wood-rush, great 175
wood-rush, hairy 203, 262, 263
wood-sage 155, 174, 181, 185, 206, 208, 209, 214, 222, 229, 235, 258, 261, 264, 271

yew 93, 94

Zephyranthes candida 243, 244